# Entropy Beyond the Second Law

Thermodynamics and statistical mechanics for equilibrium, non-equilibrium, classical, and quantum systems

# Entropy Beyond the Second Law

Thermodynamics and statistical mechanics for equilibrium, non-equilibrium, classical, and quantum systems

**Phil Attard**

**IOP** Publishing, Bristol, UK

ISBN 978-0-7503-1590-6 (ebook)
ISBN 978-0-7503-1588-3 (print)
ISBN 978-0-7503-1589-0 (mobi)

DOI 10.1088/978-0-7503-1590-6

Version: 20180401

IOP Expanding Physics
ISSN 2053-2563 (online)
ISSN 2054-7315 (print)

British Library Cataloguing-in-Publication Data: A catalogue record for this book is available from the British Library.

Published by IOP Publishing, wholly owned by The Institute of Physics, London

IOP Publishing, Temple Circus, Temple Way, Bristol, BS1 6HG, UK

US Office: IOP Publishing, Inc., 190 North Independence Mall West, Suite 601, Philadelphia, PA 19106, USA

# Contents

# Preface

'A theory is the more impressive the greater the simplicity of its premises, the more different kinds of things it relates, and the more extended its area of applicability. Therefore the deep impression that classical thermodynamics made upon me. It is the only physical theory of universal content which I am convinced will never be overthrown,'

Albert Einstein, quoted in Schlipp P A (1979) *Autobiographical Notes. A Centennial Edition* (Open Court Publishing Company) p 31

'The law that entropy always increases holds, I think, the supreme position among the laws of Nature…if your theory is found to be against the second law of thermodynamics I can give you no hope'

Arthur S Eddington (1928) *The Nature of the Physical World* (Cambridge University Press, Cambridge) ch 4

'There is no concept in the whole field of physics which is more difficult to understand than is the concept of entropy, nor is there one which is more fundamental'

Francis Sears (1950) *Principles of Physics I: Mechanics, Heat, and Sound,* (Reading, MA: Addison-Wesley) p 459

'Few physical concepts have caused as much confusion and misunderstanding as has that of entropy'

James Lovelock (1979) *Gaia: A New Look at Life on Earth,* (Oxford University Press) p 2

'our science colleagues may…help us with entropy, which to me is a more difficult concept than anything economics has to offer.'

Talling Koopmans (2005) quoted in Faber M, Niemes H and Stephan G *Entropy, Environment, and Resources: An Essay in Physico-economics* (Berlin: Springer-Verlag) p 77

**Entropy and the second law**

The Second Law of Thermodynamics—entropy increases—is arguably the most fundamental law of nature, as the above quotations from Einstein, Eddington, and Sears attest. And yet for some, entropy itself is a mystery, the cause of confusion and misunderstanding. How can it be that something so fundamental and universal, as evidenced from the interests and expertise of the quoted authors, is so hard to grasp? How can it have happened that thoughtful people appreciate the significance and universal applicability of the Second Law without understanding the nature of entropy, its sole concern?

I blame Clausius, who in the middle of the 19th century initiated the field of thermodynamics by defining entropy as the integrated heat flow divided by the temperature. This is the empirical result of measurement and observation. It is undoubtedly mathematically and quantitatively correct, but it offers nothing in the way of understanding or reason or physical basis. To my mind it obscures rather than clarifies the nature of entropy, and it is little wonder that anyone approaching entropy from this view point should be confused. This definition has made entropy an enigma, and the resultant mystique surrounding it has given rise to a belief in the power of the Second Law of Thermodynamics that is akin to magic. For many, the unquestioning acceptance of both is an act of faith.

It was Boltzmann, shortly after Clausius, who identified entropy with the logarithm of the number of molecular configurations, thereon founding the field of statistical mechanics. Here was not only a concrete physical picture for entropy, but also an explanation of the Second Law of Thermodynamics. Considering the transitions between states of a system, it is intuitively obvious that these should likely be to states with more configurations rather than less, which is precisely the direction of entropy increase. For example, a die is more likely to land showing an even number than a six because there are three of the former and only one of the latter; the state 'even' has more entropy than the state 'six'. Of course the fact that entropy is just a way of counting explains why it and the Second Law have such universal application beyond the fields of thermodynamics and statistical mechanics.

What is a little puzzling is that Boltzmann's insight into the nature of entropy and the reason for the Second Law has not settled the matter and that the confusion over entropy persists amongst some to the present day. I suspect that part of the reason for this is that the fields of thermodynamics and statistical mechanics have remained separate from the beginning. Practitioners have always specialized in one or the other, and the concerns and approaches of each are rather different. This goes deeper than the facile observation that thermodynamics is concerned with macroscopic phenomena and statistical mechanics is focussed on microscopic and molecular behavior. To some extent the two fields appeal to different aspects of the intellect. Thermodynamics has grown as a discipline based on empirical rules and quantitative relationships that have been gleaned from observation and measurement. It can be a very practical field in which practitioners are more interested in actual outcomes and real applications than in model systems or qualitative explanations. It is often taught as a collection of expressions to be memorized without necessarily recognizing the connections between different results.

Statistical mechanics is more focussed on deducing the thermodynamic laws and relationships from the probabilistic behavior of atoms and molecules. It can produce quantitative predictions of thermodynamic properties from the molecular interactions, although just as often the qualitative behavior of a system is what is sought. Whereas thermodynamics has discovered universal laws by reasoning inductively from empirical observation, statistical mechanics has focussed more on the reasons why those laws exist, and has sought to logically deduce them from the fundamental mechanical laws that govern the Universe.

Statistical theoretical approaches in broader fields such as biology, finance, economics, health, sociology, etc often use results and procedures from thermodynamics and from statistical mechanics. However, one should be aware of a serious caveat: the laws governing the behavior of atoms and molecules, which give quantitative and predictive certainty to statistical mechanics, and reproducibility to thermodynamics, generally have no analogue in these broader fields. The statistical data that these fields utilize are empirical and historical, and because there is no underlying mechanical laws, there is no certainty that the same data will hold in the future. Despite this caveat, the general concepts of entropy and of the Second Law of Thermodynamics are frequently applied beyond the disciplines of thermodynamics and statistical mechanics.

This book takes a new and unified look at thermodynamics and statistical mechanics. It attempts to identify their essential conceptual ingredients, and to derive these two disciplines from the ground up as two ends of a spectrum linked by entropy. In the equilibrium case, chapters 1 and 2, much is already known and confirmed by laboratory measurement, computer simulation, and mathematical analysis. What is new in this case is not the final results but their interpretation. For example, in chapter 1 Boltzmann's definition of entropy is generalized as the logarithm of the weight of a state. A small change perhaps, but it accommodates the case that configurations are not equally likely, it shows that probability and entropy are conceptually equivalent, and it goes beyond physical systems governed by molecular mechanics. These demonstrate the universality of the statistical thermodynamic formalism and its generic potential for application in diverse areas of human endeavor. In this vein the information entropy, often attributed to Shannon in his theory of communications, and also widely exploited by Jaynes in his 'MaxEnt' theory of probability and inductive reasoning, is here shown to be just part of the total thermodynamic entropy, as is demonstrated by results for the two-dimensional Ising model of statistical mechanics.

Similarly, what is new in chapter 2 is not so much adumbration of the conventional results of equilibrium thermodynamics, but rather that their derivation proceeds from the perspective of entropy. This shows, for example, the equivalence of free energy and total entropy, and that the minimization of the free energy is just the Second Law of Thermodynamics in a different guise. The analysis shows that the First Law of Thermodynamics is a statement about conservation laws, from which very general conclusions follow about the generic formulation of entropy for an arbitrary system. It is this idea which in part points to the way in which entropy should be developed beyond equilibrium.

## Entropy beyond the second law

> 'For in the sciences the authority of thousands of opinions is not worth as much as one tiny spark of reason in an individual man'
>
> Galileo (1612)
> *Third Letter on Sunspots*

'If different persons, not knowing each other's work, have been pursuing different clews that led to the same result, why then it shows that there may be something in it. But if this is only the same story, filtered through two channels, and reaching me in two ways, then that don't make it any stronger'

Abraham Lincoln (1861)
(Philadelphia, PA)

'A discovery must be, by definition, at variance with existing knowledge'

Albert Szent-Györgyi (1972)
Dionysians and Apollonians *Science* **176** 966

This book proceeds beyond the Second Law in several respects. In chapter 3 attention turns to non-equilibrium thermodynamics. For such time dependent systems, the Second Law of *Equilibrium* Thermodynamics is insufficient; it tells the direction of motion but not its speed. Nevertheless, looked at from the right perspective, and with the insight gleaned from the generic analysis of chapter 1, it is possible to develop an entropy for time varying systems that is the basis for a Second Law of *Non-Equilibrium* Thermodynamics.

The non-equilibrium entropy provides a driving force for motion in a generalized sense, as, for example, hydrodynamic fluxes, wave number selection, and pattern formation. Accordingly, steady heat flow chapter 3, Brownian motion chapter 5, and molecular motion in an open system chapter 5, are analyzed from this perspective. The stochastic, dissipative equations that result finally explain and justify some known results, such as the Langevin equation, and the fluctuation–dissipation theorem, and produce some new ones besides. They provide a basis for the formulation of non-equilibrium statistical mechanics in chapter 6.

Finally, chapter 7 formulates quantum statistical mechanics from first principles by showing how the wave function of an open quantum system collapses into entropy states. This gives a type of trace for the partition function and for the statistical average of an operator, which, via a formally exact transformation, is written as an integral over classical phase space. The leading order term in the expansion gives classical statistical mechanics. Hence the analysis shows that entropy is responsible for the fact that classical mechanics governs the motion of the world around us even though quantum mechanics governs the behavior of the underlying atomic and sub-atomic constituents.

The chapters that go beyond the usual formulation of entropy and of the Second Law lie at the cutting edge of contemporary scientific research. As such, many of the results reflect my own analysis and ideas rather than a consensus of opinion. As Szent-Györgyi noted above, this is necessarily the case with any discovery. Some results, such as those for non-equilibrium thermodynamics, may be described as controversial given that they differ from the very many books and papers already published on the subject. But the point of the quotations from Galileo and Lincoln is that multiple books or authors in agreement is not as convincing as independent research (surprisingly enough, even scientists copy). Similarly, authority figures are entitled to serious consideration, but no more than that; science would not be science if authorities were exempt from being questioned or from the rules of evidence. Although consensus can be consoling when it is the result of independent and critical thought, if it is merely copying, or following authority, or fitting in, then it adds nothing to the sum total of knowledge. Truth is determined by evidence alone. Evidence is not the number of proponents, or status of authority, or degree of consensus. Evidence is measurement, computation, mathematical analysis, or reasoned argument.

It is for this reason that this book is structured a little differently to most books on thermodynamics and statistical mechanics. Since the content goes beyond existing knowledge of entropy and of the Second Law of Thermodynamics, I feel an obligation, greater than usual, to justify the results with concrete evidence. I have tried to avoid simply asserting a result as commonly held, or of copying a result and citing some other source as its authority. Instead, I derive almost everything from first principles, making the book largely self-contained and the presentation uniformly coherent. This does sometimes pose a problem in that general axioms can be introduced or quantities defined before it is obvious that they are useful, but it is the price to be paid for a bottom up approach. Similarly, the novelty of the forms of entropy developed here to go beyond the Second Law means that at this stage the number of concrete examples that illustrate the formalism is rather limited. Nevertheless, I make many connections with known theorems and results for which there is a convincing consensus, and hopefully these should provide some sort of guide to the utility of the general approach taken here.

Entropy has given me understanding and insight into the nature of the world, and it has provided a basis for many of my contributions to the scientific literature. In this book I hope to pass on some of this knowledge to the benefit of readers in their own endeavors.

Phil Attard<br>(Sydney)<br>July, 2017

# Author biography

## Phil Attard

Phil Attard is a research scientist working broadly in the areas of statistical mechanics, thermodynamics, and colloid and surface science. He has held academic positions at various universities in Australia, Europe and North America, and he was a Professorial Research Fellow of the Australian Research Council. He has authored some 120 papers, 10 review articles, and 4 books, with over 6000 citations.

As an internationally recognized researcher he has made seminal contributions to the theory of electrolytes and the electric double layer, to measurement techniques for atomic force microscopy and particle interactions, and to computer simulation and integral equation algorithms for condensed matter. Attard is perhaps best known for his discovery of nanobubbles and for his role in establishing their nature.

In recent years his research has been focused on non-equilibrium thermodynamics and statistical mechanics, and on the foundations of quantum statistical mechanics. He has identified a new type of entropy—the second entropy—as the variational principle for non-equilibrium thermodynamics, and he has derived the general form of the non-equilibrium probability distribution for statistical mechanics. The theory provides a coherent approach to non-equilibrium systems and to irreversible processes, and with it can be derived many well-known empirical results and theorems. The theory has also led to the development of stochastic molecular dynamics and non-equilibrium Monte Carlo computer simulation algorithms.

Attard advocates the understanding of entropy as a physical weight, and the formulation of probability in terms of set theory. This unique perspective forms the basis for this and his other three books:

Attard P (2002) *Thermodynamics and Statistical Mechanics: Equilibrium by Entropy Maximisation* (London: Academic)
Attard P (2012) *Non-Equilibrium Thermodynamics and Statistical Mechanics: Foundations and Applications* (Oxford: Oxford University Press)
Attard P (2015) *Quantum Statistical Mechanics: Equilibrium and Non-Equilibrium Theory from First Principles* (Bristol: IOP Publishing)

**IOP** Publishing

Entropy Beyond the Second Law
Thermodynamics and statistical mechanics for equilibrium, non-equilibrium, classical, and quantum systems
**Phil Attard**

# Chapter 1

## Entropy counts

'How do I love thee, let me count the ways'

Browning (1850)

'You should call it entropy, because nobody knows what entropy really is'

von Neumann (circa 1939)

'Shannon entropy and Boltzmann entropy are completely unrelated'

Kline (1999)

The main aim of this chapter is to define entropy in its generic form. Several applications and illustrations are also given as an aid to understanding and applying entropy.

One of the issues that will be addressed in this chapter is the meaning of the Gibbs–Shannon information entropy,

$$S = -k_{\rm B} \sum_{\alpha} \wp_{\alpha} \ln \wp_{\alpha}. \tag{1.1}$$

Here $k_B$ is Boltzmann's constant and $\wp_\alpha$ is the probability of the state $\alpha$ of the system. This equation appears in almost all books on statistical mechanics or statistical thermodynamics, and it would be fair to call this the conventional expression for the entropy. It was given explicitly by Boltzmann (1866) and by Gibbs (1902). Upon it is founded Shannon's (1948) information theory and Jaynes' (1957, 2003) maximum entropy formulation of statistical mechanics and probability theory.

In my opinion, there is a great deal of misunderstanding about this expression. Most workers believe that it is 'the' entropy, and that it should be maximized. And yet there are others who believe that it is unrelated to the thermodynamic entropy. I myself believe neither, and in section 1.1.3 I will give the derivation of the full expression for the entropy and explain the proper use of this particular expression for the so-called information entropy.

## 1.1 Entropy, weight, probability, and information

One might think that any book on entropy, particularly any book that focuses on its role in thermodynamics and statistical mechanics, would begin with the Second Law of Thermodynamics,

$$\text{All systems evolve in the direction of increasing entropy.} \tag{1.2$a$}$$

Alternatively,

$$\text{The entropy increases during spontaneous changes of a system.} \tag{1.2$b$}$$

The very simplicity of the law underscores its universality and profundity.

But the problem with beginning with the Second Law is that entropy is undefined and unexplained by it. And in consequence, the reason for the law is not evident. To accept the law as merely an empirical fact confirmed by countless measurement is to sell reason short. From that limited point of view, the Second Law of Thermodynamics is nothing but superstition, a magic spell to mesmerize the credulous, an incantation to be memorized and repeated with the mind closed to further possibilities and new applications. Much better instead to understand the nature of entropy, for then one might hope to know when the law does not apply (in fact, systems where entropy has spontaneously decreased are quite common), and to develop an appropriate version of the law for new situations beyond those that form its empirical basis.

Another reason to eschew the Second Law of Thermodynamics as a starting point is that it is the *second* law. If ordinal numbers mean anything, then surely the normal precedent should be followed by beginning with the *First* Law of Thermodynamics, which is sometimes stated as

$$\text{The change in energy is the work done plus the heat flow.} \tag{1.3$a$}$$

But in many ways it is preferable to give it in the form

$$\begin{gathered}\text{Energy cannot be created or destroyed,}\\ \text{but it can be converted from one form to another.}\end{gathered} \tag{1.3$b$}$$

In short,

$$\text{Energy is conserved.} \tag{1.3$c$}$$

An orderly account would obviously discuss the First Law of Thermodynamics before the Second Law. But this raises the question of how to relate the First Law to entropy.

### 1.1.1 Conservation, counting, and linear additivity

The First Law is based upon the concept of conservation. Things that are conserved can be counted. And counting is what defines entropy.

The ability to count is a fundamental human ability that occurs in a core part of the brain, which indicates that it is very old in evolutionary terms Our ancestors no doubt counted things like tribe members, animals and possessions. These are all concrete objects that we would say in the present context share the property of being conserved. There would be little point in, say, counting the coins in your purse if the coins appeared or disappeared at random. One can shake the purse, or move it from one place to another without changing the number of coins.

This is not to say that counting is restricted to conserved objects. Immaterial things such as kisses can be counted without invoking any conservation law. But certainly in the present context it is the counting of conserved objects that is the prime focus.

Closely connected to the notion of conservation is the notion of linear additivity. For example, one can transfer some coins from one purse to another. The total number of coins is conserved, which is to say that the sum of the coins in the two purses is the same before and after the transfer.

The notion of linear additivity is not identical to that of conservation. For example, the square root of the number of coins in a purse is conserved. But the sum of the square roots of the coins in two purses is not the same after a transfer as before. Obviously, linear additive conserved variables are the easiest to deal with. And for all practical purposes we can restrict attention to the conservation of linear additive quantities. Of course it is no accident that the quantities that form the basis for scientific theories are precisely those quantities that are conserved and linear additive. Examples include energy, momentum, mass, and charge. More familiar to non-scientists, quantities like length, area, volume, and number are also conserved and linear additive.

Some quantities of value in the physical sciences are linear additive without necessarily being conserved. As we shall see, entropy is one such linear additive, non-conserved quantity. As we shall also see, it is most useful to formulate entropy as dependent on linear additive quantities, which mostly but not always are also conserved quantities. Part of the challenge in developing scientific theories from first principles is in identifying and formulating the linear additive quantities that will form its basis.

One of the direct consequences of linear additivity is the thermodynamic concept of extensivity. Essentially linear additive conserved variables scale with the size of the system, other things (intensive variables) being equal. Intensive variables are neither linear additive nor conserved; examples include pressure, temperature, voltage and surface tension. It will be argued that entropy is best formulated as dependent solely on extensive variables. This very simple idea of the scaling of extensive variables will be used profitably in several chapters below to guide the development of quite sophisticated theory.

### 1.1.2 Probability, weight, and entropy

From counting it is but a short step to probability, which leads directly to entropy. Probability theory is best formalized in terms of set theory. With the exception of parts of chapter 7, classical probability theory is invoked throughout.

We suppose that the system exists in a variety of states. These states can be microstates or macrostates. Microstates are the smallest indivisible states of the system. For example, the specified positions and momenta of all the particles. A macrostate is a set of microstates with the same value of some physical observable. A collective of macrostates correspond to the same physical observable, for example, the energy collective.

The set of all microstates is complete (the system at any one time must be in a particular microstate), and the microstates are disjoint (the system cannot be in more than one microstate at a time).

The set of all macrostates of a given collective is complete, and the macrostates of a given collective are disjoint. Degenerate macrostates are forbidden in the formalism, which means that there is a one-to-one relationship between the macrostate labels and the values of the corresponding physical observable. The system can be simultaneously in macrostates of different collectives.

We are interested in the probability of the system being in a particular state. We label microstates by a Roman letter and macrostates by a Greek letter. We suppose that each microstate $j$ has a non-negative weight $w_j$. The weights are usually defined up to a positive scale factor. Weight is the generalization of number when the objects or states being counted are not identical.

For the general formulation of the theory it does not matter how these weights come about or are calculated. However, for the application of the theory to a specific problem it does. The most convenient situation is when the microstates can be chosen such that they have equal weight, which may then be set to unity. In this case weight is the same as number. This is not necessary for either the formal development of the theory or for specific applications.

The weight of a macrostate $\alpha$ is the total weight of the microstates that it contains,

$$W_\alpha = \sum_j w_j \delta(A_j - A_\alpha) = \sum_{j \in \alpha} w_j. \tag{1.4}$$

The first equality contains a Kronecker-$\delta$ function ($\delta(0) = 1, \delta(x) = 0, \ x \neq 0$). This follows because the microstate $j$ is in the macrostate $\alpha$ if the value of the relevant observable for the microstate is equal to the value that defines the macrostate, $A_j = A_\alpha$. In the event that the microstates have equal weight and that this is set to unity, then the macrostate weight is just the number of microstates that it contains.

The total system weight is the sum of the weights of the microstates. Because a collective is complete and disjoint, this is also equal to the sum of the macrostate weights,

$$W = \sum_j w_j = \sum_\alpha W_\alpha. \tag{1.5}$$

The probability of the system being in the microstate $j$ is just

$$\wp(j) = \frac{w_j}{W}. \tag{1.6}$$

This is obviously normalized, $\sum_j \wp(j) = 1$. Similarly, the probability of the system being in the macrostate $\alpha$ is just

$$\wp(\alpha) = \frac{W_\alpha}{W}. \tag{1.7}$$

Consider now two different macrostate collectives, for example, energy and number of particles (without enforcing any conservation laws at this time). Label the macrostates of these by $\alpha$ and $\beta$. Because macrostates of different collectives are not disjointed, the probability that the system simultaneously has value $\alpha$ for the first observable and $\beta$ for the second is

$$\begin{aligned}\wp(\alpha, \beta) &= \frac{1}{W}\sum_j \delta(A_j - A_\alpha)\delta(B_j - B_\beta)w_j \\ &= \frac{1}{W}\sum_{j \in A_\alpha \cap B_\beta} w_j.\end{aligned} \tag{1.8}$$

This is proportional to the total weight of the set formed by the intersection of the two macrostates, see figure 1.1.

Given that the system is in the macrostate $\alpha$, then the probability that it is in the macrostate $\beta$ is

$$\begin{aligned}\wp(\beta|\alpha) &= \frac{1}{W_\alpha}\sum_{j \in A_\alpha \cap B_\beta} w_j \\ &= \frac{\wp(\alpha, \beta)}{\wp(\alpha)}.\end{aligned} \tag{1.9}$$

This is called the conditional probability, and it is usually written in the form of Bayes' theorem

$$\wp(\alpha, \beta) = \wp(\beta|\alpha)\wp(\alpha). \tag{1.10}$$

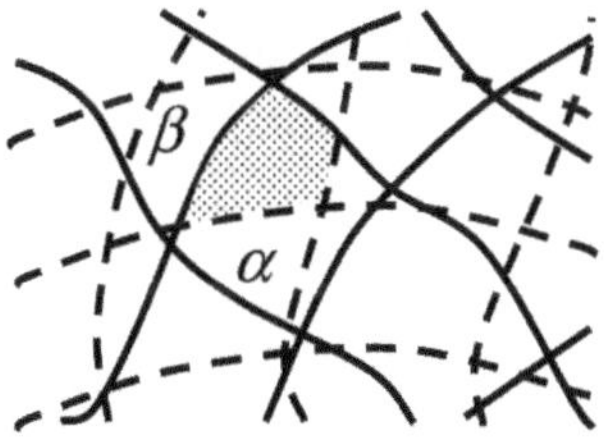


**Figure 1.1.** Two macrostate collectives. The $\alpha$ macrostate of one (solid lines) and the $\beta$ macrostate of the other (dashed lines) are labeled. The joint state $\alpha\beta$, their intersection, is shaded.

By the usual rules of set theory, the probability that the system is in the macrostate $\alpha$, or in the macrostate $\beta$ is

$$\wp\left(A_{\alpha} \bigcup B_{\beta}\right) = \wp(\alpha) + \wp(\beta) - \wp\left(\alpha, \beta\right). \tag{1.11}$$

The final term corrects for double counting of the microstates in the intersection of the two macrostates.

Now consider the case that the total system is made of two independent, non-interacting sub-systems. The macrostates of interest for the first sub-system may be labeled $\alpha$ and that for the second $\beta$. It does not matter in this case whether or not these correspond to the same physical observable because the sub-systems are independent. The weight of the macrostate $\alpha\beta$ of the total system is the product of the weights of the individual systems,

$$W_{\text{tot}}(\alpha\beta) = W_1(\alpha) W_2(\beta). \tag{1.12}$$

(This is readily checked by taking the weight equal to the number of microstates.) The total weight of the total system is also just the product of the total weights,

$$\begin{aligned} W_{\text{tot}} &= \sum_{\alpha,\beta} W_{\text{tot}}(\alpha\beta) \\ &= \sum_{\alpha} W_1(\alpha) \sum_{\beta} W_2(\beta) \\ &= W_1 W_2. \end{aligned} \tag{1.13}$$

Correspondingly, the probability for the total macrostate is the product of the individual probabilities,

$$\wp_{\text{tot}}(\alpha\beta) = \wp_1(\alpha)\, \wp_2(\beta). \tag{1.14}$$

As mentioned in the preceding subsection, the most convenient physical variables are those that are linear additive. The above example of two independent sub-systems shows that weight and probability are multiplicative rather than additive. But this deficiency is easy to overcome by taking logarithms. One simply defines the entropy as the logarithm of the weight. The total entropy is

$$S = k_{\text{B}} \ln W. \tag{1.15}$$

Here $k_{\text{B}} = 1.38 \times 10^{-23}$ J K$^{-1}$ is Boltzmann's constant. In the event that the microstates have equal unit weight, $W$ is just the total number of microstates and this corresponds to Boltzmann's original definition of entropy. Similarly, the entropy of a microstate is

$$S_j = k_{\text{B}} \ln w_j, \tag{1.16}$$

and the entropy of a macrostate is

$$S_{\alpha} = k_{\text{B}} \ln W_{\alpha}. \tag{1.17}$$

Because weights are usually defined up to a multiplicative constant, entropy is also defined up to an additive constant.

It follows from these that the probability of a state is proportional to the exponential of its entropy,

$$\wp(\alpha) = \frac{W_\alpha}{W} = \frac{e^{S_\alpha/k_B}}{W}. \tag{1.18}$$

Of course the same functional form holds also for microstates. It is important to note that there is a one-to-one relationship between entropy and probability (apart from the additive constant, which can be absorbed into the total weight anyway). This means, for example, it would make no sense to attempt to maximize the entropy with respect to the probability.

Entropy is commonly identified with disorder and unpredictability. This is a reasonable picture because entropy increases with weight, and weight can be thought of as number. Hence at the simplest level, the greater the entropy of a macrostate, the more microstates it contains. Each of these microstates is a possible configuration of the system, and it is harder to predict which one the system is actually in when there are many possibilities than when there are just a few. Furthermore, there are many more disordered arrangements of objects than there are ordered arrangements. This is illustrated in figure 1.2. One can see that there is just one microstate in the ordered 'phobic' macrostate (the first line, no identical circles touching) and just two microstates in the ordered 'philic' macrostate (the second line, all identical circles clustered

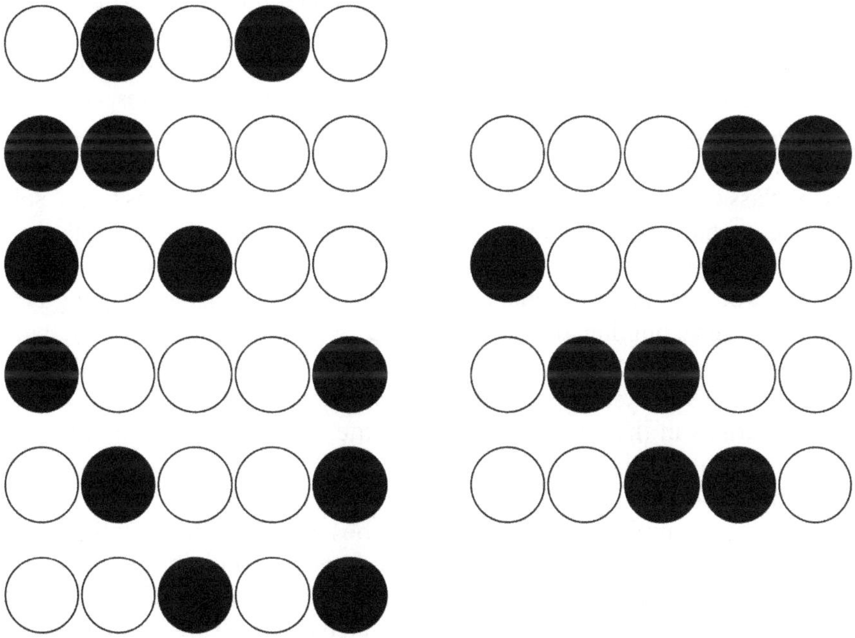

**Figure 1.2.** All ten arrangements of two filled and three empty circles in a line. The first line shows the 'phobic' and the second shows the 'philic' macrostate.

together). With such a definition of order, one can see that in general the majority of arrangements are disordered. Hence again there is a consonance between increasing entropy and increasing disorder.

### 1.1.3 Information entropy

In view of the above definitions it is straightforward to write the total entropy as a functional of the macrostate probability (Attard 2000, 2012),

$$\begin{aligned} S &= k_B \ln W \\ &= k_B \sum_\alpha \wp_\alpha \ln W \\ &= k_B \sum_\alpha \wp_\alpha \left[ \ln W_\alpha - \ln \frac{W_\alpha}{W} \right] \\ &= \sum_\alpha \wp_\alpha [S_\alpha - k_B \ln \wp(\alpha)]. \end{aligned} \tag{1.19}$$

This formula holds as well for microstates as for any collective of macrostates. Comparing this to the conventional result, equation (1.1), one must conclude that both cannot be correct.

Clearly equation (1.1) is missing the term $\sum_\alpha \wp_\alpha S_\alpha$, which can be interpreted as the average entropy per macrostate. The term that equation (1.1) does include, $-k_B \sum_\alpha \wp_\alpha \ln \wp_\alpha$, can be interpreted as the uncertainty due to the system having access to multiple macrostates. In any case, one needs both terms to give the total entropy of the system. Further, both terms are necessary for the total entropy of the system to be invariant to the representation: both the functional form of the formula and the numerical result for the total entropy are the same irrespective of the collective of macrostates or microstates that are used.

Figure 1.3 shows the exact entropy per site for the two-dimensional Ising model. This is a classic bench mark in statistical mechanics, with Onsager's solution representing perhaps the first exact solution of a non-trivial, physically realistic model that possesses a phase transition (Baxter 1982). The total entropy, which is the logarithm of the partition function, increases with increasing coupling between the spins. At high coupling this is dominated by the reservoir entropy, which is minus the average energy divided by the temperature. At low coupling it is dominated by the sub-system entropy, which is essentially the weighted number of spin configurations in the most likely energy state. When the spins interact weakly, they are free to randomly and independently orient, which increases their disorder and the entropy of the sub-system.

It can be seen in figure 1.3 that the Gibbs–Shannon information entropy, equation (1.1), evaluated in this case using the four spin probability $\wp^{2,2}(\sigma^4)$, gives only the sub-system part of the total entropy (see section 5.6 for details of the calculation). It neglects the reservoir entropy, which is effectively an internal entropy for each spin microstate. One can see that when the spins are highly correlated, the Gibbs–Shannon entropy is a negligible part of the total entropy.

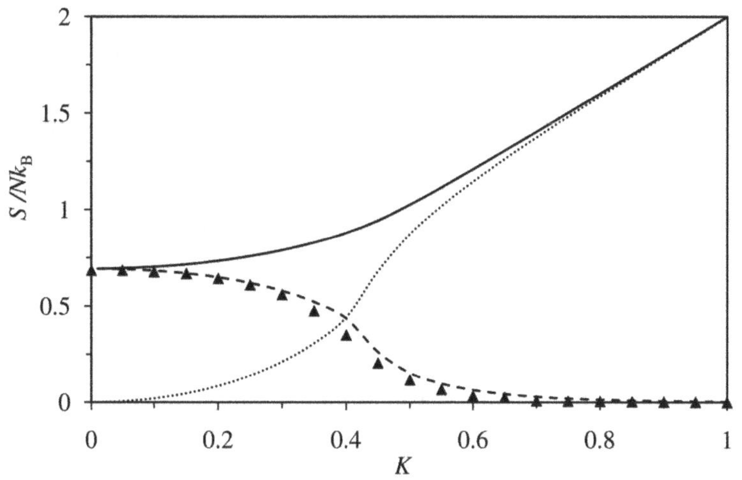


**Figure 1.3.** The entropy per site for the Ising model on a square lattice as a function of the coupling parameter (canonical equilibrium system). The curves are the exact result for the total (solid), reservoir (dotted), and subsystem (dashed) entropy. The symbols are the Gibbs–Shannon information entropy using the four spin probability function (Markov superposition formula (Attard 1999)).

One of the major uses of equation (1.1) has been in Jaynes' (1957, 2003) maximum entropy theory, in which the probability distribution that maximizes $S[\wp]$ is chosen. A similar use of $S[\wp]$ is made in Shannon's (1948) information theory. But in both cases the procedure is dubious for two reasons. First, the idea is obviously based upon the Second Law of Thermodynamics. We have not yet derived the Second Law of Thermodynamics but when we do we shall show that it holds for specific types of systems and variables. It does not hold for the probability distribution. It does not even make sense to say that the entropy is a maximum with respect to the probability distribution because there is a one-to-one relationship between the entropy of a state $S_\alpha$ and the probability of a state $\wp_\alpha$. The total entropy of the system is fixed by the sum of the entropies of the macrostates and it is pointless to attempt to use the total entropy to find the macrostate probability. The second reason that maximizing equation (1.1) with respect to the probability is pointless is that it only gives part of the entropy of the system, not the total entropy. Even if the Second Law of Thermodynamics were applicable, then it would refer to the total entropy, not part of the total entropy. What would be the use of maximizing part of the entropy?

In some books one sees it written that equation (1.1) can be shown to be true for equally likely states, as if this justifies the formula more generally. In this case the probability of a microstate is just $\wp_i = 1/n$, where the total number of microstates is $n = \sum_i$. Since the states are equally likely, their weight may be set to unity, $w_i = 1$, and their entropy may be set to zero, $S_i = k_B \ln w_i = 0$. In this case both equations (1.1) and (1.19) give the total entropy correctly as $S = k_B \ln n$. But there is no point in using equation (1.1) for equally likely states because if you know the number of equally likely states then you already know the total entropy. There is not even any point in using equation (1.19) more generally because if you know $S_\alpha$, which is required by that expression, then you already know the macrostate weight $W_\alpha$ and probability $\wp_\alpha$, the total weight $W = \sum_\alpha W_\alpha$, and the total entropy $S = k_B \ln W$. And finally, the fact that equation (1.1) is true for equally likely states does not make it true in general or useful in particular.

The reader may be interested to pinpoint where Shannon (Shannon and Weaver 1949, appendix 2) and Jaynes (2003, section 11.3) go wrong in their derivation of the entropy as a functional of the probability, equation (1.1). With the microstate probability $\wp_i$ and the macrostate probability $\wp_\alpha$ they say that the entropy of the system can be written as a functional of these with an additional term representing the uncertainty of locating the microstate within each macrostate,

$$S[\wp_i] = S[\wp_\alpha] + \sum_\alpha \wp_\alpha S_\alpha, \quad S_\alpha \equiv S[\wp(i|\alpha)]. \tag{1.20}$$

Here the macrostate entropy $S_\alpha \equiv S[\wp(i|\alpha)]$ is meant to be the same functional of the probability as $S[\wp_i]$ and $S[\wp_\alpha]$, with the conditional probability of microstates in the macrostate $\alpha$ as its argument.

As I have pointed out in greater detail elsewhere (Attard 2012), this equation has an internal contradiction. The notation in equation (1.20) implies that the microstate functional $S[\wp_i]$ and the macrostate functional $S[\wp_\alpha]$ are the same function of their arguments. But this contradicts the content of equation (1.20), which implies that $S[\wp_i]$ is the total entropy and that $S[\wp_\alpha]$ is only part of the total entropy. The problem is that while Shannon and Jaynes accept that macrostates have internal entropy $S[\wp(i|\alpha)]$ (and so $S[\wp_\alpha]$ is only part of the total entropy), they assume that microstates do not, ($S[\wp_i]$ is the entire total entropy). The last assumption is incorrect for microstates with non-uniform weight. The way to resolve this discrepancy is to rewrite equation (1.20) in terms of the macrostate probability with the total entropy explicit,

$$S_{\text{total}} = S[\wp_\alpha] + \sum_\alpha \wp_\alpha S_\alpha, \quad S_\alpha \equiv S[\wp(i|\alpha)]. \tag{1.21}$$

Using the subsequent arguments of Shannon (Shannon and Weaver 1949, appendix 2), and of Jaynes (2003, section 11.3), this can be solved to yield the correct expression for the entropy as a functional of the probability, equation (1.19) (Attard 2012), as will now be shown.

Following Shannon and Jaynes, for uniform microstates, $\wp_i^{\text{u}} = 1/n$, and uniform macrostates, $\wp_\alpha^{\text{u}} = m/n$, suppose that $S[\wp_i^{\text{u}}] = \sigma(n)$, $S[\wp_\alpha^{\text{u}}] = \sigma(n/m)$, $S[\wp^{\text{u}}(i|\alpha)] = \sigma(m)$, and that $\sigma(1) = 0$. Then one has

$$\begin{aligned} S_{\text{total}} &= \sigma(n/m) + \sum_\alpha \wp_\alpha \sigma(m) \\ &= \sigma(n) + \sum_i \wp_i \sigma(1), \end{aligned} \tag{1.22}$$

which has solution $\sigma(n) = k_B \ln n$. For the case of non-uniform macrostates, $\wp_\alpha = m_\alpha / n$, and uniform microstates $\wp_i^u = 1/n$, then $S[\wp^u(i|\alpha)] = \sigma(m_\alpha)$, and

$$\begin{aligned} S_{\text{total}} &= \sigma(n) \\ &= S[\wp_\alpha] + \sum_\alpha \wp_\alpha \sigma(m_\alpha), \end{aligned} \tag{1.23}$$

or

$$\begin{aligned} S[\wp_\alpha] &= k_B \ln n - \sum_\alpha \wp_\alpha k_B \ln m_\alpha \\ &= - k_B \sum_\alpha \wp_\alpha \ln \wp_\alpha. \end{aligned} \tag{1.24}$$

This shows that the $\wp \ln \wp$ term is just part of the total entropy, and that the full expression is

$$\begin{aligned} S_{\text{total}} &= S[\wp_\alpha] + \sum_\alpha \wp_\alpha S_\alpha \\ &= \sum_\alpha \wp_\alpha [S_\alpha - k_B \ln \wp_\alpha], \end{aligned} \tag{1.25}$$

which is of course equation (1.19).

In describing the Gibbs–Shannon expression equation (1.1) as either incorrect, or else, in the uniform probability case, as correct but pointless, one should concede that this formula has long been successfully used in communications and information theory. Perhaps the problem is one of nomenclature. If Shannon had just called *his* $H$-function the information, as he originally introduced it (Shannon 1948, section 6), rather than calling it the entropy based on its resemblance to a similar function used in statistical mechanics, then it could have been judged solely on its utility for informatics, rather than by whether or not it represented the total entropy of a physical system.

As a concrete example, consider a fair six-sided die. In the first case let the number showing be reported, which means that there are six states each with probability $\wp_i = 1/6$ and internal entropy $S_i = 0$. In this case the Gibbs–Shannon information entropy is the same as the total entropy of the system,

$$S_{GS} = S_{\text{total}} = k_B \ln 6. \tag{1.26}$$

In the second case, suppose that the state of the die is reported as 1, 3, 5 or even. The first three states each have probability $\wp_i = 1/6$ and entropy $S_i = 0$, and the final state has probability $\wp_4 = 1/2$ and entropy $S_4 = k_B \ln 3$. The Gibbs–Shannon information entropy is in this case

$$S_{GS} = k_B \left\{ \frac{3}{6} \ln 6 + \frac{1}{2} \ln 2 \right\} = k_B \ln \sqrt{12} \approx k_B \ln 3.5, \tag{1.27}$$

which is less than in the first case. The total entropy is

$$S_{\text{total}} = S_{\text{GS}} + k_{\text{B}}\left\{\frac{3}{6}\ln 1 + \frac{1}{2}\ln 3\right\} = k_{\text{B}}\ln 6, \tag{1.28}$$

which is the same as in the first case. Arguably, information has been lost in the second case because the even numbers are not individually identified, and so the Gibbs–Shannon information entropy may be regarded as a realistic quantification of the information in the two systems. Similarly, because the two systems are physically identical, the total entropy must be the same in each, which it is. This example suggests that the Gibbs–Shannon formula should be reserved for informatic applications, and that when applied to physical systems the additional contribution required for the total entropy should be added.

There has been a great deal of discussion over the years as to whether or not the Gibbs–Shannon information entropy is in any way related to the thermodynamic or physical entropy. My attitude is that on the one hand, in the physical sciences, it is incorrect to imply that the Gibbs–Shannon information entropy is 'the' entropy of a physical system. Clearly it is only part of the total entropy, as the above analysis, equation (1.19), and the results in figure 1.3 show. On the other hand, it *is* part of the total entropy of a physical system, and it offers a relatively straightforward and quickly converging way to calculate that part (see section 5.6). This part can be difficult to obtain by other methods, whereas the remaining part, the internal entropy of the macrostate, which often turns out to be the reservoir entropy of a physical system, is not difficult to obtain. The latter can be simply added to the Gibbs–Shannon information entropy to get the total entropy. Hence provided that one is explicit that the Gibbs–Shannon information entropy is only part of the total entropy, it can be quite useful in thermodynamics and statistical mechanics.

### 1.1.4 Continuum

The above analysis was implicitly for discrete states. In many cases it is worthwhile taking the continuum limit. It is quite straightforward to transform the sums to integrals in the usual fashion. Alternatively, one can derive the results for the continuum directly, as is now done.

Let the state of the system be $\mathbf{x}$, a point in a multidimensional space, and let $\omega(\mathbf{x})$ be the weight density, which is non-negative. (The quantity $\mathbf{x}$ may be regarded as the continuum analogue of microstates, although similar procedures also hold for continuum macrostates.) The total weight of the system is

$$W = \int \mathrm{d}\mathbf{x}\ \omega(\mathbf{x}). \tag{1.29}$$

The total entropy is just the logarithm of this $S = k_{\text{B}} \ln W$.

The probability density is proportional to the weight density

$$\wp(\mathbf{x}) = \frac{\omega(\mathbf{x})}{W}. \tag{1.30}$$

This is obviously normalized, $\int \mathrm{d}\mathbf{x}\, \wp(\mathbf{x}) = 1$. The probability density has the interpretation that $\wp(\mathbf{x})\mathrm{d}\mathbf{x}$ is the probability of the system being within $\mathrm{d}\mathbf{x}$ of $\mathbf{x}$.

The average of a function of the state of the system is

$$\langle f \rangle = \int \mathrm{d}\mathbf{x}\, \wp(\mathbf{x}) f(\mathbf{x}). \tag{1.31}$$

One can introduce an arbitrary volume element $\Delta(\mathbf{x})$, and define the entropy of the state of the system in terms of it as

$$S(\mathbf{x}) = k_{\mathrm{B}} \ln [\omega(\mathbf{x})\Delta(\mathbf{x})]. \tag{1.32}$$

With this the probability density is

$$\wp(\mathbf{x}) = \frac{e^{S(\mathbf{x})/k_{\mathrm{B}}}}{\Delta(\mathbf{x}) W}. \tag{1.33}$$

The volume element $\Delta(\mathbf{x})$ is solely a matter of convenience that makes the argument of the logarithm dimensionless and gives the probability density the correct dimensions. Obviously the probability density is independent of the choice of $\Delta(\mathbf{x})$, since the one implicit in the state entropy cancels with the one that appears explicitly in the probability density. Since the volume element has no physical consequences, it can be taken to be a constant or it could be ignored altogether.

The total entropy of the system may be written as a functional of the probability density,

$$\begin{aligned} S &= k_{\mathrm{B}} \ln W \\ &= \int \mathrm{d}\mathbf{x}\, \wp(\mathbf{x}) k_{\mathrm{B}} \ln W \\ &= \int \mathrm{d}\mathbf{x}\, \wp(\mathbf{x}) \left[ S(\mathbf{x}) - k_{\mathrm{B}} \ln \frac{e^{S(\mathbf{x})/k_{\mathrm{B}}}}{W} \right] \\ &= \int \mathrm{d}\mathbf{x}\, \wp(\mathbf{x}) [S(\mathbf{x}) - k_{\mathrm{B}} \ln \{\wp(\mathbf{x})\Delta(\mathbf{x})\}]. \end{aligned} \tag{1.34}$$

This is the continuum analogue of equation (1.19). Again the arbitrary volume element that appears explicitly here, cancels with that implicit in $S(\mathbf{x})$, and the total entropy is independent of the choice of $\Delta(\mathbf{x})$.

In the case of the continuum, the macrostates are represented as hypersurfaces in the space. These hypersurfaces correspond to constant values of particular observables. In general, the observables depend upon the state of the system $\mathbf{x}$.

## 1.2 A combinatorial example

In this section a simple combinatorial example is given that illustrates the use of entropy. The analysis follows Attard (2002, section 1.2).

### 1.2.1 Entropy

Suppose that the system consists of a container with $N$ cells, and that there are currently $n \leqslant N$ objects in the container, with at most one object per cell.

A microstate of the system is a specification of which cells are occupied. There are $W_N(n) = {}^N C_n \equiv N!/(N-n)!n!$ possible microstates. The entropy of the system with this number of cells and objects is therefore

$$S_N(n) = k_B \ln W_N(n) = k_B \ln \frac{N!}{(N-n)!n!}. \tag{1.35}$$

Now add a second container of $M$ cells containing $m$ objects to the system. The number of microstates for this container is $W_M(m) = {}^M C_m$, and the entropy for this second container alone is $S_M(m) = k_B \ln W_M(m)$. The number of microstates of the total system with the fixed allocation of objects in the two containers is

$$W_{N,M}(n, m) = W_N(n) W_M(m). \tag{1.36}$$

It is the product that appears because for each configuration of one container all configurations of the other container are possible. The entropy of the total system with the fixed allocation is obviously

$$\begin{aligned} S_{N,M}(n, m) &= k_B \ln W_{N,M}(n, m) \\ &= k_B \ln W_N(n) + k_B \ln W_M(m) \\ &= S_N(n) + S_M(m). \end{aligned} \tag{1.37}$$

This is an explicit example of the linear additive nature of entropy.

### 1.2.2 Equilibrium allocation

Now introduce a transition rule that allows exchange of objects between the two containers. Choose a cell at random in the first container and a cell at random in the second container and swap their contents. (One can use similar transition rules within each container to change their internal configurations, but this is not necessary for what follows.) Obviously, if both cells are occupied or if both cells are empty nothing happens. But if the first cell is occupied and the second cell is empty, then an object is transferred from the first container to the second: $n \Rightarrow n-1$ and $m \Rightarrow m+1$. The total number of objects $n_{tot} = n(t) + m(t)$ is conserved by the transfer rule.

Intuitively, one expects that on average objects will move from the container with the greater concentration to that with the lesser. It is the concentration of particles rather than number of particles that is relevant because simply doubling the size of the container and doubling the number of particles does not change the probability of a nett transfer. Eventually a steady state will be reached where there is no nett transfer, and the concentrations are the same in the two containers. (Concentration is an example of an intensive variable, whereas number is an extensive variable. It is quite common in thermodynamics that the optimum state is found by equalizing intensive variables.) The steady state will treat objects and empty cells equally, since the transfer of an object in one direction is equivalent to the transfer of an empty cell in the opposite direction.

In order to makes this intuition quantitative, one needs the expression for the transfer probability. The probability of a cell in the first container being occupied is just $n/N$, and the probability of it being empty is of course $(N - n)/N$. Analogous expressions hold for the second container. The probability for an object to be transferred from the first container to the second is therefore

$$\wp(1 \to 2) \equiv \wp(n - 1|n) = \frac{n}{N}\frac{M - m}{M}. \tag{1.38}$$

As the product of two independent normalized probabilities, this is correctly normalized. Obviously the probability of an object going from the second container to the first is

$$\wp(2 \to 1) \equiv \wp(n + 1|n) = \frac{N - n}{N}\frac{m}{M}. \tag{1.39}$$

The probability of no nett exchange is just

$$\wp(n|n) = \frac{n}{N}\frac{m}{M} + \frac{N - n}{N}\frac{M - m}{M}. \tag{1.40}$$

The steady state of no nett transfer is denoted with an over-line, $n = \bar{n}$ and $m = \bar{m}$. At this point the transfer probabilities must be equal, $\wp(\bar{n} - 1|\bar{n}) = \wp(\bar{n} + 1|\bar{n})$, which is to say

$$\frac{\bar{n}M - \bar{n}\bar{m}}{NM} = \frac{\bar{m}N - \bar{n}\bar{m}}{NM}, \quad \text{or} \quad \frac{\bar{n}}{N} = \frac{\bar{m}}{M}. \tag{1.41}$$

As was intuited above, there is no nett transfer on average when the proportion of occupied cells is the same in both containers. Clearly one has the same result for empty cells, $(N - \bar{n})/N = (M - \bar{m})/M$.

### 1.2.3 Maximum entropy

It is now shown that the entropy of the total system is a maximum in this steady state. Hence it is reasonable to call it the equilibrium state. It will be shown below that the transition rule drives the system to this state, and that it is stable to small fluctuations.

The total entropy for the system constrained to be in the equilibrium state is

$$S_{N,M}(\bar{n}, \bar{m}) = k_B \ln \frac{N!}{(N - \bar{n})!\bar{n}!} + k_B \ln \frac{M!}{(M - \bar{m})!\bar{m}!}. \tag{1.42}$$

Now consider a state on one or other side of the equilibrium state, $n = \bar{n} + p$ and $m = \bar{m} - p$, and consider a neighboring state, $n' = n + 1$, and $m' = m - 1$. If $p > 0$ then there are more than the equilibrium number of objects in the first container, $n > \bar{n}$, and if $p < 0$ then there are less than the equilibrium number of objects in the first container, $n < \bar{n}$. The difference between their entropies is

$$
\begin{aligned}
&S_{N,M}(n', m') - S_{N,M}(n, m) \\
&= k_B \ln \frac{(N-n)!n!}{(N-n-1)!(n+1)!} \frac{(M-m)!m!}{(M-m+1)!(m-1)!} \\
&= k_B \ln \frac{N-n}{n+1} \frac{m}{M-m+1} \\
&= k_B \ln \frac{\alpha - \frac{p}{M}}{\alpha + \frac{p+1}{N}} \frac{1-\alpha-\frac{p}{N}}{1-\alpha+\frac{1+p}{M}},
\end{aligned}
\tag{1.43}
$$

where $\alpha \equiv \bar{n}/N = \bar{m}/M$. If $p > 0$ ($n' = n + 1$ is further from $\bar{n}$ than $n$), both fractions are less than unity, the logarithm is negative, and the entropy difference is negative. Hence the entropy decreases moving further away from equilibrium and increases toward equilibrium. If $p < 0$ ($n' = n + 1$ is closer to $\bar{n}$ than $n$), both fractions are greater than unity, the logarithm is positive, and the entropy difference is positive. Hence the entropy increases toward equilibrium and decreases moving further away from equilibrium. One concludes that the total entropy is a concave down function that attains its maximum at equilibrium $\bar{n}$.

One has to distinguish between the constrained entropy and the unconstrained entropy. If objects are transferable between the two containers, then the total number of distinct configurations is $W_{N+M}(n+m) = {}^{N+M}C_{n+m}$, since all cells are now accessible to all objects. The total entropy is the usual logarithm of this $S_{N+M}(n+m) = k_B \ln W_{N+M}(n+m)$. This is the unconstrained entropy, where $n$ and $m$ are not individually fixed because the containers are effectively combined.

The total number of configurations in this unconstrained case must be greater than any of those of the isolated containers with a fixed allocation of the objects. The latter case is the constrained entropy case. The reason that the unconstrained entropy must be greater than the constrained entropy is that it includes all the configurations of the latter, plus all the configurations with a different allocation of the objects. That is, $W_{N+M}(n+m) \geqslant W_N(n)W_M(m)$ or in terms of the entropy, $S_{N+M}(n+m) \geqslant S_N(n) + S_M(m)$, for any $n$, $m$. (This is a strict inequality unless $N$ or $M$ is 0, or unless there are no objects, $n + m = 0$, or no empty cells, $n + m = N + M$.) One concludes that the entropy of the two containers able to exchange objects is greater than the entropy of the two isolated containers with the equilibrium allocation of objects, which is greater than the entropy of any other allocation of objects to the isolated containers,

$$
S_{N+M}(n+m) \geqslant S_N(\bar{n}) + S_M(\bar{m}) \geqslant S_N(n) + S_M(m), \tag{1.44}
$$

where $\bar{n} + \bar{m} = n + m$, and $\bar{n}/N = \bar{m}/M$. This behavior of the constrained entropy is shown in figure 1.4.

Note that it is the constrained entropy that is linear additive.

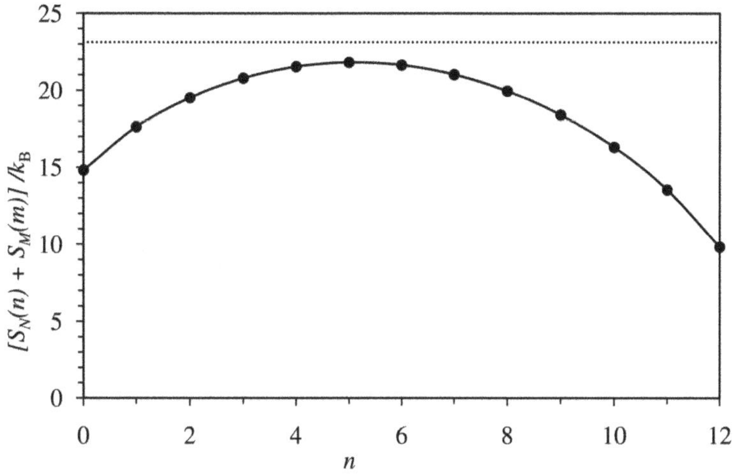


**Figure 1.4.** The constrained total entropy for two containers, $N = 18$ and $M = 24$, with a total of $n + m = 12$ particles, as a function of $n$. The maximum $S_N(\bar{n}) + S_M(\bar{m}) = 21.8k_B$ occurs at $\bar{n} = 5$. The dotted line is the unconstrained total entropy, $S_{N+M}(n + m) = 23.1k_B$.

### 1.2.4 Stability and the direction of motion

We now prove that the equilibrium state is stable, which is to say that if $n = \bar{n} + p$, $m = \bar{m} - p$, then $\wp(1 \to 2) > \wp(2 \to 1)$ if $p > 0$, and vice versa. For arbitrary $p$ we have

$$\begin{aligned}\frac{\wp(1 \to 2)}{\wp(2 \to 1)} &= \frac{n(M - m)}{(N - n)m} \\ &= \frac{(\bar{n} + p)(M - \bar{m} + p)}{(N - \bar{n} - p)(\bar{m} - p)} \\ &= \frac{\left(\alpha + \frac{p}{N}\right)}{\left(\alpha - \frac{p}{M}\right)} \times \frac{\left(1 - \alpha + \frac{p}{m}\right)}{\left(1 - \alpha - \frac{p}{N}\right)},\end{aligned} \tag{1.45}$$

where $\alpha \equiv \bar{n}/N = \bar{m}/M$. Since the number of occupied cells cannot be negative, one must have $|p| \leqslant \bar{n}$, and $|p| \leqslant \bar{m}$, and since the number of empty cells cannot be negative one must have $|p| \leqslant N - \bar{n}$, and $|p| \leqslant M - \bar{m}$. Hence, by inspection, if $p > 0$, then the two fractions are both greater than unity, and if $p < 0$, then the two fractions are both less than unity. Hence one can conclude that

$$\frac{\wp(1 \to 2)}{\wp(2 \to 1)} \begin{cases} > 1, & \text{if } n > \bar{n}, \\ < 1, & \text{if } n < \bar{n}. \end{cases} \tag{1.46}$$

These say that if $n > \bar{n}$, then there is more likely to be a decrease in $n$ than an increase, and vice versa. This shows that the equilibrium state is a stable state, which is to say that fluctuations in $n$ from $\bar{n}$ are likely to be countermanded by the subsequent transitions.

This proof of the stability of the equilibrium state shows that if the system is displaced from the equilibrium state, either by spontaneous fluctuations, initial preparation, or forces of constraint, then there is a nett driving force toward the equilibrium state. The most likely flux of particles is in the direction of increasing constrained entropy. Although the velocity has not been discussed, on symmetry grounds (vectors are proportional to vectors) one might anticipate that it is proportional to the gradient in entropy.

Finally, there is nothing in this model that precludes a spontaneous fluctuation from the equilibrium state. During such a fluctuation the constrained entropy decreases. This shows that the Second Law of Thermodynamics should not be taken overly literally. For a macroscopic system fluctuations are relatively small and can be neglected. For a macroscopic system prepared in a state away from equilibrium, with overwhelming probability it will evolve in the direction of increasing entropy.

### 1.2.5 Physical interpretation

The interpretation of these results for this simple combinatorial model gives insight into the behavior of thermodynamic and statistical mechanical systems more generally. One fundamental point is the distinction between microstates and macrostates. In the present example a microstate is a configuration of the objects, which is to say the list of occupied cells. In contrast, a macrostate is the number of occupied cells in a container, irrespective of the specific cells occupied. Obviously each macrostate contains many microstates. A specified macrostate constrains the two containers to hold the given number of objects. If no macrostate is specified, there is no such constraint, objects are free to transfer between the two containers, and the number of occupied cells in each container ranges over all possible values. This means that the number of configurations for the constrained system (i.e. specified macrostate) is less than the number of configurations without constraint (i.e. no specified macrostate). It follows that the entropy of a system constrained to be in a given macrostate (constrained or macrostate entropy) is less than the entropy of an unconstrained system (unconstrained or total entropy).

If the system is initially set up in a macrostate other than the equilibrium one, then there is likely to be a nett flux of objects toward the equilibrium allocation. This is not driven by any increase in the unconstrained or total entropy because once the total number of objects is fixed and allowed to transfer between the containers the number of possible configurations (microstates) does not change. Instead, the evolution toward equilibrium is a macroscopic flux, with each successive macrostate most likely having greater entropy than the preceding one. That is, it is the entropy constrained by the current value of the quantity in flux that increases during the approach to equilibrium. It is the constrained entropy that is a maximum in the equilibrium macrostate.

It ought to be clear that the progress toward equilibrium is a statistical one that holds on average, or in a most likely sense. It is entirely possible for a temporary reversal of the macrostate flux to occur due to the random nature of the microstate transitions. Likewise, once the equilibrium macrostate is attained, there can be temporary fluctuations to nearby macrostates, which by definition have lower entropy.

The above mathematical analysis shows the increase in the constrained entropy during the approach to the equilibrium state. The physical reason for this increase does not lie in the microstates, because the total number of unconstrained configurations is fixed (once the total number of particles and the freedom to exchange is specified) and hence so is the total unconstrained entropy. The microstates do not drive the approach to equilibrium. Transitions between microstates scrambles the configurations, but each such microstate transition is as likely as its reverse. There is no flux and no equilibration at the level of microstates.

It is the flux in macrostates that equilibrates the system. The essential point is that the number of microstates differs between macrostates. Because of this the transition between macrostates is asymmetric. If the entropy increases from macrostate 1 to 2 to 3 (i.e. the number of microstates in each macrostate increases in this sequence), then a system in a microstate belonging to macrostate 2 is more likely to make a transition to a microstate in 3 than in 1 simply because there are more target microstates in 3 than in 1.

This point is clear in the present combinatorial example where the transitions are purely stochastic, and the number of possible target microstates for a given microstate is proportional to the size of the target macrostate. An example of this is illustrated in figure 1.5. Although this model is highly idealized, the broader point

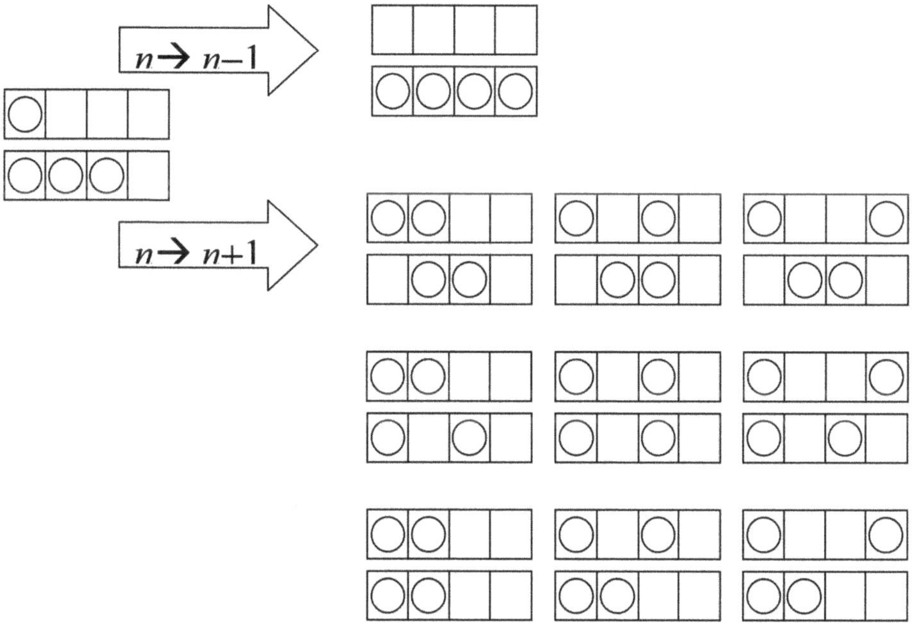


**Figure 1.5.** All microstate transitions that change the macrostate for a system of two containers, $N = M = 4$ with $n + m = 4$ particles. The initial microstate lies in the macrostate $n = 1, m = 3$ (left). There is one transition that decreases $n$ (upper right), and nine transitions that increase $n$ (lower right). The constrained entropy of the initial macrostate is $S_{4,4}(1, 3) = k_B \ln 16$, that of the $n$-decreasing final macrostate is $S_{4,4}(0, 4) = 0$, and that of the $n$-increasing final macrostate is $S_{4,4}(2, 2) = k_B \ln 36$, which in fact is the maximum constrained entropy.

also holds for the statistical contribution to more physical models with realistic dynamics.

The asymmetry in the macrostate transition probability, $\wp(2 \to 3) > \wp(2 \to 1)$ if $S_3 > S_1$, creates the irreversibility in the progress toward equilibrium. Of course the opposite transition may occur at any instant, but microstate transitions that increase macrostate entropy are more probable than the reverse, and in the long run these will occur most frequently.

The equilibrium macrostate is the one with the most microstates. Although fluctuations to nearby macrostates occur, which necessarily decrease the constrained entropy of the system, usually the constrained entropy is such a sharply peaked function of the macrostates that the effects of such fluctuations are relatively negligible.

## 1.3 Entropy and the second law

Let us return to the Second Law as enunciated by Clausius (1850),

> The entropy increases during spontaneous changes in the state of a system. (1.47)

Compare this to Boltzmann's (1895), identification of the physical basis of entropy,

> The entropy of a state is the logarithm of the number of molecular configurations in the state. (1.48)

The general definition of entropy given earlier in this chapter is slightly broader than Boltzmann's, since it uses weight rather than number, but it reduces to Boltzmann's in the case that the molecular configurations have equal weight, which they do in the phase space of an isolated system, but not more generally.

Boltzmann's insight showed the equivalence of entropy and probability, since for equally likely molecular configurations, the probability of a state is proportional to the number of configurations that it contains. This equivalence was made extant in the general definition of entropy in terms of weight given earlier in the chapter. The relationship of probability to entropy introduces a stochastic element into thermodynamics, with the Second Law now interpreted to mean that

> Spontaneous transitions will more likely occur from less probable states to more probable states. (1.49)

This does not have quite the zing of Clausius' version, but it does have greater precision. For macroscopic systems the transition probability approaches certainty. This version of the Second Law was implicitly assumed above in the discussion of the macrostate transition probability, and it seems to be intuitively obvious. But whether or not Boltzmann has indeed provided an explanation and a proof of the Second Law of Thermodynamics requires greater scrutiny.

The probabilistic nature of the Second Law of Thermodynamics can be readily seen. An equilibrium system does not change macroscopically in time, but due to molecular motion, fluctuations about the equilibrium state must occur. It will be seen that the relative magnitude of these fluctuations in macroscopic variables scales inversely with the square root of the system size, and so they are generally too small to measure experimentally. Any spontaneous fluctuation away from the equilibrium state must be to a state of lower entropy, which disproves any literal interpretation of Clausius' version of the second law. From this one can conclude that the Second Law of Thermodynamics in its literal form should be applied to systems that have been prepared in an initial state far from equilibrium, which is to say a low entropy, spontaneously improbable state. In this case the state of the system will spontaneously evolve toward the equilibrium, maximum entropy, most probable state, and the probability of such evolution approaches certainty for a macroscopic system.

Boltzmann's definition of entropy, equation (1.48), applies to the probability of a state, whereas the second law, especially in its more precise form, equation (1.49), applies to the probability of transitions between states. Obviously these are different things. Equilibrium theory depends on Boltzmann's identification of entropy and probability for a state, whereas non-equilibrium theory depends on the nature of transition probabilities. It is slightly perverse that Clausius' Second Law is always regarded as a law of equilibrium thermodynamics, whereas a more pedantic interpretation would be that it applies to not-in-equilibrium systems. In any case, in what fashion the transition probabilities are determined by the Second Law is now discussed.

The challenge is to make more precise the intuitively appealing notion that transitions to a more probable state should be more probable then transitions to a less probable state. One would like to deduce something about the transition probability from the state probability and relate this to the spontaneous evolution of entropy and the second law.

The unconditional probability of observing the transition from the state $\alpha$ to the state $\alpha'$ in the time interval $\tau > 0$ may be denoted $\wp(\alpha', \alpha|\tau)$. The probability of the reverse transition is $\wp(\alpha, \alpha'|\tau)$. Since there is no preferred direction of time in an equilibrium system, one is just as likely to observe the forward transition as the reverse, $\wp(\alpha', \alpha|\tau) = \wp(\alpha, \alpha'|\tau)$. (For simplicity, at this introductory level we assume that the states are even functions of the molecular velocities.) According to Bayes' theorem, which was given above as one of the laws of probability, the unconditional and conditional probabilities are related as $\wp(\alpha', \alpha|\tau) = \wp(\alpha'|\alpha, \tau)\wp(\alpha)$. Here, $\wp(\alpha'|\alpha, \tau)$ is the probability of the system being in the state $\alpha'$ given that it was in the state $\alpha$ a time $\tau$ earlier, and $\wp(\alpha)$ is the probability that the system is in the state $\alpha$. For an equilibrium system the latter does not depend upon time.

For a large time interval, any two states are uncorrelated. In this limit the unconditional probability reduces to the product of the singlet probabilities, $\wp(\alpha', \alpha|\tau) \to \wp(\alpha')\wp(\alpha)$, $\tau \to \infty$. Also, the conditional probability is independent of the initial state, $\wp(\alpha'|\alpha, \tau) \to \wp(\alpha')$, $\tau \to \infty$. If one now introduces a third state $\alpha''$, then in this limit

$$\frac{\wp(\alpha'|\alpha, \tau)}{\wp(\alpha''|\alpha, \tau)} \to \frac{\wp(\alpha')}{\wp(\alpha'')}, \quad \tau \to \infty. \tag{1.50}$$

If the state $\alpha'$ is more probable than the state $\alpha''$ (Boltzmann would say it has more entropy), then this says that a system in the state $\alpha$ is in the long term more likely to make the transition to $\alpha'$ than to the state $\alpha''$. In other words, transitions to states of higher entropy are more likely to be observed than the reverse. In this sense Boltzmann's physical interpretation of entropy explains Clausius' Second Law of Thermodynamics.

Although illuminating, this argument only applies in the long time limit, and so it cannot be the whole story. On shorter time intervals correlations between the initial and destination states can be expected. Although not rigorous, one can still make an informative argument for finite time intervals. Let $S(x)$ be the entropy of the state $x$. We are interested in the microstate transitions, and let $n_{\pm}(x)$ denote the number that increases or decreases the value of $x$. Now for an equilibrium system, the number of forward transitions must equal the number of reverse transitions, $n_{+}(x) = n_{-}(x)$. Let us make the intuitively appealing assumption that the number of such molecular transitions is proportional to the number of molecular configurations, $n_{\pm}(x) \propto \exp S(x)/k_{\mathrm{B}}$. The interpretation of this assumption is that larger states (i.e. those with a greater number of molecular configurations) have a larger number of transitions. Now consider that a state has finite width $\Delta_x > 0$. In this case the excess number of forward transitions is $n_{+}(x + \Delta_x/2) - n_{-}(x - \Delta_x/2) \propto \Delta_x \mathrm{d}S(x)/\mathrm{d}x$. One sees that if the states are ordered in terms of increasing entropy, (i.e. $\mathrm{d}S(x)/\mathrm{d}x \geqslant 0$), then there are more forward transitions in the direction of increasing entropy than backward transitions. This is the essence of the Second Law of Thermodynamics.

This argument provides a second link between Boltzmann's interpretation of entropy and Clausius' Second Law of Thermodynamics. The essential points are that the number of transitions between configurations is proportional to the number of configurations themselves. It is also essential that the transitions are steps of finite width, which is to say that this does not apply for infinitesimal transitions. It follows from this second argument that the driving force for the transitions is the gradient in the entropy, as was asserted above in the combinatorial example, section 1.2.4. A more sophisticated version of these arguments is given in section 3.1.5 below.

## 1.4 Nature of probability and randomness

Since entropy is in essence the same as probability (because there is a one-to-one relationship between the two), it is worthwhile discussing the conceptual basis of probability in the physical sciences. Such a fundamental understanding of probability determines not just the mathematical formulation of probability theory, but also the quantitative values of the weights that are to be applied to the microstates that occur in statistical mechanics.

Probability is also intimately connected to the notion of randomness. Of course, classical statistical mechanics, with which much of this book is concerned, is based on the classical equations of motion (Newton's, or equivalently, Hamilton's). Since these are deterministic, there is an issue as to how randomness arises in the physical sciences.

### 1.4.1 Probability

There are three philosophical positions that have been adopted for the nature of probability. The first is frequency, which holds that the probability of a state (or an event) is the frequency with which the state occurs in a long sequence of repeated trials. The second is credibility or degree of reasonable belief. This is also known as the subjective interpretation of probability, and it holds that the probability of a state is the strength with which the observer believes that the state will occur. The third is measure (or propensity), which holds that the probability of a state is the weight of the set of conditions that give rise to the state.

Of these three philosophical positions, the present author rejects the first two and embraces the third for the following reasons.

*Frequency*

To say that frequency is probability is to confuse cause and effect. The frequency with which an event or state occurs is a direct consequence of the probability of the event or state. In some cases where it is possible to make a sequence of trials, it is possible to measure probability by the frequency, But in the physical sciences we would say that the probability of the state existed whether or not the trials were carried out, whereas the frequency only exists after the trials are performed. Further, once-only events, such as the winner of a particular horse race, or the weather on a particular day, have a probability associated with them prior to the event, even though it is physically impossible to repeat these in a sequence of trials.

Of course some might say that frequency is just a mental device that allows probability to be explained to the uninitiated, or that allows it to be concretely visualized. But the problem with this defence is that it replaces the real physical origins of probability with an imaginary mental artifice. What artificial properties are created by this picture, and what real properties are overlooked?

This discussion of whether probability causes frequency, or frequency causes probability is not an empty academic exercise carried out solely for the sake of the argument. It does have at least one important consequence in the practice and formulation of statistical mechanics.

The ensemble picture of statistical mechanics is very old, having been used by both Boltzmann and Gibbs. In fact it is so widely used that many workers would be unaware that there is any alternative way of formulating statistical mechanics. The ensemble formulation asserts that the system is but one member of an ensemble of macroscopically identical systems, and that the probability of a particular state of the system is equal to the proportion of systems in the ensemble that are in that state. This ensemble interpretation is applied to classical statistical mechanics, as well as to the density matrix of quantum statistical mechanics. The evolution of the members of the ensemble in time is said to correspond to the evolution of the probability. In so far as the proportion of an ensemble in a given state is the probability of the state, the ensemble picture of statistical mechanics is obviously just a form of the frequency interpretation of probability.

There are two main objections to the ensemble picture. First, the experimental reality is that measurements are performed on a single system, not an ensemble of systems. (Repeat measurements are not the same as an ensemble: they are not essential, and there are not Avogadro's number of them.) Either the probability theory that underlies statistical mechanics works for a single system, or else the theory is a failure. If probability theory works for a single system, then ensembles are as superfluous as the aether.

Second, the members of the ensemble are implicitly conserved. This implies a conservation law for the probability during its evolution via the molecular equations of motion. Conservation laws are very serious laws of nature, and they should not be postulated without serious reason. That the members of an ensemble are conserved in someone's imagination does not make it true that probability is conserved in a real physical system. In fact it can be shown that for certain non-equilibrium systems weight is not conserved. This is not something that can be deduced from the ensemble picture. However, with hindsight, because ensembles are purely imaginary, any law, real or imaginary, can be added to ensembles to make them behave as they ought to behave. The invocation of arbitrary laws to make ensemble theory fit nature is neither convincing nor is it in accord with the way that science ought to be done.

*Credibility*

The credibility or subjectivist notion of probability holds that it is a measure of the strength of a person's belief that the state will occur. The beliefs must be rational and in accord with the rules of inductive reasoning, which happen to be the same as the rules of probability. However, two people with different information may attribute different probabilities to a particular state. Jaynes (2003) was a particularly strong advocate of the credibility interpretation of probability.

The problem with the subjectivists view is that it divorces probability from the underlying physical causes. For example, in Jaynes' treatment of the canonical probability distribution, the temperature is nothing more than a Lagrange multiplier that reflects the observer's ignorance of the energy of the system. This contrasts with the interpretation of most thermodynamicists, who view temperature as a physical attribute independent of what an observer knows or does not know. Similarly, Jaynes views entropy as a measure of the observer's ignorance, whereas I view entropy as the logarithm of the weight of relevant microstates. To most scientists who believe that physical phenomena have an objective reality, the subjectivist view is peculiarly solipsistic.

The credibility interpretation of probability provides the conceptual basis for the MaxEnt theory for obtaining probability distributions, again developed by Jaynes. This has been discussed above in connection with equations (1.1) and (1.19), where it was criticized as being unsound.

*Measure*

In this book an objective interpretation of probability is used, rather similar to the propensity interpretation of Popper (1959). Probability is taken to be a physical property of the system, namely it is proportional to the measure or weight of microstates in the specified macrostate. In the event that the microstates have equal weight, the weight of the macrostate is just the number of microstates in it.

Accepting the physical nature of probability, it remains to give a prescription for determining the microstate weights in a given system. In the simplest case the microstates are discrete and identical apart from their label, as for example, are the six faces of a true die. In such cases the microstates have equal weight, which, without loss of generality, can be assigned unit value so that the weight of a macrostate is simply the number of microstates that it contains.

One should not assume without reason that discrete microstates are equally weighted in a particular case. Similarly, in the case that the microstates form a continuum, the weight density may be uniform or non-uniform, depending on the physical characteristics of the system.

One of the most important cases, and the one upon which is focussed most attention in this book, is when the microstates represent the positions and momenta of all the particles of the system. This is called phase space. In the fundamental case when the system is isolated from the rest of the Universe, the weight density of the total phase space is uniform. This can be established (see chapter 5) by equating phase space averages (i.e. the average of a phase function over phase space using the weight density) to time averages (i.e. the simple average of the values of the phase function at discrete time intervals). This takes it as axiomatic that time is homogeneous and has uniform weight.

Having established the uniformity of classical phase space of the total isolated system, it is straightforward to derive the weight density for that phase space of the particles contained in a sub-system of the total isolated system, which, it turns out, is not uniform. Such sub-systems are the most common application of statistical mechanics.

In the case of quantum statistical mechanics, one can take a similar approach to establish the weight of the discrete quantum states that form the microstates of the system, as is detailed in chapter 7.

### 1.4.2 Randomness and irreversibility

As mentioned at the beginning of this section, there is a close connection between probability and randomness: if the system were fully determined, one would not need probability and a statistical description. In classical mechanics, the motion of the particles and the consequent evolution of the system is fully deterministic. Hence there is an issue to be addressed regarding how randomness arises from such deterministic equations of motion.

A related problem is that the classical equations of motion are time-reversible: if the velocity of all the particles in the system is reversed, then continuing forward in time the system will exactly retrace its prior history. This is called microscopic reversibility (see sections 5.1.2 and 5.4.2). This means that if the entropy had been increasing, then upon particle velocity reversal it would start decreasing. This reversibility in the equations of motion apparently contradicts the time asymmetry mandated by the Second Law of Thermodynamics, namely that a system evolves only in the direction of increasing entropy.

It turns out that the resolution of these apparent paradoxes revolves around the very formulation of statistical mechanics and thermodynamics. But first, three alternative proposals must be dismissed.

*Quantum questions?*

Some try to simply dismiss these issues as an artefact of classical mechanics that would disappear in a proper quantum treatment. In particular, some argue that quantum mechanics is inherently uncertain and unpredictable, and that this is sufficient to resolve the issue. In fact this does not solve the problem for three reasons. First, in quantum mechanics, the wave function not only evolves in a fully deterministic fashion via Schrödinger's equation, but also the complex conjugate of the wave function evolves in an exactly time-reversed fashion, both of which are analogous to the classical case. Second, most atoms and molecules behave classically due to their size; in practice for most terrestrial condensed matter systems quantum corrections contribute less than 1% for any atom larger than argon. The problems of indeterminacy and irreversibility holds for such systems and the classical equations have to be able to account for them. Third, classical statistical mechanics and thermodynamics exist as disciplines independent of quantum mechanics, as of course is clear from the fact that they were developed more than 50 years before the discovery of quantum theory. Therefore, again one must conclude that quantum mechanics cannot be essential to resolving these two paradoxes.

Of course, since quantum mechanics is the fundamental law of nature, both classical mechanics and classical statistical mechanics ought to be derived from it. This derivation is carried out in chapter 7, where the physical picture that gives rise to the classical equations turns out to be the same as that discussed here as the origin of randomness and irreversibility.

*Collective chaos?*

A second explanation proposed by some is that of ensembles, which have already been dealt with as an explanation of probability itself, section 1.4.1. Introduced by Boltzmann, an ensemble consists of identical macroscopic copies that differ in their initial microscopic or molecular configuration. The ensemble as a whole evolves according to the deterministic equations of motion of each isolated member. The randomness in this picture arises from the assigned distribution of the initial microstates of the members of the ensemble.

The ensemble picture is today more or less the standard view of probability and randomness in thermodynamics and statistical mechanics, but there are several weaknesses in it, as has already been outlined. First, ensembles are imaginary. They are nothing but a mental image that has no relationship to the reality of a single system being measured or characterized. Second, the ensemble picture does not fix the probability distribution of initial states, but this must be obtained by other physical considerations, which considerations of course could and should be instead applied directly to the real system, not the imaginary ensemble. Third, fixing the number of ensemble members implies a conservation law for probability that is not always true in reality. Just because ensembles are conserved in someone's mental picture, it does not make it true that probability is conserved in the real world. Fourth, ensembles are unnecessary, and that which is unnecessary has no place in science. Instead, one should seek the true physical origin of randomness for the single real system being studied, and one should not be bound by some imaginary scheme that obscures understanding and that blocks the mathematical formulation of randomness in statistical mechanics.

*Initially imprecise?*

As a third explanation, some suggest that randomness arises in deterministic equations of motion from imprecisely specified initial conditions. Alternatively, the precise intermolecular interactions may be unknown or too complex to calculate, and these cause indeterminacy in the system's trajectory. While these are undoubtedly practical impediments to the precise calculation and application of the equations of motion, they cannot be the physical principle that underlies randomness. For example, the irreversibility that is embodied in the Second Law of Thermodynamics is an experimental fact that does not require the specification of initial conditions or intermolecular interactions to be measured. This suggests that randomness and irreversibility must be built in to the very equations of motion that determine the evolution of the types of systems to which the Second Law refers.

*Reservoir randomness*

The quantitative approach to randomness and irreversibility that this book takes is based on the so-called reservoir formalism. The idea is sketched in figure 1.6. In this book what is called 'the system' means the total system (others might call this the Universe); sub-system means that part of the system of direct interest (others call this the system), and reservoir means the remaining part of the system (some others call this the environment).

It is generally the case that one has a detailed interest in a very small part of the Universe. For example, the shape of a macromolecule may be the focus, in which

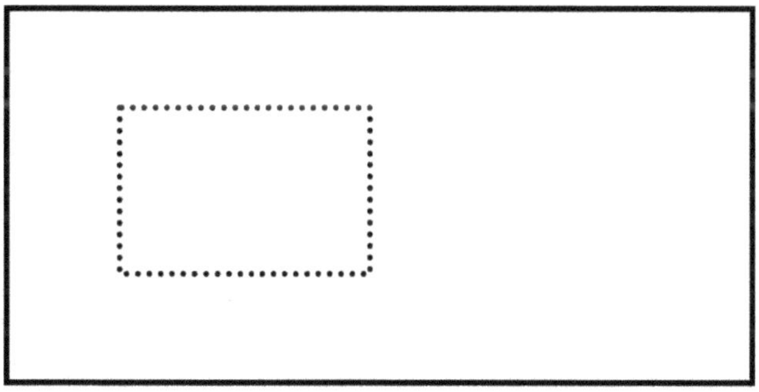

**Figure 1.6.** A system consisting of a sub-system and a (much larger) reservoir. The total system is isolated; all interactions and exchanges are forbidden across the solid boundary. The type of reservoir is determined by the specific interactions and exchanges that are allowed across the dotted boundary.

case the solvent, far-separated solutes, container walls, etc are of no interest except in so far as they affect directly the macromolecule. But one cannot ignore these completely, because, for example, the nearby solvent molecules directly affect the conformation of the macromolecule, and the far away solvent molecules act as heat, pressure, and solvent bath. In these circumstances the macromolecule and the nearby solvent molecules form the sub-system, and the far away solvent molecules, container walls, external laboratory, etc, form the reservoir. The sub-system is followed in molecular detail, and the reservoir is followed at the macroscopic level, which is to say it is characterized solely by intensive thermodynamic parameters such as temperature, pressure, chemical potential, etc.

The reservoir formalism as it gives rise to equilibrium thermodynamics will be set out in chapter 2. Here the relevance of the formalism to randomness and irreversibility will be described. The discussion is in terms of classical mechanics; the application of the formalism to quantum systems is described in chapter 7.

Let $\mathbf{\Gamma}_{\mathrm{tot}}$ denote a point in classical phase space, a microstate, of the total system (i.e. the positions and momenta of all the molecules in the system). Since the latter is isolated it evolves deterministically via Hamilton's equations of motion. The microstate of the total system at time $t$ may be denoted $\mathbf{\Gamma}_{\mathrm{tot}}(t|\mathbf{\Gamma}_{\mathrm{tot},0})$, where the initial configuration at $t = 0$ is $\mathbf{\Gamma}_{\mathrm{tot},0}$. (For simplicity, it is assumed that the Hamiltonian of the system is not explicitly time dependent.) This is called the trajectory of the total system. It is both deterministic and reversible.

Deterministic means that the initial point $\mathbf{\Gamma}_{\mathrm{tot},0}$ fully determines all subsequent points. This is to say that if the trajectory were ever to revisit one point, it would also revisit all subsequent points in order, and after the same time intervals. This also means that the trajectory cannot cross itself (i.e. each point on the trajectory is preceded by a unique point, and succeeded by a unique point).

Reversible actually means three symmetries (see section 5.4.2). Time reversibility, which says that if $\mathbf{\Gamma}_2$ transitions from $\mathbf{\Gamma}_1$ after time $\tau$, $\mathbf{\Gamma}_1 \overset{\tau}{\rightarrow} \mathbf{\Gamma}_2$, then $\mathbf{\Gamma}_1$ transitions from $\mathbf{\Gamma}_2$ after time $-\tau$, or $\mathbf{\Gamma}_2 \overset{-\tau}{\rightarrow} \mathbf{\Gamma}_1$. Conjugate reversibility, which, with a dagger denoting the state with all velocities reversed, means that the first transition just mentioned, $\mathbf{\Gamma}_1 \overset{\tau}{\rightarrow} \mathbf{\Gamma}_2$, implies the velocity reversed transition $\mathbf{\Gamma}_2^{\dagger} \overset{\tau}{\rightarrow} \mathbf{\Gamma}_1^{\dagger}$. Microscopic reversibility is the combination of these two, namely $\mathbf{\Gamma}_1 \overset{\tau}{\rightarrow} \mathbf{\Gamma}_2$, implies $\mathbf{\Gamma}_1^{\dagger} \overset{-\tau}{\rightarrow} \mathbf{\Gamma}_2^{\dagger}$.

The total phase space comprises that of the sub-system and that of the reservoir, $\mathbf{\Gamma}_{\mathrm{tot}} = \{\mathbf{\Gamma}_{\mathrm{s}}, \mathbf{\Gamma}_{\mathrm{r}}\}$. A microstate of the sub-system is a point in the phase space of the sub-system, $\mathbf{\Gamma}_{\mathrm{s}}$. This gives the position and momenta of all the molecules of the sub-system. The sub-system phase space is a subspace of the total phase space, and a microstate of the sub-system is the projection of a microstate of the total system. Many microstates of the total system, each corresponding to different reservoir microstates, project onto the same microstate of the sub-system. This projection operation has the same effect as what is often called a contracted description.

The projection of the total trajectory onto the sub-system may be written, $\mathbf{\Gamma}_{\mathrm{s}}' = \hat{\mathrm{P}}_{\mathrm{s}}\mathbf{\Gamma}(t|\mathbf{\Gamma}_{\mathrm{s},0}, \mathbf{\Gamma}_{\mathrm{r}}')$. For a different reservoir, but the same sub-system initial point, one has a different sub-system trajectory $\mathbf{\Gamma}_{\mathrm{s}}'' = \hat{\mathrm{P}}_{\mathrm{s}}\mathbf{\Gamma}(t|\mathbf{\Gamma}_{\mathrm{s},0}, \mathbf{\Gamma}_{\mathrm{r}}'')$. Hence the current sub-system configuration alone does not uniquely determine the sub-system evolution;

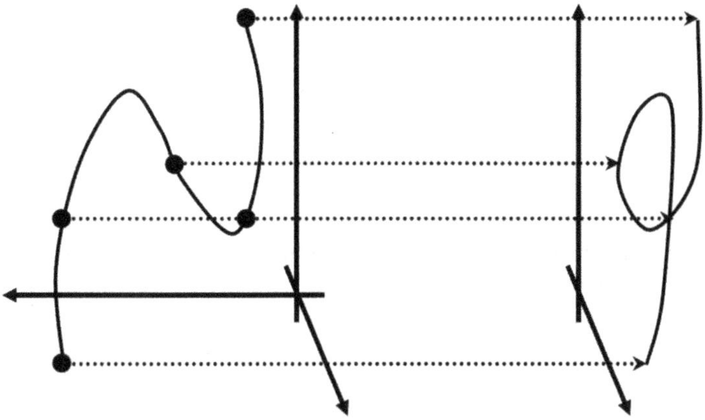

**Figure 1.7.** A spiral trajectory in three-dimensional space (left) becomes a trajectory with a crossing point and loop when projected onto the two-dimensional plane (right).

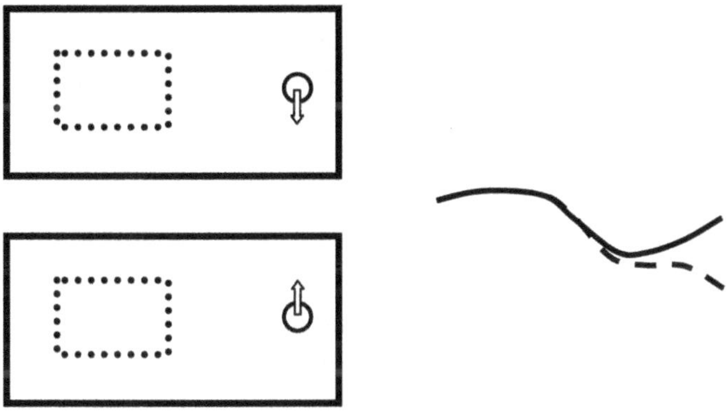

**Figure 1.8.** Two systems in which the configurations differ only by the velocity of a single reservoir particle far from the sub-system (left). The corresponding sub-system trajectories (right) bifurcate at the moment the perturbing influence of the reservoir particle reaches the sub-system. A similar bifurcation point exists in the past.

the projection operation makes the sub-system trajectory indeterminant in the sub-system. That is, revisiting the point $\mathbf{\Gamma}_{s,0}$ does not guarantee that subsequent points on the original trajectory will recur exactly, because each $\mathbf{\Gamma}_{s,0}$ could correspond to a different $\mathbf{\Gamma}_r$. Each such configuration of the reservoir molecules has a different influence on the evolution of the sub-system. This means that the projected trajectory can cross itself (figure 1.7) and bifurcate (figure 1.8).

This indeterminacy in the sub-system evolution when the reservoir configuration is not treated explicitly is manifest in stochastic terms that are added to the classical picture of Hamilton's deterministic equations of motion, and this gives rise to a probabilistic treatment of the sub-system state. Quantitative treatments of the probability follow from the physical nature of the interaction between the sub-system and the reservoir, as will be given in later chapters (chapters 4 and 6).

This interpretation of randomness in the equations of motion —it arises from the projection of the total system onto the sub-system that is being treated explicitly— also accounts for the second objection that was made to Boltzmann's identification of entropy with the number of molecular configurations, namely that the equations of motion are time reversible but the Second Law of Thermodynamics gives a preferred direction to time. It turns out that the projection operation contributes both stochastic and dissipative terms to the equations of motion, and that these are time irreversible. It will be shown that the dissipative term is a force increasing the entropy, as legislated by Clausius.

This randomness in the equations of motion leads directly to statistical mechanics and the need for a probabilistic treatment of molecular configurations. There is a direct link between randomness in statistical mechanics and thermodynamics. Different thermodynamic systems are based on the sub-system being able to exchange specified quantities (generally linear additive conserved variables such as energy, volume, number, etc.) with the reservoir. The mechanism by which this occurs is most usually via intermolecular interactions across the boundary. Therefore the current configuration of the molecules of the reservoir influences the motion of the molecules of the sub-system near the boundary immediately, and all the molecules of the sub-system ultimately. The physical interpretation of randomness as arising from the projection of external influences leads to the quantitative evaluation of probability.

## Summary

- The entropy of a system or state is the logarithm of the weight of the system or state. In the simplest case of equally likely microstates, weight is number. States with high entropy may be thought of as disordered or unpredictable.
- Macrostates are complete and disjoint. Microstates are complete, disjoint, and indivisible.
- The probability of a state is the exponential of its entropy. Probability, like entropy and weight, is a physical, measurable property.
- Conditional transitions to a more probable state are more probable than the reverse. The Second Law of Thermodynamics for conditional transitions is probably true.
- Randomness and irreversibility arise from the projection of the deterministic motion of the total system onto the sub-system of interest.

## References

Attard P 1999 Markov superposition expansion for the entropy and correlation functions in two and three dimensions *Statistical Physics on the Eve of the Twenty-First Century* eds M T Batchelor and L T Wille (Singapore: World Scientific)

Attard P 2000 The explicit density functional and its connection with entropy maximisation *J. Stat. Phys.* **100** 445

Attard P 2002 *Thermodynamics and Statistical Mechanics: Equilibrium by Entropy Maximisation* (London: Academic)

Attard P 2012 Is the information entropy the same as the statistical mechanical entropy? arXiv:1209.5500v1

Baxter R J 1982 *Exactly Solved Models in Statistical Mechanics* (London: Academic)

Boltzmann L 1866 Über die mechanische bedeutung des zweiten hauptsatzes der wärmetheorie *Wiener Berichte* **53** 195

Boltzmann L 1895 *Vorlesungen Über Gastheorie* vols 1 and 2 (Leipzig)

Browning E B 1850 *Sonnets from the Portuguese* No. 43

Brush S G (ed) 1964 *Vorlesungen Über Gastheorie* (Berkeley, CA: University of California Press) (in English)

Clausius R 1850 Über die bewegende kraft der wärme *Ann. Phys.* **79** 368, 500

Clausius R 1851 On the moving force of heat, and the laws regarding the nature of heat itself which are deducible therefrom *Phil. Mag.* **2** 1–21

Gibbs J W 1902 *Elementary Principles in Statistical Mechanics Developed with Special Reference to the Rational Foundation of Thermodynamics* (New Haven, CT: Yale University Press)

Jaynes E T 1957 Information theory and statistical mechanics I *Phys. Rev.* **106** 620

Jaynes E T 1957 Information theory and statistical mechanics II *Phys. Rev.* **108** 171

Jaynes E T 2003 *Probability Theory: The Logic of Science* ed G L Bretthorst (Cambridge: Cambridge University Press)

Kline S J 1999 *The Low-Down on Entropy and Interpretive Thermodynamics* (La Cañada, CA: DCW Industries), p 12

Popper K R 1959 The propensity interpretation of probability *Br. J. Phil. Sci.* **10** 25–42

Shannon C E 1948 A mathematical theory of communication *Bell Syst. Tech. J.* **27** 379–423

Shannon C E and Weaver W 1949 *The Mathematical Theory of Communication* (Urbana, IL: University of Illinois Press)

Tribus M and McIrvine E C 1971 Energy and information *Sci. Am.* **224** 179–88

**IOP** Publishing

Entropy Beyond the Second Law
Thermodynamics and statistical mechanics for equilibrium, non-equilibrium, classical, and quantum systems
**Phil Attard**

# Chapter 2

## Entropy varies

'Any method involving the notion of entropy, the very existence of which depends on the second law of thermodynamics, will doubtless seem to many far-fetched, and may repel beginners as obscure and difficult of comprehension'

Gibbs (1873)

'[A reformulation of the postulates] exhibits explicitly the internal consistency of the logical structure... it enables a deeper insight and intuition... and important extensions and generalizations are suggested and made practical'

Callan (1960)

'One makes a discovery when one sees what everybody else does, but thinks what nobody else had thought before'

Szent-Györgyi and Hargittai (2011)

The main aim of this chapter is to set out the general thermodynamic procedure for treating equilibrium systems. The procedures for obtaining the thermodynamic potential are given in general, with examples from common equilibrium systems derived. The novel twist on conventional thermodynamics is the relation between total entropy and free energy, from which arises the variational principle based on the Second Law of Thermodynamics, the nature of the constrained quantities that are its subject, and the relation to fluctuation theory.

The so-called reservoir formalism is used. The idea of a reservoir or bath is mainly due to Gibbs (1902), and is standard in thermodynamics (Callan 1960, Sears and Salinger 1986). The equilibrium free energies for various reservoirs derived here are the same as the conventional ones. However, the interpretation of these in terms of total entropy, and the consequent variational principle, are as developed in my previous book (Attard 2002).

doi:10.1088/978-0-7503-1590-6ch2

As a concrete example to illustrate the general ideas, the canonical equilibrium system is used, which is a sub-system in thermal contact with a heat reservoir. This will bring us to the Helmholtz free energy,

$$F = E - TS. \tag{2.1}$$

Here $E$ is the energy, $T$ is the temperature and $S$ is the entropy. This expression appears in all books on thermodynamics, but the question is whether or not it is really understood. In particular, what exactly is 'the' entropy $S$ that appears here? An example of what commonly goes wrong is the associated idea that

the Helmholtz free energy should be minimized. (2.2)

By the end of the chapter readers will understand what is wrong with this exhortation, they will appreciate the real physical import of the Helmholtz free energy, and they will be able to write and use the expression more powerfully.

## 2.1 Entropy of an isolated system

In chapter 1 the First Law of Thermodynamics was introduced as a law for the conservation of energy. Conservation lead to counting, counting to linear additivity, and linear additivity to entropy. It was pointed out that although not all linear additive quantities were conserved quantities, linear additivity was a very useful property to have and that it is widely exploited in the physical sciences.

The fundamental quantities that have proved most useful in thermodynamics are the conserved linear additive quantities energy $E$, number $N$, and volume $V$. Figure 2.1 shows an isolated system that is characterized by the values of these three quantities. The energy is composed of kinetic energy (i.e. due to particle motion) and potential energy (i.e. due to inter-particle interactions, as well as possible interactions with the walls and other fixed external fields). Despite the incessant motion of the particles that make up the system, these three quantities, $E$, $N$, and $V$, are unchanging.

For a multi-component system, one can regard $N$ as a vector whose components, $N_\alpha$ say, represent the number of each species.

Other linear additive quantities, such as the linear or angular momentum, can also be constants of the motion if the isolated system is not subject to external forces

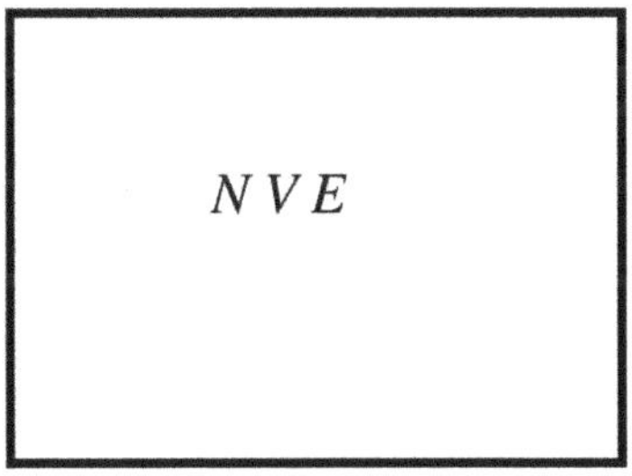


**Figure 2.1.** An isolated system with number of particles $N$, volume $V$, and energy $E$.

(and therefore enjoys translational or rotational symmetry). However, they are not widely used in thermodynamics.

Any function of these constants is also a constant of the motion. Hence one can *define* the entropy of the isolated system as dependent on them, $S(E, N, V)$. In the language of chapter 1, the particular state point with values $\{E, N, V\}$ is the macrostate formed from the intersection of the three macrostates in the three different collectives. The entropy $S(E, N, V)$ is the logarithm of the weight of the molecular microstates in that particular macrostate.

As mentioned, experience has shown the utility of formulating the entropy as dependent solely on extensive variables, and a number of results that flow directly from this will now be established. A further reason will be seen from the discussion of fluctuation theory in section 2.5, where the Gaussian distribution of the fluctuations follows directly from the central limit theorem, since by definition extensive variables are the sum of linear additive variables that themselves are randomly distributed. The same cannot be said of intensive variables.

### 2.1.1 Extensivity

Figure 2.2 shows two systems, both composed of $\lambda$ sub-systems. In both cases the total energy is $E_{\text{tot}} = \lambda E_1$, the total number is $N_{\text{tot}} = \lambda N_1$, and the total volume is $V_{\text{tot}} = \lambda V_1$. In the first system, the sub-systems are identical and isolated, which is to say each is in the macrostate $\{E_1, N_1, V_1\}$. Since entropy is linear additive, the entropy of the first system is

$$S(E_{\text{tot}}, N_{\text{tot}}, V_{\text{tot}}; \lambda) = \lambda S(E_1, N_1, V_1). \tag{2.3}$$

On the left-hand side $\lambda$ has been included explicitly to emphasize the fact that the total system comprises $\lambda$ isolated sub-systems. On the right-hand side the isolated sub-system entropy $S(E_1, N_1, V_1)$ includes contributions from the boundaries that contain and isolate it from the rest of the Universe. However in the thermodynamic limit of large sub-system size, such boundary effects scale with a boundary volume (the surface area times a molecular interaction length), which is negligible compared to the sub-system volume. Therefore, we may call $S(E_1, N_1, V_1)$ 'the' sub-system entropy in the macrostate $\{E_1, N_1, V_1\}$ irrespective of the precise nature of the sub-system boundary.

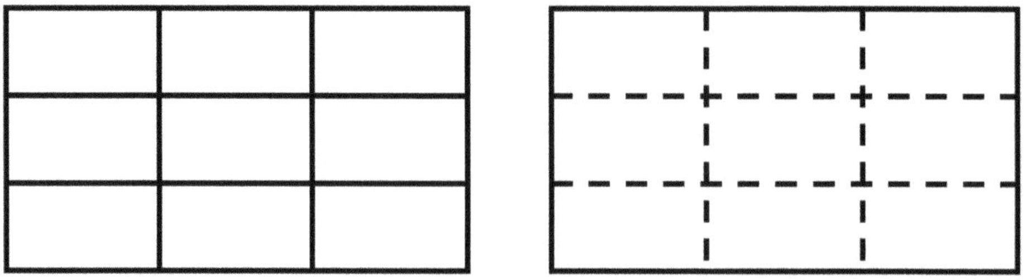

**Figure 2.2.** A system with total energy $E_{\text{tot}}$, total number $N_{\text{tot}}$, and total volume $V_{\text{tot}}$ composed of $\lambda = 9$ identical isolated sub-systems (left), and composed of $\lambda = 9$ same-sized sub-systems able to exchange energy and number (right).

The second system shown in figure 2.2 contains $\lambda$ equal sized sub-systems that can exchange energy and number with each other. These sub-systems are really imaginary cells that sub-divide the total volume. Because the cells are all the same size, symmetry arguments indicate that most likely they have the same energy $E_1 = E_{tot}/\lambda$ and number $N_1 = N_{tot}/\lambda$. (It will turn out that thermodynamics always deals with the most likely state; in the thermodynamic limit this in fact equals the average state.) These most likely values will be signified with an over-line in much of the analysis that follows. The entropy of this system is

$$S(E_{tot}, N_{tot}, V_{tot}) \gtrsim \lambda S(E_1, N_1, V_1). \tag{2.4}$$

On the left-hand side is a single isolated total system. Two assumptions are made to compare this to the right-hand side. First, for a given macrostate, the cells with 'transparent' boundaries that allow interactions and passage across them have the same entropy as isolated cells with 'opaque' boundaries that prohibit interactions and particle transport. This is justified in the thermodynamic limit where boundary conditions make negligible contribution to the total. Second, the left-hand system in figure 2.2 is constrained to be in a specific macrostate, namely the most likely one of equal energy and number in all the sub-systems, whereas the right-hand system can be in this macrostate, and in many others besides. Since the right-hand system is unconstrained in this respect, it must have more microscopic configurations available than the left-hand system and its entropy must be strictly greater than that of the constrained system. Therefore, one must have as a strict inequality

$$S(E_{tot}, N_{tot}, V_{tot}) > S(E_{tot}, N_{tot}, V_{tot}; \lambda) = \lambda S(E_1, N_1, V_1). \tag{2.5}$$

(For a detailed illustration of this point, see the combinatorial example in section 1.2.3.) The difference between the entropies of the two systems is due to the fluctuations in the energy and number of the sub-systems of the right-hand total system (ignoring, as above, other boundary effects).

The entropy due to these fluctuations is negligible compared to the entropy of the most likely macrostate, which, by symmetry, is the state in which the sub-systems all have the same energy and number. The reason that the fluctuation entropy is relatively negligible is because of the large value of the constrained entropy to which it is being added: the most likely macrostate is the state of maximum constrained entropy. This is true by definition, since probability is the exponential of the entropy, and 'most likely' means 'most probable'. More quantitatively, to obtain the unconstrained entropy, one must add to the maximum value of the constrained entropy the entropy due to fluctuations about the maximal macrostate. But, as will be shown below, the fluctuation entropy scales with the square root of the size of the sub-system, whereas the entropy of the sub-system scales with its size, as will next be shown. The former is negligible compared to the latter in the thermodynamic limit of large sub-system size. This is the reason for turning the strict inequality into an approximate inequality. In fact, in the thermodynamic limit one can write

$$S(E_{tot}, N_{tot}, V_{tot}) \approx S(E_{tot}, N_{tot}, V_{tot}; \lambda) = \lambda S(E_1, N_1, V_1). \tag{2.6}$$

This says that the unconstrained total entropy is approximately equal to the maximum value of the constrained total entropy. This is a general rule that is illustrated by the combinatorial example shown in figure 1.4.

These last three equations try to explain two things. On the one hand the total unconstrained entropy necessarily must be strictly greater than any one value of the constrained total entropy, including the maximal value of the constrained total entropy. On the other hand, in the thermodynamic limit fluctuations about the most likely state give a negligible contribution, and so the maximal constrained total entropy may be taken to equal the total unconstrained entropy.

### 2.1.2 Intensive variables

In the final form for the total entropy, $\lambda$ acts as a scale factor. It says that if one scales the extensive (i.e. linear additive) conserved variables all by the same amount, then the total entropy is also scaled by that amount. This means that the entropy is an extensive variable, which is to say that it scales with the size of the system. The extensivity of the entropy may be signified as

$$S(E_1, N_1, V_1) \sim \mathcal{O}(V_1). \tag{2.7}$$

In view of this it is useful to define the energy and number densities,

$$\varepsilon_1 \equiv \frac{E_1}{V_1}, \text{ and } \rho_1 \equiv \frac{N_1}{V_1}. \tag{2.8}$$

Because of extensivity, the entropy density is a function of these alone,

$$\sigma(\varepsilon_1, \rho_1) \equiv \frac{S(E_1, N_1, V_1)}{V_1}. \tag{2.9}$$

These densities are intensive variables, which is to say that they are independent of the size of the system, other things being equal (i.e. all extensive independent variables are scaled by the same amount). One reason intensive variables are useful is because they are localized in the sense that they do not depend on the whole system. They are often used as a first approximation for the local properties of an inhomogeneous system. Intensive variables are sometimes called field variables.

The mathematical operation of differentiation is essentially the same as forming a fraction. Hence the derivative of an extensive variable with respect to another extensive variable yields an intensive variable. This is one reason why it is useful to define the isolated system entropy as a function of the extensive variables that form the macrostates, namely the derivatives of the entropy with respect to its independent variables are intensive.

The first such derivative is with respect to energy, which is defined to yield the temperature,

$$T^{-1} \equiv \frac{\partial S(E, N, V)}{\partial E}. \tag{2.10}$$

The temperature of the isolated system is intensive and dependent on the state of the isolated system, $T(E, N, V)$. At the moment this is just a mathematical definition of the function $T$, and it remains to show that it has the same physical properties as the temperature of familiar experience.

The derivative with respect to volume yields the pressure,

$$p \equiv T \frac{\partial S(E,\, N,\, V)}{\partial V}, \tag{2.11}$$

and that with respect to number yields the chemical potential,

$$\mu \equiv -T \frac{\partial S(E,\, N,\, V)}{\partial N}. \tag{2.12}$$

For a multi-component system, there is a chemical potential $\mu_\alpha$ conjugate to the number of each species $N_\alpha$. Again, these are intensive variables that are functions of the state of the isolated system, $p(E,\, N,\, V)$ and $\mu(E,\, N,\, V)$, and these can be shown to have the same properties as the physical quantities of the same name.

With these definitions of the partial derivative of the entropy, the total derivative of the entropy of an isolated system is

$$\mathrm{d}S = \frac{1}{T}\mathrm{d}E + \frac{p}{T}\mathrm{d}V - \frac{\mu}{T}\mathrm{d}N. \tag{2.13}$$

In the thermodynamic limit, the intensive variables defined by the above entropy derivatives are functions only of the energy and number density. With the entropy density $\sigma(\varepsilon,\, \rho) \equiv S(E,\, N,\, V)/V$, this may be seen explicitly,

$$\frac{1}{T} = \left(\frac{\partial V\sigma(\varepsilon,\, \rho)}{\partial V\varepsilon}\right)_{N,V} = \left(\frac{\partial \sigma(\varepsilon,\, \rho)}{\partial \varepsilon}\right)_{\rho}, \tag{2.14}$$

$$\frac{-\mu}{T} = \left(\frac{\partial V\sigma(\varepsilon,\, \rho)}{\partial V\rho}\right)_{E,V} = \left(\frac{\partial \sigma(\varepsilon,\, \rho)}{\partial \rho}\right)_{\varepsilon}, \tag{2.15}$$

and

$$\begin{aligned} \frac{p}{T} &= \left(\frac{\partial V\sigma(\varepsilon,\, \rho)}{\partial V}\right)_{E,N} \\ &= \left(\frac{\partial V\sigma(\varepsilon,\, \rho)}{\partial V}\right)_{\varepsilon,\rho} + \left(\frac{\partial V\sigma(\varepsilon,\, \rho)}{\partial \varepsilon}\right)_{\rho,V}\left(\frac{\partial \varepsilon}{\partial V}\right)_{E} + \left(\frac{\partial V\sigma(\varepsilon,\, \rho)}{\partial \rho}\right)_{\varepsilon,V}\left(\frac{\partial \rho}{\partial V}\right)_{N} \\ &= \sigma(\varepsilon,\, \rho) - \frac{\varepsilon}{T} + \frac{\rho\mu}{T}. \end{aligned} \tag{2.16}$$

That is, in the thermodynamic limit one has $T(\varepsilon,\, \rho)$, $p(\varepsilon,\, \rho)$, and $\mu(\varepsilon,\, \rho)$.

### 2.1.3 Concavity of the entropy

For the case of the system consisting of $\lambda$ identical sub-systems able to exchange energy with each other, the right-hand part of figure 2.2, the equilibrium state is the one where the sub-systems all have the same energy, $E = E_{\text{total}}/\lambda$. (For simplicity focus on only energy in the first instance.) This follows from symmetry arguments, or from experience. Any other distribution of energy $\{E_i\}$ must have a lower entropy. Hence,

$$\begin{aligned}\lambda S(E, N, V) &\geqslant \sum_{i=1}^{\lambda} S(E_i, N, V) \\ &= \lambda S(E, N, V) + \frac{\partial S(E, N, V)}{\partial E} \sum_{i=1}^{\lambda} (E_i - E) \\ &\quad + \frac{1}{2}\frac{\partial^2 S(E, N, V)}{\partial E^2} \sum_{i=1}^{\lambda} (E_i - E)^2,\end{aligned} \tag{2.17}$$

where a second order Taylor expansion about $E$ has been performed. The first term on the right-hand side cancels with the left-hand side. The second term is zero from energy conservation, $\sum_{i=1}^{\lambda} E_i = \lambda E$. These mean that the third term must be negative semi-definite, vanishing if and only of $E_i = E$. Since it contains a sum of non-negative terms, one concludes that the second energy derivative of the entropy must be strictly negative

$$S_{EE} < 0. \tag{2.18}$$

Hence, the entropy of an isolated system is strictly concave with respect to energy.

The argument can be repeated for the other extensive variables, either one at a time or allowing simultaneous exchange. The general conclusion is

$$S_{aa} < 0, \quad \text{and} \quad S_{aa}S_{bb} - S_{ab}^2 > 0, \tag{2.19}$$

where the subscripts denote the second derivatives with respect to any of the isolated sub-system extensive variables. The second form follows because the eigenvalues of the Jacobean must be negative, and so the determinant of the Jacobean of the extensive variables taken pairwise must be positive. (One can obtain higher order conditions as well.) These are necessary conditions for any thermodynamic state to be stable.

Stable means that the system can exist with uniform densities corresponding to these values of the extensive variables. A system in a uniform state in which this concavity condition is violated can increase its total entropy by sub-dividing into coexisting phases, each with a uniform density that is stable, and with overall densities equal to the original values.

## 2.2 Heat reservoir and the Helmholtz free energy

We now turn explicitly to the canonical equilibrium system, namely a sub-system able to exchange energy with a heat reservoir, figure 2.3. As mentioned in section 1.4.2, the

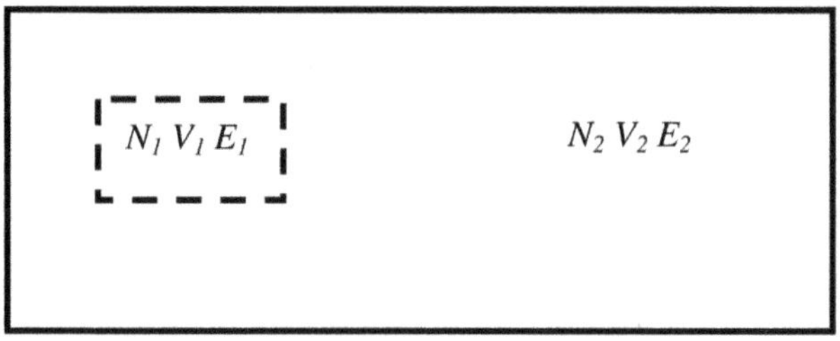


**Figure 2.3.** A system composed of a sub-system 1 and reservoir 2 able to exchange energy.

reservoir formalism is fundamental to how randomness occurs in statistical mechanics, and how statistical mechanics and thermodynamics are structured, and so it is worth defining in detail what is meant by a reservoir.

A system comprising a sub-system and a reservoir has four characteristics: the system is isolated, the reservoir is infinitely larger than the sub-system, the boundary region between the reservoir and the sub-system is infinitely smaller than the sub-system, and the reservoir and the sub-system can exchange one or more linear additive, conserved variables. (In practice, in order to establish the size of the sub-system there must be at least one linear additive, conserved variable that cannot be exchanged.)

The reason for insisting that the sub-system be large compared to its boundary region (i.e. its surface area times the thickness of direct molecular interactions) is so that boundary effects can be neglected. This condition is also known as the thermodynamic limit. The reason for the reservoir being large compared to the sub-system is so that higher order terms in a Taylor expansion can be neglected, as will be required shortly.

In what follows it will be important to keep track of the dependent and the independent variables. To date energy $E$ has been taken as an independent variable for the isolated system entropy, $S(E, N, V)$. In part of what follows it will be necessary to regard it as dependent on the temperature, at which time it will be denoted $E(N, V, T)$ or by $\overline{E}$. By the partial derivative definition, we have already seen that temperature is a dependent variable, $T(E, N, V)$. As will be discussed shortly, the properties of entropy make the relationship between energy and temperature one-to-one: $T_1 = T(E_1, N, V)$ if and only if $E_1 = E(N, V, T_1)$. In other words, $T_1 = T(E(N, V, T_1), N, V)$.

### 2.2.1 Constrained total entropy

For the system comprising a sub-system able to exchange energy with a reservoir, figure 2.3, the total energy is fixed, $E_{\text{total}} = E_1 + E_2$. Because the reservoir is infinitely larger than the sub-system, the sub-system energy is much less than the total energy, $E_1 \ll E_{\text{total}}$. This holds in practice for all conceivable configurations of the system. Taking the thermodynamic limit for the sub-system, the boundary conditions become negligible, and the entropy of the total system (in the energy macrostate $E_1$) is just the sum of the isolated sub-system and isolated reservoir entropies (each in their corresponding macrostate). These facts mean that the entropy of the total system, constrained to be in the macrostate $E_1$, is

$$\begin{aligned} S_{\text{total}}(E_1|E_{\text{total}}, N_1, V_1, N_2, V_2) \\ &= S_1(E_1, N_1, V_1) + S_2(E_{\text{total}} - E_1, N_2, V_2) \\ &= S_1(E_1, N_1, V_1) + S_2(E_{\text{total}}, N_2, V_2) - E_1\frac{\partial S_2(E_{\text{total}}, N_2, V_2)}{\partial E_{\text{total}}} + \cdots \\ &= S_1(E_1, N_1, V_1) + \text{const.} - \frac{E_1}{T_2}. \end{aligned} \tag{2.20}$$

The first equality writes the system entropy as the sum of that of the isolated sub-system and isolated reservoir. Here we have allowed for the possibility that the reservoir and sub-system are composed of different materials, and therefore their entropies may be different functions of their arguments, namely $S_1$ and $S_2$. This is a detail that is already irrelevant by the final equality.

The second equality performs a Taylor expansion of the reservoir entropy about the total energy. The leading, zeroth order, term, $S_2(E_{\text{total}}, N_2, V_2)$, is a constant completely independent of the sub-system. It therefore has no effect on the state of the sub-system and it can be neglected. It will be recalled from section 1.1.2 that entropy was only defined up to an additive constant.

The first order term is linear in the sub-system energy and the (inverse) reservoir temperature $T_2(E_{\text{total}}, N_2, V_2)$. This term, $-E_1/T_2$, is extensive with the sub-system size, $\mathcal{O}(V_1)$, as is the sub-system entropy itself, $S_1(E_1, N_1, V_1) \sim \mathcal{O}(V_1)$. The second order term in the Taylor expansion, which has been neglected, is

$$\frac{1}{2}E_1^2 \frac{\partial^2 S_2(E_{\text{total}}, N_2, V_2)}{\partial E_{\text{total}}^2} \sim \mathcal{O}\left(\frac{V_1^2}{V_2}\right). \tag{2.21}$$

In the second derivative here, the numerator scales with $V_2$ and the denominator with $V_2^2$. Hence the full second order term is $\mathcal{O}(V_1^2/V_2)$, which is a factor of $V_1/V_2$ smaller than the two terms that are explicitly retained above. (Both $S_1(E_1, N_1, V_1)$ and $E_1/T_2$ are $\mathcal{O}(V_1)$.) Since the reservoir is infinitely larger than the sub-system, this and higher order terms are completely negligible. This is one of the reasons why it is advantageous to write the entropy as a function of extensive variables only. The net effect of neglecting the higher order terms in the Taylor expansion of the reservoir is that the temperature of the reservoir is fixed and unchanging no matter how much energy is exchanged with the sub-system. In other words, $T_2(E_2, N_2, V_2) = T_2(E_{\text{total}}, N_2, V_2)$, with negligible error $\mathcal{O}(V_1/V_2)$.

As a result of these manipulations, one sees that the heat reservoir only enters the equations via its temperature $T_2$. Hence one can write the constrained total entropy as

$$S_{\text{total}}(E_1|N_1, V_1, T_2) = S_1(E_1, N_1, V_1) - \frac{E_1}{T_2}. \tag{2.22}$$

Since the temperature is that of the reservoir, and all other properties belong to the sub-system, one can abbreviate this still further and write

$$S_{\text{total}}(E|N, V, T) = S_s(E, N, V) - \frac{E}{T}. \tag{2.23}$$

Of course this concise expression is easy to misinterpret if one is unaware of its derivation. Hopefully the reader will never be misled; $T$ is the temperature of the reservoir and $S_s$ is the entropy of the isolated sub-system. The reader will also know that the left-hand side is the entropy of the total system constrained to be in the macrostate that the sub-system has energy $E$. The reader will also recognize that $-E/T$ is the sub-system-dependent part of the reservoir entropy. Finally, the reader will notice that all four of the arguments of the total entropy are independent.

Just a note on notation. The vertical bar here is the same as that used for the conditional probability. Variables to the left of it are the constrained (or exchangeable, or fluctuating) variables, and variables to the right of it are the conditioning (or fixed) variables. The exponential of the total entropy essentially gives the probability that the sub-system has energy $E$ given the conditions, as will be discussed shortly. This explains the notation that is used here and in similar circumstances throughout.

The energy derivative of the constrained total entropy is

$$\frac{\partial S_{\text{total}}(E|N, V, T)}{\partial E} = \frac{\partial S_s(E, N, V)}{\partial E} - \frac{1}{T}. \tag{2.24}$$

The first term on the right-hand side is $1/T(E, N, V)$, which is the reciprocal of the sub-system temperature, and the second term is the reciprocal the reservoir temperature.

Now one can appeal to the Second Law of Thermodynamics to determine the equilibrium state. The equilibrium state is the most likely value of the sub-system energy. The total entropy is maximized by the energy that makes its derivative vanish. This gives the equilibrium energy $\overline{E}$ as the one satisfying

$$\left.\frac{\partial S_{\text{total}}(E|N, V, T)}{\partial E}\right|_{E=\overline{E}} = 0 \Leftrightarrow T(\overline{E}, V, N) = T. \tag{2.25}$$

This says that equilibrium (i.e. the macrostate of maximum total entropy) corresponds to temperature equality between the sub-system and the heat reservoir. This is a version of the Zeroth Law of Thermodynamics. This is an implicit equation for the equilibrium energy of the sub-system, $\overline{E} = E(N, V, T)$.

In the case that the sub-system has a lower temperature than the reservoir, $\partial S(E, N, V)/\partial E = 1/T(E, N, V) > 1/T$, then the total constrained entropy increases when energy is transferred from the reservoir to the sub-system, $\partial S_{\text{total}}(E|N, V, T)/\partial E > 0$. The opposite occurs in the case that the sub-system is hotter than the reservoir. In common parlance, heat flows from a hot body to a cold body. This is consistent with the fact that an isolated system has entropy that is a concave function of energy, $S_{EE} < 0$, since this means that its temperature must increase with energy, $\partial T(E, N, V)/\partial E = -T^2 S_{EE} > 0$. Hence energy flowing to a cold body increases its temperature.

### 2.2.2 Constrained Helmholtz free energy

The constrained total entropy is the mathematical representation of the Second Law of Thermodynamics. Constraining the energy $E$ fixes the sub-system in a not-in-equilibrium state. When the constraint is relaxed, the system will move toward equilibrium by exchanging energy with the reservoir. These spontaneous changes in the sub-system energy state are in the direction of increasing constrained total entropy.

The equilibrium state is the state of maximum constrained total entropy.

It is important to note the physical basis of the above derivation. The constrained variable in this case is the sub-system energy, and it is with respect to this, and only this, that the constrained total entropy is maximized in the equilibrium state.

The present variational principle maximizes the constrained total entropy. Alternatively, one can minimize a thermodynamic potential. The general definition of this is simply the negative of the (reservoir) temperature times the constrained total entropy. The negative sign converts the maximum into a minimum. Multiplying the entropy by the temperature gives the units of energy.

In the present case the thermodynamic potential is the constrained Helmholtz free energy,

$$\begin{aligned} F(E|N, V, T) &\equiv -TS_{\text{total}}(E|N, V, T) \\ &= E - TS_{\text{s}}(E, N, V). \end{aligned} \tag{2.26}$$

By design, this is a minimum in the equilibrium state, and the energy flow is down the constrained free energy gradient.

The same rule applies in the general case: the constrained thermodynamic potential is the negative of the temperature times the constrained total entropy, $F(X|Y, x, T) = -TS_{\text{total}}(X|Y, x, T)$, where $X$ are the exchangeable variables, $x$ are the conjugate reservoir field variables, $Y$ are the fixed sub-system variables, and $T$ is the temperature of the reservoir. Concrete examples will be given in section 2.3 below.

Because the relationship between the thermodynamic potential and the total entropy is trivial, the properties of the former are essentially the same as the latter. It is a mathematical and physical fact that the thermodynamic potential is redundant. The present author's main argument against it is that it is unnecessary, and it obscures the role of entropy and its physical origin. Such rational arguments however carry little weight against convention and historical tradition. One suspects that it is more likely for energy to flow from a cold body to a hot body than it is for the free energy to be replaced by the total entropy in text books and the literature.

The *equilibrium thermodynamic potential* is defined as the minimum value of the constrained thermodynamic potential. For the present heat reservoir this is the Helmholtz free energy, and the minimum obviously occurs at $\overline{E} = E(N, V, T)$,

$$\overline{F}(N, V, T) \equiv F(\overline{E}|N, V, T) = \overline{E} - TS_{\text{s}}(\overline{E}, N, V). \tag{2.27}$$

The Helmholtz free energy is a function of just three independent variables, namely $N$, $V$, and $T$, since $\overline{E} = E(N, V, T)$ is a dependent variable. The entropy that appears on the right of the definition of the Helmholtz free energy is that of the isolated sub-system with the equilibrium energy $\overline{E}$. The overline on the Helmholtz free energy indicates that it is an equilibrium property.

Note that as a short-hand notation, sometimes we write $\overline{S}_s(N, V, T)$ instead of $S_s(\overline{E}(N, V, T), N, V)$. This is not ideal notation because entropy is fundamentally a function of extensive variables. Also it risks confusion with the unconstrained total entropy $S_{\text{total}}(N, V, T)$, which is essentially the logarithm of the partition function. This possible problem is exacerbated when the subscript s for sub-system is dropped. In any case, it should always be kept in mind that the entropy that appears explicitly in the formula for the Helmholtz free energy is the entropy of the isolated sub-system in the equilibrium energy macrostate.

The equilibrium thermodynamic potential, $F_{\text{TD}}(N, V, T)$, contains less information than the constrained thermodynamic potential, $F(E|N, V, T)$, since it is a function of three independent variables instead of four. The constrained thermodynamic potential describes the approach to energy equilibrium and energy fluctuations about the equilibrium state, whereas the equilibrium thermodynamic potential only describes the equilibrium state. Because the constrained thermodynamic potential describes the excursions of the system to not-in-equilibrium states (see section 2.5 below), it may also be called the *fluctuation potential.*

The three types of total entropy and the respective free energies are summarized in table 2.1. The partition function is the total weight (see section 2.5.3),

$$Z(N, V, T) = \sum_E e^{S_{\text{total}}(E|N,V,T)/k_B}. \tag{2.28}$$

By definition, the logarithm of the total weight is the total unconstrained entropy $S_{\text{total}}(N, V, T) = k_B \ln Z(N, V, T)$. Since the free energy in general is minus the temperature times the total entropy, the statistical mechanical free energy follows as $F_{\text{SM}}(N, V, T) = -TS_{\text{total}}(N, V, T)$. The thermodynamic free energy is the minimum value of the constrained free energy, and in the present canonical equilibrium case it is called the Helmholtz free energy, $F_{\text{TD}}(N, V, T) = -TS_{\text{total}}(\overline{E}|N, V, T) = \overline{E} - TS_s(\overline{E}, N, V)$. One has the inequalities

$$S_{\text{total}}(N, V, T) > S_{\text{total}}(\overline{E}|N, V, T) \geqslant S_{\text{total}}(E|N, V, T). \tag{2.29}$$

**Table 2.1.** Canonical equilibrium total entropies and free energies.

| | Entropy | Free Energy | Formula |
|---|---|---|---|
| Constrained | $S_{\text{total}}(E\|N, V, T)$ | $F(E\|N, V, T)$ | $= E - TS_s(E, N, V)$ |
| Max. constrained | $S_{\text{total}}(\overline{E}\|N, V, T)$ | $F_{\text{TD}}(N, V, T)$ | $= \overline{E} - TS_s(\overline{E}, N, V)$ |
| Unconstrained | $S_{\text{total}}(N, V, T)$ | $F_{\text{SM}}(N, V, T)$ | $= -k_BT \ln Z(N, V, T)$ |

In the thermodynamic limit, fluctuations are relatively negligible, and the inequality can be replaced by an equality and $S_{\text{total}}(N, V, T) \approx S_{\text{total}}(\overline{E}|N, V, T)$, or $F_{\text{SM}}(N, V, T) \approx F_{\text{TD}}(N, V, T)$.

One can compare the conventional expression for the Helmholtz free energy given at the beginning of this chapter, equation (2.1) with the equilibrium equation (2.27) or the more general not-in-equilibrium form, equation (2.26). In the equilibrium comparison, which is the most direct, one sees that the conventional expression does not avert to the dependence of the energy on the number, volume and temperature, $\overline{E} = E(N, V, T)$. In this sense equation (2.1) may be criticized as being imprecise. One also sees that it does not avert to the fact that the entropy that appears explicitly, $S(\overline{E}, N, V)$, is the entropy for the isolated sub-system in the energy macrostate $\overline{E}$. In this sense it is misleading to call this 'the' entropy. In the same vein it would be better to explicitly recognize that the Helmholtz free energy is just minus the temperature times the maximum value of the constrained total entropy. Finally, comparing the conventional expression for the Helmholtz free energy, Equation (2.1), to the more general constrained thermodynamic potential equation (2.26), one sees that the conventional expression for the Helmholtz free energy neither recognizes nor exhibits its variational nature with respect to energy.

One sometimes sees it asserted that the Helmholtz free energy is a minimum, as in the exhortation that is equation (2.2), which implies that it gives a variational principle. There are three things wrong with this assertion. First, the Helmholtz free energy is $\overline{F}(N, V, T)$, and this does not obey any variational principle. Second, the constrained thermodynamic potential $F(E|N, V, T)$ does obey a variational principle with respect to the sub-system energy, and it is essential that the variational parameter be stated and understood explicitly. This is not mere nit picking. It is not uncommon in the literature to see the Helmholtz free energy formulated for a particular problem and then for it to be minimized with respect to some parameter of interest. Unless that parameter is the energy, such a procedure is a mathematical and physical absurdity. Third, the variational principle for the constrained thermodynamic potential derives directly from the Second Law of Thermodynamics via its relationship to the total entropy. This essential physical basis and its further consequences are missed entirely by the simplistic incantation that the Helmholtz free energy should be minimized.

### 2.2.3 Derivatives of the Helmholtz free energy

A concrete example of the power of the constrained thermodynamic potential over the usual free energy expressions is in the ease with which it can be differentiated. Because it provides a variational principle for equilibrium, (i.e. it is a minimum with respect to variations in sub-system energy), the energy can effectively be regarded as a fixed independent variable rather than a dependent variable during differentiation.

For example, the temperature derivative of the Helmholtz free energy is

$$\begin{aligned}\frac{\partial \overline{F}(N, V, T)}{\partial T} &= \left(\frac{\partial F(\overline{E}(N, V, T)|N, V, T)}{\partial T}\right)_{N,V} \\ &= \left.\frac{\partial F(E|N, V, T)}{\partial T}\right|_{E=\overline{E}} + \left.\frac{\partial F(E|N, V, T)}{\partial E}\right|_{E=\overline{E}} \frac{\partial \overline{E}(N, V, T)}{\partial T} \\ &= \left.\frac{\partial F(E|N, V, T)}{\partial T}\right|_{E=\overline{E}} \\ &= -S_{\mathrm{s}}(\overline{E}, N, V).\end{aligned} \tag{2.30}$$

The result is that the temperature derivative of the Helmholtz free energy is the negative of the sub-system entropy in the equilibrium state.

The third equality follows from the variational principle for the constrained thermodynamic potential, $\partial F(E|N, V, T)/\partial E\,|_{E=\overline{E}} = 0$. Hence even though the equilibrium sub-system energy is a dependent variable, $\partial \overline{E}(N, V, T)/\partial T \neq 0$, this dependence can be ignored because it is multiplied by a derivative that vanishes at equilibrium. In other words, because the constrained thermodynamic potential is optimised at equilibrium, differentiating the Helmholtz free energy is the same as differentiating the constrained thermodynamic potential *holding* $\overline{E}$ *fixed*. This is a general feature of constrained thermodynamic potentials that can be exploited in all similar derivatives.

Now let us differentiate the Helmholtz free energy with respect to volume and number. One can use the corresponding partial derivatives of the isolated system entropy, equations (2.11) and (2.12), since one can again hold $\overline{E}$ fixed. The volume derivative gives the pressure,

$$\begin{aligned}\left(\frac{\partial \overline{F}(N, V, T)}{\partial V}\right)_{T,N} &= \left(\frac{\partial F(\overline{E}|N, V, T)}{\partial V}\right)_{\overline{E},T,N} \\ &= -T\left(\frac{\partial S_{\mathrm{s}}(\overline{E}, N, V)}{\partial V}\right)_{\overline{E},N} \\ &= -\overline{p},\end{aligned} \tag{2.31}$$

and the number derivative gives the chemical potential,

$$\begin{aligned}\left(\frac{\partial \overline{F}(N, V, T)}{\partial N}\right)_{T,V} &= \left(\frac{\partial F(\overline{E}|N, V, T)}{\partial N}\right)_{\overline{E},T,V} \\ &= -T\left(\frac{\partial S_{\mathrm{s}}(\overline{E}, N, V)}{\partial N}\right)_{\overline{E},V} \\ &= \overline{\mu}.\end{aligned} \tag{2.32}$$

It will shortly be shown that the equilibrium state is unique. Hence the pressure that appears here $\overline{p}(N, V, T)$ is equal to that of the isolated sub-system in the equilibrium energy state, $p(\overline{E}(N, V, T), N, V)$. This is necessary for thermodynamics to be internally consistent. One must have the volume derivative of the Helmholtz free energy for a sub-system in contact with a heat reservoir giving the same pressure as an isolated system with the equilibrium energy. Similar comments apply to the chemical potential, $\bar{\mu}(N, V, T) = \mu(\bar{E}(N, V, T), N, V)$ and to the sub-system entropy $\bar{S}_s(N, V, T) = S_s(\bar{E}(N, V, T), N, V)$. The overline here denotes the fact that these are properties of the sub-system in thermal equilibrium.

These results for the partial derivatives give the total differential of the Helmholtz free energy,

$$\mathrm{d}\bar{F}(N, V, T) = -\bar{S}_s(N, V, T)\,\mathrm{d}T - \bar{p}(N, V, T)\,\mathrm{d}V + \bar{\mu}(N, V, T)\,\mathrm{d}N. \quad (2.33)$$

Finally, there is one more example that exploits the variational nature of the constrained thermodynamic potential for a sub-system in contact with a heat reservoir. Dividing both sides of the Helmholtz free energy by temperature, $\bar{F}/T = \bar{E}/T - \bar{S}_s(\bar{E}, N, V)$, and differentiating with respect $T^{-1}$ one obtains

$$\left(\frac{\partial(\bar{F}(N, V, T)/T)}{\partial(1/T)}\right)_{N,V} = \bar{E}. \quad (2.34)$$

Again this follows because $\bar{E}$ and hence $S_s(\bar{E}(N, V, T), N, V)$ can be held constant.

### 2.2.4 Concavity, uniqueness, and thermal stability

*Concavity of the entropy*

It is now shown that the entropy of the isolated system is a concave function of its arguments. The derivation is based on the extensivity of the entropy and the fact that the constrained total entropy is a maximum at equilibrium.

In figure 2.4 two systems are shown, each comprising two sub-systems that are identical except for energy. In the left-hand system the sub-systems are isolated from each other and have the same volume $V$, number $N$, but different energies $E_1$ and $E_2$. The entropies of the sub-systems are $S_1 = S(E_1, N, V)$ and $S_2 = S(E_2, N, V)$. From the linear additive nature of the entropy, the total entropy with the insulating partition in place is $S_{\text{total}} = S_1 + S_2$.

If the insulating partition is replaced by a conducting partition, as in the right-hand diagram, energy irreversibly flows from the sub-system with more energy to

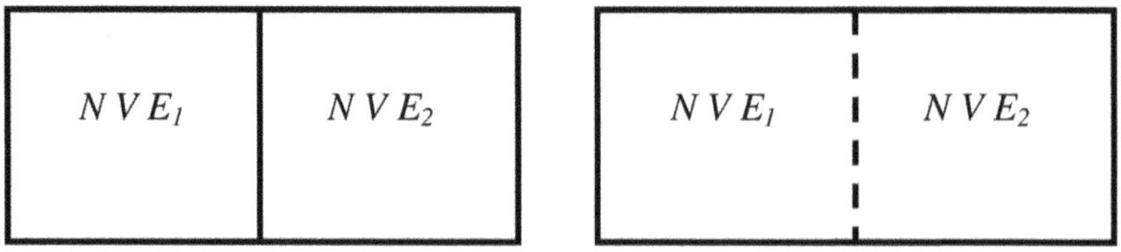


**Figure 2.4.** Two isolated systems, each composed of two sub-systems identical except for energy. In the system on the left, the sub-systems are separated by an insulating wall that prevents energy exchange. In the system on the right, the sub-systems are separated by a conducting wall that allows energy exchange.

that with less, as is obvious on symmetry grounds. The total entropy at the end of this process must be greater than at the beginning,

$$S_{\text{total}}(E_1 + E_2, 2N, 2V) \geqslant S(E_1, N, V) + S(E_2, N, V). \tag{2.35}$$

This is, of course, demanded by the Second Law of Thermodynamics: energy flows in the direction of entropy increase. At the microscopic level, the number of microstates must be greater when the constraint is relaxed. This is a strict inequality unless initially $E_1 = E_2$.

The equilibrium state is the one in which each sub-system has the same energy, namely $(E_1 + E_2)/2$, and this is the state of maximum entropy. Once equilibrated, there is no further nett energy flow, and the removal and insertion of the partition becomes a reversible process. In this case the entropy does not change and one must have

$$S_{\text{total}}(E_1 + E_2, 2N, 2V) = 2S\left(\frac{E_1 + E_2}{2}, N, V\right). \tag{2.36}$$

This expression assumes that fluctuations about the equilibrium state make relatively negligible contribution to the total entropy on the left-hand side. This result is the $\lambda = 2$ case that was derived above in the discussion of extensivity; see section 2.1.1 and figure 2.2.

These two equations combined yield

$$S\left(\frac{E_1 + E_2}{2}, N, V\right) > \frac{1}{2}[S(E_1, N, V) + S(E_2, N, V, )], \tag{2.37}$$

excluding the case $E_1 = E_2$. This shows that entropy is a concave function of energy, since any chord to the entropy curve lies below the curve. A Taylor expansion as $E_1 \to E_2$ shows that this is equivalent to

$$\left(\frac{\partial^2 S(E, N, V)}{\partial E^2}\right)_{N,V} < 0. \tag{2.38}$$

Again, a negative second derivative defines a concave function. Analogous steps can be carried out individually for volume and for particle number. One can conclude in general that the entropy is a concave function of its arguments.

It is not much harder to derive the result for simultaneous variations. For example, considering two isolated systems that are identical apart from their energies and volumes, and allowing exchange of these, one can readily show that

$$S_{\text{total}}\left(\frac{E_1 + E_2}{2}, N, \frac{V_1 + V_2}{2}\right) > \frac{1}{2}[S(E_1, N, V_1) + S(E_2, N, V_2)]. \tag{2.39}$$

Performing a Taylor expansion of the right-hand side about the equilibrium energy and volume, the zeroth and linear terms cancel, which leaves

$$S_{EE}\, \mathrm{d}E^2 + 2S_{EV}\, \mathrm{d}E\, \mathrm{d}V + S_{VV}\, \mathrm{d}V^2 < 0, \tag{2.40}$$

where the subscripts denote partial derivatives, and $\mathrm{d}E$ and $\mathrm{d}V$ are arbitrary. Again, a negative Jacobian defines a concave function of several variables. This result is consistent with equation (2.19). We shall use this result below.

*Uniqueness of the thermodynamic state*

In thermodynamics it is sometimes useful to swap dependent and independent variables. For example, to date we have described an isolated system by its entropy, which is a function of the extensive variables energy, number and volume, $S(E, N, V)$. For mathematical reasons it is sometimes convenient to instead take energy as the dependent variable, $E(S, N, V)$.

In general one can interchange variables, such as

$$x_1 = x(y_1) \Leftrightarrow y_1 = y(x_1), \tag{2.41}$$

if the relationship between the conjugate variables is one-to-one. This is the same as saying that the function is invertible,

$$x = f(y) \Leftrightarrow y = f^{-1}(x). \tag{2.42}$$

The condition for a continuous function to be invertible is that its derivative must be finite and nonzero on the interior of its domain of definition. Since entropy is a strictly concave function of its arguments, it is indeed an invertible.

First the energy–entropy swap. (We shall assume that entropy is a continuous function of energy.) As we have seen, entropy is strictly concave, $S'' < 0$. Therefore it cannot have more than one turning point, which we can assume occurs at $E_1$, with $S'(E_1) = 0$. This turning point, if it exists, must be a maximum. Obviously entropy is monotonically increasing, $S'(E) > 0$, for $E < E_1$, which means that $T(E) > 0$. It is monotonically decreasing, $S'(E) < 0$, for $E > E_1$, which is to say that $T(E) < 0$. Since systems with negative temperature are unstable, we shall only consider systems with positive temperatures here and throughout, $E < E_1$. Denote the lowest energy or ground state by $E_0$. Because the entropy is strictly concave, its gradient may possibly be infinite here, but nowhere else. An infinite gradient, $S'(E_0) = \infty$, corresponds to $T = 0$, which is absolute zero. These considerations show that on the domain $(E_0, E_1)$ there is a one-to-one relationship between the entropy and the energy. Hence the entropy function is invertible and one can write $S(E, N, V)$ or $E(S, N, V)$ as convenient.

Now the energy–temperature swap. On the physical domain $(E_0, E_1)$, the concavity of the entropy, $S_{EE} < 0$, means that $\partial(1/T)/\partial E < 0$, or $\partial T/\partial E > 0$. Since temperature is a function of energy, $T(E)$, its energy derivative cannot be infinite (otherwise there would be more than one temperature for a given energy). Hence the relationship between energy and temperature must be one-to-one, which is to say that it is invertible, $T_1 = T(E_1, N, V) \Leftrightarrow E_1 = E(T_1, N, V)$.

Entirely analogous arguments can be made for the other variables, namely pressure and volume can be swapped, as can number and chemical potential.

*Thermal stability*

For the present case of a sub-system in thermal contact with a heat reservoir, the constrained thermodynamic potential, $F(E|N, V, T) = E - TS(E, N, V)$, is a convex function of energy (because $S(E, N, V)$ is a concave function of energy, and the term $E$ is linear). Similarly, it is also a convex function of volume and number, so that we can write generically $F'' = -TS'' > 0$. As discussed at length above, the Second Law of Thermodynamics confers upon the constrained thermodynamic potential a variational principle that means that it is minimized by the equilibrium energy. Its minimum value is equal to the Helmholtz free energy.

As mentioned above, the Helmholtz free energy $\overline{F}(N, V, T) = F(\overline{E}|N, V, T)$ does not obey a variational principle. It is not a minimum with respect to number, volume, or temperature.

The thermal stability of matter may be deduced from the convexity of the constrained thermodynamic potential. Suppose that in a small region of the system there is a local increase in energy, with volume and particle number remaining constant. This increases the constrained thermodynamic potential for that region, considering it as a quasi-isolated sub-system. Equivalently, it decreases the local constrained total entropy. This is unfavorable, which is to say improbable, and the change is most likely to be counteracted by energy flowing back out of the region. This restores it to equilibrium. Conversely a local fluctuation to a lower energy is most likely followed by a countervailing back flow of energy.

It is the convexity of the constrained thermodynamic potential that makes matter thermally stable by damping such local fluctuations in the energy. Thermal stability does not arise from the Helmholtz free energy, but rather from the variational nature of the underlying constrained thermodynamic potential.

In the next section analogous optimization principles will be derived for volume and particle number reservoirs. It will be similarly shown that matter is mechanically stable to volume and particle number fluctuations.

## 2.3 Various reservoirs

The canonical equilibrium system that was just treated in detail may be called a constant temperature system, because the reservoir sets the temperature of the sub-system with which it can exchange energy. All that was required to obtain the constrained total entropy (equivalently, the constrained thermodynamic potential) for that system was to select the relevant linear additive, conserved variable, in this case energy, write the total entropy as the sum of that of the sub-system and that of the reservoir, and do a first order Taylor expansion of the latter, discarding the constant term.

In this section this procedure is repeated for several common reservoirs.

### 2.3.1 Constant pressure

An isobaric system is one in which the reservoir sets the pressure of the sub-system. The arrangement of the sub-system and reservoir remains the same as in figure 2.3, except that now the boundary is permeable to both energy and volume. This means that not only is the boundary made of conducting material, but also that it is flexible or moveable so that the volume of the sub-system can change at the expense of that of the reservoir. Thus, the total energy $E_{\text{total}} = E_1 + E_2$ and total volume $V_{\text{total}} = V_1 + V_2$ are fixed, but not their partitioning between the two systems.

As the second system is an energy and volume reservoir (i.e. it is very much larger than the sub-system), the Taylor expansion of its entropy about the total energy and volume may be terminated at the linear term. The zeroth order term, $S_2(E_{\text{total}}, N_2, V_{\text{total}})$, is a constant independent of the sub-system that may be discarded. Hence the sub-system-dependent part of the total entropy $S_{\text{total}} = S_1 + S_2$ is (dropping the subscripts)

$$S_{\text{total}}(E, V|N, p, T) = S_s(E, N, V) - \frac{E}{T} - \frac{pV}{T}. \tag{2.43}$$

This uses the volume derivative of the entropy, equation (2.11), for the pressure $p$ of the reservoir, and the energy derivative, equation (2.10), for the temperature $T$ of the reservoir. The reservoir only enters through its temperature and pressure. The constrained total entropy is extensive.

This is the constrained total entropy for the sub-system with arbitrary (i.e. not in equilibrium) values of energy or volume. By the Second Law of Thermodynamics, a spontaneous change in the volume must increase the total constrained entropy, or

$$\begin{aligned}\Delta S_{\text{total}}(E, V|N, p, T) &= \frac{\partial S_{\text{total}}(E, V|N, p, T)}{\partial V}\Delta V \\ &= \left[\frac{p(E, N, V)}{T(E, N, V)} - \frac{p}{T}\right]\Delta V \\ &\geqslant 0.\end{aligned} \tag{2.44}$$

Suppose that the system is in thermal equilibrium, $T(E, N, V) = T$, but not in mechanical equilibrium, $p(E, N, V) \neq p$. In this case this equation says that when the internal sub-system pressure is greater than the external reservoir pressure, $p(E, N, V) > p$, the sub-system spontaneously grows, $\Delta V > 0$, and vice versa. This is of course what one would expect on physical grounds and it justifies identifying the mathematical object $p$ with the physical object 'pressure'.

As usual, the equilibrium state is the one that maximizes the constrained total entropy, and is denoted $\overline{E}(N, p, T)$ and $\overline{V}(N, p, T)$. This variational procedure can be flipped into a minimization problem by defining the constrained thermodynamic potential as the negative of the temperature times the constrained total entropy,

$$\begin{aligned}G(E, V|N, p, T) &\equiv -TS_{\text{total}}(E, V|N, p, T) \\ &= E - TS_s(E, N, V) + pV \\ &= F(E|N, V, T) + pV.\end{aligned} \tag{2.45}$$

By design this constrained thermodynamic potential is a convex function of the sub-system volume and energy. It is appropriate for a sub-system with moveable conducting walls in contact with a temperature and pressure reservoir. In the event that thermal equilibrium is established, $E = \overline{E}(N, V, T)$, one can define $\tilde{G}(V|N, p, T) = \overline{F}(N, V, T) + pV$ as the constrained thermodynamic potential that determines volume fluctuations and equilibration.

Minimizing $G(E, V|N, p, T)$ with respect to energy yields $T(E, V, N) = T$, which is an implicit equation for the equilibrium sub-system energy for a given (constrained) volume, $\overline{E}(N, V, T)$. The minimization with respect to volume yields $p(E, V, N)/T(E, V, N) = p/T$, which is an implicit equation for the sub-system volume for a given sub-system energy and reservoir temperature and pressure, $\overline{V}(E|N, p, T)$. The simultaneous solution of these gives the equilibrium quantities, $\overline{E}(N, p, T)$ and $\overline{V}(N, p, T)$. The equilibrium sub-system volume also follows directly from the minimization of the thermally equilibrated constrained thermodynamic potential, $\partial\tilde{G}(V|N, p, T)/\partial V = 0$, or

$$\left.\frac{\partial \overline{F}(N, V, T)}{\partial V}\right|_{V=\overline{V}} = -p. \tag{2.46}$$

The minimum value of the constrained thermodynamic potential defines the Gibbs free energy,

$$\begin{aligned}\overline{G}(N, p, T) &\equiv G(\overline{V}, \overline{E}|N, T, p) \\ &= \tilde{G}(\overline{V}|N, T, p) \\ &= F(N, \overline{V}, T) + p\overline{V}.\end{aligned} \tag{2.47}$$

Since the Gibbs free energy must be extensive, and since it is a function of only one extensive variable, $N$, then it must be proportional to it, $\overline{G} \propto N$. The proportionality constant will be derived shortly.

Again, one can exploit the power of the constrained thermodynamic potential by invoking its variational nature when taking derivatives. That is, differentiating the Gibbs free energy is the same as differentiating the constrained thermodynamic potential while holding $V = \overline{V}$ and $E = \overline{E}$ fixed. Accordingly, the pressure derivative gives the equilibrium sub-system volume,

$$\left(\frac{\partial \overline{G}(N, p, T)}{\partial p}\right)_{N,T} = \overline{V}, \tag{2.48}$$

the number derivative gives the equilibrium sub-system chemical potential,

$$\left(\frac{\partial \overline{G}(N, p, T)}{\partial N}\right)_{p,T} = \left(\frac{\partial \overline{F}(N, \overline{V}, T)}{\partial N}\right)_{\overline{V},T} = \overline{\mu}, \tag{2.49}$$

and the temperature derivative gives the negative of the equilibrium sub-system entropy,

$$\left(\frac{\partial \overline{G}(N, p, T)}{\partial T}\right)_{N,p} = \left(\frac{\partial \overline{F}(N, \overline{V}, T)}{\partial T}\right)_{N,V} = -\overline{S}_{\mathrm{s}}. \tag{2.50}$$

Hence the total differential of the Gibbs free energy is

$$d\overline{G} = \overline{V}dp + \overline{\mu}dN - \overline{S}_s dT. \tag{2.51}$$

The second of the above derivatives combined with the above extensivity argument shows that

$$\overline{G}(N, p, T) = \overline{\mu}(p, T)N. \tag{2.52}$$

Finally, dividing both sides of the definition by $T$, one also has that

$$\left(\frac{\partial(\overline{G}/T)}{\partial(1/T)}\right)_{N,p} = \overline{E} + p\overline{V}. \tag{2.53}$$

The quantity on the right-hand side is called the enthalpy, and it will recur below.

### 2.3.2 Constant chemical potential

Now consider a reservoir that can exchange particles and energy with the sub-system of interest. This set-up is called an open system or, better, an open sub-system, or a grand canonical system. It can also be called a constant chemical potential system, as this and the temperature are fixed by the reservoir.

A usual, the total entropy is the sum of that of the sub-system and reservoir, each considered as isolated and in the designated macrostate, $S_{\text{total}} = S_1 + S_2$. The total energy $E_{\text{total}} = E_1 + E_2$ and the total particle number $N = N_1 + N_2$ are fixed, as well as the individual volumes, $V_1$ and $V_2$. Again, the relative size of the reservoir allows all terms in the Taylor expansion of the reservoir entropy to be neglected except for the linear one. (The zeroth term can again be dropped because it is a constant that is independent of the sub-system.) Hence dropping the subscripts, the constrained total entropy is

$$S_{\text{total}}(N, E|\mu, V, T) = S_s(E, V, N) - \frac{E}{T} + \frac{\mu}{T}N. \tag{2.54}$$

This gives the total entropy for a system when the sub-system of fixed volume $V$ has energy $E$ and particle number $N$, which are not necessarily the equilibrium values, while in thermal and diffusive contact with a reservoir of temperature $T$ and chemical potential $\mu$.

The constrained total entropy has derivatives

$$\frac{\partial S_{\text{total}}(N, E|\mu, V, T)}{\partial N} = \frac{-\mu(E, N, V)}{T(E, N, V)} + \frac{\mu}{T}, \tag{2.55}$$

and

$$\frac{\partial S_{\text{total}}(N, E|\mu, V, T)}{\partial E} = \frac{1}{T(E, N, V)} - \frac{1}{T}. \tag{2.56}$$

The first term on the right-hand side in each case is the derivative of the isolated sub-system entropy, and they give the chemical potential and the temperature of the sub-system in the specified macrostate. The maximum of the constrained total entropy of course corresponds to the vanishing of its derivatives. The implicit equations for the equilibrium energy $\overline{E}(\mu, V, T)$ and the equilibrium particle number $\overline{N}(\mu, V, T)$ that this gives are $T(\overline{E}, \overline{N}, V) = T$ and $\mu(\overline{E}, \overline{N}, V) = \mu$. Obviously, equilibrium corresponds to temperature and chemical potential equality between the sub-system and the reservoir.

Spontaneous changes in the sub-system particle number $\Delta N$ must lead to an increase in the total entropy

$$\begin{aligned}\Delta S_{\text{total}}(N, E|\mu, V, T) &= \frac{\partial S_{\text{total}}(N, E|\mu, V, T)}{\partial N}\Delta N \\ &= \left[\frac{-\mu(E, N, V)}{T(E, N, V)} + \frac{\mu}{T}\right]\Delta N \\ &\geqslant 0.\end{aligned} \tag{2.57}$$

Assuming thermal equilibration, $T(\overline{E}, N, V) = T$, when the chemical potential of the sub-system is greater than that of the reservoir, $\mu(\overline{E}(N, V, T), N, V) > \mu$, this says that $\Delta N < 0$, which is to say particles spontaneously flow from the sub-system to the reservoir.

In other words, this says that particles move down the chemical potential gradient. This is analogous to what happens in a thermal system: energy moves down a temperature gradient. It is the opposite of what happens in an isobaric system, where volume moves up a pressure gradient. The similarities and differences in the three cases can be accounted for by whether or not the definition of the field variable in terms of an entropy derivative involves a minus sign, or whether or not the definition refers to the reciprocal of the field variable.

The corresponding constrained thermodynamic potential is the negative of the total entropy times the temperature,

$$\begin{aligned}\Omega(E, N|\mu, V, T) &\equiv -TS_{\text{total}}(E, N|\mu, V, T) \\ &= E - TS_{\text{s}}(E, N, V) - \mu N \\ &= F(E|N, V, T) - \mu N.\end{aligned} \tag{2.58}$$

This is a convex function of $N$ and of $E$ that is minimized by their equilibrium values. Thus, this is the minimization variational principle for a sub-system able to exchange energy and particles with a reservoir of chemical potential $\mu$ and temperature $T$.

If thermal equilibration is faster than diffusive equilibration, then one can define $\tilde{\Omega}(N|\mu, V, T) \equiv \Omega(\overline{E}(N, V, T), N|\mu, V, T) = \overline{F}(N, V, T) - \mu N$. This involves the Helmholtz free energy, and it provides a variational principle for a sub-system at the same temperature as the reservoir, but not in diffusive equilibrium with it.

The concavity of the isolated sub-system entropy means that the constrained thermodynamic potential is a convex function of particle number and energy, which can be figuratively denoted $\Omega'' = -TS_s'' > 0$. The minimum value of the constrained thermodynamic potential of this open sub-system is the equilibrium free energy, which is this case is called the grand potential. It is

$$\begin{aligned}\overline{\Omega}(\mu, V, T) &\equiv \Omega(\overline{E}, \overline{N}|\mu, V, T)\\ &= \overline{F}(\overline{N}, V, T) - \mu\overline{N}.\end{aligned} \tag{2.59}$$

The properties of the variational procedure allow the otherwise dependent variables $\overline{E}$ and $\overline{N}$ to be held fixed during differentiation of the grand potential. Hence one has

$$\left(\frac{\partial\overline{\Omega}(\mu, V, T)}{\partial\mu}\right)_{V,T} = -\overline{N}, \tag{2.60}$$

$$\left(\frac{\partial\overline{\Omega}(\mu, V, T)}{\partial V}\right)_{\mu,T} = \left(\frac{\partial\overline{F}(\overline{N}, V, T)}{\partial V}\right)_{\bar{N},T} = -\bar{p}, \tag{2.61}$$

and

$$\left(\frac{\partial\overline{\Omega}(\mu, V, T)}{\partial T}\right)_{\mu,V} = \left(\frac{\partial\overline{F}(\overline{N}, V, T)}{\partial T}\right)_{\bar{N},V} = -\overline{S}_s. \tag{2.62}$$

This last quantity is the entropy of the isolated sub-system with the equilibrium energy and particle number, $\overline{S}_s(\mu, V, T) \equiv S_s(\overline{E}(\mu, V, T), \overline{N}(\mu, V, T), V)$. It follows that the total differential of the grand potential is

$$\mathrm{d}\overline{\Omega} = -\overline{N}\,\mathrm{d}\mu - \bar{p}\,\mathrm{d}V - \overline{S}_s\,\mathrm{d}T. \tag{2.63}$$

One also has

$$\left(\frac{\partial(\overline{\Omega}/T)}{\partial(1/T)}\right)_{\mu,V} = \overline{E} - \mu\overline{N}. \tag{2.64}$$

Finally, because the grand potential must be extensive, and because volume is the only one of its arguments that is extensive, the grand potential must be proportional to the volume. From the above volume derivative one concludes that

$$\overline{\Omega}(\mu, V, T) = -\bar{p}V. \tag{2.65}$$

*Multicomponent system*

It is not unusual to have several particle species present. Let $N_\alpha$ be the number of particles of type $\alpha$ in the sub-system. The chemical potential for this species is defined as

$$\mu_\alpha \equiv -T\left(\frac{\partial S_s}{\partial N_\alpha}\right)_{E,V,N_{\gamma\neq\alpha}}. \tag{2.66}$$

The formulae given above for a single species are changed to include the sum over species. For example, for $m$ different types of particles, the total differential of the isolated system entropy becomes

$$\mathrm{d}S_{\mathrm{s}}(E,\, \underline{N},\, V) = \frac{1}{T}\,\mathrm{d}E + \frac{p}{T}\,\mathrm{d}V - \frac{1}{T}\sum_{\alpha=1}^{m}\mu_\alpha\,\mathrm{d}N_\alpha. \tag{2.67}$$

It is often convenient to use vector notation, and to replace the sum by a scalar product. For example, the constrained thermodynamic potential may be written as

$$\Omega\big(E,\, \underline{N}\big|\underline{\mu},\, V,\, T\big) = E - TS_{\mathrm{s}}(E,\, \underline{N},\, V) - \underline{\mu}\cdot\underline{N}. \tag{2.68}$$

The diffusive equilibrium condition is obviously the chemical potential equality between the sub-system and the reservoir for each species, $\bar{\mu}_\alpha \equiv \mu_\alpha(\overline{E},\, \bar{\underline{N}},\, V) = \mu_\alpha$.

### 2.3.3 Constant enthalpy

We now consider an isolated sub-system with a moveable wall or piston, figure 2.5. The number of particles in the sub-system $N_1$ is fixed. Volume can be exchanged with the surrounding reservoir, $\mathrm{d}V_1 = -\mathrm{d}V_2$. Even though the walls are insulated (adiabatic, adiathermal, no heat flow), energy can still exchange between the sub-system and the reservoir via so-called $pV$-work. The total energy is fixed, $E_{\mathrm{total}} = E_1 + E_2$, which means that $\mathrm{d}E_1 = -\mathrm{d}E_2$.

The energy of the sub-system changes as its volume changes, $E_1(V_1)$. As the wall or piston moves, the change in energy of the reservoir is linearly proportional to the change in sub-system volume, $\mathrm{d}E_2 = p_2\mathrm{d}V_1$. (This can be seen in mechanical terms: if the piston has area $A_1$ and mass $M$, and if the acceleration due to gravity is $g$, then the constant external pressure is $p_2 = Mg/A_1$, and the change in energy is proportional to the change in height, $\mathrm{d}E_2 = Mg\mathrm{d}h_1$.) In any event, the external pressure $p_2$ is taken to be constant. With this it may be seen that changes in the volume and energy of the sub-system at constant particle number are related by

$$\frac{\mathrm{d}E_1}{\mathrm{d}V_1} = -p_2, \quad \text{or} \quad E_1 + p_2 V_1 = \text{const.} \tag{2.69}$$

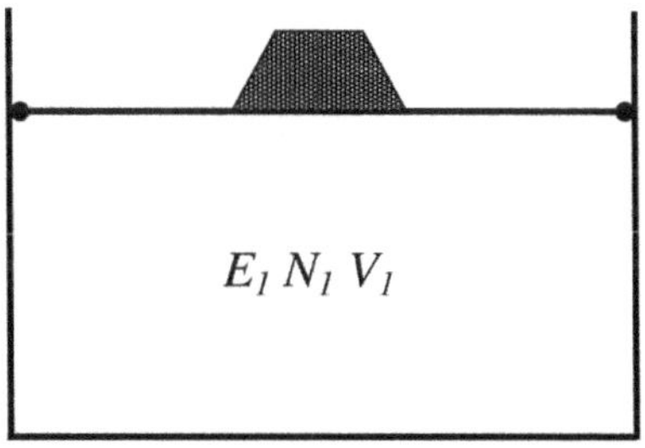


**Figure 2.5.** An isolated sub-system 1 with moveable piston.

The quantity that is constant during volume changes without heat flow is like the enthalpy,

$$H = E_1 + p_2 V_1. \tag{2.70}$$

(Usually the difference between the sub-system pressure and the reservoir pressure is ignored.) The enthalpy previously appeared as essentially the inverse temperature derivative of the Gibbs free energy, equation (2.53). The fact that this is constant in the present system of no heat flow is consistent with a form of the First Law of Thermodynamics, namely $pV$-work at constant pressure changes the heat of the system as $\mathrm{d}Q = \mathrm{d}E + p\mathrm{d}V = \mathrm{d}(E + pV)$, which is just the change in enthalpy. In the present case this is zero because the system has been defined to have no heat flow.

In a reversible change, the internal pressure balances the external pressure, $p_1(E_1, N_1, V_1) = p_2$. In this case the enthalpy of the sub-system is constant, $\mathrm{d}H_1 = \mathrm{d}E_1 + p_1\mathrm{d}V_1 = 0$. More generally, $p_1 \neq p_2$, and it is the enthalpy-like quantity $E_1 + p_2 V_1$ that is constant, where the constant externally applied pressure is used, not the internal pressure of the sub-system.

The total entropy is that of the sub-system alone, since the entropy of the reservoir is constant

$$\begin{aligned} \mathrm{d}S_2(E_2, N_2, V_2) &= \frac{1}{T_2}\mathrm{d}E_2 + \frac{p_2}{T_2}\mathrm{d}V_2 \\ &= \frac{1}{T_2}\left[p_2\mathrm{d}V_1 - p_2\mathrm{d}V_1\right] \\ &= 0. \end{aligned} \tag{2.71}$$

One could have guessed that this would be the case because the external pressure could have been made mechanical in origin. Hence the total entropy is $S_{\text{total}}(E, V|N, p) = S_1(E, V, N)$, with $E + pV =$ const., where the subscripts have been dropped. Hence the change in total entropy due to a change in sub-system volume is

$$\begin{aligned} \Delta S_{\text{total}} &= \frac{\partial S_1}{\partial E}\frac{\mathrm{d}E}{\mathrm{d}V}\Delta V + \frac{\partial S_1}{\partial V}\Delta V \\ &= \frac{p(E, N, V) - p}{T(E, N, V)}\Delta V. \end{aligned} \tag{2.72}$$

For the total entropy to increase, the volume of the sub-system has to increase when its pressure is greater than the external pressure, and the volume has to decrease when the internal pressure is less than the external pressure. This was also found to be the case for the isobaric system, which allowed heat flow, equation (2.44). Equilibrium, of course, corresponds to pressure equality.

Since the enthalpy $H = E + pV$ is constant, it may be taken as one of the independent variables for the insulated, variable volume system. The constrained thermodynamic potential may be defined as

$$\Phi(V|H, N, p; T) = -TS_s(E, N, V), \quad E \equiv H - pV. \tag{2.73}$$

The temperature here is arbitrary, as it is introduced solely to give the potential the dimensions of energy. It enters as a trivial multiplicative factor whose value has no physical consequences. By construction $\Phi$ is a convex function of volume minimized at the equilibrium volume $\overline{V}(H, N, p)$. The volume derivative is

$$\begin{aligned} \frac{1}{T}\frac{\partial \Phi(V|H, N, p; T)}{\partial V} &= -\frac{\partial S_s(E, N, V)}{\partial E}\frac{\partial (H - pV)}{\partial V} - \frac{\partial S_s(E, N, V)}{\partial V} \\ &= \frac{p - p(E, N, V)}{T(E, N, V)}, \quad E \equiv H - pV. \end{aligned} \tag{2.74}$$

Again, this vanishes at equilibrium, when the internal pressure of the sub-system equals the external pressure of the reservoir, $p(\overline{E}, \overline{V}, N) = p$, where $\overline{E} = H - p\overline{V}$.

As usual, the thermodynamic potential of this system is the minimum value of the constrained thermodynamic potential, $\bar{\Phi}(H, N, p; T) = \Phi(\overline{V}|H, N, p; T)$. And again as usual, the variational nature of $\Phi$, means that $\bar{\Phi}$ can be differentiated at fixed $\overline{V}$. Hence one has

$$\frac{1}{T}\left(\frac{\partial \bar{\Phi}}{\partial N}\right)_{H,p,T} = \frac{1}{T}\left(\frac{\partial \Phi}{\partial N}\right)_{H,p,T,\bar{V}} = \frac{\bar{\mu}}{\overline{T}}, \tag{2.75}$$

$$\frac{1}{T}\left(\frac{\partial \bar{\Phi}}{\partial H}\right)_{N,p,T} = \frac{1}{T}\left(\frac{\partial \Phi}{\partial H}\right)_{N,p,T,\bar{V}} = \frac{-1}{\overline{T}}, \tag{2.76}$$

and

$$\frac{1}{T}\left(\frac{\partial \bar{\Phi}}{\partial p}\right)_{H,N,T} = \frac{1}{T}\left(\frac{\partial \Phi}{\partial p}\right)_{H,N,T,\bar{V}} = \frac{\overline{V}}{\overline{T}}. \tag{2.77}$$

Together with the trivial result $\partial\bar{\Phi}/\partial T = \bar{\Phi}/T$, these give the total differential,

$$\mathrm{d}\bar{\Phi} = -\frac{T}{\overline{T}}\,\mathrm{d}H + \frac{\overline{V}T}{\overline{T}}\,\mathrm{d}p + \frac{\bar{\mu}T}{\overline{T}}\,\mathrm{d}N + \frac{\bar{\Phi}}{T}\,\mathrm{d}T. \tag{2.78}$$

Finally, it is worth mentioning that the change in internal energy was taken above to be equal and opposite to that of the reservoir, $\mathrm{d}E_1 = -\mathrm{d}E_2 = -p_2\mathrm{d}V_1$. However, since no heat flows, the change in energy of the sub-system is just the work done by it on the piston, $\mathrm{d}E_1 = -p_1\mathrm{d}V_1$. There is an obvious contradiction between these two in the case that $p_1 \neq p_2$.

The problem can be resolved by noting that the piston acquires kinetic energy as it accelerates due to the initial force imbalance. This has been neglected in the above analysis. Also neglected has been any heat created by friction or by viscous dissipation. If the viscous dissipation of the reservoir is comparable to that of the sub-system, one cannot say how the kinetic energy of the piston is divided between the two when it dies out. Because of the adiathermal walls, the arbitrary internal heating does not equilibrate between them. Although equilibrium still corresponds to pressure equality, one does not know the final energy, temperature, or volume of the sub-system. This indeterminacy of linked isenthalpic systems has been noted by Callan (1960, appendix C).

### 2.3.4 Constant entropy

So far physical reservoirs have been considered. These allow the exchange of energy, volume, and particles, which are the independent variables of the sub-system entropy. We now wish to consider a case in which the sub-system entropy itself is an independently specified variable. This is done more to illustrate the mathematical procedures than for any application to a real physical system.

Suppose that the sub-system can exchange energy and volume with a reservoir while keeping the sub-system entropy constant. In this case we can regard the volume as independent and write the energy as $E(S, V, N)$. As usual, the constrained total entropy is

$$S_{\text{total}}(V|S, p, N; T) = S - \frac{E(S, V, N)}{T} - \frac{pV}{T}, \tag{2.79}$$

where $T$ and $p$ are the temperature and pressure of the reservoir, respectively. Only the quantities that depend on the sub-system are kept here. Again, as usual, the constrained thermodynamic potential is

$$\begin{aligned} H^*(V|S, p, N; T) &= -TS_{\text{total}}(V|S, p, N; T) \\ &= E(S, V, N) + pV - TS. \end{aligned} \tag{2.80}$$

One can discard the final constant term since we are interested only in variations of the potential with the constrained volume at constant sub-system entropy,

$$H(V|S, p, N) = E(S, V, N) + pV. \tag{2.81}$$

This is just the enthalpy, and it describes volume fluctuations of an isentropic sub-system.

Now for an isolated system at constant entropy and number,

$$0 = \mathrm{d}S(E, N, V) = \frac{1}{T}\,\mathrm{d}E + \frac{p}{T}\,\mathrm{d}V. \tag{2.82}$$

Hence $\partial E(S, V, N)/\partial V = -p(S, V, N)$. It follows that the volume derivative of the constrained thermodynamic potential is

$$\left(\frac{\partial H}{\partial V}\right)_{S,p,N} = -p(S, V, N) + p. \tag{2.83}$$

As expected the equilibrium volume, $\overline{V}(S, p, N)$, equalizes the sub-system and reservoir pressures. Being derived from the total entropy, the constrained thermodynamic potential is a convex function of volume. Hence the extremum is a minimum,

$$\left(\frac{\partial^2 H}{\partial V^2}\right)_{S,p,N} = \left(\frac{\partial^2 E}{\partial V^2}\right)_{S,N} = -\left(\frac{\partial p(S,\, V,\, N)}{\partial V}\right)_{S,N} > 0. \tag{2.84}$$

The equilibrium enthalpy is the minimum value of the constrained thermodynamic potential,

$$\overline{H}(S, p, N) = E(S, \overline{V}, N) + p\overline{V}. \tag{2.85}$$

This is the equilibrium thermodynamic potential for this isentropic system. Holding $\overline{V}$ fixed as per the variational principle, one has

$$\left(\frac{\partial \overline{H}(S, p, N)}{\partial p}\right)_{N,S} = \overline{V}, \tag{2.86}$$

$$\left(\frac{\partial \overline{H}(S, p, N)}{\partial N}\right)_{p,S} = \left(\frac{\partial E(S, \overline{V}, N)}{\partial N}\right)_{\overline{V},S} = \bar{\mu}, \tag{2.87}$$

and

$$\left(\frac{\partial \overline{H}(S, p, N)}{\partial S}\right)_{N,p} = \left(\frac{\partial E(S, \overline{V}, N)}{\partial S}\right)_{N,\overline{V}} = \overline{T}. \tag{2.88}$$

These may be summarised by the total differential,

$$d\overline{H} = \overline{V}dp + \bar{\mu}dN + \overline{T}dS. \tag{2.89}$$

In the case of the full thermodynamic potential, $\overline{H}^*(S, p, N; T)$, the results remain the same, except that the final one is replaced by

$$\left(\frac{\partial \overline{H}^*(S, p, N; T)}{\partial S}\right)_{N,p;T} = \overline{T} - T. \tag{2.90}$$

Note that there is no requirement for the temperature of the sub-system to equal that of the reservoir. In addition one has the entropy itself

$$\left(\frac{\partial \overline{H}^*(S, p, N; T)}{\partial T}\right)_{S,p,N} = -S, \tag{2.91}$$

and the equilibrium enthalpy

$$\left(\frac{\partial(\overline{H}^*(S, p, N; T)/T)}{\partial T^{-1}}\right)_{S,p,N} = -(\overline{E} + p\overline{V}). \tag{2.92}$$

The results for the various reservoirs are summarized in table 2.2.

**Table 2.2.** Summary of the common thermodynamic systems (Attard 2002).

| System | Constrained potential<br>Parameters | Equilibrium potential<br>Differential |
|---|---|---|
| Isolated | –<br>$N, V, E$ | Entropy, $S$<br>$T\mathrm{d}S = \mathrm{d}E + p\,\mathrm{d}V - \mu\,\mathrm{d}N$ |
| Isothermal | $F = E - TS_s(E, N, V)$<br>$E\vert N, V, T$ | Helmholtz, $\overline{F}$<br>$\mathrm{d}\overline{F} = -\overline{S}_s\,\mathrm{d}T - \overline{p}\,\mathrm{d}V + \overline{\mu}\,\mathrm{d}N$ |
| Open | $\Omega = E - TS_s(E, N, V) - \mu N$<br>$N, E\vert\mu, V, T$ | Grand, $\overline{\Omega}$<br>$\mathrm{d}\overline{\Omega} = -\overline{S}_s\,\mathrm{d}T - \overline{p}\,\mathrm{d}V - \overline{N}\,\mathrm{d}\mu$ |
| Isobaric | $G = E - TS_s(E, N, V) + pV$<br>$E, V\vert N, p, T$ | Gibbs, $\overline{G}$<br>$\mathrm{d}\overline{G} = -\overline{S}_s\,\mathrm{d}T + \overline{V}\mathrm{d}p + \overline{\mu}\,\mathrm{d}N$ |
| Isenthalpic | $\Phi = -TS_s(H - pV, N, V)$<br>$V\vert N, H, p; T$ | $\overline{\Phi}$<br>$\overline{T}\mathrm{d}\overline{\Phi}/T = -\,\mathrm{d}H + \overline{V}\,\mathrm{d}p + \overline{\mu}\,\mathrm{d}N$<br>$+\,\overline{\Phi}\overline{T}\mathrm{d}T/T^2$ |
| Isentropic | $H = E(S, V, N) + pV$<br>$V\vert N, S, p$ | Enthalpy, $\overline{H}$<br>$\mathrm{d}\overline{H} = \overline{V}\,\mathrm{d}p + \overline{\mu}\,\mathrm{d}N + \overline{T}\,\mathrm{d}S$ |

## 2.4 Thermodynamic second derivatives

The second derivatives of the equilibrium thermodynamic potentials provide measurable physical parameters such as the heat capacity, compressibility, thermal expansivity, etc. These also determine the probability of the statistical fluctuations about equilibrium as will be shown in section 2.5.

The thermodynamic systems analyzed above each have a characteristic set of dependent and independent variables. However, because there is a one-to-one mapping between the equilibrium values of the dependent variables and the independent variables, which is to say that the thermodynamic state is unique, one is free to choose any set of three variables as the independent ones. To simplify the notation in this section, equilibrium quantities will not be over-lined, and the dependent and independent variables should be gleaned from the context.

### 2.4.1 Concavity of the thermodynamic potentials

The thermodynamic reservoir formalism is based on the exchange between the sub-system and the reservoir of one or more conserved, linear additive quantities. The constrained thermodynamic potential is convex with respect to these constrained extensive variables, as follows from the fact that it is the negative of the constrained total entropy. The equilibrium thermodynamic potential, which is the minimum value of the constrained thermodynamic potential, effectively replaces as independent variables these exchangeable extensive variables of the sub-system with the intensive variables of the reservoir that are their thermodynamic conjugates. The equilibrium free energy or thermodynamic potential is concave with respect to these conjugate intensive variables, as is now shown.

Let $X$ be the conserved extensive variable and let the conjugate intensive (field) variable be

$$x = -\left(\frac{\partial E}{\partial X}\right)_S, \tag{2.93}$$

or

$$x = T\frac{\partial S}{\partial X}. \tag{2.94}$$

Typically $X$ is $E$, $N$, or $V$, in which case $x$ is 1, $-\mu$, or $p$, respectively. The variable $X$ could also be some additional extensive, exchangeable system parameter. The part of the total entropy that depends on the sub-system for a reservoir at $x$ and $T$, with $X$ being exchanged, is

$$S_{\text{total}}(X|x, T) = S_s(X) - \frac{xX}{T}. \tag{2.95}$$

Now has already been shown, the total entropy is concave with respect to the extensive variable,

$$\left(\frac{\partial^2 S_{\text{total}}(X|x, T)}{\partial X^2}\right)_x < 0. \tag{2.96}$$

The equilibrium value of $X$ is $\overline{X}(x, T)$, and this gives the maximum total entropy, $\overline{S}_{\text{total}}(x, T) = S_{\text{total}}(\overline{X}(x, T)|x, T)$. This equilibrium value of the total entropy is a convex function of the reservoir field $x/T$, as is now proven.

One has

$$\begin{aligned}\frac{d^2 S_{\text{total}}(\overline{X}|x, T)}{d(x/T)^2} &= \frac{\mathrm{d}}{\mathrm{d}(x/T)}\left[\frac{\partial S_s(X)}{\partial X}\frac{\mathrm{d}\overline{X}}{\mathrm{d}(x/T)} - \frac{x}{T}\frac{\mathrm{d}\overline{X}}{\mathrm{d}(x/T)} - \overline{X}\right] \\ &= \frac{-\mathrm{d}\overline{X}(x, T)}{\mathrm{d}(x/T)}.\end{aligned} \tag{2.97}$$

The first two terms in the brackets cancel because at equilibrium, the sub-system intensive variables equal those of the reservoir, $x(\overline{X}(x, T)) = x$ and $T(\overline{X}(x, T)) = T$. For any value of the constrained exchangeable variable, say $X'$, not equal to the equilibrium value, $X' \neq \overline{X}$, then

$$S_{\text{total}}(X'|x, T) < S_{\text{total}}(\overline{X}|x, T). \tag{2.98}$$

This follows because equilibrium is defined as the maximum of the total entropy. Writing out the two sides of this inequality explicitly and re-arranging gives

$$\frac{x\overline{X}}{T} - \frac{xX'}{T} < S_s(\overline{X}) - S_s(X'). \tag{2.99}$$

The second order expansion of the sub-system entropy about $X'$ is

$$\begin{aligned} S_s(\overline{X}) &= S_s(X') + (\overline{X} - X')S_{s,X}(X') + \frac{(\overline{X} - X')^2}{2}S_{s,XX}(X') \\ &= S_s(X') + (\overline{X} - X')\frac{\bar{x}'}{\overline{T}'} + \frac{(\overline{X} - X')^2}{2}S_{s,XX}(X'). \end{aligned} \tag{2.100}$$

Inserting this into the inequality yields

$$\left[\frac{x}{T} - \frac{\bar{x}'}{\overline{T}'}\right](\overline{X} - X') < \frac{(\overline{X} - X')^2}{2}S_{s,XX}(X'). \tag{2.101}$$

Since the isolated system entropy is a concave function of its arguments, the right-hand side is negative. Hence divide both sides by $(\overline{X} - X')^2$ and take the limit $X' \to \overline{X}$ to obtain

$$\frac{\mathrm{d}(\bar{x}/\overline{T})}{\mathrm{d}X} < 0. \tag{2.102}$$

This is equivalently but more conveniently written

$$\frac{\mathrm{d}\overline{X}}{\mathrm{d}(x/T)} < 0. \tag{2.103}$$

This shows that the equilibrium total entropy (i.e. its maximum value) is a convex function of $x/T$,

$$\frac{\mathrm{d}^2 S_{\text{total}}(\overline{X}(x)|x)}{\mathrm{d}(x/T)^2} > 0. \tag{2.104}$$

Since the equilibrium thermodynamic potential is essentially the negative of the equilibrium total entropy $\overline{F}(x) \equiv F(\overline{X}(x)|x) = -TS_{\text{total}}(\overline{X}(x)|x)$, the convexity of the latter translates into concavity of the former,

$$\frac{\mathrm{d}^2(\overline{F}(x)/T)}{\mathrm{d}(x/T)^2} < 0. \tag{2.105}$$

Finally, one can readily identify $X$ and $x$ in a given case by noting that

$$\frac{\mathrm{d}(\overline{F}(x)/T)}{\mathrm{d}(x/T)} = \overline{X}(x). \tag{2.106}$$

### 2.4.2 Heat capacity

The heat capacity tells one how much energy is required to change the temperature of a sub-system. It is actually defined as the reciprocal of this, namely the rate of change of heat with temperature.

Heat flow can be thought of as the rate of microscopic or statistical energy change. Work is the rate of macroscopic energy change, the most common example being pressure–volume or $pV$ work. Neither heat not work are state functions (one cannot say that in a given thermodynamic state the system has a certain amount of heat, or a certain amount of work), but changes in the sum of the two together give the change in energy, which is a state function.

From the First Law of Thermodynamics, the change in energy of the system is the change in heat of the system plus the work done on the system, $\mathrm{d}E = \mathrm{d}Q + \mathrm{d}W$. Hence the change of energy differs if the change is at constant volume ($\mathrm{d}W = 0$) or at constant pressure ($\mathrm{d}W = -p\,\mathrm{d}V$).

The heat capacity at constant volume is

$$C_V = \left(\frac{\partial Q}{\partial T}\right)_{V,N} = \left(\frac{\partial \overline{E}}{\partial T}\right)_{V,N}, \tag{2.107}$$

where this is the equilibrium energy of a sub-system in contact with a heat reservoir of temperature $T$, $\overline{E}(N, V, T)$. Since the equilibrium energy is the derivative of the Helmholtz free energy, $\overline{F}(N, V, T)$, equation (2.34), this is the second derivative

$$C_V = \frac{-1}{T^2}\left(\frac{\partial^2(\overline{F}/T)}{\partial(1/T)^2}\right)_{V,N}. \tag{2.108}$$

The heat capacity is evidently an extensive variable.

In the preceding section, it was established that in general the equilibrium thermodynamic potentials were concave with respect to the conjugate intensive variables of the reservoir. The present case corresponds to $x \equiv 1$ (and $X \equiv E$), and so one concludes that the heat capacity at constant volume is positive,

$$C_V > 0. \tag{2.109}$$

The heat capacity at constant pressure is

$$\begin{aligned} C_p &= \left(\frac{\partial Q}{\partial T}\right)_{p,N} \\ &= \left(\frac{\partial \overline{E}}{\partial T}\right)_{p,N} + p\left(\frac{\partial \overline{V}}{\partial T}\right)_{p,N} \\ &= \frac{-1}{T^2}\left(\frac{\partial^2(\overline{G}/T)}{\partial(1/T)^2}\right)_{p,N}, \end{aligned} \tag{2.110}$$

where the enthalpy $E + pV$ has been written as the derivative of the Gibbs free energy, equation (2.53). Again invoking the concavity established in the preceding section, one identifies $x \equiv 1$ and $X = H$, and one concludes that the heat capacity at constant pressure is positive,

$$C_p > 0. \tag{2.111}$$

Further analysis shows that $C_p > C_V$. This signifies that additional heat is required to raise the temperature of a sub-system at constant pressure because extra work is required to expand the volume against the external pressure.

### 2.4.3 Compressibility

The compressibility is the rate of change of volume with pressure. In this case one has a choice of making the change at constant temperature or at constant entropy. The isothermal compressibility is defined to be

$$\chi_T = \frac{-1}{\overline{V}} \left( \frac{\partial \overline{V}}{\partial p} \right)_{T,N}. \tag{2.112}$$

The compressibility is evidently an intensive variable. The volume here is $\overline{V}(N, p, T)$, which is the pressure derivative of the Gibbs free energy, so that this may be rewritten

$$\chi_T = \frac{-1}{\overline{V}} \left( \frac{\partial^2 \overline{G}}{\partial p^2} \right)_{T,N} = \frac{-1}{\overline{V}T} \left( \frac{\partial^2 \overline{G}/T}{\partial (p/T)^2} \right)_{T,N}. \tag{2.113}$$

Invoking the concavity result and identifying $x \equiv p/T$ and $X \equiv V$, it follows that the isothermal compressibility is positive, $\chi_T > 0$.

For the case of constant entropy, the adiabatic compressibility is

$$\chi_S = \frac{-1}{\overline{V}} \left( \frac{\partial \overline{V}}{\partial p} \right)_{S,N}. \tag{2.114}$$

Since the equilibrium volume is the pressure derivative of the enthalpy at constant entropy, equation (2.86), this can be written as the second derivative of the enthalpy,

$$\chi_S = \frac{-1}{\overline{V}} \left( \frac{\partial^2 \overline{H}}{\partial p^2} \right)_{S,N}. \tag{2.115}$$

Again this is positive.

### 2.4.4 Maxwell relations

The four examples given above were 'pure' second derivatives of equilibrium thermodynamic potentials. One can also form the mixed second derivatives. Interchanging the order of differentiation gives rise to the Maxwell relations.

For example, the two ways of taking the temperature and volume derivatives of the Helmholtz free energy can be equated to each other,

$$\frac{\partial^2 \overline{F}(N, V, T)}{\partial V \partial T} = \frac{\partial^2 \overline{F}(N, V, T)}{\partial T \partial V}. \tag{2.116}$$

Performing the inner derivatives, equations (2.30) and (2.31) yield

$$\left(\frac{\partial \overline{S}_s(N, V, T)}{\partial V}\right)_{T,N} = \left(\frac{\partial \overline{p}(N, V, T)}{\partial T}\right)_{V,N}. \tag{2.117}$$

There are many such Maxwell relations. One is best advised to derive each as it is required rather than attempting to remember them all.

## 2.5 Probability and fluctuation theory

In this section the probability of fluctuations in the state of a sub-system able to exchange with a reservoir is analyzed in a generic fashion.

We shall denote the set of extensive sub-system variables that are fixed by $Y$, and the extensive sub-system variables that can be exchanged with the reservoir by $X$. The conjugate intensive variables are defined by the derivatives of the isolated sub-system entropy,

$$x = T\frac{\partial S(X, Y)}{\partial X} \text{ and } y = T\frac{\partial S(X, Y)}{\partial Y}. \tag{2.118}$$

The most usual pairs of conjugate variables are $\{x, X\} = \{1, E\}$, $\{p, V\}$, and $\{-\mu_\alpha, N_\alpha\}$. The $X$ and the $x$ are vectors in the case that more than one quantity is exchangeable with the reservoir.

### 2.5.1 Exchange with a reservoir

Using a subscript s for the sub-system and r for the reservoir, as usual the total entropy is

$$\begin{aligned} S_{\text{total}}(X_s | Y_s, Y_r, X_{\text{total}}) &= S_s(X_s, Y_s) + S_r(X_{\text{total}} - X_s, Y_r) \\ &= S_s(X_s, Y_s) - \frac{x_r}{T_r} X_s. \end{aligned} \tag{2.119}$$

Only that part of the total entropy that depends on the sub-system, and terms extensive with the sub-system, have been kept here. Dropping the subscripts since the reservoir only enters through its intensive variables this is

$$S_{\text{total}}(X | Y, x, T) = S_s(X, Y) - \frac{x}{T} X. \tag{2.120}$$

The temperature that appears here is that of the reservoir. In the event of multiple exchangeable parameters, $xX \Rightarrow \mathbf{x} \cdot \mathbf{X}$.

The probability that the sub-system is in the macrostate $X$ (i.e. has an amount $X$ of exchangeable material) is simply proportional to the exponential of the total entropy,

$$\wp(X|Y, x, T) = \frac{1}{Z(Y, x, T)} e^{S_{\text{total}}(X|Y,x,T)/k_{\text{B}}} \\ = \frac{1}{Z(Y, x, T)} e^{S_{\text{s}}(X,Y)/k_{\text{B}}} e^{-\beta x X}, \tag{2.121}$$

where $\beta \equiv 1/k_{\text{B}}T$ is often called the inverse temperature. The normalizing factor $Z(Y, x, T)$ is called the partition function.

As we have seen, the total entropy is a maximum when the conjugate sub-system field variable equals that of the reservoir,

$$\left. \frac{\partial S_{\text{total}}(X|Y, x, T)}{\partial X} \right|_{X=\bar{X}} = 0 \Leftrightarrow x_{\text{s}}(\overline{X}_{\text{s}}, Y_{\text{s}}) = x_{\text{r}}. \tag{2.122}$$

Obviously this corresponds to the maximum of the probability distribution, and $\overline{X}$ is the most likely macrostate of the system, or, equivalently, the most likely value of the exchangeable variable of the sub-system.

*Concavity of the entropy*

It was argued in section 2.1.3 that the entropy of an isolated sub-system must be a concave function of its extensive arguments in order for the thermodynamic state to be stable. This condition leads to

$$S_{ii} < 0, \quad \text{and } S_{ii}S_{jj} - S_{ij}^2 > 0, \tag{2.123}$$

where the subscripts denote the second derivatives with respect to any of the isolated sub-system $X_i$. (The second condition comes from the fact that the eigenvalues of the Jacobean matrix must all be negative, and, for two exchangeable parameters, the product of the two eigenvalues is the determinant.)

In the present case of exchange with a reservoir, for a stable equilibrium state, the total entropy must be a concave function of its constrained arguments. But this is guaranteed by the concavity of the isolated system entropy, since the part of the total entropy connected to the reservoir is a linear function of the exchangeable variables, and concavity reflects the second derivative.

The connection between the two cases is not so surprising since a macroscopic isolated system forms a reservoir for any part of itself.

*Gaussian probability*

The concavity of the total entropy means that the probability distribution for the constrained variable that the sub-system can exchange with a reservoir has a well defined peak at $\overline{X}$. One can make a second order expansion of the exponent about this peak to obtain a Gaussian distribution,

$$\wp(X|Y, x, T) \approx \frac{1}{Z_G(Y, x, T)} e^{S_s''(X-\overline{X})^2/2k_B}, \tag{2.124}$$

where $S_s'' \equiv S_s''(\overline{X}(Y, x, T), Y)$ is the Jacobian matrix of second derivatives of the isolated sub-system entropy evaluated in the equilibrium state.

Note that the contributions from higher order derivatives are negligible, as can be seen from extensivity arguments. (The $n$th entropy derivative $S_s^{(n)}$ scales with $V^{1-n}$, whereas $(X - \overline{X})^n$ scales with $V^{n/2}$, so that their product decreases in magnitude with increasing volume for $n > 2$.) Hence this Gaussian form for the probability distribution is exact in the thermodynamic limit, $V \to \infty$.

This result is essentially the central limit theorem, which in essence says that the probability distribution of variables that are the sum of random variables is Gaussian. The utility of Gaussian distributions is another strong argument for formulating the entropy as dependent solely on extensive variables.

From the easily proved properties of the Gaussian distribution, the average value is equal to the most likely value,

$$\langle X \rangle_{Y,x,T} = \overline{X}(Y, x, T). \tag{2.125}$$

This result is essential to the consistency of thermodynamics and statistical mechanics. Thermodynamics always refers to the most likely state or value. Statistical mechanics mainly deals with average values. By this result the two are the same.

### 2.5.2 Constrained and equilibrium thermodynamic potential

As we have seen, in order to make more direct contact with conventional thermodynamics, it is useful to introduce the constrained thermodynamic potential as the negative of the temperature times the constrained total entropy,

$$\begin{aligned} F(X|Y, x; T) &= -TS_{\text{total}}(X|Y, x, T) \\ &= xX - TS_s(X, Y). \end{aligned} \tag{2.126}$$

(The reservoir temperature $T$ that appears here effects the potential only trivially in the event that energy is not an exchangeable parameter.) This may also be called the fluctuation potential, for reasons that will become clearer shortly. It follows that the probability distribution is essentially the exponential of this constrained thermodynamic potential

$$\wp(X|Y, x, T) = \frac{1}{Z(Y, x, T)} e^{-\beta F(X|Y,x,T)}. \tag{2.127}$$

By design, the constrained thermodynamic potential is a convex function of the exchangeable parameters that reaches its minimum at the equilibrium value, $X = \overline{X}(Y, x, T)$. This minimum value is called the equilibrium thermodynamic potential,

$$\overline{F}(Y, x, T) = F(\overline{X}|Y, x, T) = x\overline{X} - TS_s(\overline{X}, Y). \tag{2.128}$$

The utility of the constrained thermodynamic potential is that it provides a variational principle for the exchanged variable. Amongst other things this makes differentiation of the equilibrium thermodynamic potential particularly simple because one may effectively hold the exchanged variable constant during differentiation. For example

$$\begin{aligned}\frac{\partial \overline{F}(Y, x, T)}{\partial x} &= \frac{\partial F(\overline{X}|Y, x, T)}{\partial x} + \left.\frac{\partial F(X|Y, x, T)}{\partial X}\right|_{X=\bar{X}} \frac{\partial \overline{X}(Y, x, T)}{\partial x} \\ &= \frac{\partial F(\overline{X}|Y, x, T)}{\partial x} \\ &= \overline{X}.\end{aligned} \tag{2.129}$$

For the same reason differentiation with respect to one of the non-exchangeable subsystem variables yields

$$\begin{aligned}\frac{\partial \overline{F}(Y, x, T)}{\partial Y} &= \frac{\partial F(\overline{X}|Y, x, T)}{\partial Y} + \left.\frac{\partial F(X|Y, x, T)}{\partial X}\right|_{X=\bar{X}} \frac{\partial \overline{X}(Y, x, T)}{\partial Y} \\ &= \frac{\partial F(\overline{X}|Y, x, T)}{\partial Y} \\ &= -T\frac{\partial S(\overline{X}, Y)}{\partial Y} \\ &= -T\frac{y}{T_s},\end{aligned} \tag{2.130}$$

where the conjugate intensive variable of the sub-system and also the sub-system temperature appear. Dividing by temperature and differentiating by inverse temperature yields

$$\begin{aligned}&\frac{\partial(\overline{F}(Y, x, T)/T)}{\partial T^{-1}} \\ &= \frac{\partial(F(\overline{X}|Y, x, T)/T)}{\partial T^{-1}} + \left.\frac{\partial(F(X|Y, x, T)/T)}{\partial X}\right|_{X=\bar{X}} \frac{\partial \overline{X}(Y, x, T)}{\partial T^{-1}} \\ &= \frac{\partial(F(\overline{X}|Y, x, T)/T)}{\partial T^{-1}} \\ &= x\overline{X}.\end{aligned} \tag{2.131}$$

In the event of multiple exchangeable variables, the right-hand side is $\mathbf{x} \cdot \overline{\mathbf{X}}$.

The equilibrium thermodynamic potential is a convex function of the non-exchangeable extensive variables (because its second derivative is the negative of the second derivative of the isolated sub-system entropy). It is a concave function of the reservoir field variables conjugate to the exchangeable variables. Hence

$$\frac{\partial^2 \overline{F}}{\partial x_i^2} < 0, \quad \text{and} \quad \frac{\partial^2 \overline{F}}{\partial x_i^2}\frac{\partial^2 \overline{F}}{\partial x_j^2} - \left(\frac{\partial^2 \overline{F}}{\partial x_i \partial x_j}\right)^2 > 0. \tag{2.132}$$

The curvature of the equilibrium thermodynamic potential determines the sign of physical quantities such as the heat capacity or the compressibility.

### 2.5.3 Partition function

The partition function that normalizes the probability distribution is

$$
\begin{aligned}
Z(Y, x, T) &= \sum_X e^{S_{\text{total}}(X|Y,x,T)/k_B} \\
&= \sum_X e^{S_s(X,Y)/k_B} e^{-\beta x X} \\
&= \int \mathrm{d}X e^{S_s(X,Y)/k_B} e^{-\beta x X}.
\end{aligned} \tag{2.133}
$$

In the integral form one often introduces an appropriate constant factor that makes the partition function dimensionless. This is unimportant as it has no non-trivial consequences. Since the partition function is the weighted sum over all macrostates, its logarithm is the unconstrained total entropy,

$$
S_{\text{total}}(Y, x, T) \equiv k_B \ln Z(Y, x, T). \tag{2.134}
$$

This explains why any constant factor multiplying the partition function has no physical consequences: entropy is only defined up to an additive constant.

*Averages*

The partition function acts as a generating function for the system. The average value of various properties of the sub-system may be obtained as derivatives of the logarithm of the partition function. The latter is the unconstrained total entropy of the system.

The average of an exchangeable variable of the sub-system is

$$
\begin{aligned}
\langle X \rangle_{Y,x,T} &= \int \mathrm{d}X \, \wp(X|Y, x, T) \, X \\
&= \frac{1}{Z(Y, x, T)} \int \mathrm{d}X \, e^{S_s(X,Y)/k_B} e^{-\beta x X} X \\
&= \frac{1}{Z(Y, x, T)} \int \mathrm{d}X \, e^{S_s(X,Y)/k_B} \frac{-\partial e^{-\beta x X}}{\partial(\beta x)} \\
&= \frac{-\partial \ln Z(Y, x, T)}{\partial(\beta x)}.
\end{aligned} \tag{2.135}
$$

Similarly, the sub-system field parameters conjugate to the non-exchangeable variables, $y = T_s \partial S(X, Y)/\partial Y$, have the average value

$$
\begin{aligned}
\langle \beta_s y \rangle_{Y,x,T} &= \int \mathrm{d}X\, \wp(X|Y,x,T)\, \beta_s y \\
&= \frac{1}{Z(Y,x,T)} \int \mathrm{d}X\, e^{S_s(X,Y)/k_B} e^{-\beta x X} \frac{\partial S_s(X,Y)}{k_B \partial Y} \\
&= \frac{1}{Z(Y,x,T)} \int \mathrm{d}X\, e^{-\beta x X} \frac{\partial e^{S_s(X,Y)/k_B}}{\partial Y} \\
&= \frac{\partial \ln Z(Y,x,T)}{\partial Y}.
\end{aligned}
\tag{2.136}
$$

The average of the sub-system field parameters conjugate to the exchangeable variables, is

$$
\begin{aligned}
\langle \beta_s x_s \rangle_{Y,x,T} &= \int \mathrm{d}X\, \wp(X|Y,x,T)\, \beta_s x_s \\
&= \frac{1}{Z(Y,x,T)} \int \mathrm{d}X\, e^{S_s(X,Y)/k_B} e^{-\beta x X} \frac{\partial S_s(X,Y)}{k_B \partial X} \\
&= \frac{1}{Z(Y,x,T)} \int \mathrm{d}X\, e^{-\beta x X} \frac{\partial e^{S_s(X,Y)/k_B}}{\partial X} \\
&= \frac{-1}{Z(Y,x,T)} \int \mathrm{d}X\, e^{-\beta x X} \frac{\partial e^{-\beta x X}}{\partial X} \\
&= \beta x.
\end{aligned}
\tag{2.137}
$$

An integration by parts has been performed here, with the integrated part vanishing because extreme (i.e. at the integral limits) values of the exchangeable parameter have vanishingly small probability. This says that the average of the sub-system intensive variable conjugate to the exchangeable parameter is equal to that of the reservoir. This is what one would expect from the above finding that average values equal the most likely values.

*Fluctuations*

As we have just seen, the derivative of the unconstrained total entropy, which is the logarithm of the partition function, gives the average of various properties of the sub-system. Likewise, the second derivative gives the fluctuations in those quantities. One has

$$
\begin{aligned}
\frac{\partial^2 \ln Z(Y,x,T)}{\partial(\beta x)^2} &= \frac{-\partial \langle X \rangle_{Y,x,T}}{\partial(\beta x)} \\
&= \frac{-\partial}{\partial(\beta x)} \left\{ \frac{1}{Z(Y,x,T)} \int \mathrm{d}X\, e^{S_s(X,Y)/k_B} e^{-\beta x X}\, X \right\} \\
&= \langle X^2 \rangle_{Y,x,T} - \langle X \rangle_{Y,x,T}^2 \\
&= \langle (\Delta X)^2 \rangle_{Y,x,T}.
\end{aligned}
\tag{2.138}
$$

Here the fluctuation or departure from the norm is $\Delta X \equiv X - \langle X \rangle_{Y,x,T}$. The square root of the average of the square of the fluctuation is a measure of the extent to which the system can be expected to depart at any instant from its average value.

The average of the square of the departure is evidently positive. Hence the unconstrained total entropy is a convex function of the reservoir conjugate field variables. The fluctuation in $X$ measures the width of the probability distribution $\wp(X|Y, x, T)$: a small fluctuation means a sharply peaked distribution. The numerator on the left-hand side is extensive (the total entropy scales with the size of the sub-system), and the denominator is intensive. Hence the average of the square of the fluctuation is extensive. The expected relative error in a measurement of $X$ is

$$\frac{\sqrt{\langle (\Delta X)^2 \rangle_{Y,x,T}}}{\langle X \rangle_{Y,x,T}} \sim \mathcal{O}(V^{-1/2}), \tag{2.139}$$

since the denominator is extensive. This vanishes in the thermodynamic limit, $V \to \infty$. Hence the probability distribution for the exchangeable parameter becomes infinitely sharply peaked in the thermodynamic limit. This means that the average value is the same as the most likely value. It also means that the relative statistical error in the measured value of the exchangeable quantity is negligible for a macroscopic system.

Similarly, for two exchangeable quantities, the cross second derivative is

$$\begin{aligned} \frac{\partial^2 \ln Z(Y, x, T)}{\partial(\beta x_i)\partial(\beta x_j)} &= \langle X_i X_j \rangle_{Y,x,T} - \langle X_i \rangle_{Y,x,T} \langle X_j \rangle_{Y,x,T} \\ &= \langle \Delta X_i \, \Delta X_j \rangle_{Y,x,T}. \end{aligned} \tag{2.140}$$

This is called the cross-correlation, and it can be positive or negative. If it vanishes then the two exchangeable quantities are said to be uncorrelated with each other.

It is possible to combine the cross- and self-correlation in such a way that a positive quantity is obtained,

$$\begin{aligned} &\left( \frac{\partial}{\langle X_i \rangle_{Y,x,T} \partial(\beta x_i)} - \frac{\partial}{\langle X_j \rangle_{Y,x,T} \partial(\beta x_j)} \right)^2 \ln Z(Y, x, T) \\ &= \left\langle \left( \frac{X_i}{\langle X_i \rangle_{Y,x,T}} - \frac{X_j}{\langle X_j \rangle_{Y,x,T}} \right)^2 \right\rangle_{Y,x,T}. \end{aligned} \tag{2.141}$$

The averages on the left-hand side are to be regarded here as constants that are not differentiated.

Differentiation with respect to the inverse temperature yields

$$\begin{aligned} \frac{\partial^2 \ln Z(Y, x, T)}{\partial \beta^2} &= \frac{-\partial \langle xX \rangle_{Y,x,T}}{\partial \beta} \\ &= \langle (\Delta(xX))^2 \rangle_{Y,x,T}. \end{aligned} \tag{2.142}$$

This gives the fluctuation of the exchangeable work term; for multiple quantities, $xX \Rightarrow \mathbf{x} \cdot \mathbf{X}$.

One can similarly analyze the non-exchangeable extensive variables of the sub-system. This yields

$$\begin{aligned}\frac{\partial^2 \ln Z(Y, x, T)}{\partial Y^2} &= \frac{\partial \langle \beta_s y \rangle_{Y,x,T}}{\partial Y} \\ &= \langle (\beta_s y)^2 \rangle_{Y,x,T} - \langle \beta_s y \rangle^2_{Y,x,T} + \left\langle \frac{\partial(\beta_s y)}{\partial Y} \right\rangle_{Y,x,T} \\ &= \langle (\Delta(\beta_s y))^2 \rangle_{Y,x,T} + \left\langle \frac{\partial^2 S_s(X, Y)/k_B}{\partial Y^2} \right\rangle_{Y,x,T}.\end{aligned} \tag{2.143}$$

The unconstrained total entropy is extensive, and so the left-hand side is $\mathcal{O}(V^{-1})$. The final term on the right-hand side is similarly $\mathcal{O}(V^{-1})$. Hence the average of the square of the fluctuation in the sub-system intensive variables (the ones not conjugate to the exchangeable variables) must scale as $\mathcal{O}(V^{-1})$, which in the thermodynamic limit is negligible compared to $\langle \beta_s y \rangle_{Y,x,T} \sim \mathcal{O}(V^0)$.

The first term on the right-hand side is positive, whereas the second term is negative due to the concavity of the entropy of the isolated sub-system. Hence the sign of the left-hand side is indeterminate. However the result may be rewritten as

$$\frac{\partial^2 \ln Z(Y, x, T)}{\partial Y^2} - \left\langle \frac{\partial^2 S_s(X, Y)/k_B}{\partial Y^2} \right\rangle_{Y,x,T} = \langle (\Delta(\beta_s y))^2 \rangle_{Y,x,T}, \tag{2.144}$$

which says that the left-hand side is positive.

*Partition function and thermodynamic potential*

The logarithm of the partition function is essentially the unconstrained total entropy,

$$S_{\text{total}}(Y, x, T) = k_B \ln Z(Y, x, T). \tag{2.145}$$

This plays an important role in the formalism as its derivatives generate average values and their fluctuations.

In conventional approaches instead of this it is the equilibrium thermodynamic potential that is equated to the logarithm of the partition function. There is no inconsistency here because for macroscopic systems (i.e. in the thermodynamic limit) the two are equal because fluctuations are relatively negligible. This can be seen explicitly,

$$\begin{aligned}&- k_B T \ln Z(Y, x, T) \\ &= -k_B T \ln \int \mathrm{d}X\, e^{S_s(X,Y)/k_B} e^{-\beta x X} \\ &\approx -k_B T \ln \left[ e^{S_s(\bar{X},Y)/k_B} e^{-\beta x \bar{X}} \int \mathrm{d}X\, e^{S_s''(\bar{X},Y)(X-\bar{X})^2/2k_B} \right] \\ &= \beta x \overline{X} - T S_s(\overline{X}, Y) - \frac{k_B T}{2} \ln [\mathrm{Det}\{-S_s''(\overline{X}, Y)/2\pi k_B\}] \\ &= \overline{F}(Y, x, T) - \frac{k_B T}{2} \ln [\mathrm{Det}\{-S_s''(\overline{X}, Y)/2\pi k_B\}].\end{aligned} \tag{2.146}$$

For multiple exchangeable parameters, $x\overline{X} \Rightarrow \underline{x} \cdot \underline{\overline{X}}$, and $S_s''(\overline{X}, Y)$ is the Jacobean matrix of second derivatives. In this derivation the exponent has been expanded to second order and the Gaussian integral evaluated. This gives the equilibrium thermodynamic potential or free energy plus a logarithmic correction. The latter arises from the contribution of the fluctuations to the unconstrained total entropy, and for a macroscopic system it can be neglected. (This is, in essence, the same result as was given for the combinatorial example in section 1.2.3.) One concludes that to leading order the conventional expression is correct: the equilibrium thermodynamic potential is essentially the logarithm of the partition function,

$$\overline{F}(Y, x, T) = -k_B T \ln Z(Y, x, T). \tag{2.147}$$

This result means that the logarithmic derivative of the partition function equals essentially the derivative of the equilibrium thermodynamic potential. Hence average values, which are given by the former, equal the most likely values, which are given by the latter. Similarly for the second derivatives, so that the sign of the average square fluctuation determines the sign corresponding to equilibrium derivatives.

## Summary

- The constrained total entropy is the sum of the sub-system entropy and the reservoir entropy, each isolated and in the specified macrostate. The reservoir entropy upon Taylor expansion is the negative of the sub-system constrained exchangeable parameters times the conjugate reservoir field variables. The constrained thermodynamic potential is the negative of the reservoir temperature times the constrained total entropy.
- The constrained total entropy is maximized by the equilibrium or most likely value of the constrained parameter. The corresponding minimum value of the constrained thermodynamic potential is the equilibrium thermodynamic potential, which is commonly called a free energy.
- The concavity of the isolated system entropy leads to the concavity of the constrained total entropy, the stability of matter, the uniqueness of the thermodynamic state, and the equivalence of the equilibrium state obtained with different reservoirs. Exchangeable variables flow up the entropy gradient, which makes the conjugate field variable more uniform spatially. In common parlance, energy flows from a hot body to a cold body, thereby decreasing the temperature disparity.
- The constrained total entropy, constrained thermodynamic potential, and equilibrium thermodynamic potential are all extensive.
- The equilibrium thermodynamic potential is a concave function of its field variables and a convex function of its extensive variables. These determine the sign of its second derivatives, which give common thermodynamic parameters such as the heat capacity, compressibility, etc.
- The values of the sub-system variables exchangeable with a reservoir fluctuate about their equilibrium values. The logarithm of the partition function gives the total unconstrained entropy, which is approximately equal to the

maximum value of the constrained total entropy, which is essentially the equilibrium thermodynamic potential.

- Fluctuations of extensive variables are Gaussian distributed with a variance determined by the second derivative of the equilibrium thermodynamic potential (equivalently, the isolated system entropy). Fluctuations are relatively negligible for a macroscopic system, to which extent the Second Law of Thermodynamics may be regarded as a deterministic rather than a probabilistic law.

## References

Attard P 2002 *Thermodynamics and Statistical Mechanics: Equilibrium by Entropy Maximisation* (London: Academic)

Callan H B 1960 *Thermodynamics* (New York: Wiley)

Gibbs W 1873 *Graphical Methods in the Thermodynamics of Fluids*

Gibbs J W 1902 *Elementary Principles in Statistical Mechanics Developed with Special Reference to the Rational Foundation of Thermodynamics* (New Haven, CT: Yale University Press)

Sears F W and Salinger G L 1986 *Thermodynamics, Kinetic Theory, and Statistical Thermodynamics* 3rd edn (Reading, MA: Addison-Wesley)

Szent-Györgyi A and Hargittai I 2011 *Drive and Curiosity* (New York: Prometheus) p 22

**IOP** Publishing

Entropy Beyond the Second Law
Thermodynamics and statistical mechanics for equilibrium, non-equilibrium, classical, and quantum systems
**Phil Attard**

# Chapter 3

# Driven by entropy

'The decay of a system from a given non-equilibrium state produced by a spontaneous fluctuation obeys, *on the average*, the (empirical) law for the decay from the same state back to equilibrium, when it has been produced by a constraint which is then suddenly removed'

Onsager (1953)

'"Entropy" is a word that seems to attract the crackpots of the pseudo-scientific societies… the second law is highly attractive to those who are rather more philosophic and hand-waving than is acceptable in the normal circles of the hard-sciences'

Paltridge (2005)

'The mistakes of the common herd are usually in the same direction. Like sheep, they all follow a single leader'

Fisher (1906)

The main aim of this chapter is to formulate non-equilibrium thermodynamics as a variational principle of the appropriate entropy. The particular theory has been developed by me (Attard 2005a), and the presentation follows previous work (Attard 2012). The approach is illustrated by treating in detail the canonical non-equilibrium system, namely steady heat flow.

The most wide-spread competing approach to non-equilibrium thermodynamics is a variational principle based on the rate of entropy production,

$$\delta \dot{S}\,|_{\bar{J}} = 0. \tag{3.1}$$

This asserts that the optimum non-equilibrium state (symbolized by the flux $\bar{J}$) is the one that extremizes the rate of entropy production, $\dot{S}$, which is also called the

doi:10.1088/978-0-7503-1590-6ch3

dissipation. There are also variants of this based on extremizing different functions of the dissipation. This is claimed to be the variational principle upon which the field of non-equilibrium thermodynamics must be based. It is meant to be the generalization of the Second Law of Thermodynamics to non-equilibrium systems. This formula has quite a long history (see below), and it has been advocated and used by at least three Nobel Laureates (Rayleigh, Onsager, Prigogine), and other authorities besides.

## 3.1 Entropy for non-equilibrium systems

In this section the way to formulate an entropy for non-equilibriums systems is given in the generality of sets and states. But first is briefly discussed the need to develop a principle beyond the Second Law as a basis for non-equilibrium thermodynamics, which leads to a historical review of the above canard.

### 3.1.1 The second law is not enough

Chapter 2 analyzed the variational principle for equilibrium thermodynamics, namely the Second Law of Thermodynamics that entropy is maximized. An equilibrium system does not change macroscopically with time, although of course microscopic fluctuations occur due to the unceasing molecular motion.

A non-equilibrium system is one that changes macroscopically with time. Such changes can be due to an applied time-dependent mechanical field, or to an applied thermodynamic gradient. An example of the latter is steady heat flow, where the energy of the hot reservoir decreases and that of the cold reservoir increases over time.

In fact equilibrium and non-equilibrium systems are more closely connected than is implied by the above distinction. As we shall see, the fluctuations in time of an equilibrium system are intimately related to the flux or motion in a non-equilibrium system.

The reader may be puzzled by the opening paragraph of this sub-section in which the Second Law of Thermodynamics was referred to as the basis of equilibrium thermodynamics. Surely the fact that it refers to spontaneous changes in a system is a signifier of a non-equilibrium system?

It is certainly true that the Second Law refers to macroscopic changes, and that these are the stuff of non-equilibrium thermodynamics. It is also undeniably true that the Second Law provides the variational principle that underlies the quantitative treatment of equilibrium systems, as evidenced by the results in chapter 2. The point is, however, that while the Second Law sets the direction of change for a system that is not in equilibrium, it does not quantitatively determine the rate of change. In short, the Second Law gives the future direction of time's arrow but not its speed. Any thermodynamic theory for non-equilibrium systems must involve time in a quantitative not just qualitative sense.

### 3.1.2 Principle of extreme dissipation

> 'These [reciprocal] relations can be summarized in a variation-principle… an extension of Lord Rayleigh's '*principle of the least dissipation of energy*'… the rate of increase of the entropy plays the role of a potential' Onsager (1931)

This brings us to the history and motivation for the principle of extreme dissipation, equation (3.1). The absence of quantitative time in the Second Law of Thermodynamics has led many scientists to the same general conclusion: the Second Law must be modified for non-equilibrium systems.

In order to insert time quantitatively, almost the only approach that has been tried is to replace the entropy that appears there by its time rate of change. The latter is also known as the rate of entropy production, or the rate of dissipation, or just the dissipation.

In all cases the argument has been made by analogy with the Second Law rather than from first principles, and so workers pursuing such an approach have had to guess at the variational principle. As a result two contradictory versions have been promulgated. A substantial fraction of the advocates of this approach assert that the general principle that underlies non-equilibrium thermodynamics is that the rate of entropy production is a *minimum*. Or by analogy to the Second Law,

The rate of entropy production decreases during spontaneous changes of the system. (3.2)

Onsager, in his Noble-prize winning paper on the reciprocal relations, calls this 'The Principle of Least Dissipation'. He characterizes the dissipation by generalizing the energy dissipation functions given earlier by Rayleigh (Strutt 1871), and by Kelvin (Onsager 1931). Rayleigh (Strutt 1913) attributes it to Helmholtz (1869) and to Korteweg (1883). This principle of minimal entropy production has also been asserted by Prigogine (1967), de Groot and Mazur (1984), Biot (1955, 1975), Paltridge (1979), and many others besides.

Contradicting this, more than half of the advocates of extremizing the rate of entropy production assert the exact opposite. They say that the rate of entropy production is a *maximum*, which, by analogy to the Second Law, can be formulated as

The rate of entropy production increases during spontaneous changes of the system. (3.3)

Examples of scientists arguing for this include Prigogine (1967), Paltridge (1979), Swenson and Turvey (1991), Schneider and Kay (1994), and Dewar (2003). For other examples and reviews see Kleidon and Lorenz (2005) and Martyushev and Seleznev (2006). Some scientists, including the Noble laureate Prigogine, have advocated both principles.

The variational functional given by Onsager and Machlup (1953) has been of some influence, and is based on different quadratic forms for the entropy production, and whose extremum is designed to give the optimum steady state of a non-equilibrium system. They chose the extremum to be a minimum. The variational functionals given by Hashitsume (1952, 1956), Gyarmati (1968, 1970), and Bochkov and Kuzovlev (1980), are in essence identical to that given by Onsager and Machlup (1953). Some of the proponents of this functional say the extremum should be a maximum, and others say it should be a minimum.

The reader will doubtless be troubled by the fact that two directly opposed criteria have been proposed and argued for with equal fervor. Surely there should be a unique general principle that determines the optimum non-equilibrium state, and this principle should be self-evident, incontrovertible, and backed by evidence. It is perhaps surprising that those who argue for a principle based on the rate of entropy production cannot agree amongst themselves whether it should be a maximum or a minimum. Despite the evident contradiction and the dearth of convincing evidence for either, there has been little criticism in the literature of the principle of extreme dissipation. Doubtless most workers have chosen discretion over valor, preferring to follow each other rather than to question authority. The few honorable exceptions that I am aware of are Gage *et al* (1966), Keizer and Fox, (1974), Lavenda (1985), Hunt *et al* (1987), Ross and Vlad (2005), and of course myself (Attard 2006).

I will not give here a detailed critique of the principle of extreme dissipation (for such, see Attard (2006)) since it is preferable to devote this chapter to the correct non-equilibrium variational principle rather than to be distracted by a fundamentally flawed approach. Briefly however, it can be noted that the rate of change of an extensive variable $\dot{X}$ is a flux, and multiplying by the conjugate field variable $x/T$, a thermodynamic force, gives the rate of entropy production,

$$\dot{S} = \frac{x}{T}\dot{X}. \tag{3.4}$$

There are many details that are skipped here, but the essence of the argument can be readily appreciated. Since this is a function bilinear in the flux and the force, extremizing it with respect to either can only ever yield $\pm\infty$. Onsager and Machlup (1953) get around this problem by adding quadratic terms in the force and in the flux, which also give the entropy dissipation in the optimum state, and these serve to produce an extremum at finite values (for detailed analysis, see Attard 2006). There are of course an infinite number of variational functionals that one can construct to give the known optimum steady state. What is required is the unique variational formula that comes from the physical principles that generalize the Second Law of Thermodynamics. This matter is further discussed following equation (3.96) below.

### 3.1.3 Entropy of joint macrostates

In section 1.1.2, entropy and probability were defined in terms of the weight of macrostates and microstates, which is basically set theory. It will be recalled that microstates are disjoint, indivisible, and form a complete set, and that macrostates are disjoint and also form a complete set. With $w_i$ the weight of microstate $i$, the weight of the macrostate $\alpha$ is

$$W_\alpha = \sum_i w_i \delta(A_i - A_\alpha) = \sum_{i\in\alpha} w_i, \tag{3.5}$$

and the total weight of the system is

$$W = \sum_i w_i = \sum_\alpha W_\alpha. \tag{3.6}$$

The entropy is essentially the logarithm of the weight, so that the microstate, macrostate, and total entropy is respectively

$$S_i = k_B \ln w_i, \qquad S_\alpha = k_B \ln W_\alpha, \qquad \text{and} \qquad S = k_B \ln W. \tag{3.7}$$

Since macrostates from different collectives, labeled by say $\alpha$ and $\beta$, are not disjoint, their joint weight can be calculated as the weight of their intersection,

$$W(\alpha\beta) = \sum_i w_i \delta(A_i - A_\alpha)\delta(B_i - B_\beta) = \sum_{i \in A_\alpha \bigcap B_\beta} w_i. \tag{3.8}$$

This is sketched in figure 1.1. This is the unconditional weight. If $\alpha$ and $\beta$ belong to the same collective, this is $W(\alpha\beta) = W_\alpha \delta(\alpha - \beta)$.

Because the collectives are complete, summing over one of the macrostates must reduce the joint weight to the weight of the remaining macrostate,

$$\sum_\beta W(\alpha\beta) = W_\alpha. \tag{3.9}$$

This is a type of conservation law for weight. The idea can again be seen in figure 1.1, where it is evident that the $\alpha$ macrostate is covered by all the intersecting $\beta$ macrostates.

It follows that summing over the joint weight must yield the total weight

$$\sum_{\alpha,\beta} W(\alpha\beta) = W. \tag{3.10}$$

The entropy of the joint macrostate is of course the logarithm of its weight, and the joint probability is

$$\wp(\alpha\beta) = \frac{W(\alpha\beta)}{W} = \frac{e^{S(\alpha\beta)/k_B}}{W}. \tag{3.11}$$

This is the unconditional probability, which is to say the probability of the system being in the state $\alpha$ and in the state $\beta$. The conditional probability $\wp(\alpha|\beta)$, which is to say the probability that the system is in the state $\alpha$ given that it is in the state $\beta$, was discussed in section 1.1.2. It is

$$\wp(\alpha|\beta) = \frac{\wp(\alpha\beta)}{\wp(\beta)} = \frac{W(\alpha\beta)}{W(\beta)} = e^{[S(\alpha\beta) - S(\beta)]/k_B}. \tag{3.12}$$

### 3.1.4 Second entropy for transitions

Consider the transition between two microstates in the time interval $\tau$, $i \xrightarrow{\tau} j$. This is the key to the analysis. Just as the microstates $i$ themselves provided the elements for the analysis of equilibrium systems, it is these transitions between microstates that are the elements of non-equilibrium theory.

As the microstates have weight, so do the transitions between them, and the manipulation of these transition weights is based on set theory rather similar to the analysis of the microstate weights above and in section 1.1.2. The state with the system in the microstate $i$ at time $t_1$ and $j$ at time $t_2$ may be described as a transition microstate. Attached to it is the weight, $w(j,\ t_2;\ i,\ t_1) = w(j,\ i|t_{21})$, where $t_{21} \equiv t_2 - t_1$. That this depends only on the time interval is true for an equilibrium system, which is the sole concern here.

This is the unconditional weight, and one can just as well reverse the order of the arguments and times,

$$w(j,\ t_2;\ i,\ t_1) = w(i,\ t_1;\ j,\ t_2) \Leftrightarrow w(j,\ i|t_{21}) = w(i,\ j|t_{12}). \tag{3.13}$$

These all give the weight for the system to be in the microstate $i$ at time $t_1$ and $j$ at time $t_2$. This is the unconditional weight, and the unconditional transition probability is proportional to it.

In common parlance, if $t_2 > t_1$, this would normally be called the transition from $i$ to $j$. However, it is important in the present theory to also consider transitions backward in time. Because of the symmetry of the unconditional weight, forward and backward unconditional transitions have equal weight. However, we know from the Second Law of Thermodynamics that there must be an asymmetry or irreversibility in time. For this reason we shall try to avoid using the word transition in association with the unconditional weight, and instead restrict it to conditional weights.

Conditional weights are defined in the usual way, via

$$w(j,\ i|\tau) = w(j|i,\ \tau)\ w(i). \tag{3.14}$$

That the microstate weight $w(i)$ is independent of time is true for the present equilibrium system. The conditional weight $w(j|i,\ \tau)$ is to be read as the weight associated with the fact the system is in the microstate $j$ given that it was in the microstate $i$ a time $\tau$ earlier. Here $i$ is the initial microstate and $j$ is the destination microstate. This is the weight for the unconditional transition $i$ to $j$ in time $\tau$, which is denoted $i \xrightarrow{\tau} j$. This may be a forward or a backward transition, according to the sign of $\tau$.

The conditional weight is in general asymmetric in time. This follows since $w(j,\ i|\tau) = w(i,\ j|-\tau)$, which may be rearranged as

$$\frac{w(j|i,\ \tau)}{w(i|j,\ -\tau)} = \frac{w(j)}{w(i)}. \tag{3.15}$$

This says that the weights of the forward and backward conditional transitions between a pair of microstates are in inverse ratio to the weights of the initial microstates.

In general microstate transitions are stochastic, and their probability is reflected in their weight. Of course saying that the transition is stochastic does not rule out the possibility that it has a deterministic component, or even that the random component may be a small perturbation on the deterministic component. In the event that the transition is fully deterministic, then the weight becomes a $\delta$-function. In general, the stochastic contribution to the microstate transitions of a sub-system

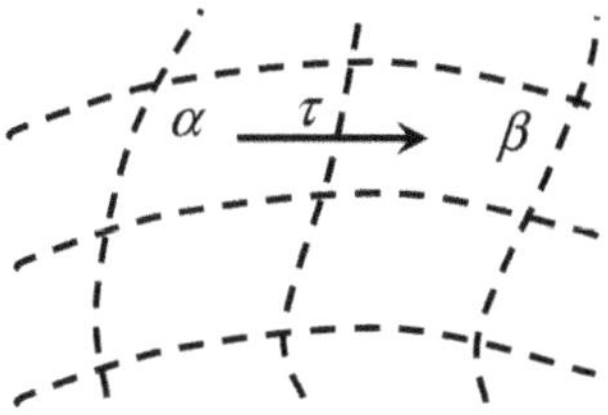


**Figure 3.1.** A transition in time $\tau$ from the $\alpha$ macrostate to the $\beta$ macrostate of a collective.

comes from the interactions with the reservoirs, which are treated in projection. It may also come from internal motion of the microstates that is not treated explicitly. (Although the microstates are defined to be indivisible, this is a statement of how the system is to be treated mathematically rather than a statement of the physical reality.)

Continuing in the same vein as the equilibrium formulation, one defines the macrostate transition, $\alpha \xrightarrow{\tau} \beta$ (see figure 3.1). These can belong to the same or different collectives. The unconditional transition weight for this is

$$\begin{aligned} W(\beta, \alpha|\tau) &= \sum_{i,j} w(j, i|\tau)\delta(A_i - A_\alpha)\delta(B_j - B_\beta) \\ &= \sum_{i\in\alpha}\sum_{j\in\beta} w(j, i|\tau). \end{aligned} \tag{3.16}$$

It ought to be clear that even if the microstate transitions were fully deterministic, the macrostate transitions must contain a stochastic element because specifying the initial macrostate does not fix the initial microstate that is required to determine the subsequent transition.

In equation (3.9) above, the weight of joint (i.e. simultaneous) macrostates was reduced to the weight of one of them by summing over the other, $\sum_\beta W(\alpha\beta) = W_\alpha$. The present transition weights can be taken to obey a similar reduction condition. We take it as a theorem (shortly proven by a time average) that for an equilibrium system the unconditional microstate weight is conserved in a microstate transition,

$$\sum_j w(j, i|\tau) = w(i) \text{ and } \sum_i w(j, i|\tau) = w(j). \tag{3.17}$$

The first says that all the weight in microstate $i$ is distributed without loss or gain amongst the possible target microstates $j$. (It is possible to scale the right-hand side by a positive constant, but this is only material in some non-equilibrium cases.) The second says that all the weight in the target microstates come from the possible initial microstates. The second is the same as the first with $\tau$ changed to $-\tau$. For the conditional microstate weight the conservation rule is equivalent to

$$\sum_j w(j|i, \tau) = 1. \tag{3.18}$$

This is plausible because the system must be in one and only one microstate at a time.

The proof of the conservation of weight during a transition follows from the axiom that time is uniform, at least in an equilibrium system. By this is meant that a time average over nodes of equal weight gives the same result as a statistical average using the microstate weights. Since the system must be in one and only one microstate at a time, one can define its trajectory in time through the microstates as $i(t)$. Let $a$ label the nodes, $t_a = a\tau$, and let the $a$th microstate on the trajectory be $i_a \equiv i(a\tau)$. The microstate weight is proportional to the number of times the trajectory visits the microstate,

$$w(i) \propto n(i) = \sum_a \delta(i_a - i). \tag{3.19}$$

The unconditional transition weight is proportional to the number of times the transition occurs on the trajectory,

$$w(j, i|\tau) \propto n(j, i|\tau) = \sum_a \delta(i_a - i)\, \delta(i_{a+1} - j). \tag{3.20}$$

Summing over the destination microstates gives

$$\begin{aligned} \sum_j n(j, i|\tau) &= \sum_j \sum_a \delta(i_a - i)\, \delta(i_{a+1} - j) \\ &= \sum_a \delta(i_a - i) \\ &= n(i). \end{aligned} \tag{3.21}$$

The penultimate equality follows because the set of microstates is complete and disjoint (and hence non-degenerate). Hence the weight is conserved during a transition. Obviously one gets the second result in equation (3.17) by summing over $i$ instead of $j$.

For certain time-varying non-equilibrium systems it is necessary to include a time dependent scale factor in this conservation law (see chapter 6).

Each microstate $i$ has a conjugate microstate $i^\dagger$. The physical interpretation of this is that in the conjugate state all the velocities of the particles in the system have their velocities reversed (and also any velocity-derived external fields like magnetic fields or Coriolis forces, if present). For an equilibrium system, the weight of a microstate and its conjugate are equal,

$$w(i^\dagger) = w(i). \tag{3.22}$$

Since reversing all the velocities is the same as reversing the direction of time, this says that time has no preferred direction in an equilibrium system. Note that in the sum over microstates, $\sum_i$, (for example, to obtain the total weight) both $j$ and $j^\dagger$ each appear once.

Because reversing the velocities is the same as reversing time, one must have for the unconditional transition weight

$$w(j, i|\tau) = w(j^\dagger, i^\dagger|-\tau). \tag{3.23}$$

This might be called microscopic equilibrium reversibility. Combining these two with equation (3.15) yields

$$\frac{w(j|i, \tau)}{w(j^\dagger|i^\dagger, -\tau)} = \frac{w(i^\dagger)}{w(i)} = 1, \text{ or } w(j|i, \tau) = w(j^\dagger|i^\dagger, -\tau). \tag{3.24}$$

This means that the forward transition has the same conditional weight as the backward transition between the conjugate microstates, This is called microscopic reversibility.

Microscopic reversibility may be confirmed explicitly for an isolated system evolving adiabatically with Hamilton's equations of motion. It holds also for stochastic dissipative transitions for an equilibrium system (see section 5.4.2).

Each macrostate $\alpha$ has a conjugate macrostate $\alpha^\dagger$. Consider the physical observable $A$, with $A_\alpha$ its value in the macrostate $\alpha$. If $A$ is an even function of the velocities (for example, energy, number, etc.), then $A_\alpha = A_{\alpha^\dagger}$. Since degenerate macrostates are precluded in the formalism (there is a one-to-one relationship between the macrostate labels and the values of the physical observable), this means that $\alpha = \alpha^\dagger$, which is to say that in this even parity case the macrostates are self-conjugate. If the physical observable is an odd function of the velocities, for example, the total momentum of the system, then $A_\alpha = -A_{\alpha^\dagger}$, and $\alpha \neq \alpha^\dagger$. If an observable is not of even parity (i.e. it is of odd or mixed parity), then both $\alpha$ and $\alpha^\dagger$ occur in the sum over macrostates. In view of this, whether the parity is even, odd, or mixed, the reverse transition between the conjugate macrostates has weight

$$\begin{aligned} W(\alpha^\dagger, \beta^\dagger|\tau) &= \sum_{i^\dagger \in \alpha^\dagger} \sum_{j^\dagger \in \beta^\dagger} w(i^\dagger, j^\dagger|\tau) \\ &= \sum_{i \in \alpha} \sum_{j \in \beta} w(i^\dagger(i), j^\dagger(j)|\tau) \\ &= \sum_{i \in \alpha} \sum_{j \in \beta} w(i^\dagger, j^\dagger|\tau) \\ &= \sum_{i \in \alpha} \sum_{j \in \beta} w(i, j|-\tau) \\ &= W(\alpha, \beta|-\tau). \end{aligned} \tag{3.25}$$

The second equality follows because for every $i^\dagger \in \alpha^\dagger$ there is an $i \in \alpha$ and for every $j^\dagger \in \beta^\dagger$ there is a $j \in \beta$. The third equality follows because $i^\dagger(i) = i^\dagger$ and $j^\dagger(j) = j^\dagger$. The fourth equality results from microscopic reversibility. The first and last equalities are just the definition of the macrostate transition weight. The derivation shows that macroscopic equilibrium reversibility, $W(\alpha^\dagger, \beta^\dagger|\tau) = W(\alpha, \beta|-\tau)$, is a direct consequence of microscopic equilibrium reversibility $w(i^\dagger, j^\dagger|\tau) = w(i, j|-\tau)$. As in that case it says that the forward and reverse transitions between macrostates have the same conditional weight,

$$W(\alpha^{\dagger}|\beta^{\dagger}, \tau) = W(\alpha|\beta, -\tau). \tag{3.26}$$

If the macrostates have even parity, $\alpha^{\dagger} = \alpha$ and $\beta^{\dagger} = \beta$ then this becomes $W(\alpha, \beta|\tau) = W(\alpha, \beta|-\tau)$. This is a stronger result but it will seldom be required for what follows.

Macroscopic reversibility appears to contradict the Second Law of Thermodynamics. Macroscopic reversibility says that there is no preference between forward and reverse transitions, whereas the Second Law says that there is a preferred direction for time. This apparent contradiction is addressed below.

The conservation law for the microstate transition weight, equation (3.17) also follows for the macrostate transition weights,

$$\sum_{\alpha} W(\alpha, \beta|\tau) = W(\beta), \quad \text{and} \quad \sum_{\beta} W(\alpha, \beta|\tau) = W(\alpha). \tag{3.27}$$

From these conservation laws it follows that the total unconditional transition weight is just the total weight itself,

$$\sum_{i,j} w(j, i|\tau) = \sum_{i} w(i) = W. \tag{3.28}$$

This is independent of the time interval $\tau$. Evidently the same result holds for the total macrostate transition weight,

$$\sum_{\alpha,\beta} W(\beta, \alpha|\tau) = \sum_{\beta} W(\beta) = W. \tag{3.29}$$

As in the formulation of entropy for the equilibrium case, one can define the second entropy as the logarithm of the transition weight,

$$S^{(2)}(\beta, \alpha|\tau) \equiv k_{\mathrm{B}} \ln W(\beta, \alpha|\tau). \tag{3.30}$$

There is an analogous expression for the microstate transitions. This could also be called the transition entropy, or the two-time entropy.

Obviously the unconditional transition probability is essentially the exponential of the second entropy,

$$\wp(\beta, \alpha|\tau) = \frac{W(\beta, \alpha|\tau)}{W} = \frac{1}{W} e^{S^{(2)}(\beta,\alpha|\tau)/k_{\mathrm{B}}}. \tag{3.31}$$

The conditional transition probability is the probability that the system will be in the macrostate $\beta$ at a time $\tau$, given that it is currently in the macrostate $\alpha$. It is given by

$$\wp(\beta|\alpha, \tau) \equiv \frac{\wp(\beta, \alpha|\tau)}{\wp(\alpha)} = \frac{W(\beta, \alpha|\tau)}{W(\alpha)}. \tag{3.32}$$

If $\tau > 0$ this a forward transition, and if $\tau < 0$ this a backward transition.

The macroscopic reversibility condition on the transition weight, equation (3.25), yields a similar result for the unconditional transition probability,

$$\wp(\alpha^{\dagger}, \beta^{\dagger}|\tau) = \wp(\alpha, \beta|-\tau) = \wp(\beta, \alpha, |\tau). \tag{3.33}$$

The interpretation of this is that the forward transition between two macrostates is as likely to occur as the reverse transition between their conjugate states. Applying this to the conditional transition probability yields

$$\wp(\alpha^{\dagger}|\beta^{\dagger}, \tau) = \frac{\wp(\alpha^{\dagger}, \beta^{\dagger}|\tau)}{\wp(\beta^{\dagger})} = \frac{\wp(\beta, \alpha|\tau)}{\wp(\beta)} = \wp(\beta|\alpha, \tau)\frac{\wp(\alpha)}{\wp(\beta)}. \tag{3.34}$$

This may be rearranged as

$$\frac{\wp(\alpha^{\dagger}|\beta^{\dagger}, \tau)}{\wp(\beta|\alpha, \tau)} = \frac{\wp(\alpha)}{\wp(\beta)} = e^{[S(\alpha)-S(\beta)]/k_{\mathrm{B}}}. \tag{3.35}$$

Recall that from microscopic reversibility, $\wp(\alpha^{\dagger}|\beta^{\dagger}, \tau) = \wp(\alpha|\beta, -\tau)$. One can conclude that the ratio of forward and reverse conditional transition probabilities equals the exponential of the entropy difference of the two macrostates.

The conditional transition probability in terms of the second entropy is

$$\wp(\beta|\alpha, \tau) = e^{[S^{(2)}(\beta,\alpha|\tau)-S(\alpha)]/k_{\mathrm{B}}}. \tag{3.36}$$

One could define the exponent to be the conditional second entropy, but this does not appear to be particularly useful.

A quantity that will prove of some interest is the most likely conditional transition state. In particular, given the current macrostate $\beta$, the most likely macrostate a time $\tau$ distant may be denoted $\bar{\alpha} \equiv \bar{\alpha}(\tau|\beta)$. This is the macrostate that maximizes the second entropy,

$$\left.\frac{\partial S^{(2)}(\alpha, \beta|\tau)}{\partial \alpha}\right|_{\alpha=\bar{\alpha}} = 0. \tag{3.37}$$

Maximizing the second entropy $S^{(2)}(\alpha, \beta|\tau)$ is the same as maximizing the transition weight $W(\alpha, \beta|\tau)$. This definition of $\bar{\alpha}$, together with the conservation condition on the transition weight, equation (3.27), can be used to 'reduce' the second entropy to the first or ordinary entropy. One has

$$\begin{aligned} S(\beta) &= k_{\mathrm{B}} \ln W(\beta) \\ &= k_{\mathrm{B}} \ln \sum_{\alpha} W(\alpha, \beta|\tau) \\ &\approx k_{\mathrm{B}} \ln W(\bar{\alpha}(\tau|\beta), \beta|\tau) \\ &= S^{(2)}(\bar{\alpha}(\tau|\beta), \beta|\tau). \end{aligned} \tag{3.38}$$

The third equality approximates the logarithm of a sum by the logarithm of the largest term in the sum. This holds when the distribution of target macrostates $\alpha$ is sharply peaked about the most likely macrostate, $\bar{\alpha}$. It is the same concept as that which sets the unconstrained total entropy equal to the maximum value of the constrained total entropy (see figure 1.4, and also the derivation leading to equation (2.147)).

This expression may be called the reduction condition (more precisely, the type 2 reduction condition; the conservation law for the transition weight may be called the reduction condition of type 1), and it turns out to be of central importance in the theory that follows. It says that the maximum value of the second entropy over the future macrostates (if $\tau > 0$, or over the past macrostates if $\tau < 0$) is equal to the first entropy of the current macrostate.

### 3.1.5 Another proof of the second law

In section 1.3 an argument was given that purported to show that Boltzmann's identification of entropy with the logarithm of the macrostate weight (number of molecular configurations in simplest terms) justifies the Second Law of Thermodynamics as enunciated by Clausius. That argument was necessarily approximate, and it seems worthwhile to revisit the issue in the light of the present results for the transition probability.

Consider macrostates in a single collective, and order them in terms of (first) entropy. Suppose that $\gamma$ is a macrostate approximately half way between the macrostates $\alpha$ and $\beta$. In this case equation (3.35) gives the product of the ratio of the conditional transition probabilities from the mid-state $\gamma$ is

$$e^{[S(\alpha)-S(\beta)]/k_{\mathrm{B}}} = \frac{\wp(\alpha|\gamma,\tau)}{\wp(\gamma^{\dagger}|\alpha^{\dagger},\tau)}\frac{\wp(\gamma|\beta,\tau)}{\wp(\beta^{\dagger}|\gamma^{\dagger},\tau)} \approx \frac{\wp(\alpha|\gamma,\tau)^2}{\wp(\beta^{\dagger}|\gamma^{\dagger},\tau)^2}. \tag{3.39}$$

By ordering the states and choosing $\gamma$ to be in the middle, one can expect that the transition $\gamma \xrightarrow{\tau} \alpha$ is approximately the same as the transition $\beta \xrightarrow{\tau} \gamma$, and similarly for the conjugate transitions. This gives the final approximation.

For simplicity one can look at even parity macrostates, $\alpha^{\dagger} = \alpha$, etc, in which case this becomes

$$\frac{\wp(\alpha|\gamma,\tau)}{\wp(\beta|\gamma,\tau)} \approx e^{[S(\alpha)-S(\beta)]/2k_{\mathrm{B}}}. \tag{3.40}$$

The interpretation of this is that if the states are ordered $S(\alpha) > S(\gamma) > S(\beta)$, then the likelihood of observing an entropy increasing transition over an entropy decreasing transition is just the exponential of half the total difference in entropy. The factor of one half here is more correct than its absence in the argument given in section 1.3. One can invoke the mid-point theorem to say that there always exists a $\gamma$ that makes this result correct.

This result again shows the connection between the entropy of Boltzmann and the Second Law of Thermodynamics of Clausius. It also again shows the probabilistic nature of the Second Law.

For even parity states, macroscopic reversibility is $\wp(\alpha, \beta|\tau) = \wp(\beta, \alpha|\tau)$. This means that the forward transition and the backward transition have equal probability in an unconditional sense. In consequence, if one state is far more likely than the other, $\wp(\alpha) \gg \wp(\beta)$, then the conditional transition to the less likely state is far less likely than the reverse, $\wp(\beta|\alpha, \tau) \ll \wp(\alpha|\beta, \tau)$. The result means that one is just as likely to observe a fluctuation *from* equilibrium as to observe the reverse fluctuation *to* equilibrium, in an unconditional sense. In other words, an entropy increasing transition $\beta \xrightarrow{\tau} \alpha$, is just as unconditionally likely an entropy decreasing transition, $\alpha \xrightarrow{\tau} \beta$.

The assertion in the Second Law of Thermodynamics that spontaneous transitions increase the entropy refers to the conditional future transition probability from an unlikely (not-in-equilibrium) state. The Second Law does not apply to the unconditional transition probability. In general, the Second Law is probabilistic, not deterministic, and so spontaneous transitions from the equilibrium macrostate to states of lower entropy in the conditional sense may well be unlikely, but in general they are not impossible.

In macroscopic systems, fluctuations in the transitions are relatively negligible, and the transitions become effectively deterministic. Hence if the system is set up in a not-in-equilibrium macrostate, it will spontaneously evolve toward macrostates of higher entropy. If the system is set up in the equilibrium or most likely macrostate, it will stay there. In this macroscopic case the Second Law holds for future conditional transitions with probability approaching unity.

### 3.1.6 Two points on the arrow of time

> 'If as we follow the arrow we find more and more of the random element in the state of the world, then the arrow is pointing towards the future; if the random element decreases the arrow points towards the past.' Eddington (1928)

There is an interesting twist on the arrow of time argument popularized by Eddington. For a system currently observed in a not-in-equilibrium macrostate, the usual formulation asks 'where will the system go to?'. But what if the obverse question were asked, namely 'from whence came the system?'. Since an answer is impossible if the previous states were set by some artificial unspecified process, one must assume that the current state arose by a spontaneous fluctuation of the system itself. In this case the answer is 'most likely the system came from a state of higher entropy'.

This answer is entirely consistent with the quantitative analysis above, where it can be noted that nothing was said about the sign of $\tau$.

One must conclude that the usual formulation of the Second Law of Thermodynamics,

$$\text{All systems evolve to states with greater entropy,} \tag{3.41a}$$

is valid, and equally valid is its obverse,

$$\text{All systems have evolved from states with greater entropy.} \tag{3.41b}$$

Alternatively, the formulation

$$\text{The entropy increases during spontaneous transitions from the current state,} \tag{3.41c}$$

is as correct as

$$\text{The entropy decreases during spontaneous transitions to the current state.} \tag{3.41d}$$

These statements hold in a probabilistic sense (i.e. they are most likely true), they are conditional on the current state alone, and they hold for a system undergoing spontaneous fluctuations (i.e. without outside interference or specific preparation). From these alternative formulations it is clear that the notion that the arrow of time points in the direction of increasing entropy is a selective interpretation of the Second Law that entirely ignores history.

The fact of the matter is that the arrow of time is symmetric with regard to the change in entropy. If one defines a fixed arrow of time as pointing from what we call the past to the future, then entropy most likely increases in a forward transition to the future (i.e. in the direction of the arrow of time), and it most likely increases in a backward transition to the past (i.e. against the arrow of time). The present results contradict the quote from Eddington given at the beginning of this sub-section. In view of this symmetry, one must conclude that there is no such thing as the arrow of time as defined by Eddington (1928).

## 3.2 Pure parity fluctuation analysis

### 3.2.1 Gaussian distribution

As was shown in section 2.5.3, for an equilibrium system the probability distribution for an extensive exchangeable parameter becomes infinitely sharply peaked in the thermodynamic limit. This means it can be accurately modeled as a Gaussian distribution, with the variance determined by the second derivative of the isolated system entropy.

In fact, the prevalence of Gaussian distributions in physical systems is more general than this. Gaussian distributions are stable because any variable that is the sum of variables that are Gaussian distributed is itself Gaussian distributed. And if the variables are arbitrarily distributed, then by the central limit theorem their sum tends toward a Gaussian distribution as their number increases. Since the physical observables that define macrostates have multiple stochastic influences, it is no surprise that they have a Gaussian distribution.

Let $X$ be a vector of extensive variables. The present analysis is simplified by taking all the components to be of the same parity, $X^\dagger = X$ or else $X^\dagger = -X$. That is, upon reversal of the velocities of all particles in the Universe, the values of the observables comprising $X$ are either all unchanged or else all negated. The more general mixed parity case will be analyzed in section 3.3 below.

We are primarily interested in non-conserved, extensive variables, which is to say variables whose value can change with time in an isolated system. The spatial distribution of energy or number are examples. Furthermore, the variable is defined such that its most likely or average value in the isolated system vanishes,

$$\overline{X} = 0. \tag{3.42}$$

For example, the difference in energy between the two halves of an isolated system most likely vanishes, because a uniform energy distribution is the one that maximizes the entropy. Obviously one can always subtract off any non-zero most likely value and interpret $X$ as the departure from the most likely value. A non-zero value of $X$ is called a fluctuation. Because $X$ is extensive, higher order contributions to its fluctuations can be expected to be relatively negligible, and its probability distribution can be expected to be Gaussian.

The focus of this section is on an isolated system. The value of the non-conserved variable can evolve in time, $\dot{X}^0(t) \neq 0$, where $t$ is time, the superscript dot signifies the time derivative, and the superscript 0 signifies the adiabatic (Hamiltonian) evolution. In this isolated system, $\overline{X} = 0$. An initial non-zero fluctuation, $X_0 \neq 0$, most likely evolves adiabatically back to zero, $\overline{X^0(t|X_0)} \to 0, t \to \infty$. (The evolution is not deterministic even in this isolated system because $X_0$ is a macrostate and the initial microstate conditioned on this macrostate is statistically distributed.) What we expect, and it remains to prove, is that there is a regime in which the rate of regression is constant and proportional to the initial fluctuation, $\overline{\dot{X}^0} = \tilde{\Lambda} X_0$.

Further, the formalism is designed to accommodate reservoirs that can exchange material related to $X$ in such a way that the most likely value is non-zero, $\overline{X}(x_{\mathrm{r}}) \neq 0$, where $x_r$ is a reservoir parameter. In this case one expects, and again it remains to prove, that the adiabatic rate of change of $X$ in the presence of the reservoir is the same as if the non-zero value had arisen from a spontaneous fluctuation of the isolated system, $\overline{\dot{X}^0} = \tilde{\Lambda}\overline{X}(x_{\mathrm{r}})$. The arrangement of the reservoir induces a thermodynamic non-equilibrium system in a steady state, and this idea is one aspect of Onsager's regression hypothesis (Onsager 1931).

For the present no reservoirs are included so that the results pertain to an equilibrium isolated system. The time correlation matrix for $X$ is defined to be

$$Q(\tau) \equiv k_{\mathrm{B}}^{-1}\langle X(\tau)\, X(0)\rangle. \tag{3.43}$$

The averand is a dyadic matrix; some authors write it as $X(\tau)\, X(0)^{\mathrm{T}}$, where the superscript T means transpose. The average is over the configurations of the isolated system. The auto- or self-correlation functions for the elements of $X$ correspond to the diagonal elements of this matrix, and the cross-correlation functions correspond to the off-diagonal elements. Note the unconventional appearance of Boltzmann's constant in the definition of the time correlation matrix, which simplifies many of the following results.

An equilibrium system is homogeneous in time, which means that

$$\langle X(\tau)\, X(0)\rangle = \langle X(0)\, X(-\tau)\rangle, \text{ or } Q(\tau) = Q(-\tau)^{\mathrm{T}}. \tag{3.44}$$

The condition of macroscopic reversibility is manifest as

$$\begin{aligned} \langle X(\tau)\, X(0)\rangle &= \langle X(-\tau)^{\dagger}\, X(0)^{\dagger}\rangle \\ &= \langle X(-\tau)\, X(0)\rangle, \text{ or } Q(\tau) = Q(-\tau). \end{aligned} \tag{3.45}$$

This result only holds for the present pure parity case. Together these two results say that the pure parity time correlation matrix is symmetric, $Q(\tau) = Q(\tau)^{\mathrm{T}}$, and even in the time interval.

The entropy of the isolated system may be expanded to second order about $X = 0$,

$$S(X) = S(0) + \frac{1}{2} S'' : XX. \tag{3.46}$$

The linear term vanishes because $X = 0$ is the most likely state, which is to say the state of maximum entropy, $S'(0) = 0$. One can discard the constant term $S(0)$ because we are only interested in the variation with $X$. The Taylor expansion can be terminated at the quadratic term, because higher order terms scale with successive higher powers of the inverse of the square root of the system size. The quadratic term is intensive on average, since $\langle XX \rangle \sim \mathcal{O}(V)$.

The second derivative matrix is

$$S'' \equiv \left. \frac{\partial^2 S(X)}{\partial X \partial X} \right|_{X=0}. \tag{3.47}$$

This can be called the fluctuation matrix, and it is obviously symmetric $S''^{\mathrm{T}} = S''$, since $\partial^2 S / \partial X_\alpha \partial X_\beta = \partial^2 S / \partial X_\beta \partial X_\alpha$. Since the entropy is a maximum at $X = 0$, the fluctuation matrix must be negative definite.

The probability distribution for the fluctuation is the exponential of the entropy. By the above expansion the exponent is a quadratic form, which is to say that the distribution is Gaussian

$$\wp(X) = \frac{1}{W} e^{S'':XX/2k_{\mathrm{B}}}, \quad W = \sqrt{\mathrm{Det}\{-2\pi k_{\mathrm{B}} S''^{-1}\}}. \tag{3.48}$$

This is in fact the total fluctuation entropy. The negative of the inverse of the fluctuation matrix is the static correlation function,

$$Q_0 \equiv Q(0) = -S''^{-1}. \tag{3.49}$$

This is readily derived from its definition,

$$\begin{aligned} Q(0) &= k_{\mathrm{B}}^{-1} \langle X(0)\, X(0) \rangle \\ &= \frac{1}{k_{\mathrm{B}} W} \int \mathrm{d}X\; XX e^{S'':XX/2k_{\mathrm{B}}} \\ &= \frac{1}{k_{\mathrm{B}} W} \int \mathrm{d}X\; X k_{\mathrm{B}} S''^{-1} \frac{\partial}{\partial X} e^{S'':XX/2k_{\mathrm{B}}} \\ &= \frac{1}{W} \left\{ [X S''^{-1} e^{S'':XX/2k_{\mathrm{B}}}]_{-\infty}^{\infty} - S''^{-1} \int \mathrm{d}X\; e^{S'':XX/2k_{\mathrm{B}}} \right\} \\ &= -\, S''^{-1}. \end{aligned} \tag{3.50}$$

In the integration by parts, the integrated part is zero because extreme fluctuations at the limits of integration have vanishing probability.

In the same fluctuation spirit we shall expand the second entropy of the isolated system to quadratic order,

$$S^{(2)}(X', X|\tau) = \frac{1}{2}A(\tau) : XX + B(\tau) : X'X + \frac{1}{2}A'(\tau) : X'X'. \tag{3.51}$$

This is for the unconditional transition $X \xrightarrow{\tau} X'$. Here $X$ is the current macrostate and $X'$ is the destination macrostate. If $\tau > 0$ then $X'$ is the future macrostate, and if $\tau < 0$ then $X'$ is the past macrostate. The coefficient matrices are the second derivatives of the second entropy evaluated in the most likely state, $X = X' = 0$,

$$A(\tau) \equiv \left. \frac{\partial^2 S^{(2)}(0, X|\tau)}{\partial X \partial X} \right|_{X=0}, \quad A'(\tau) \equiv \left. \frac{\partial^2 S^{(2)}(X', 0|\tau)}{\partial X' \partial X'} \right|_{X'=0},$$

and

$$B(\tau) \equiv \left. \frac{\partial^2 S^{(2)}(X', X|\tau)}{\partial X \partial X'} \right|_{X'=X=0}. \tag{3.52}$$

Clearly the first two matrices are symmetric, $A(\tau) = A(\tau)^{\mathrm{T}}$ and $A'(\tau) = A'(\tau)^{\mathrm{T}}$. The fluctuation second entropy has maximum value $S^{(2)}(0, 0|\tau) = 0$.

For the present equilibrium isolated system one has time homogeneity, which implies $S^{(2)}(X', X|\tau) = S^{(2)}(X, X'|-\tau)$. This yields

$$A(-\tau) = A'(\tau), \qquad \text{and} \qquad B(\tau) = B(-\tau)^{\mathrm{T}}. \tag{3.53}$$

Macroscopic reversibility says that the second entropy must have the symmetry

$$S^{(2)}(X', X|\tau) = S^{(2)}(X'^{\dagger}, X^{\dagger}|-\tau) = S^{(2)}(X', X|-\tau). \tag{3.54}$$

The final equality, which only holds in the present pure parity case, can be confirmed by inspection of the quadratic form. It follows that

$$A(\tau) = A'(\tau), \qquad \text{and} \qquad B(\tau) = B(\tau)^{\mathrm{T}}. \tag{3.55}$$

Maximizing the second entropy with respect to $X'$ yields the most likely destination state given the present state,

$$\left. \frac{\partial S^{(2)}(X', X|\tau)}{\partial X'} \right|_{X'=\overline{X}'} = 0. \tag{3.56}$$

Performing the derivative on the quadratic form yields the most likely destination explicitly,

$$\overline{X}' \equiv \overline{X}(\tau|X) = -A(\tau)^{-1}B(\tau)X. \tag{3.57}$$

The destination may be in the future, $\tau > 0$, or in the past, $\tau < 0$. If in the future, this result answers the question 'where will the system most likely go to?'. If in the past it answers the question 'where did the system most likely come from?'. In either case this result says that the most likely destination fluctuation is linearly proportional to the current fluctuation. This is a direct consequence of the Gaussian form for the fluctuations. Because the fluctuation form for the first or normal entropy is quadratic in the fluctuation, $S(X) = S'': XX/2$, its gradient is linear in the fluctuation, $\nabla S(X) \equiv S'(X) = S''X$. The gradient in the first entropy can be called the thermodynamic force. The fact that the most likely destination is linearly proportional to the fluctuation $X$ is the same as saying that it is linearly proportional to the thermodynamic force. Hence the present result for the most likely destination provides the justification for an aspect of the regression hypothesis (Onsager 1931), namely that the rate of return to equilibrium is linearly proportional to the gradient in the entropy.

It is useful to rewrite the second entropy as a function of the departure from the most likely destination state,

$$\begin{aligned} S^{(2)}(X', X|\tau) = & \frac{1}{2}A(\tau) : [X' + A(\tau)^{-1}B(\tau)X]^2 \\ & + \frac{1}{2}[A(\tau) - B(\tau)A(\tau)^{-1}B(\tau)] : XX. \end{aligned} \tag{3.58}$$

Here and often below the notation of a vector squared is shorthand for the corresponding dyadic matrix.

Above, the reduction condition on the second entropy was given, equation (3.38). It says that the second entropy for the transition to the most likely destination state equals the first entropy of the current state,

$$S^{(2)}(\overline{X}', X|\tau) = S(X). \tag{3.59}$$

Physically, the reduction condition arises from the sharply peaked nature of the probability distribution, which means that fluctuations are relatively negligible. In the above form for the second entropy, the first term vanishes when $X' = \overline{X}(\tau|X)$, leaving only the final term that is quadratic in $X$. Hence by the reduction condition, the coefficient of this must be equal to the first entropy fluctuation matrix,

$$A(\tau) - B(\tau)A(\tau)^{-1}B(\tau) = S''. \tag{3.60}$$

This is an exact relation between the two fluctuation transition matrices, the third matrix already having been eliminated, $A'(\tau) = A(\tau)$. Because this holds for all time intervals, it will prove to be the key equation in determining the coefficients of a small $\tau$ expansion of the fluctuation transition matrices.

The transition probability is the exponential of the second entropy,

$$\wp(X', X|\tau) = \frac{1}{W^{(2)}} e^{S^{(2)}(X',X|\tau)/k_B}. \tag{3.61}$$

The exponent is a quadratic form, which means that the transition probability is Gaussian, by design. Using the second entropy rearranged as above, and also the reduction condition, the normalization factor is readily seen to be

$$W^{(2)} = \sqrt{\mathrm{Det}\{-2\pi k_B A(\tau)^{-1}\}}\sqrt{\mathrm{Det}\{-2\pi k_B S''^{-1}\}}. \tag{3.62}$$

For a Gaussian distribution, the average value equals the most likely value. This allows the easy evaluation of the time correlation matrix, and gives its relationship with the transition fluctuation matrices,

$$\begin{aligned} Q(\tau) &= k_B^{-1}\langle X'X\rangle \\ &= k_B^{-1}\langle \overline{X'}X\rangle \\ &= -\, A(\tau)^{-1}B(\tau)k_B^{-1}\langle XX\rangle \\ &= A(\tau)^{-1}B(\tau)S''^{-1}. \end{aligned} \tag{3.63}$$

Using these, the most likely destination, equation (3.57), can be rewritten in terms of the time correlation function,

$$\overline{X'} = -A(\tau)^{-1}B(\tau)X = -Q(\tau)S''X. \tag{3.64}$$

The final two factors give the gradient in the first entropy, $S''X = S'(X) \equiv \nabla S(X)$. This is the thermodynamic force that drives the fluctuation back to its equilibrium value. This result is a general form of the regression hypothesis that holds for arbitrary time intervals $\tau$. If one multiplies both sides of this by $X^{\mathrm{T}}$ on the right and takes the average then one obtains the identity,

$$\langle[\nabla S(X)]X^{\mathrm{T}}\rangle = -k_B \mathrm{I}. \tag{3.65}$$

One can eliminate the two fluctuation transition matrices in favor of the time correlation matrix. From the reduction condition, equation (3.60), and the time correlation function expression equation (3.63), one obtains

$$A(\tau)^{-1}S'' = \mathrm{I} - A(\tau)^{-1}B(\tau)A(\tau)^{-1}B(\tau) = \mathrm{I} - Q(\tau)S''Q(\tau)S''. \tag{3.66}$$

Hence

$$A(\tau) = [\mathrm{I} - S''Q(\tau)S''Q(\tau)]^{-1}S''. \tag{3.67}$$

This is evidently symmetric, as it must be. Taking the transpose of the time correlation expression equation (3.63), and using the result of macroscopic reversibility, $B(\tau)^{\mathrm{T}} = B(\tau)$, one also obtains

$$B(\tau) = [\mathrm{I} - S''Q(\tau)S''Q(\tau)]^{-1}S''Q(\tau)S'' = A(\tau)Q(\tau)S''. \tag{3.68}$$

Hence the time correlation function matrix completely determines the fluctuation transition matrices. It follows that it also determines the second entropy and the transition probability.

### 3.2.2 Exponential Markovian decay

So far the analysis describes exactly fluctuations, with the coefficients having been related to the time correlation function. The general result for the most likely destination state, equation (3.64), is a form of the regression hypothesis, since it showed that this was linearly proportional to the current state. That result was valid for arbitrary time intervals. Now the aim is to elucidate the time dependence of the time correlation function and of the fluctuation transition matrices, which form the proportionality constant for the most likely destination.

The time correlation function can of course be obtained quantitatively and in molecular detail using statistical mechanics for any particular system. The aim of thermodynamics, however, is to abstract from any particular system and to obtain a general description of matter in terms of a few macroscopic parameters that can be measured, which is pursued here.

The starting point is Markovian systems, which in many cases are a good model of real world behavior. It will now be shown that the time correlation functions in these systems are exponentially decaying. Following that derivation, the small time limit will be analyzed, in which limit it will be shown that both Markovian and non-Markovian time correlation functions decay linearly. This limit turns out to be sufficient to develop the formalism of non-equilibrium thermodynamics of steady state systems.

For the transition $X \xrightarrow{\tau} X'$, for sufficiently long time intervals the current and the destination macrostate are uncorrelated. This means that the transition probability becomes the product of the probability of each fluctuation independently, and the second entropy becomes the sum of the first entropies, $S^{(2)}(X', X|\tau) \to S(X') + S(X)$, $\tau \to \infty$. This means that the transition fluctuation matrices must behave as

$$A(\tau) \to S'', \quad B(\tau) \to 0, \quad \tau \to \infty. \tag{3.69}$$

These limiting expressions can also be derived from equations (3.67) and (3.68) using the fact that the time correlation function vanishes in the long time limit, $Q(\tau) \to 0$, $\tau \to \infty$.

For small time intervals, the most likely destination state must be the same as the current state, $\overline{X}(\tau|X) \to X, \tau \to 0$. This must be so since it takes time to depart from the current macrostate. Inserting this into equation (3.57) yields

$$A(\tau)^{-1}B(\tau) \to -\mathrm{I}, \quad \tau \to 0. \tag{3.70}$$

Again, I is the identity matrix.

The present analysis is for fluctuations all with the same time parity. In this case equations (3.53) and (3.55) show that the fluctuation transition matrices are even functions of time, $A(\tau) = A(-\tau)$, and $B(\tau) = B(-\tau)$. From this one can see that the time correlation function is also an even function of time, $Q(\tau) = Q(-\tau)$, which in fact was shown in equations (3.44) and (3.45).

The results for $\tau \to \infty$ are formally exact. However, this is not the limit that is required for the formal thermodynamic analysis of steady state non-equilibrium systems. For these one has to ascertain the applicable time regime, and the qualitative behavior of the time correlation function therein. This can be done by looking at a specific system, and reasoning inductively that the behavior is universally applicable.

The regression of a fluctuation, $\overline{X}(\tau|X)$, is shown in figure 3.2. The particular macrostate illustrated is the first energy moment, $X \equiv E_1$, which is a non-conserved extensive variable that is the simplest measure of the non-uniformity of the energy

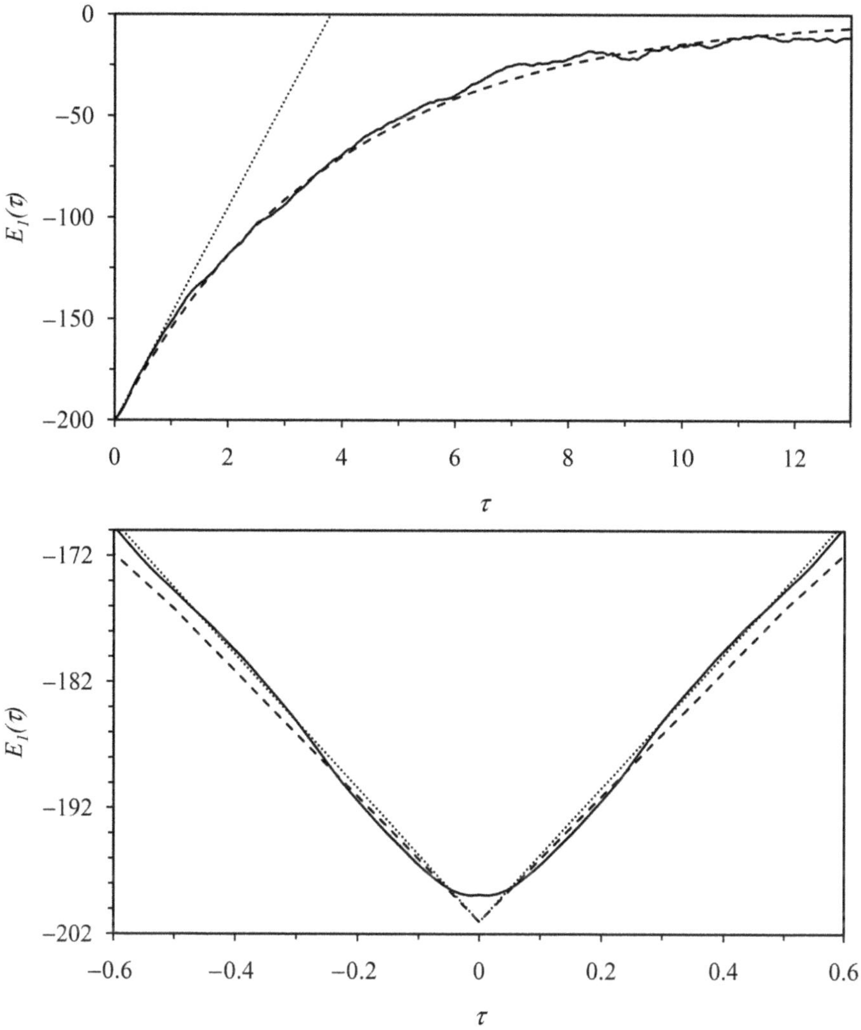


**Figure 3.2.** Adiabatic decay of the first energy moment (dimensionless units). The solid curve is the average over dynamically disordered initial states of molecular dynamics trajectories of a Lennard-Jones fluid. The dotted line is fitted, $E_1(\tau) = a + b|\tau|$, and the dashed curve is an exponential, $E_1(\tau) = ae^{|\tau|b/a}$. The lower chart magnifies the data at small times. Data from Attard (2005a).

distribution in the system. The negative values shown indicate that initially there is excess energy in the negative half of the system.

The behavior shown in figure 3.2 typifies the regression of all fluctuations of the same time parity. In this case the regression is an even function of time. There are three more or less distinct time regimes. In the molecular or infinitesimal time regime, for the data of figure 3.2 this is $|\tau| \lesssim 0.1$, the fluctuation is smooth about $\tau = 0$, as can be seen in the lower part of the figure. Hence it has a Taylor expansion in even powers of $\tau$, $\overline{X}(\tau|X) \sim X + c(X)\tau^2 + d(X)\tau^4 + \cdots$.

The small time regime is $0.1 \lesssim |\tau| \lesssim 1$ in figure 3.2, and it is identified by linear or steady decay of the fluctuation. Because of the pure parity, the regression must be an even function of time, which means that there must be non-analytic terms in its expansion in this regime, $\overline{X}(\tau|X) \sim a(X) + b(X)|\tau|$, with $a(X) \approx X$. Terms of order $\mathcal{O}(\tau^2, |\tau|^3)$ have been neglected here. This non-analytic expansion is the only expansion that is both linear in $\tau$ and even in $\tau$. It cannot apply in the molecular regime, where the expansion must be analytic, as the lower part of figure 3.2 shows. The non-analytic terms in the small-$\tau$ expansion sum an infinity of terms in the molecular regime Taylor expansion.

The long time regime is here $|\tau| \gtrsim 1$. In this particular system the fluctuation regresses exponentially to zero, as the fit in figure 3.2 confirms. However, such exponential behavior is not the case for all fluctuations, and sometimes the regression can be power law or even non-monotonic.

The molecular regime can be thought of as an inertial regime where the regression of the fluctuation comes up to speed from a standing start. The extent and nature of this regime varies with the type of fluctuation and the details of the intermolecular potential. More universal is the small time regime in which the rate of change of the fluctuation is constant. The actual rate of change is proportional to the magnitude of the initial fluctuation, with the proportionality constant depending upon the specific system. The latter are related to the hydrodynamic transport coefficients, as will shortly be clarified. This is the reason why the short time regime is the relevant regime for steady state non-equilibrium thermodynamics. Since the magnitude of the fluctuation decreases with increasing time, so must the driving force and the rate of change. This sets an upper limit on the small time regime which is defined by a constant rate of change. That is, in the small time regime, the change in magnitude of the fluctuation must be relatively negligible, $|X(\tau) - X(0)| \ll |X(0)|$.

The exponential decay evident in figure 3.2 is an indicator of a Markov system. Such a system is defined by the present state solely determining the future state irrespective of the previous history of the system. In a Markov system the transition probability for three states at times $t_1 < t_2 < t_3$ factorizes

$$\begin{aligned} \wp(X_3, X_2, X_1|t_{32}; t_{21}) &= \wp(X_3|t_{32}, X_2; t_{21}, X_1)\wp(X_2, X_1|t_{21}) \\ &\approx \wp(X_3|t_{32}, X_2)\wp(X_2, X_1|t_{21}). \end{aligned} \tag{3.71}$$

The first equality is the formally exact expression for the conditional probability: the probability of $X_3$ depends upon the preceding two states. The final equality is true if the intermediate state $X_2$ removes any influence of the state $X_1$ on the final state $X_3$. One can say that the final state has no memory of the earlier state. This is the Markov approximation.

Markovian systems have a greatly simplified form of regression. The most likely transition $X_1 \to X_3$ in time $t_{31}$ according to equation (3.64) is

$$\overline{X_3} = -Q(t_{31})S''X_1. \tag{3.72}$$

Similar to the analysis of Fox and Uhlenbeck (1970), one can divide the time interval with an intermediate time, $t_1 < t_2 < t_3$, so that the total transition is the product of a pair of successive transitions,

$$\overline{X}_3 = -Q(t_{32})S''\overline{X}_2 = Q(t_{32})S''Q(t_{21})S''X_1. \tag{3.73}$$

The first equality invokes the Markov approximation: the transition $X_2 \to X_3$ is not influenced directly by the prior state $X_1$. For these two expressions to be equal one must have

$$-Q(t_{31})S'' = Q(t_{32})S''Q(t_{21})S'', \tag{3.74}$$

which implies that

$$Q(\tau) = -e^{|\tau|\Lambda S''/2}S''^{-1}. \tag{3.75}$$

Here $\Lambda$ is a constant matrix, and exponentiation is defined by the power law expansion. This is known as Doob's theorem, and it applies to any random process that is both Gaussian and Markov (Doob 1942). The absolute value of the time interval appears here because the time correlation function must be an even function of time in the present case of pure parity.

This general analysis proves the exponential decay of the time correlation functions in Markov systems. The decay rate is determined by $\Lambda$, which will turn out to be related to the transport coefficients of the system.

For small enough time intervals the exponential can be linearized,

$$Q(\tau) = -S''^{-1} - \frac{|\tau|}{2}\Lambda, \quad |\tau|\text{Det}\{\Lambda S''\} \ll 1. \tag{3.76}$$

A more precise delineation of the regime of validity of this linearization would invoke the eigenvalue of $\Lambda S''$ that is smallest in magnitude.

### 3.2.3 Small time expansion

The preceding result that the time correlation function of a Markov system was linear in $|\tau|$ on short time scales cannot hold on infinitesimal time scales. Instead, in that regime the Taylor expansion, which is analytic, must involve only even powers of the time interval,

$$Q(\tau) = -S''^{-1} + \frac{1}{2}\tau^2\ddot{Q}(0) + \cdots, \quad \tau \to 0. \tag{3.77}$$

This and the preceding two equations give different expressions for the time dependence of the time correlation function. The expressions apply respectively in each of the three time regimes identified in figure 3.2.

The physical origin of these three time regimes is readily understood. Initially, $\tau = 0$, the system is positionally ordered, $X(0) \neq 0$, and dynamically disordered, $\dot{X}(0) = 0$. Over molecular timescales, say $\tau_{\text{relax}}$, the system becomes dynamically ordered, $\dot{X}(\tau_{\text{relax}}) \neq 0$. Even though dynamic order is a lower entropy state than dynamic disorder, we shall soon show that dynamic order that leads to increasing position disorder has greater second entropy than dynamic disorder.

The amount of dynamic order induced after this molecular time interval is proportional to the extent of the fluctuation, $\dot{X}(\tau) \propto X(\tau)$, $\tau > \tau_{\rm relax}$. This result is manifest in the exponential decay of $\overline{X}(\tau|X)$ in a Markov system. However, the result is more general than Markov systems, both on symmetry grounds (the vector $X$ is the only vector present, and so it must be the one that sets the direction of the vector $\dot{X}$, which is the so-called Curie principle), and on the grounds that for fluctuations higher order terms in $X$ can be neglected.

For time intervals longer than the infinitesimal time $\tau_{\rm relax}$, but not too much longer, the changes in the fluctuation from its initial value are negligible, $X(\tau) \approx X(0)$, and so one has $\dot{X}(\tau) \propto X(0)$. Integrating this over this so-called short time regime leads to equation (3.76), where the absolute value of the time interval is required from the symmetry of the present pure parity system. We shall call the time intervals where this is valid short time intervals.

From the nature of the above argument, the short time expansion equation (3.76), is likely to be universally applicable. It is not restricted to the Markov systems that were used to derive it.

The following results apply to the small time regime. Expansions in powers of $|\tau|$ will now be obtained for the second entropy fluctuation matrices.

One can rearrange the fluctuation form of the second entropy as

$$\begin{aligned} S^{(2)}&(X', X|\tau) \\ &= \frac{1}{2}A(\tau) : X^2 + B(\tau) : X'X + \frac{1}{2}A(\tau) : X'^2 \\ &= \frac{-1}{2}B(\tau) : [X' - X]^2 + \frac{1}{2}[A(\tau) + B(\tau)] : [X'^2 + X^2]. \end{aligned} \tag{3.78}$$

This is for a pure parity system. This uses the inside-out convention for the double scalar product, $A : XY \equiv Y \cdot AX$, and writes the symmetric dyadic matrix as the square of a vector, $X^2 \equiv XX$.

As mentioned above, in the limit $\tau \to 0$ the destination state must be the same as the current state, $\overline{X'} = X$. Hence the transition probability must become a delta-function, $\wp(X', X|\tau) \to \delta(X' - X)\wp(X)$, $\tau \to 0$. In view of this, the first term in the second equality of the second entropy above must diverge,

$$B(\tau) = \frac{1}{|\tau|}\Lambda^{-1} + B + \mathcal{O}(\tau), \quad |\tau| \to 0. \tag{3.79}$$

The constant matrix $\Lambda$ is written to be consistent with the Markov analysis, as will soon be made clear explicitly. Since the second term in the second equality of the second entropy must remain finite, the leading term must cancel so that the expansion for $A(\tau)$ must be of the form

$$A(\tau) = \frac{-1}{|\tau|}\Lambda^{-1} + A + \mathcal{O}(\tau), \quad |\tau| \to 0. \tag{3.80}$$

To leading-order these two expansions give $A(\tau)^{-1}B(\tau) = -\mathrm{I}$, $|\tau| \to 0$, for which the result was already given as equation (3.70). By design, in the present case of pure parity, only terms even in time appear in these expansions.

Because the second entropy gives the transition probability, the quadratic form must be negative, which means that $A(\tau)$ must be negative definite, which means that $\Lambda$ must be positive definite. Since $A(\tau)$ is symmetric, so is $\Lambda$.

Inserting these two expansions into the reduction condition, equation (3.60), yields

$$\begin{aligned} S'' &= A(\tau) - B(\tau)A(\tau)^{-1}B(\tau) \\ &= \frac{-1}{|\tau|}\Lambda^{-1} + A - \left[\frac{1}{|\tau|}\Lambda^{-1} + B\right]\left[\frac{-1}{|\tau|}\Lambda^{-1} + A\right]^{-1}\left[\frac{1}{|\tau|}\Lambda^{-1} + B\right] \\ &= 2[A + B] + \mathcal{O}(\tau). \end{aligned} \tag{3.81}$$

Only the leading order term, $\mathcal{O}(|\tau|^0)$ has been retained here, linear and higher order terms having been neglected. One sees from this that the sum of the two constant coefficient matrices in the small time expansion of the transition fluctuation matrices is determined to equal half the first entropy fluctuation matrix. This will turn out to be a remarkably powerful result.

For a Markov system one has the stronger result, $B = 0$ and $A = S''/2$. This is not required for what follows.

With this result one can give the explicit proof of the regression hypothesis. Given the current state $X$, the most likely destination state of a transition, equation (3.57), for short time intervals $\tau$, is

$$\begin{aligned} \overline{X}(\tau|X) &= -A(\tau)^{-1}B(\tau)X \\ &= -\left[\frac{-1}{|\tau|}\Lambda^{-1} + A\right]^{-1}\left[\frac{1}{|\tau|}\Lambda^{-1} + B\right]X \\ &= [\mathrm{I} + |\tau|\Lambda(A + B)]X \\ &= X + \frac{|\tau|}{2}\Lambda S''X + \mathcal{O}(\tau^2). \end{aligned} \tag{3.82}$$

This says that the most likely change or regression in a fluctuation is linear in time and linearly proportional to the thermodynamic force $\nabla S(X) \equiv S'(X) = S''X$. The regression is symmetric in the future and in the past. This holds for systems of pure parity and is the general form of the regression hypothesis.

Multiplying both sides of this equation by $k_B^{-1}X$ on the right, the average yields the time correlation function, $Q(\tau) = -S''^{-1} - |\tau|\Lambda/2$. This is just the Markov result in the small time limit, equation (3.76).

For such a Markov system, one can simply piece together this result for a small time interval to obtain the result for larger time intervals, which is exponential decay of the time correlation function. For a non-Markov system such a procedure would be invalid, because successive time intervals would be correlated with each other, whereas the present second entropy expression neglects all conditions prior to the current state.

In the small time limit the second entropy is explicitly

$$\begin{aligned} S^{(2)}(X', X|\tau) &= \frac{-1}{2|\tau|}\Lambda^{-1} : [X' - \overline{X}(\tau|X)]^2 + \frac{1}{2}S'' : X^2 \\ &= \frac{-1}{2|\tau|}\Lambda^{-1} : [X' - X]^2 + \frac{1}{2}S : X'X - \frac{|\tau|}{8}X \cdot S''\Lambda S''X \\ &= \frac{-1}{2|\tau|}\Lambda^{-1} : [X' - X]^2 + \frac{1}{4}S'' : [X'^2 + X^2] \\ &\quad -\frac{|\tau|}{8}X \cdot S''\Lambda S''X + \mathcal{O}(\tau^2). \end{aligned} \tag{3.83}$$

This neglects $-B : [X' - X]^2/2$ and other terms of order $\tau^2$. By design, the first equality satisfies the reduction condition exactly. That is, the maximum value of the second entropy with respect to the destination state given the current state equals the first entropy of the current state,

$$\overline{S^{(2)}}(\tau|X) \equiv S^{(2)}(\overline{X'}, X|\tau) = \frac{1}{2}S'' : X^2. \tag{3.84}$$

However, the final equality in equation (3.83) is more approximate in the sense that it satisfies the reduction condition to $\mathcal{O}(\tau)$.

### 3.2.4 Stochastic dissipative equations of motion

The difference between the actual destination and the most likely destination is the stochastic part of the transition,

$$X' - \overline{X}(\tau|X) = \tilde{R}. \tag{3.85}$$

Hence the transition in full is

$$X(t + \tau) = X(t) + \frac{|\tau|}{2}\Lambda S''X(t) + \tilde{R}(t) + \mathcal{O}(\tau^2). \tag{3.86}$$

This is the stochastic dissipative equation of motion.

The second entropy in the small time limit given by the first equality in equation (3.83) is

$$S^{(2)}(X', X|\tau) = S(X) - \frac{1}{2|\tau|}\Lambda^{-1} : [X' - \overline{X}(\tau|X)]^2. \tag{3.87}$$

Since $\wp(X', X|\tau) = \wp(X'|X, \tau)\wp(X)$, and $\wp(X) = e^{S(X)/k_B}/Z^{(1)}$, in the small time limit the conditional transition probability is just the Gaussian

$$\begin{aligned} \wp(X'|X, \tau) &= \frac{1}{Z'}e^{-\Lambda^{-1}:[X'-\bar{X}(\tau|X)]^2/2k_B|\tau|} \\ &= \frac{1}{Z'}e^{-\Lambda^{-1}:\tilde{R}\tilde{R}/2k_B|\tau|} \\ &= \wp(\tilde{R}), \end{aligned} \tag{3.88}$$

where $Z' \equiv [\text{Det}\,\{2\pi k_B|\tau|\Lambda\}]^{1/2}$. Since $\tilde{R} = X' - \overline{X}(\tau|X)$, the conditional probability distribution for $X'$ is equal to the probability distribution for $\tilde{R}$ because at constant $X$, $\mathrm{d}X' = \mathrm{d}\tilde{R}$. From this it follows that the stochastic change on average vanishes, $\langle \tilde{R} \rangle = 0$, and it has variance

$$\langle \tilde{R}\tilde{R} \rangle = |\tau| k_B \Lambda. \tag{3.89}$$

This last result is the fluctuation dissipation theorem.

### 3.2.5 Regression theorem and the reciprocal relations

The small time expansion is all that is required to treat steady state, thermodynamic, non-equilibrium systems, as will shortly be made clear. It also gives several related results that have been long known, albeit by different methods.

It has just been established that in the small time limit to linear order in $\tau$, given the current state $X$, the most likely destination state is

$$\overline{X'} \equiv \overline{X}(\tau|X) = -Q(\tau)S''X = X + \frac{1}{2}|\tau|\Lambda S''X. \tag{3.90}$$

One may define the coarse velocity as the average velocity over the interval. Its most likely value is

$$\overline{\overset{\circ}{X}}(X) \equiv \frac{\overline{X'} - X}{\tau} = \frac{\hat{\tau}}{2}\Lambda S''X. \tag{3.91}$$

Here and throughout the sign of the time interval is $\hat{\tau} \equiv \text{sign}(\tau) = \tau/|\tau|$. This says that the coarse velocity is proportional to the thermodynamic force, $S'(X) \equiv \partial S(X)/\partial X = S''X$.

This result is essentially the Onsager regression hypothesis: the most likely flux is linearly proportional to the thermodynamic driving force (Onsager 1931). Of course the present derivation makes it a theorem rather than a hypothesis. Flux is here the coarse velocity $\overset{\circ}{X}$, as will become clearer in the explicit treatment of heat flow in section 3.4 below. The 'internal' thermodynamic driving force is here the gradient of the sub-system entropy. This will be shown below to equal the 'external' thermodynamic driving force that is imposed by the reservoirs.

An interesting feature of the present result is that the coarse velocity is independent of the length of the time interval over which it is measured, which one would expect of any velocity, but it does depend upon the sign of $\tau$. The most likely flux reverses sign for the future, $\hat{\tau} > 0$, compared to the past, $\hat{\tau} < 0$. This is an obvious consequence of a fluctuation in an equilibrium system: the system will most likely return to equilibrium, and the system most likely came from equilibrium.

The transport matrix $\Lambda$ is the coefficient matrix for the leading order term in the small time expansion of $A(\tau)$ and $B(\tau)$, equations (3.79) and (3.80). Since both of these are symmetric, the transport matrix itself must be symmetric

$$\Lambda = \Lambda^{\mathrm{T}}. \tag{3.92}$$

This says that the off-diagonal transport coefficients, which couple different forces and fluxes, are equal. This result is just the reciprocal relations, the explanation for which (Onsager 1931) culminated in a Noble prize.

There are a number of comparatively well-known examples of reciprocal relations. One simple example is multi-component isothermal diffusion. The regression theorem in this case is equivalent to Fick's law, since it says that the mass fluxes are proportional to the chemical potential gradients. There are two relevant diffusion constants for species $\alpha$ and $\beta$. One gives the proportionality of the flux $\alpha$ to the chemical potential gradient $\beta$, and the other gives the obverse, the proportionality of the flux $\beta$ to the chemical potential gradient $\alpha$. Because the transport matrix is symmetric, these are equal to each other.

Another example is the thermoelectric effect, which couples electric and heat currents. The Seebeck effect applies to an electric current driven by a temperature gradient, and the Peltier effect applies to a heat flux driven by a voltage difference, or electro-chemical gradient. The symmetry of the transport matrix $\Lambda$ ensures that the respective proportionality constants for these two effects are equal to each other.

Probably the most widely used example of a reciprocal relation, albeit one in which the origin of the phenomenon is rarely acknowledged, is chemical reaction rates. Even undergraduate students learn that the rate coefficient for the forward reaction is equal to that of the reverse reaction. This result in fact can only be proven by the symmetry of the transport matrix. This at that time unexplained example of a reciprocal relation was one of the original motivations for Onsager's (1931) work, for which he was subsequently awarded a Noble prize.

Onsager's original derivation of the reciprocal relations consisted of combining the regression hypothesis (that the rate of regression of a fluctuation was proportional to the fluctuation itself) with macroscopic reversibility. In the present case the regression hypothesis has been derived as a theorem. Macroscopic reversibility was used in the present derivation to show that the cross fluctuation transition matrix is symmetric, $B(\tau) = B(\tau)^{\mathrm{T}}$, and that the two 'pure' fluctuation transition matrices are equal, $A(\tau) = A'(\tau)$. Equivalently, the time correlation function is symmetric, $Q(\tau) = Q(\tau)^{\mathrm{T}}$.

These symmetries, as well as the symmetry of the transport matrix, only hold for the pure parity case. Mixed parity is treated in section 3.3.

### 3.2.6 Green–Kubo expression

The Green–Kubo relations give the hydrodynamic transport coefficients as time integrals of the time correlation function of an equilibrium system (Green 1954, Kubo 1966, Kubo *et al* 1978, Keizer 1987, Zwanzig 2001). The above result for the small time expansion of the most likely destination equation (3.90), is equivalent to the Green–Kubo expressions, as is now shown. Since Onsager (1931) gave the relationship between the time correlation function and the transport coefficient in his paper on the reciprocal relations, it may be observed that he had given the Green–Kubo expressions some 20 years before either Green or Kubo.

The result for the most likely destination, equation (3.90), may be re-written in Green–Kubo form as

$$\begin{aligned}\Lambda &= \frac{-2}{|\tau|}[Q(\tau) + S''^{-1}] \\ &= \frac{-2}{k_B|\tau|}[\langle X(\tau)X(0)\rangle - \langle X(0)X(0)\rangle] \\ &= \frac{-2}{k_B|\tau|}\int_0^\tau dt' \langle \dot{X}(t')X(0)\rangle.\end{aligned} \tag{3.93}$$

For small but non-infinitesimal $\tau$, where the present expansion is valid, the right-hand side must be independent of $\tau$. In practice this defines the regime of validity for the small $\tau$ expansion. Typically, simulations show the right-hand side has a broad peak as a function of $\tau$, the maximum of which is generally taken as 'the' transport coefficient. The average is an equilibrium average, and it is typically taken for an isothermal rather than an isolated system. One can use integration by parts and various symmetry arguments to rewrite this is in several mathematically (but not necessarily computationally) equivalent ways.

### 3.2.7 Physical meaning of the second entropy

In order to provide a physical interpretation of the second entropy, it is useful to rewrite it in terms of the flux (i.e. the coarse velocity, $\overset{\circ}{X} \equiv (X' - X)/\tau$) and the transport matrix. Using equation (3.90), the departure from the most likely destination is

$$X' - \overline{X'} = \tau\overset{\circ}{X} - \frac{|\tau|}{2}\Lambda S''X + \mathcal{O}(\tau^2). \tag{3.94}$$

Inserting this into the expression for the second entropy, the first equality in equation (3.83), to leading-order in $\tau$ one has

$$\begin{aligned}&S^{(2)}(X', X|\tau) \\ &= S(X) + \frac{1}{2}\left[\frac{-\Lambda^{-1}}{|\tau|} + \mathcal{O}(\tau^0)\right] : [X' - \overline{X'}]^2 \\ &= S(X) - \frac{|\tau|}{2}\Lambda^{-1} : \left[\overset{\circ}{X} - \overline{\overset{\circ}{X}}\right]^2 + \mathcal{O}(\tau^2) \\ &= S(X) - \frac{|\tau|}{2}\Lambda^{-1} : \overset{\circ}{X}{}^2 + \frac{\tau}{2}\overset{\circ}{X}\cdot S''X - \frac{|\tau|}{8}X\cdot S''\Lambda S''X + \mathcal{O}(\tau^2).\end{aligned} \tag{3.95}$$

This expression writes the second entropy explicitly in terms of the coarse velocity; as a quadratic form, it gives the entropy due to the fluctuations in flux. By construction the flux, $\overset{\circ}{X}$, does not depend upon the length of the time interval, and as such it is an intensive function of time. In contrast, the second entropy as a linear function of the time interval is extensive in time.

*Dynamic order versus structural disorder*

Writing the second entropy in terms of the fluctuation in the flux allows for a straightforward physical interpretation. The four terms on the right-hand side of the final equality each contribute to the understanding of it.

The first term is the ordinary entropy $S(X) = S'': X^2/2$. Once the optimum flux $\overline{\overset{\circ}{X}}(X)$ has been established, this is all that remains of the second entropy. Given the optimum flux, maximizing $S(X)$ yields the optimum static structure $\overline{X}$. For the case of an isolated system this vanishes, but when exchange with a reservoir is allowed, the reservoir first entropy added to this term gives an optimum structure that is non-zero, $\overline{X}(x_r) \neq 0$. This case will be analyzed below. The appearance of the first entropy is a direct consequence of the reduction condition, equation (3.38). On the basis of this discussion one can say that the first or ordinary entropy determines the static structure of a non-equilibrium system.

Turning to the second term, one sees that it is quadratic in the flux. Since the transport matrix must be positive definite (see the discussion following equation (3.80)), this term is negative and acts to reduce the second entropy. This is the term that prevents the flux becoming infinitely large, which is a crucial difference between the second entropy approach and the approaches based upon the principle of extreme dissipation (see the discussion in section 3.1.2 above and the following equation (3.96) below). The reason that the optimum flux must have an upper limit is that it represents a dynamically ordered state, which is a state of low entropy. Of course the challenge, and the point of the second entropy, is to balance quantitatively the unfavorable aspects of dynamic order with the favorable aspect of reducing the structural order over time.

This brings us to the third term, which is proportional to the thermodynamic force, $\nabla S(X) = S'(X) = S''X$. In fact it is proportional to the scalar product of the flux and the force, $\tau\overset{\circ}{X} \cdot \nabla S(X)/2$. In the future, $\tau > 0$, this term is positive when the flux is in the same direction as the force, which is the direction that increases the first entropy. (The case $\tau < 0$, and the meaning of the appearance of terms with $\tau$ and $|\tau|$ in the second entropy is discussed along with the irreversibility of the fluctuation below.) Because $S''$ is negative definite, the direction of increasing first entropy is opposite to the direction of the fluctuation $X$, which is to say that the first entropy increases when the flux reduces $X$ (going forward in time). This term is half the entropy dissipated over the time interval. It is this second term that drives the flux to be non-zero and that balances the entropic cost of the dynamic order that is the flux.

The final term does not appear to have a straightforward physical interpretation. Obviously it is negative and depends on the static structure, not the flux. Mathematically the reduction condition demands it so that the second and third terms can be canceled when the flux is optimum.

One can see that the physical basis for the variational principle embodied by the second entropy is rather simple, namely that the optimum flux arises from two competing effects. The (static) order of the system decreases over time as entropy is produced by the flux being aligned with the thermodynamic gradient. This term is favorable, linear in the flux, and dominates at small fluxes. Competing against this is the dynamic order of the system, which increases as the magnitude of the flux increases. This term unfavorable, quadratic in the flux, and dominates at large fluxes.

*Dissipation*

The variational nature of the second entropy stands in stark contrast to the dissipation, the extremization of which, it will be recalled from section 3.1.2, has been advocated by some authorities as the principle for determining the optimum non-equilibrium state. In general the rate of entropy production is a function of both the fluctuation and the flux,

$$\dot{S}(\dot{X}, X) = \dot{X} \cdot \frac{\partial S(X)}{\partial X} = \dot{X} \cdot S'' X. \tag{3.96}$$

In this formally exact expression, the fluctuation $X$ and the flux $\dot{X}$ are independent variables.

According to the second entropy, for a given fluctuation $X$, the optimal rate of entropy production is therefore

$$\overline{\overset{\circ}{\dot{S}}}(X) \equiv \dot{S}(\overline{\overset{\circ}{\dot{X}}}, X) = \frac{\hat{\tau}}{2} X \cdot S'' \Lambda S'' X. \tag{3.97}$$

The optimal flux that appears here was determined by maximizing the second entropy, not by extremizing the dissipation. Since the transport matrix $\Lambda$ is positive definite, the optimal dissipation is evidently positive.

The rate of entropy production, equation (3.96), is neither maximized nor minimized by the optimum flux. At a glance one can see from equation (3.96), taking a one component system and $X > 0$ to make the point, that the maximum dissipation occurs at $\dot{X} = -\infty$, and the minimum dissipation occurs at $\dot{X} = +\infty$. A function that is linear in the flux has no non-trivial extrema; it can be a maximum or a minimum only at the boundaries of its range. It is the quadratic form of the second entropy that enables it to be used as a variational principle, which is in stark contrast to the dissipation. As was remarked in section 3.1.2, this is fundamentally why the dissipation does not give the variational principle for non-equilibrium systems.

*Irreversible trajectory*

As already mentioned, the second entropy contains terms that depend on $\tau$ and on $|\tau|$. (Actually terms dependent on $\tau$ only occur in the mixed parity case; see section 3.3.) The former distinguishes between the future and the past, whereas the latter does not. It is these latter terms that make the trajectory irreversible.

Irreversibility means that if one goes backward from the most likely destination, most likely one does not regain the original state. If $X$ is the original fluctuation then this reversal operation corresponds to

$$\begin{aligned} \overline{X}(-\tau|\overline{X}(\tau|X)) &= \left[\mathrm{I} + \frac{|\tau|}{2}\Lambda S''\right]\left[\mathrm{I} + \frac{|\tau|}{2}\Lambda S''\right]X \\ &= X + |\tau|\Lambda S'' X + \mathcal{O}(\tau^2). \end{aligned} \tag{3.98}$$

One sees that reversing the time step after the first step gives more of the same.

Consider the sequence of transitions $X_1 \xrightarrow{\tau} X_2 \xrightarrow{-\tau} X_3$. The initial state is $X_1$, the most likely intermediate state is $X_2 \equiv \overline{X}(\tau|X_1)$, and the most likely final state is $X_3 \equiv \overline{X}(-\tau|X_2)$. This says that in the future, $\tau > 0$ the system will move from $X_1$ toward equilibrium, so that $S(X_2) > S(X_1)$. However, the most likely prior state for $X_2$ is $X_3$ not $X_1$, since the former is even closer to equilibrium, $S(X_3) > S(X_2) > S(X_1)$. Reversibility and irreversibility is further discussed in section 5.4.2 below.

### 3.2.8 Third entropy

Although only the second entropy is necessary to deal with steady state, thermodynamic, non-equilibrium systems, it is of interest to discuss the third entropy because this shows how non-Markovian behavior can arise. Since Gaussian fluctuations of all orders are the sum of pair-wise products, the third entropy in its most general form is

$$\begin{aligned} S^{(3)}(X_3, X_2, X_1|t_{32}, t_{21}) = & \frac{1}{2}C : X_3^2 + \frac{1}{2}C' : X_1^2 + \frac{1}{2}D : X_2^2 \\ & + E : X_2X_3 + E' : X_2X_1 + F : X_1X_3. \end{aligned} \tag{3.99}$$

For simplicity the dependence on the time intervals of the fluctuation transition matrices has been suppressed. In the symmetric case that $t_{32} = t_{21}$, the primed and unprimed matrices are equal. By rearranging this in terms of $(X_3 - X_2)^2$ and $(X_2 - X_1)^2$, in the limit $t_{31} \to 0$ one would conclude that to leading order the fluctuation transition matrices $C$, $D$, and $E$ would go like $|t_{21}|^{-1}$. The fluctuation transition matrix $F$ contains non-Markovian behavior. It goes like $\mathcal{O}(|t_{21}|^0)$ to leading order.

This matrix is zero in a Markovian system, $F = 0$. In this case the third entropy is just the sum of the consecutive second entropies,

$$S^{(3)}(X_3, X_2, X_1|t_{32}, t_{21}) = S^{(2)}(X_3, X_2|t_{32}) + S^{(2)}(X_2, X_1|t_{21}) - S(X_2). \tag{3.100}$$

The final term cancels the first entropy that is counted twice for the intermediate state.

### 3.2.9 Beyond fluctuations

The above analysis was for fluctuations in an isolated system, and for these the quadratic forms for the first and second entropy are valid. One can expect that the fluctuations in an extensive variable are relatively small. For heat flow below, in section 3.4.4 we shall treat the case of energy exchange with heat reservoirs. This raises the general possibility of how to treat the problem when the relevant variable $X$ is not relatively small.

Suppose that the reservoirs impose a non-zero most likely value, $\overline{X}(x_r)$. The fluctuation about this may be denoted $\delta X = X - \overline{X}(x_r)$. This fluctuation may be considered small, and the first and second entropy may be expanded to quadratic order in it. All the above theory now holds with $X \Rightarrow \delta X$. The first entropy fluctuation matrix $S''$ is evaluated at $X = \overline{X}(x_r)$ rather than at $X = 0$, and similarly for the second entropy fluctuation transition matrices $A(\tau)$ and $B(\tau)$. The time correlation function is define in terms of this fluctuation,

$$Q(\tau) = k_{\mathrm{B}}^{-1}\langle \delta X(\tau)\, \delta X(0)\rangle_{X_{\mathrm{r}}}. \tag{3.101}$$

## 3.3 Mixed parity fluctuations

### 3.3.1 Second entropy

Up to now, fluctuations of pure time parity have been treated, which is to say either $X^\dagger = X$, or else $X^\dagger = -X$. (Recall that the dagger denotes the macrostate with the same molecular positions but with the molecular velocities reversed.) Now our attention turns to the mixed parity case, where the components of $X$ each have pure parity, but they do not all necessarily have the same parity. This is not a serious restriction because any mixed parity macrostate can be decomposed into its pure parity components, $X_\alpha^\pm = [X_\alpha \pm X_\alpha^\dagger]/2$, and these individually can be included in $X$.

The diagonal parity matrix $\varepsilon$ is defined with elements $\varepsilon_{ij} = \pm\delta_{ij}$ such that

$$X^\dagger = \varepsilon X, \tag{3.102}$$

Obviously $\varepsilon^2 = \mathrm{I}$, and $X_i^\dagger = \varepsilon_{ii} X_i$.

In general the time correlation function is the equilibrium average of the dyadic matrix of the fluctuations, $Q(\tau) \equiv k_{\mathrm{B}}^{-1}\langle X(\tau)X(0)\rangle$. The time homogeneity of an equilibrium system holds as well for mixed parity systems as it does for pure parity systems,

$$\langle X(\tau)X(0)\rangle = \langle X(0)X(-\tau)\rangle, \text{ or } Q(\tau) = Q(-\tau)^{\mathrm{T}}. \tag{3.103}$$

However, macroscopic reversibility in the mixed parity case reads,

$$\langle X(\tau)X(0)\rangle = \langle X(-\tau)^\dagger X(0)^\dagger\rangle, \text{ or } Q(\tau) = \varepsilon Q(\tau)^{\mathrm{T}}\varepsilon. \tag{3.104}$$

The first entropy in fluctuation approximation is unchanged,

$$S(X) = \frac{1}{2}S'' : XX, \tag{3.105}$$

again with the fluctuation matrix related to the time correlation matrix,

$$S'' \equiv \left.\frac{\partial^2 S(X)}{\partial X \partial X}\right|_{X=0} = -k_{\mathrm{B}}\langle XX\rangle^{-1} = -Q(0)^{-1}. \tag{3.106}$$

Macroscopic reversibility at $\tau = 0$ reads $Q(0) = \varepsilon Q(0)\varepsilon$, which means that

$$\varepsilon S''\varepsilon = S'', \text{ or } \varepsilon S'' = S''\varepsilon. \tag{3.107}$$

Macroscopic reversibility at $\tau = 0$, means the coupling between variables of different time parity instantaneously vanishes,

$$\left\langle X_i(0)X_j(0)\right\rangle = 0, \text{ if } \varepsilon_{ii} \neq \varepsilon_{jj}. \tag{3.108}$$

From this it follows that $\{S''^{-1}\}_{ij} = 0$ if $\varepsilon_{ii}\varepsilon_{jj} = -1$. This result says that if all variables of the same parity are grouped together then the time correlation matrix is 'block' diagonal at $\tau = 0$, as is the fluctuation matrix. This result does not imply that variables of different parity are uncorrelated for all times.

The quadratic form for the second entropy in this mixed parity case can most generally be written as

$$S^{(2)}(X', X|\tau) = \frac{1}{2}X \cdot A(\tau)X + X \cdot B(\tau)X' + \frac{1}{2}X' \cdot C(\tau)X'. \tag{3.109}$$

Here $X \equiv X(0)$ is the current state, and $X' \equiv X(\tau)$ is the destination state. The matrices $A(\tau)$ and $C(\tau)$ are second derivative matrices of the same variable,

$$A(\tau) \equiv \left.\frac{\partial^2 S^{(2)}(0, X|\tau)}{\partial X \partial X}\right|_{X=0}, \quad \text{and} \quad C(\tau) \equiv \left.\frac{\partial^2 S^{(2)}(X', 0|\tau)}{\partial X' \partial X'}\right|_{X'=0}. \tag{3.110}$$

These are clearly symmetric, $A(\tau) = A(\tau)^{\mathrm{T}}$ and $C(\tau) = C(\tau)^{\mathrm{T}}$. The matrix $B(\tau)$ is the cross second derivative,

$$B(\tau) \equiv \left.\frac{\partial^2 S^{(2)}(X', X|\tau)}{\partial X \partial X'}\right|_{X'=X=0}. \tag{3.111}$$

The second entropy must be negative $S^{(2)}(X', X|\tau) \leqslant S^{(2)}(0, 0|\tau) = 0$. This follows because fluctuations are unfavorable, and constant contributions have been set to zero. This means that the matrices $A(\tau)$ and $C(\tau)$ must be negative definite, and the double matrix formed from the three coefficient matrices must also be negative definite.

The second entropy has similar symmetries to the time correlation function. From time homogeneity, $S^{(2)}(X', X|\tau) = S^{(2)}(X, X'|-\tau)$, it follows that

$$A(-\tau) = C(\tau), \quad \text{and} \quad B(\tau) = B(-\tau)^{\mathrm{T}}. \tag{3.112}$$

Macroscopic reversibility applied to the second entropy reads $S^{(2)}(X', X|\tau) = S^{(2)}(X'^{\dagger}, X^{\dagger}|-\tau) = S^{(2)}(\varepsilon X', \varepsilon X|-\tau)$. Hence the fluctuation transition matrices have the symmetries

$$\varepsilon A(\tau)\varepsilon = C(\tau), \quad \text{and} \quad \varepsilon B(\tau)\varepsilon = B(\tau)^{\mathrm{T}}. \tag{3.113}$$

The most likely destination state given the present state maximizes the second entropy over $X'$. This gives

$$\overline{X'} \equiv \overline{X}(\tau|X) = -C(\tau)^{-1}B(\tau)^{\mathrm{T}}X. \tag{3.114}$$

Similarly, the most likely initial state that would lead to the destination $X'$ maximizes the second entropy over $X$,

$$\overline{X} \equiv \overline{X}(-\tau|X') = -A(\tau)^{-1}B(\tau)X'. \tag{3.115}$$

This can also be written $\overline{X}(-\tau|X') = -C(-\tau)^{-1}B(-\tau)^{\mathrm{T}}X'$.

It is useful to rewrite the second entropy in terms of the departure from the most likely destination state,

$$\begin{aligned} S^{(2)}(X', X|\tau) = &\frac{1}{2}[X' + C(\tau)^{-1}B(\tau)^{\mathrm{T}}X]^{\mathrm{T}}C(\tau)[X' + C(\tau)^{-1}B(\tau)^{\mathrm{T}}X] \\ &+ \frac{1}{2}X^{\mathrm{T}}[A(\tau) - B(\tau)C(\tau)^{-1}B(\tau)^{\mathrm{T}}]X. \end{aligned} \tag{3.116}$$

Alternatively, it can be written in terms of the departure from the most likely initial state,

$$\begin{aligned} S^{(2)}(X', X|\tau) = &\frac{1}{2}[X + A(\tau)^{-1}B(\tau)X']^{\mathrm{T}}A(\tau)[X + A(\tau)^{-1}B(\tau)X'] \\ &+ \frac{1}{2}X'^{\mathrm{T}}[C(\tau) - B(\tau)^{\mathrm{T}}A(\tau)^{-1}B(\tau)]X'. \end{aligned} \tag{3.117}$$

The reduction condition, equation (3.38), for the present mixed parity case reads

$$S^{(2)}(\overline{X}', X|\tau) = S(X), \text{ or } S^{(2)}(X', \overline{X}|\tau) = S(X'). \tag{3.118}$$

It follows from the preceding expressions that

$$A(\tau) - B(\tau)C(\tau)^{-1}B(\tau)^{\mathrm{T}} = S'', \tag{3.119}$$

and

$$C(\tau) - B(\tau)^{\mathrm{T}}A(\tau)^{-1}B(\tau) = S''. \tag{3.120}$$

These two expressions are equivalent to each other, since one can be turned into the other by multiplying before and after by the parity matrix, and applying macroscopic reversibility, equation (3.113). This result of the reduction condition may be compared to the result in the pure parity case, equation (3.60), in which case $A(\tau) = C(\tau)$ and $B(\tau) = B(\tau)^{\mathrm{T}}$.

The transition fluctuation matrices can be related to the time correlation matrix by invoking the fact that for Gaussian statistics, the average value and the most likely value are the same. Hence the time correlation function is given by

$$\begin{aligned} Q(\tau) &= k_{\mathrm{B}}^{-1}\langle X'X\rangle \\ &= k_{\mathrm{B}}^{-1}\langle \overline{X}'X\rangle \\ &= -\, C(\tau)^{-1}B(\tau)^{\mathrm{T}}k_{\mathrm{B}}^{-1}\langle XX\rangle \\ &= C(\tau)^{-1}B(\tau)^{\mathrm{T}}S''^{-1}. \end{aligned} \tag{3.121}$$

Similarly, $Q(\tau) = S''^{-1}B(\tau)^{\mathrm{T}}A(\tau)^{-1}$.

Using these and the reduction condition, equation (3.119), the transition fluctuation matrices can each be expressed of the time correlation matrix and the entropy fluctuation matrix. One can readily show that

$$A(\tau) = [\mathrm{I} - S''Q(\tau)^{\mathrm{T}}S''Q(\tau)]^{-1}S''. \tag{3.122}$$

This reduces to the pure parity result, equation (3.67), when the time correlation function matrix is symmetric.

Inserting this into the second form for the time correlation function, one obtains for the cross fluctuation matrix

$$B(\tau) = [\mathrm{I} - S''Q(\tau)^{\mathrm{T}}S''Q(\tau)]^{-1}S''Q(\tau)^{\mathrm{T}}S'' = A(\tau)Q(\tau)^{\mathrm{T}}S''. \tag{3.123}$$

As in the pure parity case, the time correlation function completely determines the fluctuation transition matrices, the second entropy, and the transition probability. In particular, the most likely destination, equation (3.114), may be written

$$\overline{X}(\tau|X) = -Q(\tau)S''X. \tag{3.124}$$

This is formally unchanged from the result in the case of pure parity, equation (3.64) (although of course $Q(\tau)$ is not now symmetric). The term $S''X = S'(X) \equiv \nabla S(X)$ remains the thermodynamic driving force toward the equilibrium state. Again, as in the pure parity case, multiplying by $X^{\mathrm{T}}$ on the right, the subsequent average yields an identity, since $\langle[\nabla S(X)]X\rangle = -k_{\mathrm{B}}\mathrm{I}$.

### 3.3.2 Small time expansion

The difference between the small time expansion for the previous pure parity systems and the present mixed parity system, is that the time correlation function is no longer an even function of time. This means that now both $\tau$ and $|\tau|$ appear in the expansion. Because of this, we shall need the sign of the time interval, $\hat{\tau} \equiv \operatorname{sign} \tau \equiv \tau/|\tau|$.

The pure parity expansions (3.79) and (3.80) carry over directly to the present mixed parity case by including additional terms,

$$A(\tau) = C(-\tau) = \frac{A_{-1}}{|\tau|} + \frac{A'_{-1}}{\tau} + A_0 + A_0'\hat{\tau} + \mathcal{O}(\tau), \tag{3.125}$$

and

$$B(\tau) = \frac{B_{-1}}{|\tau|} + \frac{B'_{-1}}{\tau} + B_0 + B_0'\hat{\tau} + \mathcal{O}(\tau). \tag{3.126}$$

Since $A(\tau) = A(\tau)^{\mathrm{T}}$, all the coefficient matrices that appear in its expansion are symmetric. Since $B(-\tau) = B(\tau)^{\mathrm{T}}$, in its expansion the unprimed coefficient matrices are symmetric, and the primed coefficient matrices are antisymmetric.

As in the pure parity case, equation (3.70), since $\overline{X}(\tau|X) \to X$, $\tau \to 0$ still holds, one must still have that $A(\tau)^{-1}B(\tau) \to -\mathrm{I}$, $\tau \to 0$. Hence the leading-order coefficients satisfy

$$B_{-1} + \hat{\tau}B'_{-1} = -A_{-1} - \hat{\tau}A'_{-1}. \tag{3.127}$$

Now $A'_{-1}$ is symmetric whereas $B'_{-1}$ is antisymmetric. Since the two parts of this equation must separately vanish, it follows that $A'_{-1} = B'_{-1} = 0$. With this the expansions become

$$A(\tau) = \frac{-\Lambda^{-1}}{|\tau|} + A_0 + A_0'\hat{\tau} + \mathcal{O}(\tau), \tag{3.128}$$

and

$$B(\tau) = \frac{\Lambda^{-1}}{|\tau|} + B_0 + B_0'\hat{\tau} + \mathcal{O}(\tau). \tag{3.129}$$

As in the pure parity case, because $A(\tau)$ is symmetric and negative definite, the transport matrix must be symmetric and positive definite.

The two remaining symmetric coefficient matrices $A_0$ and $B_0$, can be related to each other by invoking the reduction condition. Expanding this to zeroth order in the time interval one has

$$\begin{aligned}
S'' &= A(\tau) - B(\tau)C(\tau)^{-1}B(\tau)^{\mathrm{T}} \\
&= [-\Lambda^{-1}|\tau|^{-1} + A_0 + A_0'\hat{\tau}] - [\Lambda^{-1}|\tau|^{-1} + B_0 + B_0'\hat{\tau}] \\
&\quad \times [-\Lambda^{-1}|\tau|^{-1} + A_0 - A_0'\hat{\tau}]^{-1}[\Lambda^{-1}|\tau|^{-1} + B_0 - B_0'\hat{\tau}] \\
&= [-\Lambda^{-1}|\tau|^{-1} + A_0 + A_0'\hat{\tau}] - [\Lambda^{-1}|\tau|^{-1} + B_0 + B_0'\hat{\tau}] \\
&\quad \times [-\mathrm{I} + |\tau|\Lambda A_0 - \tau\Lambda A_0']^{-1}[\mathrm{I} + |\tau|\Lambda B_0 - \tau\Lambda B_0'] \\
&= [-\Lambda^{-1}|\tau|^{-1} + A_0 + A_0'\hat{\tau}] - [\Lambda^{-1}|\tau|^{-1} + B_0 + B_0'\hat{\tau}] \\
&\quad \times [-\mathrm{I} + |\tau|\Lambda(A_0 + B_0) - \tau\Lambda(A_0' + B_0') + \mathcal{O}(\tau^2)] \\
&= [-\Lambda^{-1}|\tau|^{-1} + A_0 + A_0'\hat{\tau}] + [\Lambda^{-1}|\tau|^{-1} + A_0 + 2B_0 - A_0'\hat{\tau} + \mathcal{O}(\tau)] \\
&= 2(A_0 + B_0) + \mathcal{O}(\tau).
\end{aligned} \tag{3.130}$$

The time correlation function may also be expanded for small times,

$$\begin{aligned}
Q(\tau) &= C(\tau)^{-1}B(\tau)^{\mathrm{T}}S''^{-1} \\
&= [-\Lambda^{-1}|\tau|^{-1} + A_0 - A_0'\hat{\tau}]^{-1}[\Lambda^{-1}|\tau|^{-1} + B_0 - B_0'\hat{\tau}]S''^{-1} \\
&= [\mathrm{I} - |\tau|\Lambda A_0 + \tau\Lambda A_0']^{-1}[-\mathrm{I} - |\tau|\Lambda B_0 + \tau\Lambda B_0']S''^{-1} \\
&= [-\mathrm{I} - |\tau|\Lambda(A_0 + B_0) + \tau\Lambda(A_0' + B_0')]S''^{-1} \\
&= -S''^{-1} - \frac{|\tau|}{2}\Lambda + \tau\Theta + \mathcal{O}(\tau^2).
\end{aligned} \tag{3.131}$$

The final equality defines $\Theta$, the coefficient matrix of $\tau$. Since $Q(-\tau) = Q(\tau)^{\mathrm{T}}$, this must be antisymmetric, which yields

$$\Theta \equiv \Lambda(A_0' + B_0')S''^{-1} = -S''^{-1}(A_0' - B_0')\Lambda. \tag{3.132}$$

The fluctuation transition coefficient matrices that couple to $\hat{\tau}$ can be written in terms of the antisymmetric transport matrix,

$$A_0' = \frac{1}{2}[\Lambda^{-1}\Theta S'' - S''\Theta\Lambda^{-1}], \quad \text{and} \quad B_0' = \frac{1}{2}[\Lambda^{-1}\Theta S'' + S''\Theta\Lambda^{-1}]. \tag{3.133}$$

In view of equation (3.131), one can *define* 'the' transport matrix in the mixed parity case as $L(\hat{\tau}) \equiv \Lambda - 2\hat{\tau}\Theta$. This is defined in this way to retain the same proportionality between flux and thermodynamic force as in the pure parity case, as will shortly be made explicit. The transport matrix $L(\hat{\tau})$ consists of a symmetric and an antisymmetric matrix. These give an irreversible and a reversible contribution to the trajectory, respectively. The anti-symmetric matrix changes sign if the transport matrix is used to predict the past rather than the more usual case of predicting the future. This is a real physical effect.

The time correlation matrix has the symmetry $Q(-\tau) = \varepsilon Q(\tau)\varepsilon$, or, in component form, $Q_{ij}(-\tau) = \varepsilon_{ii}\varepsilon_{jj}Q_{ij}(\tau)$. From the small time expansion, equation (3.131), it follows that the symmetric part of the transport matrix has elements that vanish for states of different parity,

$$\Lambda_{ij} = 0 \text{ if } \varepsilon_{ii}\varepsilon_{jj} = -1. \tag{3.134}$$

We have already seen the equivalent result that $\{S''^{-1}\}_{ij} = 0$ if $\varepsilon_{ii}\varepsilon_{jj} = -1$. It also follows that the antisymmetric part of the transport matrix has elements that vanish for states of the same parity,

$$\Theta_{ij} = 0 \text{ if } \varepsilon_{ii}\varepsilon_{jj} = 1. \tag{3.135}$$

In view of this behavior it is often convenient to group the components of $X$ of the same parity together. In such a representation the various coefficient matrices subdivide into four square blocks, with those linking components of like parity on the main diagonal, and those linking unlike parity components on the off-diagonal. One sees that in this representation, the matrices $\Lambda$, $S''^{-1}$ (and hence $S''$) are symmetric and block diagonal, and the matrix $\Theta$ is antisymmetric and block adiagonal.

Inserting the above expansions and definitions into equation (3.124), one obtains the optimum flux as

$$\overline{\overset{\circ}{X}} = \frac{\hat{\tau}}{2}[\Lambda - 2\hat{\tau}\Theta]S''X = \frac{\hat{\tau}}{2}L(\hat{\tau})S'(X). \tag{3.136}$$

As in the pure parity case, the most likely flux is linearly proportional to the thermodynamic driving force. The term in brackets is the origin of the definition of 'the' transport matrix used above. In the present case there are both reversible and irreversible contributions to the most likely flux, whereas in the pure parity case the contributions were only irreversible.

The second entropy may be rewritten in terms of the flux. Invoking the departure of $X'$ from its most likely value and the reduction condition one has

$$
\begin{aligned}
&S^{(2)}(X', X|\tau)\\
&= S(X) + \frac{1}{2}C(\tau) : [X' - \overline{X}']^2\\
&= S(X) + \frac{1}{2}\left[\frac{-\Lambda^{-1}}{|\tau|} + \mathcal{O}(\tau^0)\right] : \left[X' - X - \frac{|\tau|}{2}L(\hat{\tau})S''X + \mathcal{O}(\tau^2)\right]^2\\
&= S(X) - \frac{|\tau|}{2}\Lambda^{-1} : \left[\overset{\circ}{X} - \frac{\hat{\tau}}{2}L(\hat{\tau})S''X\right]^2 + \mathcal{O}(\tau^2)\\
&= S(X) - \frac{|\tau|}{2}\Lambda^{-1} : \overset{\circ}{X}{}^2 + \frac{\tau}{2}\overset{\circ}{X} \cdot S''X - |\tau|\overset{\circ}{X} \cdot \Lambda^{-1}\Theta S''X\\
&\quad - \frac{|\tau|}{2}\Lambda^{-1} : [L(\hat{\tau})S''X]^2 + \mathcal{O}(\tau^2).
\end{aligned}
\tag{3.137}
$$

As in the pure parity case, section 3.2.7, there is a term quadratic in the flux, which is negative and therefore unfavorable. There are now two terms linear in the flux, which also couple to the thermodynamic gradient, again as in the pure parity case. The second of these terms includes the anti-symmetric transport matrix $\Theta$, which does not appear in the pure parity case.

The antisymmetric part of the transport matrix $\Theta$ gives a reversible contribution to the most likely evolution of the fluctuation. It will be recalled from the pure parity case that the symmetric transport matrix $\Lambda$ gave an irreversible contribution. For the present mixed parity case one has,

$$
\begin{aligned}
\overline{X}(\tau|X) - \overline{X}(-\tau|X) &= -[Q(\tau) - Q(-\tau)]S''X\\
&= -2\tau\Theta S''X + \mathcal{O}(\tau^2).
\end{aligned}
\tag{3.138}
$$

This would be zero for the pure parity case. This is analogous to the result that one would obtain for an ordinary mechanical evolution such as that given by Hamilton's equations, which is why it is called a reversible contribution.

It should be noted that although $\Theta$ affects the most likely evolution of the fluctuation in the present mixed parity case, it has no effect on the optimum rate of entropy production. This follows from the symmetric nature of the entropy fluctuation matrix, since one has

$$
\begin{aligned}
\overline{\overset{\circ}{S}}(X) &= \frac{1}{\tau}[\overline{X}(\tau|X) - X] \cdot S''X\\
&= X \cdot \left[\frac{\hat{\tau}}{2}S''\Lambda + S''\Theta\right]S''X\\
&= \frac{\hat{\tau}}{2}X \cdot S''\Lambda S''X + \mathcal{O}(\tau).
\end{aligned}
\tag{3.139}
$$

The contribution from $\Theta$ vanishes because it is an antisymmetric matrix, and the double scalar product is symmetric. Hence the optimum dissipation is formally the same as for a pure parity system, equation (3.97). One might have predicted this because by the Second Law of Thermodynamics the entropy production must be irreversible, which is to say it must be proportional to the sign of the time interval $\hat{\tau}$, as this result shows.

## 3.4 Steady heat flow

The canonical non-equilibrium system is that of steady heat flow. This is now analyzed in detail. Analogous methods can be applied to other types of thermodynamic fluxes (driven diffusion, volume changes, electrical currents, etc). The remaining generic type not treated here is a mechanical non-equilibrium system, which typically has an applied potential that varies with time. Both thermodynamic and mechanical non-equilibrium systems will be treated at the level of statistical mechanics in chapter 6.

### 3.4.1 First energy and first temperature

*Canonical equilibrium system*

Heat is a form of energy, and so the latter is the focus of this chapter. It will be recalled from chapter 2 that the derivative of the isolated system entropy with respect to energy is essentially the temperature,

$$\frac{1}{T} = \frac{\partial S}{\partial E}. \tag{3.140}$$

Because the number and volume of the sub-system will be held constant throughout, these will not be shown and we can simply write $S(E)$.

The temperature is an intensive variable and as such it is also given by the energy density derivative of the entropy density, equation (2.14),

$$\frac{1}{T} = \frac{\partial \sigma}{\partial \varepsilon}, \tag{3.141}$$

where $\sigma \equiv S/V$ and $\varepsilon = E/V$.

The canonical equilibrium system consists of a sub-system able to exchange energy with a reservoir of temperature $T$. The constrained total entropy is

$$S_{\text{total}}(E|T) = S(E) - \frac{E}{T}. \tag{3.142}$$

This is maximized by the equilibrium energy $\overline{E}(T)$,

$$\left.\frac{\partial S_{\text{total}}(E|T)}{\partial E}\right|_{\overline{E}} = 0 \Leftrightarrow T_s(\overline{E}) = T, \tag{3.143}$$

at which point the temperature of the sub-system equals that of the reservoir. The constrained thermodynamic potential is minus the temperature times the constrained total entropy,

$$F(E|T) \equiv -TS_{\text{total}}(E|T) = E - TS(E). \tag{3.144}$$

The Helmholtz free energy is the minimum value of this,

$$\overline{F}(T) = F(\overline{E}(T)|T) = \overline{E}(T) - TS(\overline{E}(T)). \tag{3.145}$$

*Canonical non-equilibrium system*

The canonical non-equilibrium system consists of a sub-system sandwiched between, and in thermal contact with, two heat reservoirs of different temperatures (figure 3.3). The heat flux $J$, which is uniform in the optimum non-equilibrium state, is the rate of energy change of a reservoir per unit cross-sectional area.

The treatment of the canonical non-equilibrium system first requires a detailed analysis of the static and dynamic properties of the sub-system in isolation. This is where we begin, and reservoirs will not be introduced until section 3.4.4.

For the analysis of equilibrium thermodynamics, chapter 2, the thermodynamic limit was invoked. This says that the reservoir is infinitely larger than the sub-system, and that the boundary region is infinitely smaller than the sub-system. The first of these ensures that exchange with the sub-system does not change the field variables of the reservoir, and the second of these ensures that the precise boundary conditions have no effect on the properties of the sub-system, so that the latter may be analyzed as isolated.

For the analysis of non-equilibrium thermodynamics, the same thermodynamic limit has to be invoked for both reservoirs and both boundaries. For this reason the boundary conditions parallel to the flux need not be specified as they have negligible effect on the sub-system. An additional condition is imposed, namely that the gradient in the reservoir field variables, essentially their difference divided by the width of the sub-system, is vanishingly small. This means that all quantities can be expanded in the gradient and all terms beyond the first order can be neglected.

It is also necessary to insist that the conductivity of the reservoirs be much greater than that of the sub-system. This means that the temperature within each reservoir is equal to that at its boundary with the sub-system. Or at least the width of any region

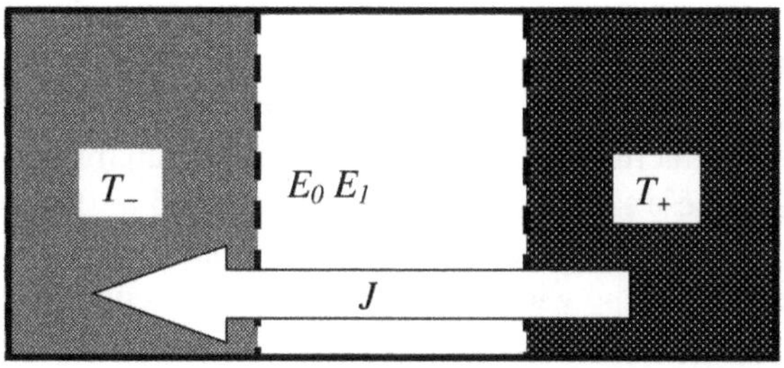


**Figure 3.3.** The canonical non-equilibrium system consists of two heat reservoirs of temperatures $T_+ > T_-$, in thermal contact with a sub-system with zeroth $E_0$ and first $E_1$ energy moments. The uniform heat flux is $J$.

of temperature non-uniformity in the reservoirs is small compared to the width of the sub-system.

The reservoirs impose a temperature gradient on the sub-system. Because of the local properties of the entropy density, there will be a corresponding gradient in energy density. One can specify the energy density $\varepsilon(\mathbf{r})$ as a macrostate of the constrained system. Choosing the $z$-axis to be aligned with the gradient, one need only consider spatial variations in this direction, $\varepsilon(z)$.

The energy density may be expanded to first order (since the gradient is vanishingly small, all higher order contributions can be neglected),

$$\varepsilon(z) = \varepsilon_0 + \varepsilon_1 z. \tag{3.146}$$

The macrostate is now specified by two parameters, $\{\varepsilon_0, \varepsilon_1\}$.

For a sub-system of width $L$ and cross-sectional area $A$, the total energy is just the zeroth energy moment,

$$E_0 = A\int_{-L/2}^{L/2} \mathrm{d}z\ \varepsilon(z) = AL\varepsilon_0. \tag{3.147}$$

The first energy moment is

$$E_1 = A\int_{-L/2}^{L/2} \mathrm{d}z\ \varepsilon(z)z = \frac{AL^3}{12}\varepsilon_1. \tag{3.148}$$

With these the macrostate of the isolated sub-system can be specified in a global sense by the zeroth and first energy moments and one can write

$$S(E_0, E_1) = A\int_{-L/2}^{L/2} \mathrm{d}z\ \sigma(\varepsilon(z)). \tag{3.149}$$

Note that it does not matter whether or not the optimum number density is spatially varying as this is a second order effect that can be neglected.

It is important to note that both of these energy moments are extensive with the cross-sectional area of the sub-system, $A$. In chapter 2 where the theory for equilibrium thermodynamics was developed, it was pointed out that extensivity played a key role, so that entropy was best formulated in terms of extensive variables. An explicit benefit of this is that in the thermodynamic limit the Taylor expansion of the reservoir entropy could be terminated at the linear term. A second benefit was that the fluctuations in the extensive exchangeable variables were relatively negligible. These same benefits can be exploited in the present formulation of non-equilibrium thermodynamic systems. In this case the thermodynamic limit of relevance is $A \to \infty$. Since the width $L$ is fixed, this also corresponds to the volume of the sub-system becoming infinite.

The local reciprocal temperature of the sub-system may also be expanded to linear order

$$\begin{aligned}\frac{\partial\sigma(\varepsilon(z))}{\partial\varepsilon(z)} &= \frac{1}{T(z)}\\ &= \frac{1}{T(0)} + z\frac{\mathrm{d}T(z)^{-1}}{\mathrm{d}z}\bigg|_{z=0} + \mathcal{O}(T'', (T')^2)\\ &\equiv \frac{1}{T_0} + z\frac{1}{T_1}.\end{aligned} \tag{3.150}$$

This defines the zeroth temperature $T_0$ as the temperature at the mid-plane, and the first temperature $T_1$ as the reciprocal of the gradient of the reciprocal temperature at the mid-plane, $T_1^{-1} = -T^{-2}\nabla T$. These are properties of the sub-system, which will sometimes be emphasized with the subscript s.

A change in the energy density is $\Delta\varepsilon(z) = \Delta\varepsilon_0 + z\Delta\varepsilon_1$. This changes the isolated sub-system entropy in the specified macrostate by

$$\begin{aligned}\Delta S(E_0, E_1) &= A\int_{-L/2}^{L/2} \mathrm{d}z\frac{\partial\sigma(\varepsilon(z))}{\partial\varepsilon(z)}\Delta\varepsilon(z)\\ &= A\int_{-L/2}^{L/2} \mathrm{d}z\left[\frac{1}{T_0} + z\frac{1}{T_1}\right][\Delta\varepsilon_0 + z\Delta\varepsilon_1]\\ &= AL\frac{\Delta\varepsilon_0}{T_0} + \frac{AL^3}{12}\frac{\Delta\varepsilon_1}{T_1}\\ &= \frac{\Delta E_0}{T_0} + \frac{\Delta E_1}{T_1}.\end{aligned} \tag{3.151}$$

Hence one can conclude that the zeroth temperature is the thermodynamic conjugate of the zeroth energy moment,

$$\frac{\partial S(E_0, E_1)}{\partial E_0} = \frac{1}{T_0}, \tag{3.152}$$

and that the first temperature is the conjugate of the first energy moment,

$$\frac{\partial S(E_0, E_1)}{\partial E_1} = \frac{1}{T_1}. \tag{3.153}$$

This last one says in brief that the first energy moment is conjugate to the temperature gradient, a result that will play a key role in the analysis of heat flow. These give the temperatures as functions of the moments, $T_0(E_0, E_1)$ and $T_1(E_0, E_1)$. We shall assume without proof that these relationships are invertible, $E_0(T_0, T_1)$ and $E_1(T_0, T_1)$.

It is somewhat idiosyncratic to analyze heat flow in terms of the zeroth and first temperatures. The relationship with conventional results is that the zeroth temperature corresponds to the temperature, $T_0 \Rightarrow T$, and the first temperature to the temperature gradient, $T_1^{-1} \Rightarrow -T^{-2}\nabla T$. Like the present treatment, the conventional treatment assumes that the temperature gradient is small so that terms beyond first order can be neglected.

### 3.4.2 Second entropy

The second entropy formalism, section 3.1.4, was based on extensive non-conserved variables of the isolated system that most likely vanished, $\overline{X} = 0$. For the present case of heat flow, the corresponding variable is the first energy moment, $E_1$, since this is not conserved, and it is extensive with the system cross-sectional area $A$. Further, it is most likely zero, since the entropy is a maximum when the energy is uniformly distributed throughout the system. As we have just seen, the field variable conjugate to the first energy moment is the first temperature, $T_1$. From now on the zeroth energy moment, $E_0$, and also the zeroth temperature, $T_0$, will be suppressed and not treated explicitly.

For heat flow we can fix the direction (e.g. parallel to the $z$-axis), and take $E_1$ to be a scalar. This means that $X$ is a scalar because it is the only fluctuating variable that we shall consider. Hence all of the matrices that were defined in the general first and second entropy analysis, section 3.1.4, become scalars here.

Accordingly, the part of the first entropy of the isolated system relevant to heat flow depends upon the first energy moment, with $\overline{E_1} = 0$ being the macrostate of maximum entropy. Invoking the fluctuation approximation the first entropy is

$$S(E_1) = \frac{1}{2} S'' E_1^2. \tag{3.154}$$

As just mentioned, the fluctuation coefficient is a scalar,

$$S'' \equiv \left. \frac{\partial^2 S(E_1)}{\partial E_1^2} \right|_{E_1=0} = \frac{-k_B}{\langle E_1(0)^2 \rangle}. \tag{3.155}$$

The averages here and below are canonical equilibrium ones. By the uniqueness of the thermodynamic state, these are the same as in an equilibrium isolated system at the energy $E_0 = \overline{E_0}(T_0)$.

The thermodynamic force is the inverse first temperature,

$$\frac{1}{T_1} \equiv \frac{\partial S(E_1)}{\partial E_1} = S'' E_1. \tag{3.156}$$

One can also write the thermodynamic force as $\nabla S(E_1) = S'(E_1) = S'' E_1$.

For the transition $E_1 \xrightarrow{\tau} E_1'$, the second entropy is

$$S^{(2)}(E_1', E_1|\tau) = \frac{A(\tau)}{2}[E_1'^2 + E_1^2] + B(\tau) E_1' E_1. \tag{3.157}$$

(Obviously as a scalar $X$ has pure parity, in this case it is even.) From the expansions of the fluctuation transition coefficients for short times, equations (3.79) and (3.80), in the present scalar, even parity case one has

$$A(\tau) = \frac{-1}{|\tau|\Lambda} + A + \mathcal{O}(\tau), \quad \text{and} \quad B(\tau) = \frac{1}{|\tau|\Lambda} + B + \mathcal{O}(\tau), \tag{3.158}$$

with $A + B = S''/2$ from the reduction condition, equation (3.81). Inserting the expansion into the most likely destination gives

$$\begin{aligned}\overline{E_1'} &= \frac{-B(\tau)}{A(\tau)} E_1 \\ &= E_1 + \frac{|\tau|\Lambda}{2} S'' E_1 + \mathcal{O}(\tau^2).\end{aligned} \tag{3.159}$$

Writing the second entropy in terms of the departure from this most likely destination, in the small time limit one has

$$\begin{aligned}S^{(2)}(E_1', E_1|\tau) &= \frac{1}{2} S'' E_1^2 - \frac{1}{2\Lambda|\tau|}\left[E_1' - E_1 - \frac{|\tau|\Lambda}{2} S'' E_1\right]^2 + \mathcal{O}(\tau^2) \\ &= \frac{1}{2} S'' E_1^2 - \frac{|\tau|}{2\Lambda}\left[\mathring{E}_1 - \frac{\hat{\tau}\Lambda}{2} S'' E_1\right]^2 \\ &= \frac{1}{2} S'' E_1^2 - \frac{|\tau|}{2\Lambda}\left(\mathring{E}_1\right)^2 + \frac{\tau}{2}\mathring{E}_1 S'' E_1 - \frac{|\tau|\Lambda}{8}(S'' E_1)^2,\end{aligned} \tag{3.160}$$

where $\hat{\tau} = \text{sign}(\tau)$. This is the particular form of equation (3.95) for heat flow. The first term on the right-hand side is the first entropy, and so this satisfies the reduction condition for the most likely destination.

The coarse velocity, $\mathring{E}_1 \equiv [E_1' - E_1]/\tau$, which is the average rate of change over the time interval, has most likely value $\hat{\tau}\Lambda S'' E_1/2$. The rate of change of the first energy moment is directly related to the energy flux in the isolated system. Because the system is isolated, the dynamics are adiabatic, which fact is signified by appending the superscript 0.

The energy flux, $J_E^0$, is defined as the amount of energy crossing a plane, per unit area, per unit time. Because we are presently considering the energy flux alone, the number flux is necessarily assumed zero, and so this corresponds to the conductive energy flux.

To be definite, we assume that the flux is in the $z$-direction. We also assume that it is uniform, $J_E^0(\mathbf{r}) = J_E^0$, which can be expected to be the average or most likely flux. This means that the only change in energy density occurs at the $z$-boundaries of the isolated system; the same amount of energy per unit time flows across each and every plane perpendicular to the $z$-axis throughout the system. Because we take the thermodynamic limit, we need not be overly concerned with the nature of the boundary regions beyond quantifying the rate of energy change in them.

Such a uniform flux removes energy at a rate $AJ_E^0$ from the boundary region at $z = -L/2$, and it adds energy at the same rate to the boundary region at $z = L/2$. The rate of change of the first energy moment due to this is

$$\mathring{E}_1 = \frac{-L}{2}(-AJ_E^0) + \frac{L}{2}(AJ_E^0) = VJ_E^0. \tag{3.161}$$

Here $V = AL$ is the volume of the isolated sub-system.

We are interested in time intervals $\tau$ that are longer than molecular times, so that following a fluctuation the rate of change of energy moment can accelerate from zero to a steady value. We also want non-uniformities in the energy profile to decay away (or to average to zero), which can be expected since in general short wavelength inhomogeneities decay more quickly than long wave length ones, and the first moment is the non-uniformity with the longest wavelength. However, we also require the time interval to be sufficiently short so that the change in energy moment is relatively negligible, and so that the short time expansion can be terminated at the linear term. The data in figure 3.2 show the regression of the first energy moment in a particular system, and it can be seen that there exist a range of values for the time interval $\tau$ in which both requirements are satisfied.

Under these circumstances, we expect a steady decay of the first energy moment, in which case the coarse velocity and the instantaneous velocity are equal. Equating the most likely energy flux to the most likely coarse rate of change of energy moment per unit volume, one has

$$\begin{aligned} \overline{J}_{\mathrm{E}}^{0} &= \frac{\hat{\tau}\Lambda}{2V} S'' E_1 \\ &= \frac{\hat{\tau}\Lambda}{2V} \frac{1}{T_1} \\ &= \frac{-\hat{\tau}\Lambda}{2VT^2} \nabla T. \end{aligned} \tag{3.162}$$

Hydrodynamics almost invariably refers to the future state. In this case $\hat{\tau} = 1$, and this result says that the most likely energy flux is linearly proportional to the temperature gradient, and in the opposite direction (i.e. energy flows from hot to cold). This is Fourier's law, and one can identify the conventional thermal conductivity as $\lambda \equiv \Lambda/2VT^2$.

### 3.4.3 Green–Kubo thermal conductivity

Green–Kubo theory relates the hydrodynamic transport coefficients to an integral of the time correlation function. The general approach was derived in section 3.2.6 from the regression theorem. For the present problem of heat transport, the thermal conductivity is related to the transport coefficient by $\lambda \equiv \Lambda/2VT^2$. One can rearrange the most likely regression of the first energy moment, equation (3.159), in several ways to give $\Lambda$ and hence $\lambda$. If one multiplies by $E_1(t)$ and averages then one obtains

$$\begin{aligned} \lambda(\tau) &= \frac{-1}{Vk_{\mathrm{B}}T^2|\tau|} \langle [\overline{E}_1(t+\tau|E_1(t)) - E_1(t)]E_1(t)\rangle \\ &= \frac{-1}{Vk_{\mathrm{B}}T^2|\tau|} \langle [E_1(t+\tau) - E_1(t)]E_1(t)\rangle. \end{aligned} \tag{3.163}$$

Terms $\mathcal{O}(\tau)$ have been neglected here. In the second equality, the actual value replaces the most likely value, which is permissible for Gaussian statistics.

In principle, $\lambda(\tau)$ ought to be independent of the time interval. Multiplying both sides by $\tau$, differentiating with respect to $\tau$, and ignoring $\mathrm{d}\lambda(\tau)/\mathrm{d}\tau$, yields

$$\lambda(\tau) = \frac{-\hat{\tau}}{Vk_{\mathrm{B}}T^2}\langle \dot{E}_1(t+\tau)E_1(t)\rangle. \tag{3.164}$$

The averages that appear here are equilibrium averages. One can evaluate them for a canonical equilibrium system. The procedure is to generate points in phase space, $\mathbf{\Gamma}_0$, according to the Maxwell–Boltzmann distribution at the zeroth temperature, $\wp_{\mathrm{MB}}(\mathbf{\Gamma}_0|T_0)$. Then the adiabatic trajectory (i.e. Hamiltonian trajectory for an isolated system) is generated, $\mathbf{\Gamma}^0(\tau|\mathbf{\Gamma}_0)$. The first moments, $E_1(0) \equiv E_1(\mathbf{\Gamma}_0)$ and $E_1(\tau) \equiv E_1(\mathbf{\Gamma}^0(\tau|\mathbf{\Gamma}_0))$ are then calculated and multiplied together. This is then averaged over the initial points. Equivalently, the change in first moment $E_1(\tau) - E_1(0)$ can be calculated by integrating its rate of change $\dot{E}_1(t)$ on the adiabatic trajectory over the time interval. Specifically, equation (3.163) can be written

$$\lambda(\tau) = \frac{-1}{Vk_{\mathrm{B}}T^2|\tau|}\int \mathrm{d}\mathbf{\Gamma}_0\, \wp_{\mathrm{MB}}(\mathbf{\Gamma}_0|T_0)\int_0^{\tau} \mathrm{d}t\; \dot{E}_1(\mathbf{\Gamma}^0(t|\mathbf{\Gamma}_0))E_1(\mathbf{\Gamma}_0). \tag{3.165}$$

To the extent that this or equation (3.164) is independent of $\tau$, then the short time regime can be said to exist and the theory is applicable.

The two formulae above are tested in a simulation of a Lennard-Jones fluid in figure 3.4. It is evident that $\lambda(\tau)$ converges more quickly to the steady state value $\lambda$ using the terminal velocity of the first energy moment, equation (3.164), than it does using the coarse velocity, equation (3.163). This makes sense because the latter includes the conductivity of the inertial phase, which is low or zero, and the relative

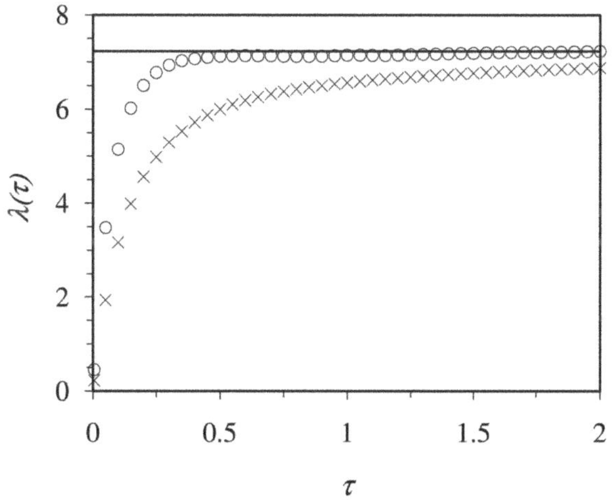


**Figure 3.4.** Thermal conductivity from Green–Kubo expressions for a Lennard-Jones fluid at $\rho = 0.8$ and $T_0 = 2$ (dimensionless units). The simulations use the coarse velocity, equation (3.163) (crosses), or the instantaneous velocity, equation (3.164) (circles). The solid line is a guide to the eye. Data from Attard (2005b).

contribution of this to the overall conductivity decays as $\tau^{-1}$, which is quite slow. The instantaneous terminal velocity, after sufficient time, gives the conductivity of the system in the steady state.

### 3.4.4 Reservoirs of different temperatures

The analysis of the static and dynamic structure of the isolated sub-system for heat flow is now complete. It has been shown that the first energy moment of the sub-system and its rate of change were the extensive variables essential to the treatment of heat flow. Now the analysis is extended to include energy exchange with heat reservoirs. Two reservoirs of different temperatures are used, sandwiching the sub-system as in figure 3.3. It will be shown that the reservoirs induce a non-zero first energy moment in the sub-system. It will also be shown that the energy flux through the sub-system due to the reservoirs is equal to the rate of decay of the energy moment as if it had arisen from a spontaneous fluctuation in an isolated sub-system. Finally, it will be shown how the two apparently contradictory requirements of constant energy moment and constant rate of decay of the energy moment are enforced by the reservoirs to give a steady state system.

*First entropy*

With reference to figure 3.3, the reservoir and sub-system boundaries are at $z = \pm L/2$, and the temperature of each respective reservoir is $T_{r\pm}$. Let $\Delta E_{r\pm}$ be the change in energy of the respective reservoir due to exchange with the sub-system at the boundary. It follows that the total change in the reservoir entropy is

$$\begin{aligned} \Delta S_r &= \frac{\Delta E_{r+}}{T_{r+}} + \frac{\Delta E_{r-}}{T_{r-}} \\ &= \frac{\Delta E_{r0}}{T_{r0}} + \frac{\Delta E_{r1}}{T_{r1}}. \end{aligned} \tag{3.166}$$

The second equality is exact and no approximation has been invoked here. The change in the zeroth and first energy moments of the reservoirs are defined as

$$\Delta E_{r0} \equiv \Delta E_{r+} + \Delta E_{r-}, \text{ and } \Delta E_{r1} \equiv \frac{L}{2}\Delta E_{r+} - \frac{L}{2}\Delta E_{r-}. \tag{3.167}$$

The zeroth and first reservoir temperatures are defined as

$$\frac{1}{T_{r0}} \equiv \frac{1}{2}\left[\frac{1}{T_{r+}} + \frac{1}{T_{r-}}\right], \text{ and } \frac{1}{T_{r1}} \equiv \frac{1}{L}\left[\frac{1}{T_{r+}} - \frac{1}{T_{r-}}\right]. \tag{3.168}$$

These definitions are obviously analogous to those used for the isolated sub-system. It is implicitly assumed that the energy exchange takes place at the boundary and that the reservoir temperature $T_{r\pm}$ applies here because any temperature gradients within a reservoir are negligible.

Let $\Delta^r E_{s\pm}$ be the reservoir-induced change in the sub-system energy at each respective boundary. This change is due to exchange with the reservoir, and so from energy conservation one must have $\Delta^{\mathrm{r}} E_{\mathrm{s}\pm} = -\Delta E_{\mathrm{r}\pm}$. This can be written in terms of the change in zeroth and first moment,

$$\Delta^{\mathrm{r}} E_{\mathrm{s}0} = -\Delta E_{\mathrm{r}0}, \quad \text{and} \quad \Delta^{\mathrm{r}} E_{\mathrm{s}1} = -\Delta E_{\mathrm{r}1}. \tag{3.169}$$

Obviously the energy exchange is assumed to occur at each boundary over a region with small width compared to the length of the sub-system.

In addition to this reservoir-induced change, the energy moments also change adiabatically. Adiabatic evolution means that of the isolated sub-system, as determined by Hamilton's equations of motion, without any direct influence of the reservoir (or artificial thermostats, or dissipative or stochastic forces). The adiabatic change can also be called the internal change because it is determined solely by the sub-system forces and it would occur if the sub-system was isolated from the reservoirs. The total change in the sub-system energy moments is

$$\Delta E_{\mathrm{s}0} = \Delta^0 E_{\mathrm{s}0} + \Delta^{\mathrm{r}} E_{\mathrm{s}0}, \quad \text{and} \quad \Delta E_{\mathrm{s}1} = \Delta^0 E_{\mathrm{s}1} + \Delta^{\mathrm{r}} E_{\mathrm{s}1}. \tag{3.170}$$

The superscript 0 denotes the adiabatic change, and the superscript r denotes the reservoir induced change. The total energy of an isolated system is conserved, which means that the adiabatic change of the zeroth energy moment vanishes, $\Delta^0 E_{\mathrm{s}0} = 0$.

In the canonical equilibrium case the reservoir energy was written $E_{\mathrm{r}} = E_{\mathrm{total}} - E_{\mathrm{s}}$ and the reservoir entropy was written $S_{\mathrm{r}} = -E_{\mathrm{s}}/T_{\mathrm{r}}$, the constant independent of the sub-system being discarded. We shall use similar notions for the present canonical non-equilibrium case, with two adjustments. First, the reservoirs exchange energy over time, and so we have to keep this in mind, at least notionally. Second, the change in reservoir energy cannot be equated to the entire change in the sub-system energy, but only the reservoir-induced part of it. In view of these considerations, and of the definitions of moments and temperatures above, the reservoir entropy can be written as

$$S_{\mathrm{r}}(t) = \frac{E_{\mathrm{r}0}(t)}{T_{\mathrm{r}0}} + \frac{E_{\mathrm{r}1}(t)}{T_{\mathrm{r}1}}. \tag{3.171}$$

The time-dependent moments defined here are meant to be only the part due to exchanges with the sub-system. Although we expect a monotonic energy flux from the hot reservoir to the cold via the sub-system, we assume that the reservoirs are so large that over the time of any measurement the reservoir temperatures are constant. These moments are defined such that their changes are the changes in reservoir moment given above,

$$\Delta E_{\mathrm{r}0}(t) = \Delta E_{\mathrm{r}0} = -\Delta^{\mathrm{r}} E_{\mathrm{s}0}, \tag{3.172}$$

and

$$\Delta E_{\mathrm{r}1}(t) = \Delta E_{\mathrm{r}1} = -\Delta^{\mathrm{r}} E_{\mathrm{s}1}. \tag{3.173}$$

With these, the constrained entropy of the total system may be written as

$$S_{\mathrm{total}}(E_{\mathrm{s}0}, E_{\mathrm{s}1} | T_{\mathrm{r}0}, T_{\mathrm{r}1}) = S_{\mathrm{s}}(E_{\mathrm{s}0}, E_{\mathrm{s}1}) + \frac{E_{\mathrm{r}0}(t)}{T_{\mathrm{r}0}} + \frac{E_{\mathrm{r}1}(t)}{T_{\mathrm{r}1}}. \tag{3.174}$$

As above, we shall write the sub-system entropy as a quadratic form in the first energy moment. This is valid provided that the temperature gradient between the reservoirs is small.

The total entropy here has two constraints: the zeroth and first energy moments. Because the zeroth moment is a conserved variable, its treatment does not materially differ from the canonical equilibrium case. To find its optimum value can maximize the total entropy with respect to $E_{s0}$, which is the same as differentiating with respect to $\Delta^r E_{s0}$. The first energy moment is not conserved, and it can change by either $\Delta^0 E_{s1}$ or by $\Delta^r E_{s1}$. In maximizing the constrained total entropy with respect to $E_{s1}$, one has to specify which of these two possible changes is meant. Because the reservoir first energy moment $E_{r1}(t)$, it is clear that the reservoir-induced change is the required one. The justification for this is that in this structural part of the problem the reservoir is the determining factor. It is only in the dynamic part of the problem that the adiabatic change becomes an essential element, as will be seen.

In view of this discussion, the derivatives of the constrained total entropy with respect to the reservoir-induced change in sub-system moments are

$$\frac{\partial S_{\text{total}}}{\partial^r E_{s0}} = \frac{1}{T_{s0}} - \frac{1}{T_{r0}}, \tag{3.175}$$

and

$$\frac{\partial S_{\text{total}}}{\partial^r E_{s1}} = \frac{1}{T_{s1}} - \frac{1}{T_{r1}}. \tag{3.176}$$

The reservoir terms here follow from the results above, $\Delta E_{r0}(t) = -\Delta^r E_{s0}$, and $\Delta E_{r1}(t) = -\Delta^r E_{s1}$.

This formally gives the result intuitively expected: the constrained total entropy is a maximum with respect to reservoir-induced changes of energy when the zeroth temperature of the sub-system equals that of the reservoir, and similarly for the first temperature,

$$\overline{T}_{s0} = T_{r0}, \quad \text{and} \quad \overline{T}_{s1} = T_{r1}. \tag{3.177}$$

Of course if one unpacks the definitions, these simply say that the sub-system temperatures at the boundaries most likely equal those of the reservoirs. This result gives the most likely static structure for this steady state thermodynamic non-equilibrium system.

In obtaining this static structure, the adiabatic change $\Delta^0$ was not required or used. It will now be shown that maximizing the second entropy with respect to this adiabatic change is what gives the most likely heat flux.

*Second entropy*

We now turn to the energy flux, and for this we need the second entropy. Since the zeroth energy moment is a conserved variable with no adiabatic change, $\Delta^0 E_{s0} = 0$, we can suppress both the zeroth energy moment and zeroth temperature. The equality of the zeroth temperature of the sub-system and the reservoir, $\overline{T}_{s0} = T_{r0}$, determines the most likely zeroth energy moment $\overline{E}_{s0} = E_{s0}(\overline{T}_{s0})$.

In terms of notation, above $\Delta^{\mathrm{r}}$ denoted the mathematical variation in a constraint. Such a variation was used to optimize the first entropy. In this section $\Delta_\tau^0$ and $\Delta_\tau^{\mathrm{r}}$ will be used to denote the adiabatic and reservoir-induced physical change over the time interval $\tau$. Dividing by $\tau$ these give the adiabatic and reservoir-induced energy fluxes. In the formalism $\Delta_\tau^0$ and $\Delta_\tau^{\mathrm{r}}$ are constrained variables with respect to which the second entropy will be maximized.

Above, the second entropy for an isolated system was given in terms of fluxes in general, equation (3.95), and in particular for heat flow, equation (3.160). These were in the small time limit, valid to $\mathcal{O}(\tau)$. The four terms in the expressions have obvious physical interpretations. First is the first or ordinary entropy of the sub-system, which is quadratic in the first energy moment. Second is the term quadratic in the flux, which acts to limit the flux. Third is the rate of entropy production, which is linear in the flux. Fourth is a term quadratic in the first entropy gradient which is required by the reduction condition. By flux in equation (3.160) is meant $\overset{\circ}{E}_1$, the coarse rate of change of the first energy moment, which for an isolated system is the internal or adiabatic rate of change.

The present case differs from the earlier one by the addition of reservoirs. Hence one has to modify equation (3.160) in several respects. First, the first entropy of the sub-system has to be replaced by that of the total system. Second, the term quadratic in the internal flux remains unchanged, because this remains the cost of dynamic order with or without reservoirs. Third, to the internal rate of entropy production is added the rate of entropy production due to the reservoirs. The latter consists of the dissipation of the reservoir entropy, plus the reservoir-induced dissipation of the sub-system entropy. Fourth, the term independent of the flux is determined by the reduction condition for the total system, namely that the maximum value of the second entropy (i.e. its value in the state of optimum flux), equals the total first entropy.

These give the second entropy for the case of heat reservoirs as

$$\begin{aligned} S_{\text{total}}^{(2)}&\left(\Delta_\tau^0 E_{\text{s1}},\, \Delta_\tau^{\text{r}} E_{\text{s1}},\, E_{\text{s1}} \middle| T_{\text{r1}},\, \tau\right) \\ &= \frac{1}{2} S_{\text{s}}'' E_{\text{s1}}^2 + \frac{E_{\text{r1}}(t)}{T_{\text{r1}}} - \frac{1}{2\Lambda|\tau|}\left[\Delta_\tau^0 E_{\text{s1}}\right]^2 + \frac{1}{2}\Delta_\tau^0 E_{\text{s1}} S_{\text{s}}'' E_{\text{s1}} \\ &\quad + \frac{1}{2}\Delta_\tau^{\text{r}} E_{\text{s1}}\left[S_{\text{s}}'' E_{\text{s1}} - \frac{1}{T_{\text{r1}}}\right] + \frac{|\tau|\Lambda}{8}\left[S_{\text{s}}''^2 E_{\text{s1}}^2 - \frac{2 S_{\text{s}}'' E_{\text{s1}}}{T_{\text{r1}}}\right]. \end{aligned} \tag{3.178}$$

This is valid in the small time limit, with terms $\mathcal{O}(\tau^2)$ neglected. The thermodynamic force of the sub-system is $S_{\text{s}}'' E_{\text{s1}} = T_{\text{s1}}^{-1}$. The various terms here can be matched with the required modifications of equation (3.160) that were just discussed. The fourth term here is the internal rate of entropy production, and is the same as the third term in equation (3.160). The fifth term here is the reservoir-induced production of total first entropy, which is the sum of that of the sub-system and that of the reservoir. The sixth term is determined by the reduction condition, the fact that it must comprise terms quadratic and linear in the first energy moment, and the need for its derivative to vanish at the optimum moment, as is discussed shortly.

It will be noted that this is linear in the reservoir-induced change in moment, $\Delta^{\rm r}_\tau E_{\rm s1}$. Hence the second entropy cannot be maximized with respect to the latter. Physically, since this is boundary driven flow, and since the conductivity of the reservoir is much greater than that of the sub-system, any change in sub-system moment from its optimum value can be immediately counteracted by flow from the reservoir. Hence one expects

$$\overline{\Delta^{\rm r}_\tau E_{\rm s1}} = -\Delta^0_\tau E_{\rm s1}. \tag{3.179}$$

We shall discuss this further shortly, and below we shall rewrite the second entropy imposing this condition explicitly.

Differentiating the second entropy with respect the internal flux gives

$$\frac{\partial S^{(2)}_{\rm total}}{\partial \Delta^0_\tau E_{\rm s1}} = \frac{-1}{\Lambda|\tau|}\Delta^0_\tau E_{\rm s1} + \frac{1}{2}S''_{\rm s}E_{\rm s1}. \tag{3.180}$$

In this $\Delta^{\rm r}_\tau E_{\rm s1}$ has been held constant; see below for the contrary case. Setting this to zero yields the most likely internal flux,

$$\overset{\circ}{\overline{E}}{}_1^0 \equiv \frac{\overline{\Delta^0_\tau E_{\rm s1}}}{\tau} = \frac{\hat{\tau}\Lambda}{2}\frac{1}{T_{\rm s1}}. \tag{3.181}$$

This is identical to that obtained for the isolated system, equation (3.162), as might have been expected because it arises from the adiabatic optimization of the flux. Again in essence, it gives Fourier's law, that the sub-system flux is proportional to the sub-system temperature gradient.

Differentiating with respect to the reservoir-induced change in moment gives

$$\begin{aligned}\frac{\partial S^{(2)}_{\rm total}}{\partial^{\rm r} E_{\rm s1}} = S''_{\rm s}E_{\rm s1} - \frac{1}{T_{\rm r1}} + \frac{1}{2}S''_{\rm s}\left[\Delta^0_\tau E_{\rm s1} + \Delta^{\rm r}_\tau E_{\rm s1}\right] \\ + \frac{|\tau|\Lambda}{8}\left[2S''^{2}_{\rm s}E_{\rm s1} - \frac{2S''_{\rm s}}{T_{\rm r1}}\right].\end{aligned} \tag{3.182}$$

Setting this to zero determines $\overline{E}_{\rm s1}$, the optimum sub-system moment at the start of the transition. For this purpose one need only consider the reservoir-induced derivative, $\partial^{\rm r}E_{\rm s1}$, assuming that adiabatic changes are negligible in comparison. This is equivalent to assuming that the thermal conductivity of the reservoirs is much greater than that of the sub-system. This is a necessary assumption for the reservoir formalism of non-equilibrium thermodynamics, akin to the thermodynamic limit in defining what a reservoir is.

From the results of maximizing the first entropy, equation (3.177), the optimum first moment is given by

$$S''_{\rm s}\overline{E}_{\rm s1} \equiv \frac{1}{\overline{T}_{\rm s1}} = \frac{1}{T_{\rm r1}}. \tag{3.183}$$

In order for the preceding derivative to vanish at this point, and hence for the second entropy to be consistent with the first entropy, one must have

$$\Delta_\tau^{\mathrm{r}} E_{\mathrm{s}1} = -\Delta_\tau^0 E_{\mathrm{s}1}. \tag{3.184}$$

This is the result anticipated above, namely that the reservoir-induced change in moment must cancel exactly the adiabatic change in moment.

One can insert this result into the expression for the second entropy and eliminate the reservoir-induced change in moment,

$$\begin{aligned} &S^{(2)}_{\mathrm{total}}(\Delta_\tau^0 E_{\mathrm{s}1}, E_{\mathrm{s}1}|T_{\mathrm{r}1}, \tau) \\ &= \frac{1}{2} S_{\mathrm{s}}'' E_{\mathrm{s}1}^2 + \frac{E_{\mathrm{r}1}(t)}{T_{\mathrm{r}1}} - \frac{1}{2\Lambda|\tau|}\left[\Delta_\tau^0 E_{\mathrm{s}1}\right]^2 + \frac{1}{2T_{\mathrm{r}1}}\Delta_\tau^0 E_{\mathrm{s}1} - \frac{|\tau|\Lambda}{8T_{\mathrm{r}1}^2}. \end{aligned} \tag{3.185}$$

The final term has also been changed in order to satisfy the reduction condition. With this expression, the derivative with respect to the internal flux now gives

$$\frac{\partial S^{(2)}_{\mathrm{total}}}{\partial \Delta_\tau^0 E_{\mathrm{s}1}} = \frac{-1}{\Lambda|\tau|}\Delta_\tau^0 E_{\mathrm{s}1} + \frac{1}{2T_{\mathrm{r}1}}, \tag{3.186}$$

which vanishes when

$$\overline{\overset{\circ}{E}_1{}^0} \equiv \frac{\overline{\Delta_\tau^0 E_{\mathrm{s}1}}}{\tau} = \frac{\hat{\tau}\Lambda}{2}\frac{1}{T_{\mathrm{r}1}}. \tag{3.187}$$

This is practically equivalent to the original result, equation (3.181), since $\overline{T}_{\mathrm{s}1} = T_{\mathrm{r}1}$.

The second entropy evaluated for the most likely adiabatic change is

$$\begin{aligned} &S^{(2)}_{\mathrm{total}}\left(\overline{\Delta_\tau^0 E_{\mathrm{s}1}}, E_{\mathrm{s}1}\middle|T_{\mathrm{r}1}, \tau\right) \\ &= S_{\mathrm{total}}(E_{\mathrm{s}1}) - \frac{1}{2\Lambda|\tau|}\left[\frac{|\tau|\Lambda}{2T_{\mathrm{r}1}}\right]^2 + \frac{1}{2T_{\mathrm{r}1}}\frac{|\tau|\Lambda}{2T_{\mathrm{r}1}} - \frac{|\tau|\Lambda}{8T_{\mathrm{r}1}^2} \\ &= S_{\mathrm{total}}(E_{\mathrm{s}1}). \end{aligned} \tag{3.188}$$

By design, this satisfies the reduction condition exactly.

The original second entropy expression, equation (3.178), evaluated at $\overline{\Delta_\tau^0 E_{\mathrm{s}1}} = |\tau|\Lambda S_{\mathrm{s}}'' E_{\mathrm{s}1}/2$ and $\Delta^{\mathrm{r}} E_{\mathrm{s}1} = -\overline{\Delta_\tau^0 E_{\mathrm{s}1}}$, is

$$\begin{aligned} &S^{(2)}_{\mathrm{total}}\left(\overline{\Delta_\tau^0 E_{\mathrm{s}1}}, \Delta_\tau^{\mathrm{r}} E_{\mathrm{s}1}, E_{\mathrm{s}1}\middle|T_{\mathrm{r}1}, \tau\right) \\ &= S_{\mathrm{total}}(E_{\mathrm{s}1}) - \frac{1}{2\Lambda|\tau|}\left[\frac{|\tau|}{2}\Lambda S_{\mathrm{s}}'' E_{\mathrm{s}1}\right]^2 + \frac{|\tau|}{4}\Lambda S_{\mathrm{s}}'' E_{\mathrm{s}1} S'' E_{\mathrm{s}1} \\ &\quad - \frac{|\tau|}{4}\Lambda S_{\mathrm{s}}'' E_{\mathrm{s}1}\left[S_{\mathrm{s}}'' E_{\mathrm{s}1} - \frac{1}{T_{\mathrm{r}1}}\right] + \frac{|\tau|\Lambda}{8}\left[S_{\mathrm{s}}''^2 E_{\mathrm{s}1}^2 - \frac{2S_{\mathrm{s}}'' E_{\mathrm{s}1}}{T_{\mathrm{r}1}}\right] \\ &= S_{\mathrm{total}}(E_{\mathrm{s}1}). \end{aligned} \tag{3.189}$$

By design, this also satisfies the reduction condition exactly.

*Cancelation of adiabatic flux*

Let us further discuss the condition $\Delta_\tau^{\mathrm{r}} E_{\mathrm{s1}} = -\Delta_\tau^0 E_{\mathrm{s1}}$. This says that the adiabatic change in the sub-system energy moment is exactly canceled by the reservoir-induced change. Once the sub-system has reached the optimum static structure induced by the reservoirs $T_{\mathrm{s1}}(\overline{E}_{\mathrm{s1}}) = T_{\mathrm{r1}}$, then most likely there is no further change over time in that structure, $\overline{\Delta_\tau E_{\mathrm{s1}}} = 0$.

This cancelation between the internally and externally induced changes in first energy moment has the effect of holding it constant. Hence there is no longer an inertial period over molecular time scales in which the flux comes up to speed. Nor is there decay over longer time scales as the fluctuation and hence driving force are reduced. In this steady state system there is no distinction between the coarse velocity $\overline{\overset{\circ}{E}_1{}^0}$ and the instantaneous velocity $\overline{\dot{E}_1{}^0}$.

The total change in the sub-system first energy moment over the time interval is of course the sum of the adiabatic change and the reservoir-induced change, which are individually non-zero. By energy conservation, the reservoir-induced change in the sub-system first energy moment is equal and opposite to the change in the reservoir first energy moment,

$$\overline{\Delta_\tau^0 E_{\mathrm{s1}}} = -\overline{\Delta_\tau^{\mathrm{r}} E_{\mathrm{s1}}} = \overline{\Delta_\tau E_{\mathrm{r1}}}. \tag{3.190}$$

Although the steady state is defined as the state in which the sub-system is macroscopically constant in time, $\overline{\Delta_\tau E_{\mathrm{s1}}} = 0$, this result says that there remains a steady energy flow between the reservoirs. In fact the energy discrepancy manifest in their temperature difference is steadily diminishing. (However, the reservoirs are so large that this effect is immeasurably small.) In a steady state non-equilibrium system, the entropy of the system steadily increases while the entropy (i.e. structure and dynamics) of the sub-system is constant.

The important quantitative meaning of this last result is that given the sub-system temperature gradient equal to that between the reservoirs, then the adiabatic energy flow in the sub-system is equal to the energy flow between the reservoirs. Since the adiabatic energy flow is equal to the rate of regression of the fluctuation of the isolated system (with the same energy moment), equation (3.181), this proves another aspect of Onsager's regression hypothesis: the flux in an isolated system is the same as the flux in the same system between two reservoirs if the spontaneous structural fluctuation in the isolated system is the same as that induced by the reservoirs.

## 3.5 Variational hydrodynamics

This treatment of heat flow can be extended to hydrodynamic fluxes in general. The second entropy provides a variational principle for these fluxes, and it gives in the optimum state the conventional hydrodynamic equations such as the Navier–Stokes and energy equations. The derivation is a little lengthy (Attard 2012, chapter 5) and in this section only the final variational principle is given. This is then applied to the case of steady heat flow in slab geometry, where the results reduce to those derived above from the behavior of the first energy moment.

The present variational formulation of hydrodynamics treats the material fluxes as constrained or fluctuating quantities that can take values independent of the thermodynamic driving forces. A similar fluctuation approach was pursued by Landau and Lifshitz (1957, 1959). Other researchers have extended and applied this approach to a range of different hydrodynamic phenomena (Fox and Uhlenbeck 1970, Keizer 1987, Schöpf and Rehberg 1994, Ortiz de Zárate and Sengers 2006). Such variational approaches to hydrodynamics possibly have computational and numerical advantages over conventional methods of evolving the system with the usual hydrodynamic equations. The present approach is directly based on the second entropy, and as such it makes a direct connection between non-equilibrium thermodynamics and hydrodynamic phenomena.

### 3.5.1 General result

For a system consisting of a sub-system and a reservoir, the total dissipation in the steady state is (Attard 2012)

$$\dot{S}_{\text{total}} = \int \mathrm{d}\mathbf{r} \left\{ \mathbf{J}_{\mathrm{E}}^{0} \cdot \nabla \frac{1}{T} - \frac{1}{T} \underline{\underline{\Pi}} : \nabla \mathbf{v} + \frac{1}{T} \sum_{\alpha} \dot{\xi}_{\alpha} A_{\alpha} - \sum_{k} \mathbf{J}_{\mathrm{N},k}^{0} \cdot \left[ \frac{\nabla \psi_k}{T} + \nabla \frac{\mu_k}{T} \right] \right\}. \quad (3.191)$$

This is valid for arbitrary fluxes and fields; everything in the integrand is a function of position and time. This assumes that if energy can cross the boundary, then the temperature of the sub-system equals that of the reservoir at every point on the boundary, and similarly for any thermodynamic field variable conjugate to any flux across the boundary (Attard 2012, section 5.2). This is for the steady state; in the contrary case the boundary integrals are non-zero.

Here $T(\mathbf{r}, t)$ is the temperature, $\mathbf{v}(\mathbf{r}, t)$ is the local barycentric or center of mass velocity, $\mu_k$ is the chemical potential of species $k$, and $\psi_k(\mathbf{r}, t)$ is the external potential. For chemical reaction $\alpha$, $A_\alpha = \sum_k \mu_k \nu_{\alpha k}$ is the chemical affinity, $\nu_{\alpha k}$ is the stoichiometric coefficient. These are the specified field variables of the sub-system that give the thermodynamic driving forces. For the constrained fluxes, $\dot{\xi}_\alpha$ is the rate of reaction, and $\mathbf{J}_{\mathrm{E}}(\mathbf{r}, t)$ and $\mathbf{J}_{\mathrm{N}}(\mathbf{r}, t)$ are the energy and number fluxes, respectively. The superscript 0 on the latter denotes the diffusive part of the flux, which is the flux in excess of the convective flux carried by the local barycentric velocity. The diffusive part of the momentum flux, $\underline{\underline{\Pi}}(\mathbf{r}, t)$, is also called the viscous pressure tensor.

In general the second entropy consists of four terms: the first entropy of the structure, the term quadratic in the fluxes, the term bilinear in the fluxes and the thermodynamic forces that represents half the rate of total entropy production, and the term independent of the fluxes that ensures that the reduction condition is satisfied by the optimum fluxes. Since the rate of dissipation is a scalar, because of the Curie symmetry principle (de Groot and Mazur 1984 section VI.2) for an isotropic system there can be no coupling between scalars, vectors, and traceless second rank tensors, nor between vector components in different directions. Hence the three types of fluxes decouple and the second entropy is

$$S^{(2)} = S_{\text{tot}} + |\tau|\left[S_0^{(2)} + S_1^{(2)} + S_2^{(2)}\right]. \tag{3.192}$$

The total first entropy is

$$S_{\text{tot}} = \int \mathrm{d}\mathbf{r}\; \sigma_0(\varepsilon^{\text{int}}(\mathbf{r}, t), \underline{n}(\mathbf{r}, t)) + \oint \mathrm{d}\mathbf{r} \frac{E_{\text{r}}(\mathbf{r}, t)}{T_{\text{r}}(\mathbf{r})}. \tag{3.193}$$

This the sum of the sub-system entropy and the reservoir entropy, with the latter being the boundary integral form of the reservoir terms in equation (3.174), assuming that only energy is exchangeable. The first entropy density, $\sigma_0$, is a function of the local species number density, $\underline{n}(\mathbf{r}, t)$, and the internal entropy density, $\varepsilon^{\text{int}}(\mathbf{r}, t) = \varepsilon(\mathbf{r}, t) - \underline{n}(\mathbf{r}, t) \cdot \underline{\psi}(\mathbf{r}, t) - \rho(\mathbf{r}, t)v(\mathbf{r}, t)^2/2$, where $\rho$ is the local total mass density, and $v^2 = \mathbf{v} \cdot \mathbf{v}$ is the square of the local barycentric velocity. When the total first entropy is optimized with respect to the internal energy density, the sub-system temperature on the boundary equals the reservoir temperature.

The remaining three terms in the second entropy come from the independently coupled scalar, vector, and traceless second rank tensor fluxes. The viscous pressure tensor is decomposed into its scalar and traceless parts, $\underline{\underline{\Pi}} = \pi \underline{\underline{I}} + \underline{\underline{\Pi}}^*$, with $\pi = \text{TR}[\underline{\underline{\Pi}}]/3$. Similarly the shear rate tensor may be decomposed as $\nabla\mathbf{v} = [\nabla \cdot \mathbf{v}/3]\underline{\underline{I}} + [\nabla\mathbf{v}]^*$. Since the viscous pressure tensor is symmetric, one has the result that $\underline{\underline{\Pi}} : \nabla\mathbf{v} = \pi\nabla \cdot \mathbf{v} + \underline{\underline{\Pi}}^* : [\nabla\mathbf{v}]^{\text{sym},*}$.

The second entropy for the scalar fluxes is a mixed parity case,

$$\begin{aligned} S_0^{(2)} = \int \mathrm{d}\mathbf{r} \frac{-1}{2}\Lambda_0^{-1} : \{\pi, \underline{\dot{\xi}}\}^2 - \frac{1}{2}\Lambda_0^{-1} : \{\bar{\pi}, \underline{\bar{\dot{\xi}}}\}^2 \\ + \frac{\hat{\tau}}{2T}\{\pi, \underline{\dot{\xi}}\} \cdot \Lambda_0^{-1}L_0(\hat{\tau})\{-\nabla \cdot \mathbf{v}, \underline{A}\}. \end{aligned} \tag{3.194}$$

One sees here and below the usual terms: a term quadratic in the flux, a term bilinear in the flux and thermodynamic force, and a term independent of the flux designed to satisfy the reduction condition for the optimum fluxes.

The second entropy for the vector fluxes is a pure parity case,

$$\begin{aligned} S_1^{(2)} = \int \mathrm{d}\mathbf{r} \frac{-1}{2}\Lambda_1^{-1} : \left\{\mathbf{J}_{\text{E}}^0, \underline{\mathbf{J}}_{\text{N}}^0\right\}^2 - \frac{1}{2}\Lambda_1^{-1} : \left\{\bar{\mathbf{J}}_{\text{E}}^0, \underline{\bar{\mathbf{J}}}_{\text{N}}^0\right\}^2 \\ + \frac{\hat{\tau}}{2}\left\{\mathbf{J}_{\text{E}}^0, \underline{\mathbf{J}}_{\text{N}}^0\right\} \cdot \left\{\nabla\frac{1}{T}, \left[\frac{-1}{T}\nabla\underline{\psi} - \nabla\frac{\underline{\mu}}{T}\right]\right\}. \end{aligned} \tag{3.195}$$

In an isotropic system, there is no coupling between different Cartesian components, and so $\Lambda_1$ is composed of three identical blocks on the main diagonal.

The second entropy for the traceless part of the viscous pressure tensor (i.e. diffusive momentum flux tensor) is also a pure parity case,

$$S_2^{(2)} = \int \mathrm{d}\mathbf{r} \frac{-1}{2}\Lambda_2^{-1} : \underline{\underline{\Pi}}^*\underline{\underline{\Pi}}^* - \frac{1}{2}\Lambda_2^{-1} : \underline{\underline{\bar{\Pi}}}^*\underline{\underline{\bar{\Pi}}}^* - \frac{\hat{\tau}}{2T}\underline{\underline{\Pi}}^* : [\nabla\mathbf{v}]^{\text{sym},*}. \tag{3.196}$$

In an isotropic system there is a single transport constant, $\Lambda_2^{-1} : \underline{\underline{\Pi}}^* \underline{\underline{\Pi}}^* = \lambda_2^{-1} \underline{\underline{\Pi}}^* : \underline{\underline{\Pi}}^*$. This completes the variational formulation of hydrodynamics in the most general case.

Maximization of this most general form for the second entropy with respect to the diffusive fluxes gives the standard equations of hydrodynamics (Attard 2012, chapter 5). This allows the transport constants that appear here, the various Λ, to be identified with the conventional hydrodynamic transport coefficients.

### 3.5.2 Conductive heat flow

For the case of conductive heat flow, the only non-zero flux is that of energy. In the slab geometry invoked earlier in this chapter, this related to the rate of change of the first energy moment of the sub-system, $\mathbf{J}_\mathrm{E}^0 = \dot{E}_1^0 \hat{\mathbf{z}}/V$.

The total first entropy is as given earlier, equation (3.174),

$$S_\mathrm{total}(E_{s0}, E_{s1}|T_{r0}, T_{r1}) = S_s(E_{s0}, E_{s1}) + \frac{E_{r0}(t)}{T_{r0}} + \frac{E_{r1}(t)}{T_{r1}}. \tag{3.197}$$

As usual, maximizing this with respect to the first energy moment of the sub-system shows that the sub-system's first temperature in the optimum state is equal to that of the reservoir, $\overline{T}_{s1} = T_{r1}$. In the remainder of the second entropy only the term that couples the vector fluxes is non-zero, and it is

$$\begin{aligned} S_1^{(2)} &= \int \mathrm{d}\mathbf{r} \left\{ \frac{-1}{2} \Lambda_1^{-1} : \mathbf{J}_\mathrm{E}^0 \mathbf{J}_\mathrm{E}^0 - \frac{1}{2} \Lambda_1^{-1} : \overline{\mathbf{J}}_\mathrm{E}^0 \overline{\mathbf{J}}_\mathrm{E}^0 + \frac{\hat{\tau}}{2} \mathbf{J}_\mathrm{E}^0 \cdot \nabla \frac{1}{T} \right\} \\ &= A \int_{-L/2}^{L/2} \mathrm{d}z \left\{ \frac{-1}{2\Lambda_1 V^2} (\dot{E}_1^0)^2 - \frac{1}{2\Lambda_1 V^2} \left( \overline{\dot{E}_1^0} \right)^2 + \frac{\hat{\tau}}{2V} \dot{E}_1^0 \frac{1}{T_{s1}} \right\} \\ &= \frac{-1}{2\Lambda_1 V} \left( \dot{E}_1^0 \right)^2 - \frac{1}{2\Lambda_1 V} \left( \overline{\dot{E}_1^0} \right)^2 + \frac{\hat{\tau}}{2} \dot{E}_1^0 \frac{1}{T_{s1}}. \end{aligned} \tag{3.198}$$

Combined with the expression for the total first entropy, this is identical to the expression for the second entropy earlier derived directly for the transitions in the first energy moment, equation (3.185). That expression enforced the condition that the reservoir-induced change in first energy moment was equal and opposite to the adiabatic change, which in physical terms ensures temperature equality between the sub-system and the reservoir on each boundary.

In the present notation, the optimum heat flux is given by $\partial S^{(2)}/\partial \dot{E}_1^0 \big|_{\overline{\dot{E}_1^0}} = 0$, which gives

$$\overline{\dot{E}_1^0} = \frac{\hat{\tau} V \Lambda_1}{2} \frac{1}{T_{s1}}, \text{ or } \mathbf{J}_\mathrm{E}^0 = \frac{-\hat{\tau}\Lambda_1}{2T^2} \nabla T. \tag{3.199}$$

Comparing this to Fourier's law, one can identify the thermal conductivity in terms of the present transport coefficient as $\lambda = \Lambda_1/2T^2$.

## 3.6 Stochastic, dissipative hydrodynamic equations

An important question in hydrodynamics is how to include noise in the evolution equations. While it is understood that many systems require such a stochastic contribution to initiate or to modify a transition, there is little consensus on the way to characterize such a random element, or how to include it in the evolution equations, or what its probability distribution is.

In actual fact this problem was addressed and solved by Landau and Lifshitz (1957, 1959). Although most conventional hydrodynamicists do not recognize the fluctuating hydrodynamics approach of Landau and Lifshitz, a number of more enlightened workers have variously applied and developed it (Fox and Uhlenbeck 1970, Keizer 1987, van Beijeren and Cohen 1988, Schöpf and Rehberg 1994, Ortiz de Zárate and Sengers 2006). The present second entropy approach to non-equilibrium systems provides a basis for, and a way of systematizing, fluctuating hydrodynamics.

The stochastic dissipative hydrodynamic equations in general begin with the conventional equations of hydrodynamics, as given, in any standard book on hydrodynamics. To be definite, here the notation used in section 5.4 of Attard (2012) is followed. The conventional equations give the most likely evolution, to which must be added the stochastic contributions. The latter are just the Gaussian distributed diffusive fluxes, as given by the exponential of the three contributions to the second entropy, equations (3.194)–(3.196).

For simplicity, I shall illustrate the idea in the Boussinesq approximation, which is well established (Reid and Harris 1958, Busse 1967, Bodenschatz *et al* 2000, Attard 2012). The conventional hydrodynamic equations for convective flow in Boussinesq approximation comprise the incompressibility equation,

$$0 = \nabla \cdot \mathbf{v}(\mathbf{r}, t), \tag{3.200}$$

the Navier–Stokes equation (infinite Prandtl number, instantaneous velocity relaxation),

$$0 = \mathcal{R}\nabla_{\parallel}^{2} T\big(\mathbf{r}, t\big) + \nabla^2\nabla^2 v_z(\mathbf{r}, t). \tag{3.201}$$

and the energy equation is

$$\frac{\partial T(\mathbf{r}, t)}{\partial t} = v_z(\mathbf{r}, t) - \mathbf{v}(\mathbf{r}, t) \cdot \nabla T(\mathbf{r}, t) + \nabla^2 T(\mathbf{r}, t). \tag{3.202}$$

These are in dimensionless form with $\mathcal{R}$ being the Rayleigh number. The temperature here is the departure from that in the conducting state.

For the stochastic, dissipative version of these, one simply returns the energy equation to its exact form in terms of the diffusive energy flux explicitly, and writes the latter as the sum of its most likely and stochastic parts,

$$\begin{aligned} \mathbf{J}_{\mathrm{E}}^{0}(\mathbf{r}, t) &= \overline{\mathbf{J}_{\mathrm{E}}^{0}}(\mathbf{r}, t) + \frac{1}{\Delta_t}\tilde{\mathbf{R}}(\mathbf{r}, t) \\ &= -\lambda\nabla T(\mathbf{r}, t) + \frac{1}{\Delta_t}\tilde{\mathbf{R}}(\mathbf{r}, t). \end{aligned} \tag{3.203}$$

This is written in SI units, $\Delta_t$ is the time step, and the heat flux and temperature are the departure from conduction. With this the stochastic, dissipative energy equation is (dimensionless)

$$\frac{\partial T(\mathbf{r}, t)}{\partial t} = v_z(\mathbf{r}, t) - \mathbf{v}(\mathbf{r}, t) \cdot \nabla T(\mathbf{r}, t) + \nabla^2 T(\mathbf{r}, t) - \frac{1}{\Delta_t^*}\tilde{\mathbf{R}}^*(\mathbf{r}, t). \tag{3.204}$$

The probability distribution of the stochastic heat flux increment is given by the exponential of the vector contribution to the second entropy, equation (3.195),

$$\begin{aligned} S_1^{(2)}/k_B &= \frac{-1}{4T_{00}^2 \lambda k_B |\Delta_t|} \int \mathrm{d}\mathbf{r}\, \tilde{\mathbf{R}}(\mathbf{r}, t) \cdot \tilde{\mathbf{R}}(\mathbf{r}, t) \\ &= \frac{-c_p^*}{4T_{00}^{*2} |\Delta_t^*|} \int \mathrm{d}\mathbf{r}^*\, \tilde{\mathbf{R}}^*(\mathbf{r}, t) \cdot \tilde{\mathbf{R}}^*(\mathbf{r}, t). \end{aligned} \tag{3.205}$$

Here the dimensionless heat capacity is $c_p^* \equiv c_p L_z^3 / k_B$, the unit of time is $\tau \equiv L_z^2 c_p / \lambda$, and the dimensionless stochastic heat increment is $\tilde{\mathbf{R}}^* = \tilde{\mathbf{R}}/(-\Delta_T) c_p L_z$, with $L_z$ being the height of the convective cell, $T_{00}$ being the mid-plane temperature, and $\Delta_T < 0$ being the temperature difference. The stochastic heat flux increment is independently distributed at each time step.

It can be mentioned that this is equivalent in fluctuating hydrodynamics to determining the variance of the stochastic heat flux from the fluctuation–dissipation theorem (Landau and Lifshitz 1957, 1959, Fox and Uhlenbeck 1970, Keizer 1987, van Beijeren and Cohen 1988, Schöpf and Rehberg 1994, Ortiz de Zárate and Sengers 2006).

## 3.7 Non-equilibrium pattern formation

In this chapter it has been argued that the second entropy, which is the entropy of transitions, is the appropriate entropy for non-equilibrium systems. The first entropy, which is essentially the entropy of structure or of state, is not so much irrelevant or non-existent for non-equilibrium systems, but rather it is insufficient to answer the important questions that one can ask about a non-equilibrium system.

An important and interesting example that underscores this point occurs in the phenomena of transitions between non-equilibrium states, or, relatedly, the formation of patterns in non-equilibrium systems. (The present author deprecates calling these non-equilibrium phases, primarily because their thermodynamic behavior is qualitatively different to that of equilibrium phases, especially in the nature of their stability and the transitions between them.) The general principle for non-equilibrium systems is that it is the transitions not the states that are the basic elements. For the present discussion we select a subset of all non-equilibrium systems, namely steady state non-equilibrium systems, in which this general principle may be clearly elucidated. In steady state systems the macroscopic state of the sub-system is constant in time, while at the same time entropy is produced continually in the reservoir or environment.

A possibly confusing point about steady state systems is that the 'state' of the sub-system appears constant in time, and therefore it is tempting to ask: what is the optimum state? In fact, because entropy is continually produced by the reservoir, such a question is meaningless. The correct question is: what is the optimum current state given the initial state? Or, what is the optimal transition from a given initial state? Such questions can be answered even if the final state, once established, is steady and constant in time.

A common and relatively simple example is the convective rolls that can occur in a fluid transporting heat. The challenge is to articulate the principle that determines the optimum shape and arrangement of such convective rolls for a given temperature gradient. This is an open question, long-standing in hydrodynamics, where it is called the wave number selection problem (see e.g. Bodenschatz *et al* (2000), and references therein).

In an equilibrium system, in which both the sub-system and the reservoir are macroscopically constant in time, the optimum state (or structure, or pattern) is the one with maximum total first entropy, (equivalently, minimum free energy). Of course there can often be barriers between the different states of the sub-system, and so sometimes more than one state may be observed. Nevertheless, *the* equilibrium state is well-defined as the state of maximum total entropy, and other possible states with lower total entropy are described as meta-stable states and, by definition, they occur with a lower likelihood.

One cannot apply the same principle to a state or pattern of a non-equilibrium system, even of a steady state non-equilibrium system, because entropy is continually produced. Hence it would make little sense to maximize the sub-system first entropy (both because it is only part of the total entropy, and because it would eventually be swamped by the amount of entropy produced by the reservoirs). Nor would it be meaningful to maximize the total entropy, because this monotonically increases with time. One cannot salvage the situation by maximizing (or extremizing) the rate of entropy production, because there is no law in thermodynamics that says that the rate of dissipation should be a maximum (see section 3.1.2).

As argued above, it is the second entropy rather than the first entropy that is useful for non-equilibrium systems. Instead of maximizing the first entropy, one should maximize the second entropy. Since the latter is the entropy of transitions, the question that should be posed is *not* what is the optimum non-equilibrium state?, *but rather* what is the optimum transition? Or, equivalently, given the current state, what is the optimum state after a given time?

Here I give two quantitative examples from convective heat flow that answer this question. The first concerns the conductive–convective transition, and the second concerns the cross-roll transition.

I compute the optimum transitions comparing the deterministic and stochastic versions of the hydrodynamic equations in the Boussinesq approximation. I also compare a semi-analytic theory, developed in section 3.7.2, that gives the mode with the fastest initial growth rate from conduction. A numerical version of this method, combined with white noise added to the initial state, is applied in section 3.7.4 to the cross-roll transition.

### 3.7.1 Stochastic method

Numerical results for the conduction–convection transition are given shortly. In this case the convective state is modeled as ideal straight rolls parallel to the $y$-axis, or $y$-rolls. Following Busse (1967), a discrete Fourier expansion for the fields is used, with the temperature being expanded as

$$\begin{aligned} T(\mathbf{r}, t) = & \sum_{n=1}^{N} T_{n0}^{\mathrm{s}} \sin \pi_n z \\ & + \sum_{n=1}^{N} \sum_{j=j_{\min}}^{j_{\max}} \left\{ T_{nj}^{\mathrm{s}} \sin \pi_n z \cos ja_1 x + T_{nj}^{\mathrm{c}} \cos \pi_n' z \cos ja_1 x \right\}, \end{aligned} \tag{3.206}$$

where $\pi_n = 2n\pi$ and $\pi_n' = (2n - 1)\pi$. Boussinesq symmetry is not enforced, but emerges spontaneously in the course of the transition. A similar expansion holds for the $z$-component of velocity (but without the $j = 0$ term). The $x$-component of velocity interchanges sines and cosines.

For the deterministic evolution, the small wave number $a_1 \approx 0.1$ is replaced by the wave number of the specified fundamental mode $a_x = 2.5 - 6$, and only about 6 harmonics are used.

The (dimensionless) stochastic heat increment has expansion

$$\begin{aligned} \tilde{R}_z(\mathbf{r}, t) = & \sum_{n=1}^{N} R_{z,n0}^{\mathrm{c}} \cos \pi_n z \\ & + \sum_{n=1}^{N} \sum_{j=j_{\min}}^{j_{\max}} \left\{ R_{z,nj}^{\mathrm{c}} \cos \pi_n z \cos ja_1 x + R_{z,nj}^{\mathrm{s}} \sin \pi_n' z \cos ja_1 x \right\}, \end{aligned} \tag{3.207}$$

and

$$\tilde{R}_x(\mathbf{r}, t) = \sum_{n=1}^{N} \sum_{j=j_{\min}}^{j_{\max}} \left\{ R_{x,nj}^{\mathrm{s}} \sin \pi_n z \sin ja_1 x + R_{x,nj}^{\mathrm{c}} \cos \pi_n' z \sin ja_1 x \right\}. \tag{3.208}$$

The (dimensionless) second entropy is

$$\begin{aligned} S_1^{(2)}/k_{\mathrm{B}} = & \frac{-c_{\mathrm{p}}^*}{4T_{00}^{*2}|\Delta_t^*|} \frac{L_x^* L_y^*}{4} \left[ \sum_{n=1}^{N} 2\left(R_{x,n0}^{\mathrm{c}}\right)^2 \right. \\ & \left. + \sum_{n=1}^{N} \sum_{j=j_{\min}}^{j_{\max}} \left\{ \left(R_{x,nj}^{\mathrm{s}}\right)^2 + \left(R_{z,nj}^{\mathrm{c}}\right)^2 + \left(R_{x,nj}^{\mathrm{c}}\right)^2 + \left(R_{z,nj}^{\mathrm{s}}\right)^2 \right\} \right]. \end{aligned} \tag{3.209}$$

Based on the experimental configuration of Busse and Whitehead (1971), $L_x = L_y = 30 - 80$ cm, $L_z = 0.5 - 1$ cm, and $a_1 \approx 2\pi/L_x \approx 0.1 - 0.2\ \mathrm{cm}^{-1}$. It was found that $N = 6$, $j_{\min} = 1 - 25$, $j_{\max} = 175$ and $\Delta_t^* = 10^{-4} - 10^{-3}$ yielded reliable results. With these parameters, the standard deviation in the (dimensionless) stochastic heat increment coefficients is typically $\mathcal{O}(10^{-12})$.

Note that in addition to the stochastic contribution to the rate of change of temperature (and the other fields), there is also a stochastic contribution to the initial state at the commencement of the transition. This is governed by the exponential of the static part of the total first entropy. The fluctuation in the internal energy contribution of the static part of the total entropy is (Attard 2012, section 6.2)

$$\begin{aligned}\Delta S_{\rm tot,st}^{(1),\rm int} &= \int_V \mathrm{d}\mathbf{r}\ \sigma_{\rm tot,st}^{\rm int}(\mathbf{r};\ \Delta T)\\ &= \int_V \mathrm{d}\mathbf{r}\ \frac{-c_{\rm p}}{2T_{00}^2}\Delta T(\mathbf{r})^2.\end{aligned}\tag{3.210}$$

This gives independent Gaussian distributions for the Fourier coefficients for the fluctuation about the initial state.

The gravitational contribution is

$$\sigma_{\rm tot,st}^{\rm g}(\mathbf{r};\ \Delta T) = \frac{\alpha m n_{00} g z}{T_{00}}\left\{1 - \frac{z\Delta_T}{L_z T_{00}}\right\}\Delta T(\mathbf{r}),\tag{3.211}$$

and the kinetic energy contribution is

$$\sigma_{\rm tot,st}^{\rm KE}(\mathbf{r};\ \Delta\mathbf{v}) = \frac{-m n_{00}}{2T_{00}}\left\{1 - \frac{z\Delta_T}{L_z T_{00}}\right\}\Delta\mathbf{v}(\mathbf{r})\cdot\Delta\mathbf{v}(\mathbf{r}).\tag{3.212}$$

Consistent with the Boussinesq approximation, these have here been neglected.

The stochastic algorithm used in Attard (2012, section 6.6) solved the deterministic hydrodynamic equations with white noise added to the initial state. The power added by the noise was $\mathcal{O}(10^5)$ times the power that would have been added by the present Gaussian distribution. Nevertheless, as will be seen, it does not appear to have affected the result for the optimum wave number following the conduction–convection or the cross-roll transition.

### 3.7.2 Linear perturbation growth rate

A plausible alternative approach to find the optimum transition from conduction to convection is based on the mode with the fastest initial growth rate. A small amplitude, periodic, convective perturbation is imposed on the conductive state and its rate of growth or decay is determined. (Linear perturbation theory more generally in convection is reviewed by Bodenschatz *et al* 2000.) It is plausible that the wave number of the perturbation with the largest positive growth rate is the wave number of the most likely final state for the conduction–convection transition.

The temperature and velocity perturbations are written with a time-dependent amplitude

$$T(\mathbf{r},\ t) = A(t)T(\mathbf{r}),\quad \text{and}\quad v_z(\mathbf{r},\ t) = A(t)v_z(\mathbf{r}),\tag{3.213}$$

so that the energy equation becomes

$$\frac{\dot{A}(t)}{A(t)} = \frac{v_z(\mathbf{r})}{T(\mathbf{r})} + \frac{1}{T(\mathbf{r})}\nabla^2 T(\mathbf{r}) = \xi. \tag{3.214}$$

Here the non-linear term has been neglected because the perturbation can be taken to be infinitesimal in its initial growth phase. The two sides of the first equality must individually be constant because they depend upon different independent variables. It follows that $A(t) = A_0 e^{\xi t}$, and $\xi$ is the growth rate of the perturbation.

The perturbation is taken to be periodic with wave number $a$,

$$\nabla_{\|}^2 T(\mathbf{r}) = -a^2 T(\mathbf{r}), \quad \text{and} \quad \nabla_{\|}^2 v_z(\mathbf{r}) = -a^2 v_z(\mathbf{r}). \tag{3.215}$$

With this the Navier–Stokes equation becomes

$$0 = -a^2 \mathcal{R} T(\mathbf{r}) + [\partial_z^2 - a^2]^2 v_z(\mathbf{r}). \tag{3.216}$$

Putting this into the energy equation and rearranging gives

$$0 = a^2 \mathcal{R} v_z(\mathbf{r}) - \xi[\partial_z^2 - a^2]^2 v_z(\mathbf{r}) + [\partial_z^2 - a^2]^3 v_z(\mathbf{r}). \tag{3.217}$$

This is the same as the result given by Reid and Harris (1958) except for the inclusion of the growth rate term, $\xi \neq 0$. The solution is evidently

$$v_z(\mathbf{r}) = \{A_1 \cosh p_1 z + A_2 \cosh p_2 z + A_3 \cosh p_3 z\} f(x, y), \tag{3.218}$$

where $f$ is periodic with wave number $a$, $\nabla_{\|}^2 f = -a^2 f$. The $p_j$ are the roots of the cubic equation

$$0 = a^2 \mathcal{R} - \xi[p^2 - a^2]^2 + [p^2 - a^2]^3. \tag{3.219}$$

The three boundary conditions at $z = \pm 1/2$ are

$$v_z = \partial_z v_z = [\partial_z^2 - a^2]^2 v_z = 0. \tag{3.220}$$

The first says that the normal component of the velocity must vanish at the upper and lower rigid surfaces. The second comes from the fact that the horizontal velocity vanishes at the surfaces (stick boundary conditions) and the compressibility equation. The third comes from the vanishing of the temperature perturbation at the surfaces and the Navier–Stokes equation. Since the boundary conditions are homogeneous, a non-trivial solution exists only if the determinant of the linear system of equations vanishes. The characteristic equation for this is

$$\begin{vmatrix} C_1 & C_2 & C_3 \\ p_1 S_1 & p_2 S_2 & p_3 S_3 \\ [p_1^2 - a^2]^2 C_1 & [p_2^2 - a^2]^2 C_2 & [p_3^2 - a^2]^2 C_3 \end{vmatrix} = 0. \tag{3.221}$$

Here $C_j \equiv \cosh p_j/2$ and $S_j \equiv \sinh p_j/2$, $j = 1, 2, 3$. Given that the roots $p_j(\mathcal{R}, a, \xi)$ have been found, this is a transcendental equation that determines the rate of growth

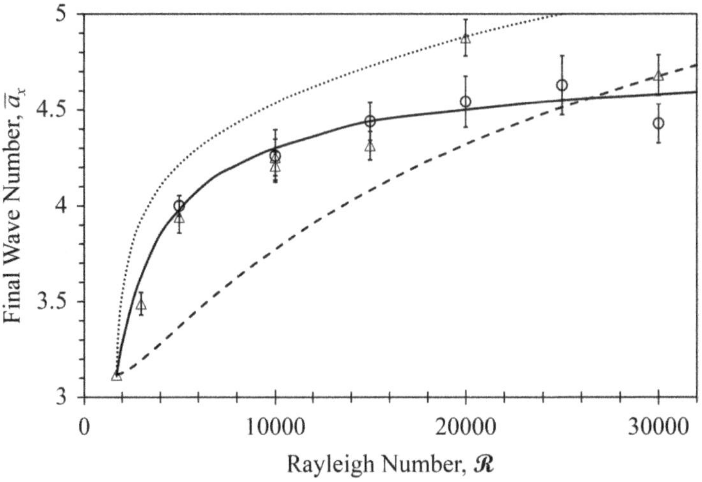


**Figure 3.5.** Wave number of the spontaneous conduction–convection transition as a function of Rayleigh number. The circles result from the stochastic, dissipative hydrodynamics equations, section 3.6, the triangles result from the conventional deterministic hydrodynamics equations with uniform white noise initially added to the temperature (Attard 2012, section 6.6), and the solid curve is the linear perturbation theory based on the largest initial growth rate, section 3.7.2. Also shown is the wave number of maximum heat flux (maximum Nusselt number, dashed curve), and the wave number of maximum sub-system first entropy (dotted curve) (Attard 2012, section 6.6).

of the mode, $\xi(\mathcal{R}, a)$. It proved relatively straightforward to solve this equation numerically. It is a reasonable proposition—absent contrary evidence in specific cases—that the preferred final wave number $\bar{a}(\mathcal{R})$ for any transition is the one that has the largest positive initial growth rate.

### 3.7.3 Wave number selection: conduction to convection

Figure 3.5 shows the most likely wave number of straight roll convection following a transition from conduction. The circles are the stochastic, dissipative hydrodynamics equations, section 3.6, the triangles are deterministic hydrodynamic equations with added initial white noise (Attard 2012, section 6.6), and the solid curve is the wave number of the initially fastest growing mode given by the linear perturbation theory, section 3.7.2. It can be seen that the mode with the initial fastest growth rate obtained from the perturbation analysis agrees with both the stochastic results. As mentioned above, the fluctuations in the initial temperature are given by equation (3.210), which is Gaussian white noise. There is no fundamental justification for instead adding uniform white noise to the initial state and then using conventional deterministic hydrodynamics equations for the transition, which is the method used in section 6.6 of Attard (2012). Moreover, the total power that was added as white noise turns out to be several orders of magnitude larger than what would have been given by equation (3.210). Nevertheless, the white noise does not appear to have done much damage, and both stochastic methods in figure 3.5 show that the initially fastest growing mode from conduction most likely establishes itself as the final steady state. The non-linear coupling in the hydrodynamic equations that comes into play as the amplitudes of the modes increases during the transition is, with rare exceptions, insufficient to change the dominant mode once it has begun to be established.

The results in figure 3.5 show a wave number that increases with increasing Rayleigh number, at first rapidly, and then, beyond about $\mathcal{R} \approx 10^4$, slowly. In contrast, the experimental measurements of Willis *et al* (1972) for a large Prandtl number silicone oil show a wave number for convection that strongly decreases with increasing Rayleigh number, in apparent contradiction of the present results. One possible reason for the discrepancy is that the present model of perfect straight rolls is too great an idealization of the experiments, and that the defects, curved rolls, and finite domains included in the experimental averages need to be accounted for. Another possibility, more germane to the present point, is that the experimental protocol of Willis *et al* corresponds to a convection–convection transition that is induced by a change in Rayleigh number, $a_x \xrightarrow{\Delta\mathcal{R}} a_x'$, whereas the present results are for the conduction–convection transition at fixed Rayleigh number. Arguably, the contradiction between the two results actually confirms the point being made: there is no such thing as an optimum non-equilibrium state, but there is an optimum transition from a given non-equilibrium state.

The figure also shows the wave number of maximum sub-system first entropy and of maximum first entropy production (equivalently maximum heat flow, also known as the Nusselt number). These are one-time properties that are a function only of the convecting state, not of the transition. It can be seen that neither the first entropy nor the rate of first entropy production determine the wave number of the convecting state selected by the transition.

It should be mentioned that at each Rayleigh number the range of wave numbers that yield stable steady state solutions to the hydrodynamic equations can be quite broad. For example, at $\mathcal{R} = 10\,000$, the present perturbation theory yields a positive growth rate in the range $0.75 \leqslant \alpha \leqslant 8.97$, and at $\mathcal{R} = 6000$ the range is $0.99 \leqslant \alpha \leqslant 7.54$, which agrees with the region of stability estimated by Reid and Harris (1958).

In the present case of the conduction–convection transition, one can conclude that the initial growth rate is responsible for wave number selection of the non-equilibrium pattern. This is a combined property of the initial state (the calculations are a perturbation carried out in the conducting state) and of the final state (the growth rate is of a trial wave number for the final steady state). It is the entropy of the transition, the second entropy, not the entropy of the final state, that is maximized by the optimum non-equilibrium pattern.

### 3.7.4 Wave number selection: cross-roll transition

The principle being argued here—that the objects of non-equilibrium systems are the transitions not the states—is exemplified by the calculated and measured data shown in figure 3.6. In this case the transition is between two linear straight roll convective states that are mutually orthogonal.

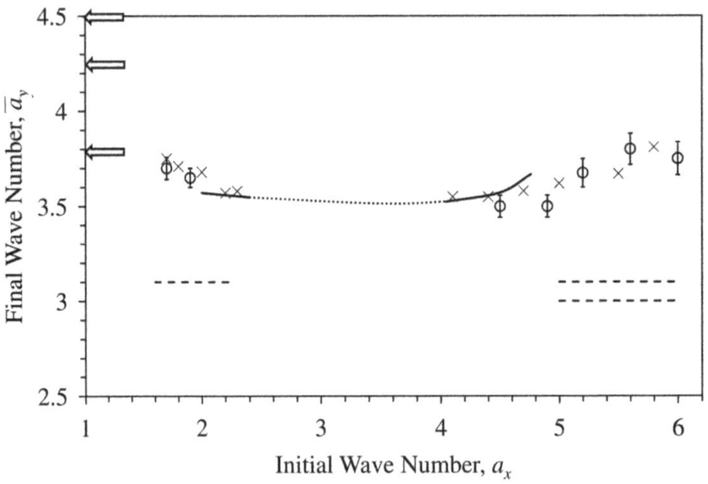


**Figure 3.6.** Cross-roll transition at a Rayleigh number $\mathcal{R} = 10^4$. The circles are the most likely final wave number for the full transition, and the crosses are the wave number $a_y$ with initial fastest growth rate, both from the hydrodynamics equations in Boussinesq approximation (see text) and initially in a stable straight roll state with wave number $a_x$. The curve is a linear perturbation theory (Busse 1967) giving the wave number with largest initial growth rate (solid portion positive, dotted portion negative) (Busse and Whitehead 1971). The horizontal dashed lines approximate the range of measured laboratory data (Busse and Whitehead 1971). The horizontal arrows on the vertical axis denote for straight rolls at $\mathcal{R} = 10^4$ the calculated wave number of maximum heat flow ($a_y = 3.77$), maximum sub-system entropy ($a_y = 4.53$), and the most likely wave number following a spontaneous transition from the conducting state ($\bar{a}_y = 4.25$).

In the experiments the system was initially given a wave number by a shadow mask and intense light source, and the transition to an orthogonal state upon removal of the light source was observed (Busse and Whitehead 1971). Although the final state was generally dominated by straight rolls, systems with defects, curved rolls, and finite domains were also used in the reported results.

The circles represent the full cross-roll transition, which were obtained using typically $N = 6$ vertical $z$-modes, $Q = 6$ horizontal $x$-modes (harmonics of $a_x$), and $j_{\max} = 150$ horizontal $y$-modes (harmonics of $a_1 = 0.2$, with $j_{\min} = 1$). Both the conventional deterministic hydrodynamic equations and the stochastic, dissipative hydrodynamic equations of section 3.6 were used. The initial conditions were either no added noise, Gaussian white noise (section 3.6), or uniform white noise that was several orders of magnitude larger (Attard 2012, section 6.6). No systematic differences between the various cases were found.

The initial growth rate was also estimated by again solving both the deterministic and the stochastic hydrodynamic equations in Boussinesq approximation, with or without added white noise. Again no systematic differences between the various cases were found. A particular fundamental mode $a_y$ and five harmonics was specified, and the exponential growth rate during the initial, linear part of the transition was obtained. This was repeated with different specified fundamental modes to find the one with the largest growth rate.

The calculations for the initial growth rate are in principle identical to the linear perturbation theory of Busse (1967), notwithstanding slightly different computational parameters. The latter essentially solves the deterministic hydrodynamic equations in Boussinesq approximation using a Fourier expansion with $N + Q = 6 - 12$ modes, and finds the wave number of the perturbation to the initial straight roll steady state with the largest initial growth rate. If this is negative the initial state is stable and no transition is expected to occur; if it is positive the initial state is unstable and a cross-roll transition is expected to occur. In the latter case one expects that at least the initial growth will be dominated by the transverse straight roll state with that wave number. Numerical results from this perturbation approach are given by Busse and Whitehead (1971) and are shown as the curves in figure 3.6.

It can be seen in figure 3.6 that the two calculations of the mode with the initial fastest growth rate are in good agreement with each other. They are also in relatively good agreement with the most likely final wave number for the full transition. In some ways this agreement is surprising because in the full transitions it was often observed that, due to the stochastic nature of the process, the dominant wave number at small times was not always the calculated most likely fastest growing mode. It was also observed that the wave number that dominated most of the transition was sometimes overtaken by another mode toward the end of the transition, due no doubt to non-linear coupling that becomes important in the late stages. The fact that there is statistical noise in the final wave number also indicates that the fastest initial growth rate is not the *sole* determinant of the final state. Despite these phenomena, it can be seen in figure 3.6 that the linear perturbation theory for the fastest initial growth rate gives a quite good estimate of the final wave number of the cross-roll transition. Computationally, the linear perturbation theory obtains the initially fastest growing mode many, many orders of magnitude more efficiently than the full calculation of the final transition.

Figure 3.6 shows that the calculated results overestimate the experimental measured final state (Busse and Whitehead 1971) by about half a wave number. In the experiments, the cross-roll transition often begins independently over small domains, which grow and join up to occupy the entire convection box, often with defects (Busse and Whitehead 1971). In the calculations, combinations of ideal straight rolls are used that occupy the entire convection box. Also, in the computations, the initial $y$-rolls are constrained to the fundamental wave number $a_x$ and up to five harmonics. Any transition pathway that proceeds by occupying a spectrum of nearby wave numbers is precluded in the computations. Such pathways may or may not contribute to the experimentally measured cross-roll transition. Finally, it is likely that mechanical noise from water flow in the heat bath in the experiments is much larger than the statistical noise used here. These differences may possibly account for the discrepancy between the calculated final wave numbers in figure 3.6 and the measured ones.

In figure 3.6, as in figure 3.5, it can be seen that the optimum final state is not equal to the state that maximizes the rate of (first) entropy production (in this case maximum heat flow), nor to the state that maximizes the sub-system first entropy. It is also not equal to the optimum state that follows a transition from conduction at this Rayleigh number.

It can be mentioned that the total entropy monotonically increased during these spontaneous transitions (Attard 2012, section 6.6). The sub-system entropy and the sub-system dependent part of the total entropy (i.e. the static entropy) were higher in the final state than in the initial state, as was the rate of entropy production (the heat flux, or the Nusselt number). The increase in these was not monotonic during a spontaneous transition.

The various approaches in figure 3.6 agree that the final steady state depends on the initial steady state. All three theories show that the final wave number varies with the initial wave number, $\bar{a}_y(a_x)$. Also, both the experiments and the theories identify a range of initial wave numbers for which no transition occurs.

The fact that the final wave number varies with the initial wave number precludes the existence of *any* one-time property that is optimized by the final state. The data in figure 3.6 are a graphic illustration of the general principle that for a non-equilibrium system there is no such thing as an optimum state. Rather, there is an optimum transition, or, equivalently, an optimum final state conditional upon a given initial state.

## Summary

- The Second Law of Thermodynamics characterizes equilibrium systems. For these it gives a variational principle, namely that the constrained entropy for the corresponding not-in-equilibrium system is a maximum. Non-equilibrium systems are *not* characterized by any analogy of the Second Law based on extremizing the rate of change of entropy.
- The second entropy characterizes transitions between two states over a time interval, and hence also rates of change and fluxes. For a constrained non-equilibrium system, the second entropy is a maximum for the optimum rate or flux. The reduction condition says that the maximum value of the second entropy equals the first entropy for the initial state.
- Fluctuations of non-conserved extensive variables are Gaussian distributed according to a quadratic form for the second entropy. The most likely regression of such a fluctuation maximizes the second entropy. A small time expansion of the second entropy shows that a fluctuation most likely regresses at a constant rate in proportion to the gradient of the first entropy. The proportionality constant comes from the leading coefficient of the small time expansion of the second entropy and may be identified with the hydrodynamic transport matrix.
- The hydrodynamic transport matrix is symmetric, which is known as the Onsager reciprocal relations, and it can be related to the time correlation function, which is known as the Green–Kubo relations.
- For the case of a steady state thermodynamic non-equilibrium system, the appropriate extensive non-conserved variable to be constrained is typically the first moment of an extensive conserved variable. Its spontaneous fluctuation when isolated can also be induced by two spatially separated reservoirs with a difference between the respective conjugate field variables. The

adiabatic (isolated) regression of the fluctuation is compensated by the rate of exchange with the reservoirs, which means that the structure of the induced moment is constant, as is the driven flux.

- Hydrodynamics can be formulated as a variational principle in which the second entropy is maximized by the optimum fluxes for the specified thermodynamic forces. Stochastic contributions to the conventional hydrodynamic equations are distributed according to the exponential of the second entropy.
- The conditionally optimum state or pattern of a non-equilibrium system is characterized by maximizing the second entropy for the transition, and it depends upon the given initial state.

## References


Attard P 2005a Statistical mechanical theory for steady state systems. II. Reciprocal relations and the second entropy *J. Chem. Phys.* **122** 154101

Attard P 2005b Statistical mechanical theory for steady state systems. III. Heat flow in a Lennard-Jones fluid *J. Chem. Phys.* **122** 244105

Attard P 2006 Statistical mechanical theory for steady state systems. VI. Variational principles *J. Chem. Phys.* **125** 214502

Attard P 2012 *Non-Equilibrium Thermodynamics and Statistical Mechanics: Foundations and Applications* (Oxford: Oxford University Press)

Biot M A 1955 Variational principles in irreversible thermodynamics with application to viscoelasticity *Phys. Rev.* **97** 1463

Biot M A 1955 1975 A virtual dissipation principle and Lagrangian equations in nonlinear irreversible dynamics *Bull. Acad. R. de Belgique* **61** 6–30

Bochkov G N and Kuzovlev Y E 1980 On the variational principle of non-equilibrium non-linear statistical thermodynamics *JETP Lett.* **30** 46

Bodenschatz E, Pesch W and Ahlers G 2000 Recent developments in Rayleigh–Bénard convection *Annu. Rev. Fluid Mech.* **32** 709

Busse F H 1967 On the stability of two-dimensional convection in a layer heated from below *J. Math. Phys.* **46** 140

Busse F H and Whitehead J A 1971 Instabilities of convection rolls in a high Prandtl number fluid *J. Fluid Mech.* **47** 305

de Groot S R and Mazur P 1984 *Non-Equilibrium Thermodynamics* (New York: North-Holland)

Dewar R 2003 Information theory explanation of the fluctuation theorem, maximum entropy production and self-organized criticality in non-equilibrium stationary states *J. Phys. Math. Gen.* **36** 631

Doob J L 1942 The Brownian movement and stochastic equations *Ann. Math.* **43** 351

Eddington A S 1928 *The Nature of the Physical World* (New York: Macmillan)

Fox R F and Uhlenbeck G E 1970 Contributions to non-equilibrium thermodynamics. I. Theory of hydrodynamical fluctuations *Phys. Fluids* **13** 1893–2881

Fisher I 1906 *The Nature of Capital and Income* (New York: Macmillan)

Gage D H, Schiffer M, Kline S J and Reynolds W C 1966 *Non-Equilibrium Thermodynamics: Variational Techniques and Stability* ed R J Donnelly (Chicago, IL: University of Chicago Press)

Green M S 1954 Markoff random processes and the statistical mechanics of time-dependent phenomena. II. Irreversible processes in fluids *J. Chem. Phys.* **22** 398

Gyarmati I 1968 On the most general form of the thermodynamic integral principle *Z. Phys. Chem.* **239** 133

Gyarmati I 1970 *Non-Equilibrium Thermodynamics: Field Theory, and Variational Principles* (Berlin: Springer)

Hashitsume N 1952 A statistical theory of linear dissipative systems I *Prog. Theor. Phys.* **8** 461

Hashitsume N 1956 A statistical theory of linear dissipative systems II *Prog. Theor. Phys.* **15** 369

Hunt K L C, Hunt P M and Ross J 1987 Dissipation in steady states of chemical systems and deviations from minimum entropy production *Physica* A **147** 48

Keizer J and Fox R F 1974 Qualms regarding the range of validity of the Glansdorff–Prigogine criterion for stability of non-equilibrium states *Proc. Natl Acad. Sci. USA* **71** 192

Keizer J 1987 *Statistical Thermodynamics of Non-Equilibrium Processes* (New York: Springer)

Kleidon A and Lorenz R D 2005 *Non-Equilibrium Thermodynamics and the Production of Entropy: Life, Earth, and Beyond* (Berlin: Springer)

Kortweg D J 1883 On a general theorem of the stability of the motion of a viscous fluid *Phil. Mag.* **16** 112

Kubo R 1966 The fluctuation–dissipation theorem *Rep. Prog. Phys.* **29** 255

Kubo R, Toda M and Hashitsume N 1978 *Statistical Physics II. Non-Equilibrium Statistical Mechanics* (Berlin: Springer)

Landau L D and Lifshitz E M 1957 Hydrodynamic fluctuations *Sov. Phys. JETP* **5** 512

Landau L D and Lifshitz E M 1959 *Fluid Mechanics* (Oxford: Pergammon Press)

Lavenda B H 1985 *Nonequilibrium Statistical Thermodynamics* (Chichester: Wiley)

Martyushev L M and Seleznev V D 2006 Maximum entropy production principle in physics, chemistry and biology *Phys. Rep.* **426** 1–45

Onsager L 1931 Reciprocal relations in irreversible processes I *Phys. Rev.* **37** 405

Onsager L 1931 Reciprocal relations in irreversible processes II *Phys. Rev.* **38** 2265

Onsager L and Machlup S 1953 Fluctuations and irreversible processes *Phys. Rev.* **91** 1505

Ortiz de Zárate J M and Sengers J V 2006 *Hydrodynamic Fluctuations in Fluids and Fluid Mixtures* (Amsterdam: Elsevier)

Paltridge G W 1979 Climate and thermodynamic systems of maximum dissipation *Nature* **279** 630

Prigogine I 1967 *Introduction to Thermodynamics of Irreversible Processes* (New York: Interscience)

Paltridge G W 2005 *Non-equilibrium Thermodynamics and the Production of Entropy: Life, Earth, and Beyond* ed A Kleidon and R D Lorenz (Berlin: Springer) p 33

Reid W H and Harris D L 1958 Some further results on the Bénard problem *Phys. Fluids* **1** 102

Ross J and Vlad M O 2005 Exact solutions for the entropy production rate of several irreversible processes *J. Phys. Chem.* **109** 10607

Schneider E D and Kay J J 1994 Life as a manifestation of the second law of thermodynamics *Math. Comput. Modelling* **19** 25

Schöpf W and Rehberg I 1994 The influence of thermal noise on the onset of travelling-wave convection in binary fluid mixtures: an experimental investigation *J. Fluid Mech.* **271** 235

Strutt J W 1871 Some general theorems relating to vibrations *Proc. Math. Soc. London* **4** 357

Strutt J W 1913 On the motion of a viscous fluid *Phil. Mag.* **26** 776

Swenson R and Turvey M T 1991 Thermodynamic reasons for perception-action cycles *Ecol Psychology* **3** 317

van Beijeren H and Cohen E G D 1988 The effects of thermal noise in a Rayleigh–Bénard cell near its first convective instability *J. Stat. Phys.* **53** 77
von Helmholtz H L F 1969 *Collected Works* (Leipzig) vol I 223
Willis G E, Deardorff J W and Somerville R C 1972 Roll-diameter dependence in Rayleigh convection and its effect upon the heat flux *J. Fluid Mech.* **54** 351
Zwanzig R 2001 *Non-Equilibrium Statistical Mechanics* (Oxford: Oxford University Press)

**IOP** Publishing

Entropy Beyond the Second Law
Thermodynamics and statistical mechanics for equilibrium, non-equilibrium, classical, and quantum systems

**Phil Attard**

# Chapter 4

## Entropy in motion

'All chance [is] direction which thou canst not see'

Pope (1734)

'We ought to regard the present state of the Universe as the effect of its antecedent state and as the cause of the state that is to follow'

Laplace (Weinert 2005)

The main aim of this chapter is to derive the equations of motion for macroscopic variables. Such variables range from the traditional Brownian particles to the more abstract thermodynamic fluctuations. Some of the results also apply to molecular coordinates, and to the evolution of the probability density. It will be shown in general that because the reservoir coordinates are projected out of the problem, the equations of motion are both stochastic and dissipative.

This dual consequence of the projection of the reservoir onto the sub-system is seldom taken into account in contemporary treatments of statistical systems. In particular, deterministic thermostatted equations of motion are widely used in computer simulations, as exemplified by the isokinetic thermostat,

$$\dot{\mathbf{p}} = -\nabla_{\mathbf{q}}\mathcal{H}(\mathbf{p}, \mathbf{q}) + \alpha\mathbf{p}, \quad \alpha(\mathbf{\Gamma}) \equiv \frac{1}{\mathbf{p}\cdot\mathbf{p}}\mathbf{p}\cdot\nabla_{\mathbf{q}}\mathcal{H}(\mathbf{p}, \mathbf{q}). \tag{4.1}$$

Here $\mathbf{q}$ are the positions, $\mathbf{p}$ are the momenta, and $\mathcal{H}$ is the Hamiltonian. Because this neglects the stochastic part of the projection operation, it violates the fluctuation–dissipation theorem, one of the most fundamental theorems of statistical mechanics.

A related misunderstanding invokes Liouville's theorem for the evolution of the probability density together with such deterministic, non-Hamiltonian equations of motion,

$$\frac{\mathrm{d}\wp(X, t)}{\mathrm{d}t} = \frac{\partial\wp(X, t)}{\partial t} + \dot{X}^{\mathrm{det}}\cdot\nabla\wp(X, t). \tag{4.2}$$

doi:10.1088/978-0-7503-1590-6ch4

Here $X$ is the macrostate, $\dot{X}^{\mathrm{det}}$ is its deterministic velocity, and $\wp(X, t)$ is its probability. It will be shown that in general this is an approximation rather than an exact theorem, and that its accuracy and reliability depends on the circumstances of each particular application.

## 4.1 Brownian motion

Brownian motion refers to the random movements of a microscopic solute or colloid particle in a solvent (Brown 1828, Haw 2002). This is a prototype of all stochastic processes. Einstein (1905) and Smoluchowski (1906) gave the earliest statistical treatments of Brownian motion, and these provided more or less direct evidence for the existence of molecules and the validity of statistical mechanics. The diffusion of the probability density that was the basis of their treatment was further developed by Fokker (1914) and Planck (1916). Langevin (1908) gave a stochastic dissipative equation that described Brownian motion itself, and this has become the general model for almost all the different types of random processes.

Here the stochastic dissipative equation for Brownian motion is derived from the second entropy, first in the Einstein–Smoluchowski form for position macrostates, and then in the more useful Langevin form for velocity macrostates. Following these the Fokker–Planck equation is discussed.

### 4.1.1 Position macrostates and Einstein's treatment

In this and the following sub-section we regard the Brownian particle's position $\mathbf{r}$ as labeling the macrostate of the system. The formalism for the time evolution of the macrostate is just the pure parity fluctuation case of chapter 3. The solvent in which the particle is embedded is treated as a reservoir and its microstates are not specified.

We shall suppose that the Brownian particle experiences a potential trap $U(\mathbf{r})$ centered on the origin, $U'(\mathbf{0}) = 0$ and $U''(\mathbf{0}) > 0$, so that the average position vanishes,

$$\langle \mathbf{r} \rangle = \mathbf{0}. \tag{4.3}$$

The reason for requiring this is so that non-zero values of $\mathbf{r}$ can be treated as a fluctuation. With the pinning potential, $\langle r^2 \rangle < \infty$.

The limit $U(\mathbf{r}) \to 0$ gives results for the free Brownian particle. It will turn out that in this section the final results are independent of $U(\mathbf{r})$, and so this is a conceptually clear way of obtaining the free particle results even in the thermodynamic limit.

For a solvent of temperature $T$, the reservoir entropy for the macrostate $\mathbf{r}$ is

$$S_{\mathrm{r}}(\mathbf{r}) = \frac{-U(\mathbf{r})}{T}. \tag{4.4}$$

For a Brownian particle, the reservoir entropy is the total entropy. The fluctuation matrix is the matrix of second derivatives of the total entropy,

$$S'' = \frac{-1}{T} U''(\mathbf{0}). \tag{4.5}$$

The probability is just the exponential of the entropy,

$$\wp(\mathbf{r}) = \frac{1}{Z(T)} e^{S_r(\mathbf{r})/k_B T}. \tag{4.6}$$

For a parabolic potential, this is a Gaussian in the position. Hence the average of the square of the position is

$$\langle \mathbf{rr} \rangle = -k_B S''^{-1}. \tag{4.7}$$

The second entropy in the pure parity case is now applied to the transition between positions of the Brownian particle. For this purpose equation (3.86) is

$$\mathbf{r}' = \mathbf{r} + \frac{|\tau|}{2} \Lambda_r S'' \mathbf{r} + \tilde{\mathbf{R}}, \quad |\tau| \gg \tau_{\text{molec}}. \tag{4.8}$$

Here $\tau_{\text{molec}}$ is a molecular time scale. Assuming a spherical particle, the transport coefficient is diagonal with the same values on the diagonal. It can be written as

$$\Lambda_r = 2k_B^{-1} D \, \mathrm{I}. \tag{4.9}$$

Here and below I is the $3 \times 3$ identity matrix. The quantity $D$ is the diffusion constant, as will be justified by comparison with Einstein's result below.

This equation does not include explicitly any adiabatic contribution to the transition. For a small time step this would be $\mathbf{r}^0(\tau|\mathbf{r}) = \mathbf{r} + \tau\mathbf{v}$. In the present case this would vanish since the transition is not conditioned on the initial velocity, and its most likely or average value is zero, $\mathbf{v} = 0$. The direct adiabatic force due to the applied potential causes a change of particle velocity. But the particle velocity is projected out of the problem, taking with it the adiabatic force. These effects are subsumed in the reservoir entropy. That is, the effects of the external force are included in a cumulative manner in the dissipative term, $\nabla S_r = S''\mathbf{r}$.

The probability distribution for the stochastic part is the Gaussian, equation (3.88),

$$\begin{aligned} \wp(\tilde{\mathbf{R}}) &= \frac{1}{Z'} e^{-\Lambda_r^{-1}:\tilde{\mathbf{R}}\tilde{\mathbf{R}}/2k_B|\tau|} \\ &= \frac{1}{(4\pi D|\tau|)^{3/2}} e^{-\tilde{\mathbf{R}}\cdot\tilde{\mathbf{R}}/4D|\tau|}. \end{aligned} \tag{4.10}$$

This is effectively the probability distribution of the step taken by a free Brownian particle over the time interval $\tau$. It is the same as a step in a random walk. It is evidently a Gaussian, as sketched in figure 4.1, with the width increasing as the square root of the time interval.

From this it follows that the stochastic change on average vanishes, $\langle \tilde{\mathbf{R}} \rangle = 0$, and it has variance

$$\langle \tilde{\mathbf{R}}\tilde{\mathbf{R}} \rangle = |\tau| k_B \Lambda_r = 2|\tau| D \, \mathrm{I}. \tag{4.11}$$

This last result is a form of the fluctuation–dissipation theorem: the variance of the fluctuations is proportional to the strength of the dissipative term.

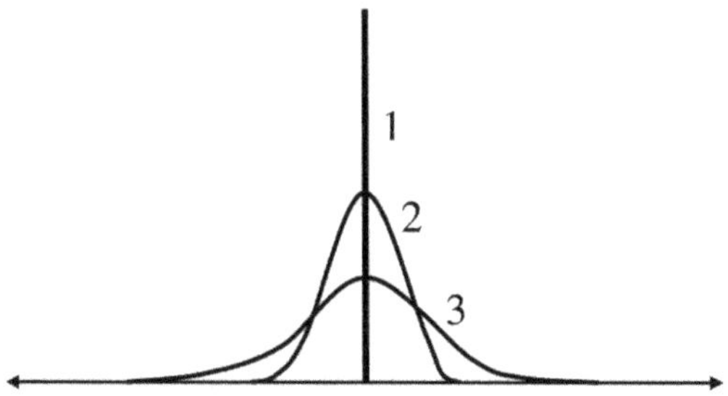


**Figure 4.1.** The backward or forward evolution of the position probability of a free Brownian particle. The central spike is the $\delta$-function for the current time.

Note that there is no correlation between the stochastic change at successive time steps, $\langle \tilde{\mathbf{R}}(t+\tau)\tilde{\mathbf{R}}(t)\rangle = 0$. This means that this is a Markov system in which one can string successive transitions together to form a trajectory.

The correlation between the macrostate change over the transition and the initial macrostate is

$$\begin{aligned}\left\langle [\mathbf{r}(t+\tau)-\mathbf{r}(t)]\mathbf{r}(t)\right\rangle &= \frac{|\tau|}{2}\Lambda_r S''\langle \mathbf{r}(t)\mathbf{r}(t)\rangle \\ &= -k_B\frac{|\tau|}{2}\Lambda_r S'' S''^{-1} \\ &= \frac{-k_B|\tau|}{2}\Lambda_r \\ &= -|\tau| D \mathrm{I}.\end{aligned} \tag{4.12}$$

This result is independent of the external potential $U(\mathbf{r})$. Hence taking the limit $U(\mathbf{r}) \to 0$ shows that this result holds also for the free Brownian particle.

One can write this in a more symmetric form as

$$\begin{aligned}&\left\langle [\mathbf{r}(t+\tau)-\mathbf{r}(t)][\mathbf{r}(t+\tau)-\mathbf{r}(t)]\right\rangle \\ &= \langle [\mathbf{r}(t+\tau)-\mathbf{r}(t)]\mathbf{r}(t+\tau)\rangle - \langle [\mathbf{r}(t+\tau)-\mathbf{r}(t)]\mathbf{r}(t)\rangle \\ &= 2|\tau| D\mathrm{I}.\end{aligned} \tag{4.13}$$

Taking the trace of this dyadic matrix gives the result in the form originally given by Einstein (1905),

$$\left\langle [\mathbf{r}(t+\tau)-\mathbf{r}(t)]\cdot[\mathbf{r}(t+\tau)-\mathbf{r}(t)]\right\rangle = 6|\tau| D. \tag{4.14}$$

The agreement justifies the identification of $D$ as the diffusion constant for the particle in the particular solvent. One notes that the root mean square displacement of the Brownian particles grows with the square root of time. This is the well-known result for a random walk.

Differentiating the Einstein result with respect to $\tau$ yields

$$\left\langle \mathbf{v}(t+\tau)\cdot[\mathbf{r}(t+\tau)-\mathbf{r}(t)]\right\rangle = -\langle \mathbf{v}(t+\tau)\cdot\mathbf{r}(t)\rangle = 3\hat{\tau} D. \tag{4.15}$$

The first equality follows because variables of opposite time parity are instantaneously uncorrelated in an equilibrium system, $\langle \mathbf{v}(t+\tau)\cdot\mathbf{r}(t+\tau)\rangle = 0$. The result says that from the starting position, the particle moves toward the origin at constant velocity.

### 4.1.2 Langevin equation

We now turn to the stochastic dissipative equation given for Brownian motion by Langevin (1908). We shall derive the equations of motion from the pure parity second entropy result, equation (3.86), and show their equivalence to Langevin's result, giving an interpretation of his physical justification.

The Langevin equation treats the Brownian particle's position $\mathbf{r}$ and velocity $\mathbf{v}$ explicitly, and so we take $\{\mathbf{r}, \mathbf{v}\}$ as a macrostate of the system. The Hamiltonian is the sum of the kinetic and potential energies of the particle,

$$\mathcal{H}(\mathbf{r}, \mathbf{v}) = \mathcal{K}(\mathbf{v}) + U(\mathbf{r}), \quad \mathcal{K}(\mathbf{v}) = \frac{m}{2}v^2 = \frac{1}{2m}p^2, \tag{4.16}$$

where $\mathbf{p} = m\mathbf{v}$ is the particle's momentum and $m$ is its mass.

The reservoir entropy for a position-velocity macrostate is just

$$S_r(\mathbf{r}, \mathbf{v}) = \frac{-\mathcal{H}(\mathbf{r}, \mathbf{v})}{T}, \tag{4.17}$$

and the fluctuation or second derivative matrix is

$$S'' = \frac{-1}{T}\begin{pmatrix} U''(\mathbf{0}) & 0 \\ 0 & m \end{pmatrix}. \tag{4.18}$$

As has been mentioned, for the Brownian particle the reservoir entropy is the same as the total entropy.

Denote the trajectory of the Brownian particle in position-velocity space by $\{\mathbf{r}(t), \mathbf{v}(t)\}$. A transition over the time interval $\tau$ has the form

$$\begin{aligned} \mathbf{r}(t+\tau) &= \mathbf{r}(t) + \tau\mathbf{v}(t) \\ \mathbf{v}(t+\tau) &= \mathbf{v}(t) - \frac{\tau}{m}\nabla U(\mathbf{r}) + \frac{|\tau|}{m}\overline{\mathbf{F}}_r(\mathbf{r}, \mathbf{v}) + \frac{1}{m}\tilde{\mathbf{F}}_r(\tau). \end{aligned} \tag{4.19}$$

This assumes that $\tau$ is of molecular or infinitesimal dimensions.

This transition contains both adiabatic and reservoir (solvent) contributions. The adiabatic contribution is

$$\begin{aligned} \mathbf{r}^0(\tau|\mathbf{r}, \mathbf{v}) &= \mathbf{r} + \tau\mathbf{v}, \\ \mathbf{v}^0(\tau|\mathbf{r}, \mathbf{v}) &= \mathbf{v}(t) - \frac{\tau}{m}\nabla U(\mathbf{r}). \end{aligned} \tag{4.20}$$

These are just Hamilton's equations of motion for the applied force $-\nabla U(\mathbf{r})$.

The reservoir contributions consist of a deterministic or most likely force $\overline{\mathbf{F}}_r(\mathbf{r}, \mathbf{v})$, and a stochastic force $\tilde{\mathbf{F}}_r(\tau)$. (The dependence of the stochastic force on the time interval will be obtained below.) These arise from the random (but statistically predictable) forces of the solvent molecules on the Brownian particle. These are real physical forces. These solvent forces effect the position evolution via the acceleration at $\mathcal{O}(\tau^2)$, which for infinitesimal $\tau$ is negligible over a single time step. The fact that the reservoir forces are irreversible (i.e. proportional to $|\tau|$ rather than to $\tau$) is justified by the symmetry arguments used to derive the second entropy for the pure parity case.

Comparing this functional form to the second entropy result, equation (3.86), one sees several similarities and differences. First, the position evolution is purely adiabatic and is not treated as a stochastic transition governed by the second entropy. This is justified for infinitesimal time intervals, as argued above. The position is a 'slave' to the velocity over the time interval, which explains why the problem is treated as stochastic in the velocity only. This is, therefore, a pure parity case, rather than a mixed parity case of position and velocity.

Second, whereas the present equation for the velocity contains both adiabatic and statistical contributions, the pure parity second entropy result, equation (3.86), contains only statistical contributions. This difference is explained by the fact that the fluctuations treated in chapter 3 were for macrostates with no external potential acting on the macrostate. Their evolution had to be treated statistically because the microstates necessary to determine adiabatic motion were unspecified. In the present case the applied potential acts directly on the transition between macrostates and this has to be directly accounted for.

For the present problem there is complete separation between the applied force acting on the Brownian particle, which gives the adiabatic contribution to the infinitesimal transition, and the solvent forces, which are characterized statistically by the second entropy. The adiabatic forces are reversible, $\propto \tau$, whereas the dissipative and stochastic forces are irreversible, $\propto |\tau|$, as the second entropy analysis showed.

Third, the present and former problem are treated at different time scales. The second entropy result, equation (3.86), was derived for small time intervals, which are much larger than the present infinitesimal time intervals. This difference in time scales raises the question of whether it is valid to apply the second entropy analysis over infinitesimal time scales as here. The original justification for requiring small time scales, say $|\tau| \gg |\tau_{\rm molec}|$, was that a fluctuation had an initial inertial period in which the velocities became organized into a steady flux. Time symmetry arguments showed that the term giving this steady flux was proportional to $|\tau|$ in the pure parity case. There are three arguments for extrapolating this result down to infinitesimal time scales in the present case. First, the present velocity macrostate is the end point of a continuous evolution, and the solvent already has had adequate time to become organized. This is particularly true in the case that the mass of the Brownian particle is much greater than the mass of the solvent molecules. Second, the relaxation time for velocity is much shorter than for position. And finally third, any term linearly proportional to $|\tau|$ can be broken down into a series of infinitesimal transitions each proportional to $|\tau_{\rm molec}| \ll |\tau|$, and over the longer term these in series return the original expression.

The fourth point in the comparison is that the second entropy approach explicitly determines the dissipative and stochastic forces in equation (3.86), whereas without additional assumptions these are undetermined in the Langevin equation. As a historical fact Langevin assumed that the dissipative force was the hydrodynamic drag force, and that the variance of the stochastic force was such that the equipartition theorem was satisfied. Such assumptions are quite plausible on physical grounds, and have been proven correct over a century of application. Nevertheless, it would be better to have a derivation of these from first principles, if for no other reason than to explain how it is possible that the hydrodynamic drag force can be applicable on molecular length scales. Also, a first principles derivation likely indicates the full generality of the stochastic dissipative equation, how to extend it to non-equilibrium systems, and how to apply them beyond the physical sciences to fields such as sociology, economics, health, etc. Obviously such extensions cannot invoke hydrodynamic drag forces or the equipartition theorem.

This last point can be made explicit by applying equation (3.86) to the present problem. Adding the adiabatic transition, the stochastic dissipative equation for Brownian motion is

$$\begin{aligned} \mathbf{r}(t+\tau) &= \mathbf{r}(t) + \tau\mathbf{v}(t) \\ \mathbf{v}(t+\tau) &= \mathbf{v}(t) - \frac{\tau}{m}\nabla U(\mathbf{r}) + \frac{|\tau|}{2}\Lambda_v S''_{vv}\mathbf{v} + \tilde{\mathbf{R}}_\mathrm{v} \\ &= \mathbf{v}(t) - \frac{\tau}{m}\nabla U(\mathbf{r}) - \frac{|\tau|\gamma}{m}\mathbf{v} + \tilde{\mathbf{R}}_\mathrm{v}. \end{aligned} \tag{4.21}$$

Here the entropy for the velocity macrostate is $S_\mathrm{r}(\mathbf{v}) = -mv^2/2T$, which has second derivative, $S''_{vv} = -m\mathrm{I}/T$, the transport matrix is proportional to the identity matrix, $\Lambda_v = 2m^{-2}T\gamma\mathrm{I}$, and the stochastic force is Gaussian distributed, equation (3.88), $\wp(\tilde{\mathbf{R}}_\mathrm{v}) = Z'^{-1}e^{-\Lambda_v^{-1}:\tilde{\mathbf{R}}_\mathrm{v}\tilde{\mathbf{R}}_\mathrm{v}/2k_\mathrm{B}|\tau|} = Z'^{-1}e^{-m^2\tilde{R}_v^2/4\gamma|\tau|k_\mathrm{B}T}$. Shortly $\gamma$ will be identified with the friction or drag coefficient.

One sees that in the second entropy theory the dissipative force in general takes the form of a drag force, $\overline{\mathbf{F}}_\mathrm{r} \propto \mathbf{v}$. It is not that the dissipative force arises from hydrodynamic drag, but rather that hydrodynamic drag forces themselves are dissipative forces. The very general statistical considerations of the second entropy show that dissipative forces are proportional to the gradient of the entropy. Since the entropy contains the kinetic energy, which is quadratic in the velocity, its velocity gradient is proportional to the velocity. The present result for a Brownian particle in a solvent would also apply to a Brownian particle in a solid, where no hydrodynamic forces are present. (Friction forces are also dissipative forces that arise from the gradient of the entropy.) Even though hydrodynamic forces and friction forces only hold on macroscopic length scales, dissipative forces are statistical in origin and apply on molecular and even sub-molecular length scales.

In the second entropy analysis there is a single parameter—the transport coefficient—that determines both the variance of the stochastic force and the magnitude of the dissipative force. In contrast, conventional treatments of the theory of Langevin (1908) regard the two forces as separate and determine the variance of the stochastic force from the equipartition theorem (see Pathria 1972 section 13.4). Figure 4.2 shows the most likely trajectory of a Brownian particle.

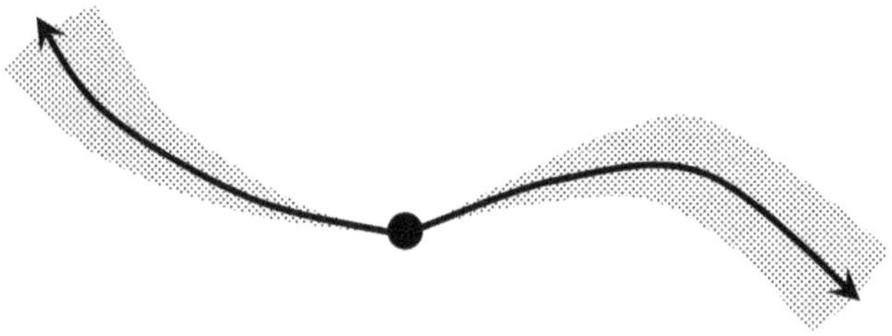

**Figure 4.2.** The most likely trajectory of a Brownian particle forward and backward in time from the present time (circle). The shaded areas indicate the stochastic contribution within a one-standard deviation envelope.

This contains the adiabatic contributions, which are time reversible, and the dissipative contributions, which are time irreversible. Hence the discontinuous first derivative in the trajectory at $t = 0$. The stochastic contributions perturb the trajectory, increasingly so as time progresses or regresses from the present.

The discontinuity in the derivative of the trajectory means that one has to specify whether the time derivative of a function on the trajectory is the forward or the backward derivative,

$$\begin{aligned}\frac{\mathrm{d}^{\pm} f(\mathbf{v}(t), t)}{\mathrm{d}t} &= \frac{\partial f(\mathbf{v}(t), t)}{\partial t} + \dot{\mathbf{v}}^{\pm}(t) \cdot \nabla_{\mathrm{v}} f(\mathbf{v}(t), t) \\ &= \frac{\partial f(\mathbf{v}(t), t)}{\partial t} + \frac{\mathbf{v}(t \pm |\tau|) - \mathbf{v}(t)}{\pm|\tau|} \cdot \nabla_{\mathrm{v}} f(\mathbf{v}(t), t) \\ &= \frac{\partial f(\mathbf{v}(t), t)}{\partial t} + \left[\frac{-1}{m}\nabla U(\mathbf{r}) \mp \frac{\gamma}{m}\mathbf{v} \pm \frac{1}{|\tau|}\tilde{\mathbf{R}}_{\mathrm{v}}\right] \cdot \nabla_{\mathrm{v}} f(\mathbf{v}(t), t).\end{aligned} \tag{4.22}$$

The upper sign refers to the forward derivative and the lower sign to the backward derivative.

### 4.1.3 Markov behavior

The second entropy analysis of Brownian motion in terms of position (i.e. Einstein's result) is consistent with the analysis in terms of velocity (i.e. Langevin's result). This may be seen as follows.

For a free Brownian particle, $U(\mathbf{r}) = 0$, the most likely rate of change of velocity given by equation (4.21) is

$$\dot{\overline{\mathbf{v}}}(t) = \frac{-\hat{\tau}\gamma}{m}\mathbf{v}(t). \tag{4.23}$$

Since $\tau$ is molecular scale, this takes the coarse velocity to equal the instantaneous velocity, $[\mathbf{v}(t + \tau) - \mathbf{v}(t)]/\tau = \dot{\mathbf{v}}(t)$.

According to this, the rate of change of velocity depends only on the current velocity, and so this is a Markov process. The solution is

$$\overline{\mathbf{v}}(t) = e^{-|t|\gamma/m}\mathbf{v}(0). \tag{4.24}$$

This says that the velocity decays exponentially from the current value (and to the current value).

The time correlation of the velocity is also exponential,

$$\begin{aligned}\langle \mathbf{v}(t)\cdot\mathbf{v}(0)\rangle &= \langle \overline{\mathbf{v}}(t)\cdot\mathbf{v}(0)\rangle \\ &= e^{-|t|\gamma/m}\langle \mathbf{v}(0)\cdot\mathbf{v}(0)\rangle \\ &= \frac{3k_BT}{m}e^{-\gamma|t|/m}.\end{aligned} \tag{4.25}$$

The final equality follows because $\langle \mathbf{vv}\rangle = -k_B S_{vv}''^{-1} = (k_BT/m)\mathrm{I}$, which is equivalent to the equipartition theorem.

Now in the analysis of the Brownian particle from the perspective of position, the Einstein result linked the diffusion constant to the correlation of velocity and position, equation (4.15). Writing $\tau_r$ as the time step used for the position equation (the interval for velocity is much smaller, as sketched in figure 4.3), that result may be written as

$$\begin{aligned}D &= \frac{\hat{\tau}_r}{3}\langle \mathbf{v}(t+\tau_r)\cdot[\mathbf{r}(t+\tau_r)-\mathbf{r}(t)]\rangle \\ &= \frac{\hat{\tau}_r}{3}\int_0^{\tau_r}\mathrm{d}t'\langle \mathbf{v}(t+\tau_r)\cdot\mathbf{v}(t+t')\rangle \\ &= \frac{\hat{\tau}_r}{3}\int_0^{\tau_r}\mathrm{d}t'\ \frac{3k_BT}{m}e^{-\gamma|\tau_r-t'|/m} \\ &= \frac{\hat{\tau}_r k_BT}{m}\frac{m\hat{\tau}_r}{\gamma}[1-e^{-\gamma|\tau_r|/m}] \\ &= \frac{k_BT}{\gamma},\quad |\tau_r| \gg \frac{m}{\gamma}.\end{aligned} \tag{4.26}$$

This says that the diffusion constant is inversely proportional to the drag constant provided that the time interval for positional diffusion is relatively large. This shows the consistency of the Einstein result for free particle diffusion with the Langevin result for the evolution of the free particle velocity.

One can also find the regime of validity for a pinned Brownian particle. For the case of a parabolic pinning potential,

$$U(\mathbf{r}) = \frac{\kappa}{2}r^2, \tag{4.27}$$

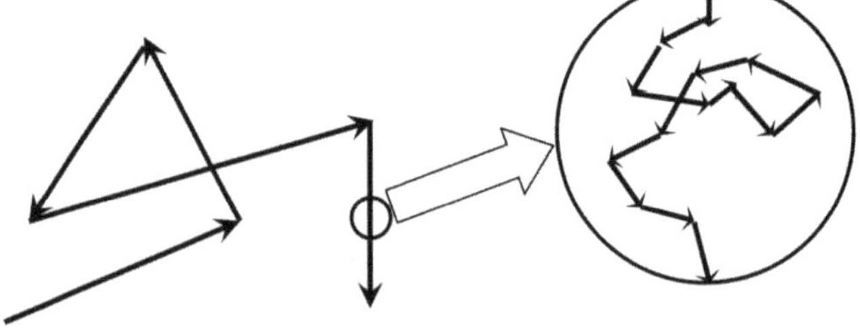

**Figure 4.3.** A Brownian trajectory in position (left) and in velocity (right).

the reservoir entropy is

$$S(\mathbf{r}, \mathbf{v}) = \frac{-m}{2T}v^2 - \frac{\kappa}{2T}r^2. \tag{4.28}$$

In this case the stochastic dissipative equation of motion for the position, equation (4.8), becomes

$$\mathbf{r}' = \mathbf{r} - |\tau_r|\frac{\kappa D}{k_B T}\mathbf{r} + \tilde{\mathbf{R}}, \quad |\tau_r| \gg \tau_{\text{molec}}. \tag{4.29}$$

In order for the discretization to be valid, the change in position must be small compared to the position itself, which means

$$\frac{m}{\gamma} \ll |\tau_r| \ll \frac{k_B T}{\kappa D} = \frac{\gamma}{\kappa}. \tag{4.30}$$

The lower limit is that obtained above for the free particle from the Markov analysis of velocity. This says that if the drag coefficient is large, or the curvature of the potential is small, then there is a time interval in which the stochastic dissipative equation of motion for the position can validly be applied in the case of an applied potential.

## 4.2 Fokker–Planck equation

We shall now derive the Fokker–Planck equation in general for stochastic dissipative equations of motion, and then use it to show that the Langevin equation yields a stationary equilibrium probability distribution when the fluctuation–dissipation theorem is satisfied. We shall also show that the stochastic dissipative equation derived from the second entropy analysis also yields a stationary probability distribution. For a more specialised treatment of the Fokker–Planck equation, see Risken (1984).

In the present book, the probability is taken to be the exponential of the entropy, and so it does not make much sense to speak of the evolution of an arbitrary probability density in the equilibrium case. Similarly, the transition probability is taken to be fixed by the exponential of the second entropy, which in turn fixes the stochastic, dissipative equations of motion. So again it is strange to speak of arbitrary equations of motion.

Nevertheless, one of the benefits that arise from the following derivation of the Fokker–Planck equation is a consistency rule that arbitrary stochastic dissipative equations of motion have to obey in order for the equilibrium probability density to be stationary. This leads to the fluctuation–dissipation theorem independent of the second entropy. It is shown that Langevin's equation not only has to have such a quantitative relationship, but that the qualitative form of the equation is fixed and unchangeable. Also the Fokker–Planck equation provides a basis for discussing Liouville's theorem.

### 4.2.1 General expression

We consider a generic stochastic dissipative transition,

$$X' = X + \tau \dot{X}^{\text{det}} + \tilde{R}. \tag{4.31}$$

Here the macrostate $X$ is to be considered quite generally. It can be a single coordinate, or a vector of multiple coordinates, possibly of mixed parity. The deterministic part of the transition consists of the adiabatic evolution and the most likely contribution due to the reservoir. For future use, the stochastic part is the difference between the actual and the most likely destination, $X' - \overline{X'} = \tilde{R}$. (We habitually use the word 'stochastic' to mean the zero mean part of the random reservoir contribution. Strictly speaking, the entire reservoir contribution is stochastic; it can be divided into a most likely deterministic part, $\overline{R} = \tau[\dot{X}^{\text{det}} - \dot{X}^{0}]$, and a zero mean random part, $\tilde{R}$. It is the latter that we mean by the word 'stochastic'.)

We shall suppose that the probability distribution for the stochastic part of the transition is Gaussian

$$\wp(\tilde{R}) = \frac{1}{Z'} e^{-\Lambda^{-1}:\tilde{R}\tilde{R}/2k_{\text{B}}|\tau|}, \quad Z' \equiv \{\text{Det}\,[2\pi k_{\text{B}}|\tau|\Lambda]\}^{1/2}. \tag{4.32}$$

Clearly $\langle \tilde{R} \rangle = 0$. This Gaussian is just the conditional transition probability, which means that the unconditional transition probability can be written as

$$\wp(X', X|\tau) = \wp(X'|X, \tau)\wp(X) = \wp(\tilde{R})\wp(X). \tag{4.33}$$

With these, and using the conservation law for probability in the equilibrium case, equation (3.17), one has

$$\begin{aligned}
\wp(X', t+\tau) &= \int \mathrm{d}X\, \wp(X', X|\tau) \\
&= \frac{1}{Z'} \int \mathrm{d}X\, e^{-\Lambda^{-1}:[X'-\overline{X'}]^2/2k_{\text{B}}|\tau|} \wp(X, t) \\
&= \frac{1}{Z'} \int \mathrm{d}\overline{X'} \left| \frac{\mathrm{d}\overline{X'}}{\mathrm{d}X} \right|^{-1} e^{-\Lambda^{-1}:[X'-\overline{X'}]^2/2k_{\text{B}}|\tau|} \\
&\quad \times \Big\{ \wp(X', t) + (X - X') \cdot \nabla \wp(X', t) \\
&\quad + \frac{1}{2}(X - X')(X - X') : \nabla\nabla \wp(X', t) + \cdots \Big\}.
\end{aligned} \tag{4.34}$$

The evolution of a volume element is sketched in figure 4.4. With this and a little thought (see equation (5.7)) one can see that the compressibility of the equations of motion is

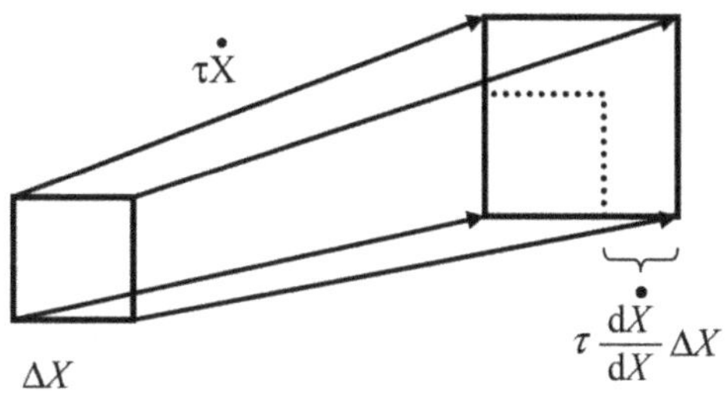


**Figure 4.4.** The evolution of a volume element in time $\tau$.

$$\begin{aligned}\left|\frac{\mathrm{d}\overline{X}'}{\mathrm{d}X}\right| &= \left|\mathrm{I} + \tau\frac{\mathrm{d}\dot{X}^{\mathrm{det}}}{\mathrm{d}X}\right| \\ &= 1 + \tau\mathrm{Tr}\frac{\mathrm{d}\dot{X}^{\mathrm{det}}}{\mathrm{d}X} + \mathcal{O}(\tau^2) \\ &= 1 + \tau\nabla\cdot\dot{X}^{\mathrm{det}}.\end{aligned} \tag{4.35}$$

Note that the adiabatic equations of motion are incompressible, $\nabla\cdot\dot{X}^0 = 0$, and so only the dissipative part contributes to this. Writing $X' - X = \tau\dot{X}^{\mathrm{det}} + \tilde{R}$, and neglecting quadratic powers of $\tau\dot{X}^{\mathrm{det}}$ and odd powers of $\tilde{R}$, the evolution becomes

$$\begin{aligned}&\wp(X', t+\tau) \\ &= \frac{1}{Z'}\Big[1 - \tau\nabla\cdot\dot{X}^{\mathrm{det}}\Big]\int \mathrm{d}\tilde{R}e^{-\Lambda^{-1}:\tilde{R}\tilde{R}/2k_{\mathrm{B}}|\tau|}\Big\{\wp(X', t) \\ &\quad - \Big(\tau\dot{X}^{\mathrm{det}} + \tilde{R}\Big)\cdot\nabla\wp(X', t) + \frac{1}{2}\Big(\tau\dot{X}^{\mathrm{det}} + \tilde{R}\Big)\Big(\tau\dot{X}^{\mathrm{det}} + \tilde{R}\Big) : \nabla\nabla\wp(X', t)\Big\} \\ &= \frac{1}{Z'}\Big[1 - \tau\nabla\cdot\dot{X}^{\mathrm{det}}\Big]\int \mathrm{d}\tilde{R}e^{-\Lambda^{-1}:\tilde{R}\tilde{R}/2k_{\mathrm{B}}|\tau|}\Big\{\wp(X', t) \\ &\quad - \tau\dot{X}^{\mathrm{det}}\cdot\nabla\wp(X', t) + \frac{1}{2}\tilde{R}\tilde{R} : \nabla\nabla\wp(X', t)\Big\} \\ &= \wp(X', t) - \tau\nabla\cdot\dot{X}^{\mathrm{det}}\wp(X', t) - \tau\dot{X}^{\mathrm{det}}\cdot\nabla\wp(X', t) + \frac{k_{\mathrm{B}}|\tau|}{2}\Lambda : \nabla\nabla\wp(X', t).\end{aligned} \tag{4.36}$$

The neglected terms are $\mathcal{O}(\tau^2)$. The leading order $\tau\dot{X}^{\mathrm{det}}$ can be evaluated at $X'$, which is why it has been taken outside of the integral twice here. This gives the partial time derivative as

$$\frac{\partial\wp(X, t)}{\partial t} = -[\nabla\cdot\dot{X}^{\mathrm{det}}]\wp(X, t) - \dot{X}^{\mathrm{det}}\cdot\nabla\wp(X, t) + \frac{\hat{\tau}k_{\mathrm{B}}}{2}\Lambda : \nabla\nabla\wp(X, t). \tag{4.37}$$

This is the Fokker–Planck equation.

Note that this partial time derivative on the left-hand side does not distinguish notationally between the forward and backward time derivatives. In contrast, the right-hand side contains irreversible terms in the dissipative contribution to the deterministic velocity and in the stochastic contribution, with $\hat{\tau} \equiv \tau/|\tau| = +1$ for the forward time derivative and $\hat{\tau} = -1$ for the backward time derivative. It will be shown next (and also in section 5.4.3) that for equilibrium systems, the right-hand side is zero with the irreversible terms canceling each other. (In chapter 6 below the same cancelation will be shown to occur for non-equilibrium steady state systems.)

### 4.2.2 Second entropy

The equilibrium probability is just the exponential of the total entropy,

$$\wp(X, t) = \frac{1}{Z} e^{S(X)/k_B}. \tag{4.38}$$

In the second entropy case, the deterministic part of the transition is

$$\dot{X}^{\text{det}} = \dot{X}^0 + \frac{\hat{\tau}}{2} \Lambda \nabla S(X). \tag{4.39}$$

Because in the equilibrium case $S(X) = -\mathcal{H}(X)/T$ it follows that $\dot{X}^0 \cdot \nabla S(X) = 0$. This assumes that the sub-system state $X$ has no internal entropy, so that the total entropy is the same as the reservoir entropy. This is a valid assumption for a Brownian particle, or for a point in classical phase space, but it does not hold, for example, for a fluctuation of a non-conserved variable in the sub-system, such as the first energy moment treated in section 3.4.

A second result required is that the adiabatic equations of motion are incompressible, $\nabla \cdot \dot{X}^0 = 0$. These two results mean that there is no adiabatic contribution to the Fokker–Planck equation. What remains is

$$\begin{aligned} \frac{\partial \wp(X, t)}{\partial t} &= \left\{ \frac{-\hat{\tau}}{2} \Lambda : \nabla\nabla S(X) - \frac{\hat{\tau}}{2k_B} \Lambda : \nabla S(X) \nabla S(X) \right. \\ &\quad \left. + \frac{\hat{\tau}}{2k_B} \Lambda : \nabla S(X) \nabla S(X) + \frac{\hat{\tau}}{2} \Lambda : \nabla\nabla S(X) \right\} \wp(X, t) \\ &= 0. \end{aligned} \tag{4.40}$$

This vanishes identically, as it must in the present equilibrium case.

*Mixed parity*

In the second entropy, the mixed parity case, the deterministic part of the transition is given by equation (3.136) with the adiabatic evolution added,

$$\dot{X}^{\text{det}} = \dot{X}^0 + \frac{\hat{\tau}}{2}[\Lambda - 2\hat{\tau}\Theta] S'' X \equiv \dot{X}^0 + \frac{\hat{\tau}}{2} L(\hat{\tau}) \nabla S(X). \tag{4.41}$$

The random forces remain Gaussian distributed with the variance $\Lambda k_B |\tau|$ unchanged from the pure parity case.

One can use the above pure parity result for the partial time derivative of the probability with the first two transport matrices replaced by $L(\hat{\tau})$,

$$\begin{aligned}\frac{\partial \wp(X,t)}{\partial t} &= \Big\{ \frac{-\hat{\tau}}{2} L(\hat{\tau}) : \nabla\nabla S(X) - \frac{\hat{\tau}}{2k_B} L(\hat{\tau}) : \nabla S(X) \nabla S(X) \\ &\quad + \frac{\hat{\tau}}{2k_B} \Lambda : \nabla S(X)\nabla S(X) + \frac{\hat{\tau}}{2} \Lambda : \nabla\nabla S(X) \Big\} \wp(X,t) \\ &= \Big\{ \frac{-\hat{\tau}}{2} \Lambda : \nabla\nabla S(X) - \frac{\hat{\tau}}{2k_B} \Lambda : \nabla S(X)\nabla S(X) \\ &\quad + \frac{\hat{\tau}}{2k_B} \Lambda : \nabla S(X)\nabla S(X) + \frac{\hat{\tau}}{2} \Lambda : \nabla\nabla S(X) \Big\} \wp(X,t) \\ &= 0. \end{aligned} \tag{4.42}$$

The second equality follows because any symmetric scalar product with the antisymmetric matrix $\Theta$ vanishes. Hence for both the pure and mixed parity case, the second entropy formulation of the stochastic dissipative equations of motion yield a stationary equilibrium probability distribution.

### 4.2.3 Dissipation in the Langevin equation

Now it is shown that the solvent force, $\mathbf{F}_r$, in the Langevin equation must necessarily be of the form of a drag force in order for the equilibrium probability distribution to be stable.

In the case of the Langevin equation, $X \equiv \{\mathbf{r}, \mathbf{v}\}$, the deterministic rates of change are

$$\begin{aligned} \dot{\mathbf{r}}^{\text{det}} &= \dot{\mathbf{r}}^0 \\ \dot{\mathbf{v}}^{\text{det}} &= \dot{\mathbf{v}}^0 + \frac{\hat{\tau}}{m} \mathbf{F}_r(\mathbf{r}, \mathbf{v}). \end{aligned} \tag{4.43}$$

The velocity gradient of the total entropy is $\nabla_v S = -m\mathbf{v}/T$. Again using the equilibrium probability, $\wp(\mathbf{r}, \mathbf{v}; t) = e^{S(\mathbf{r},\mathbf{v})/k_B}/Z$, the Fokker–Planck equation in this case is

$$\begin{aligned}\frac{\partial \wp(X,t)}{\partial t} &= \Big\{ \frac{-\hat{\tau}}{m} \nabla_v \cdot \mathbf{F}_r - \frac{\hat{\tau}}{mk_B} \mathbf{F}_r \cdot \nabla_v S(X) \\ &\quad + \frac{\hat{\tau}}{2k_B} \Lambda_v : \nabla_v S(X) \nabla_v S(X) + \frac{\hat{\tau}}{2} \Lambda_v : \nabla_v \nabla_v S(X) \Big\} \wp(X,t) \\ &= \Big\{ \frac{-\hat{\tau}}{m} \nabla_v \cdot \mathbf{F}_r + \frac{\hat{\tau}}{k_B T} \mathbf{F}_r \cdot \mathbf{v} + \frac{\hat{\tau} m^2}{2k_B T^2} \Lambda_v : \mathbf{vv} - \frac{\hat{\tau} m}{2T} \mathrm{Tr}\, \Lambda_v \Big\} \wp \\ &\quad (X,t) \\ &= \Big\{ \frac{-\hat{\tau}}{m} \nabla_v \cdot \mathbf{F}_r + \frac{\hat{\tau}}{k_B T} \mathbf{F}_r \cdot \mathbf{v} + \frac{\hat{\tau} m^2 \lambda_v}{2k_B T^2} \mathbf{v} \cdot \mathbf{v} - \frac{3\hat{\tau} m \lambda_v}{2T} \Big\} \wp(X,t), \end{aligned} \tag{4.44}$$

since $\Lambda_v = \lambda \mathrm{I}$. To make this vanish for all $\mathbf{v}$, it is clear that one must take

$$\mathbf{F}_r = -\gamma \mathbf{v}, \tag{4.45}$$

which gives

$$\frac{\partial \wp(X, t)}{\partial t} = \left\{ \frac{3\hat{\tau}}{m}\gamma - \frac{\hat{\tau}}{k_B T}\gamma \mathbf{v} \cdot \mathbf{v} + \frac{\hat{\tau} m^2 \lambda_v}{2 k_B T^2} \mathbf{v} \cdot \mathbf{v} - \frac{3\hat{\tau} m \lambda_v}{2T} \right\} \wp(X, t). \tag{4.46}$$

This vanishes when the strength of the dissipation, the drag force, is related to the strength of the fluctuations via

$$\gamma = \frac{m^2 \lambda_v}{2T}. \tag{4.47}$$

This is another version of the fluctuation–dissipation theorem. This theorem was originally derived by Langevin (1908) by demanding that the equipartition theorem be satisfied. Since the equilibrium probability yields the equipartition theorem, the present derivation based on the stationarity of the former includes the latter as a special case.

### 4.2.4 No dissipation without fluctuation

The Langevin equation for velocity, equation (4.19), may be rewritten in terms of momentum $\mathbf{p} = m\mathbf{v}$ as

$$\dot{\mathbf{p}} = \frac{-\partial \mathcal{H}(\mathbf{p}, \mathbf{q})}{\partial \mathbf{q}} - \frac{\gamma}{m}\mathbf{p} + \frac{1}{\tau}\tilde{\mathbf{F}}_r(\tau). \tag{4.48}$$

Here the position is written $\mathbf{q} \equiv \mathbf{r}$, and it has been assumed that $\tau$ is infinitesimal. This may be interpreted as applying to the coordinates of a single Brownian particle, or to the phase space of the molecules of a system.

The fluctuation–dissipation theorem says that the variance of the stochastic force is

$$\left\langle \tilde{\mathbf{F}}_r \tilde{\mathbf{F}}_r \right\rangle = m^2 k_B |\tau| \lambda_v \mathrm{I} = \frac{2 k_B T}{\gamma} |\tau| \mathrm{I}. \tag{4.49}$$

Comparing the above stochastic dissipative Langevin equation with the dissipative equations of motion with the isokinetic thermostat given at the start of the chapter, equation (4.1), one sees that the latter has the same dissipative term, $\alpha \equiv \gamma/m$, but it completely neglects the stochastic term, $\langle \tilde{\mathbf{F}}_r \tilde{\mathbf{F}}_r \rangle = 0$. This is a violation of the fluctuation–dissipation theorem.

The consequence of this violation is that the equilibrium probability distribution *cannot* be stationary under the isokinetic equations of motion despite (or because of) the fact that the kinetic energy is a constant of the motion and the equipartition theorem is obeyed. It follows that any property of the system (apart from the kinetic energy) that is obtained as a time average over the isokinetic trajectory cannot be equal to its correct average value obtained with the proper probability distribution.

The approach of using deterministic equations of motion—dissipation without fluctuation—is very commonly used for non-equilibrium molecular dynamics simulations. Although we shall not analyze non-equilibrium systems until chapter 6, one can anticipate that any dissipative equations of motion for the non-equilibrium system based upon an isokinetic or any similar deterministic thermostat, and any other deterministic term designed to mimic a particular thermodynamic flow, in the absence of a stochastic contribution that obeys a generalized fluctuation–dissipation theorem, will likewise yield erroneous results.

## 4.3 Stochastic calculus

The stochastic calculus refers to that branch of mathematics that deals with integrals and derivatives of functions of stochastic variables. The issue addressed here is whether or not the stochastic calculus is relevant to statistical mechanics in the physical sciences. For the conventional view of the stochastic calculus and the Fokker–Planck equation, see Gardener (1983) or Risken (1984); for a more detailed discussion of what follows, see Attard (2012, section 11.3.3).

There are two commonly used versions of the stochastic calculus, one due to Itô, and the other to Stratonovitch, as is now explained. In the simplest case, consider a trajectory with a random displacement added at each discrete node, $t_n = n\Delta_t$,

$$x_n = \bar{x}(t_n) + \tilde{R}_n, \tag{4.50}$$

or

$$x_{n+1} = x_n + \Delta_t \dot{\bar{x}}(t_n) + [\tilde{R}_{n+1} - \tilde{R}_n]. \tag{4.51}$$

We shall return to the variance of the increment in the random displacement below. For the present purpose of connecting with the mathematics literature on the stochastic calculus, one can consider it to be linear in the time interval, $\langle [\tilde{R}_{n+1} - \tilde{R}_n]^2 \rangle = 2D|\Delta_t|$, as in Einstein's treatment of a free Brownian particle.

The point is that the above stochastic equation specifies the trajectory at the nodes, but it is silent about what happens between the nodes. One is free to make any convenient choice, but any such choice has consequences.

The Itô calculus says that the random displacement is applied entirely at the beginning of the node, and so between nodes one has

$$x(t) = \bar{x}(t) + \tilde{R}_n, \quad t_n < t < t_{n+1}. \tag{4.52}$$

The Stratonovitch calculus says that between nodes the average random displacement at the termini is applied,

$$x(t) = \bar{x}(t) + \frac{1}{2}[\tilde{R}_{n+1} + \tilde{R}_n], \quad t_n < t < t_{n+1}. \tag{4.53}$$

These are sketched in figure 4.5. These are not the only choices that can be made, but they are the most common.

The choice of calculus makes a difference for the integral of a function over a trajectory. One has

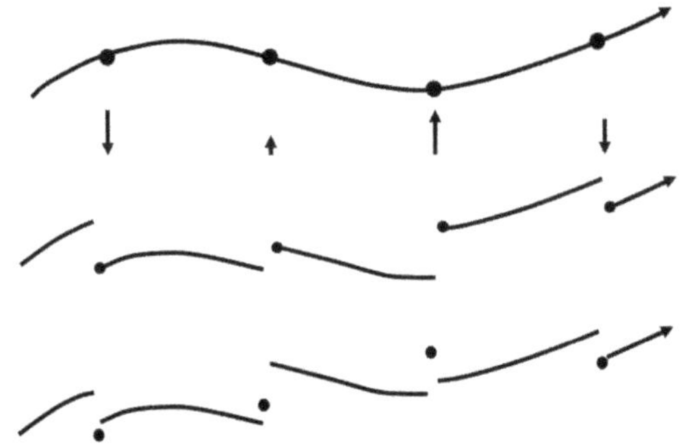


**Figure 4.5.** From top to bottom: (a) the most likely trajectory with four nodes, (b) four random displacements, (c) the Itô trajectory, and (d) the Stratonovitch trajectory.

$$\begin{aligned} G(t) &= \int_{x_0}^{x(t)} \mathrm{d}x(t')\, g(x(t')) \\ &= \sum_{i=0}^{n-1} [x_{i+1} - x_i]\, g(x(t')), \\ &= \sum_{i=0}^{n-1} [\Delta_t \dot{\bar{x}}(t_i) + \tilde{R}_{i+1} - \tilde{R}_i]\, g(x(t')). \end{aligned} \tag{4.54}$$

Here $t' = t_i + \alpha\Delta_t$, and $\alpha \in [0, 1)$. Ordinarily the precise value of $\alpha$ is immaterial; different choices of $\alpha \in [0, 1)$ change $g(x(t'))$ by $\mathcal{O}(\Delta_t)$, and hence $G(t)$ by $\mathcal{O}(n\Delta_t^2) = \mathcal{O}(\Delta_t)$, which is negligible in the limit $\Delta_t \to 0$.

In the stochastic calculus the choice of $\alpha$ makes a difference to the final result. To see this note that the value of the function within the interval differs between the two versions,

$$g(x(t')) = \begin{cases} g(x_i) + [\alpha\Delta_t \dot{\bar{x}}(t_i) + \tilde{R}_i] g'(x_i) + \cdots, & \text{Itô}, \\ g(x_i) + [\alpha\Delta_t \dot{\bar{x}}(t_i) + (\tilde{R}_i + \tilde{R}_{i+1})/2] g'(x_i) + \cdots, & \text{Strat.} \end{cases} \tag{4.55}$$

Subtracting one version of the integral from the other and averaging yields a result proportional to the variance of the increment in the random displacement

$$\begin{aligned} \langle G^{\text{Strat.}}(t) - G^{\text{Ito}}(t) \rangle &= \sum_{i=0}^{n-1} \langle [\Delta_t \dot{\bar{x}}(t_i) + R_{i+1} - R_i]\, [(\tilde{R}_{i+1} - \tilde{R}_i)/2] g'(x_i) \rangle \\ &= \frac{1}{2} \sum_{i=0}^{n-1} \langle [\tilde{R}_{i+1} - \tilde{R}_i]^2 \rangle g'(x_i) \\ &= D|\Delta_t| \sum_{i=0}^{n-1} g'(x_i). \end{aligned} \tag{4.56}$$

Since there are $n = t/\Delta_t$ terms here, the difference between the two versions of the integral is $\mathcal{O}(1)$, which cannot be neglected even for an infinitesimal time step. This shows that one has to make a particular choice for the stochastic calculus, either Itô or Stratonovitch, or some other, and the value of any integral over a stochastic trajectory is dependent on the choice.

In the physical sciences this situation is unacceptable because one cannot have any such ambiguity in the value of physical variables.

The resolution of the problem comes from noticing that in the final equality above it was assumed that the variance is twice the diffusion constant times the length of the time step. This corresponds to the Einstein result, that the variance in the displacement of a free Brownian particle scales with the length of the time interval. In the mathematics literature, and particularly in financial applications of stochastic difference equations, this is taken as an axiomatic definition of stochastic behavior. It is also assumed that the time interval can be regarded as infinitesimal on the scale of the considered motion (otherwise the discretization of the trajectory would yield inaccurate results).

In fact, however, in statistical mechanics as applied to real physical systems, the variance must equal the position self-correlation function, and this has different asymptotic behavior in different time regimes. As discussed in connection with figure 3.2, one has

$$\begin{aligned}\left\langle [\tilde{R}_{n+1} - \tilde{R}_n]^2 \right\rangle &= 2k_B q_0^s(\Delta_t) \\ &\sim \begin{cases} k_B \ddot{q}_0^s(0)\Delta_t^2, & \Delta_t \to 0, \\ 2D|\Delta_t|, & \Delta_t \gg \tau_{\text{molec}}. \end{cases}\end{aligned} \tag{4.57}$$

As in equation (4.13), the symmetrized position autocorrelation function is

$$q_0^s(\tau) \equiv \frac{1}{2k_B}\langle [x(t+\tau) - x(t)]\, [x(t+\tau) - x(t)]\rangle_0, \tag{4.58}$$

and its 'acceleration' is

$$\begin{aligned}\ddot{q}_0^s(\tau) &= \frac{-1}{2k_B}[\langle \ddot{x}(t+\tau)x(t)\rangle_0 + \langle x(t)\ddot{x}(t+\tau)\rangle_0] \\ &= \frac{1}{k_B}\langle \dot{x}(t+\tau)\dot{x}(t)\rangle_0.\end{aligned} \tag{4.59}$$

The sub-script 0 signifies a free Brownian particle, in which case $\ddot{q}_0^s(0) = T/m$, but the analysis is in fact more general than this (see section 4.6.2 below).

This result shows that in physical applications, the behavior of the variance of the increment in the random displacement is quadratic in the time interval for intervals of the order of molecular time scales, and is linear in the time interval for intervals longer than the relaxation time of the system.

If in a particular application one uses an infinitesimal time step, (because, for example, the function of interest $g(x(t))$ is rapidly varying), then the variance is quadratic in the time step, and the difference between the Itô and the Stratinovich calculus is $\mathcal{O}(n\Delta_t^2) \sim t\Delta_t$, which is negligible, and the choice of the stochastic calculus makes no difference. If, on the other hand, one uses a relatively large time step beyond the system relaxation time (because $|\Delta g(x(t'))| \ll |g(x(t'))|$ over the time step), then the variance would indeed be linear in the time step, as mathematicians invariably assume. However, since the whole theory requires that the stochastic influence be a small perturbation on the adiabatic evolution (otherwise the reservoir influence would have to be treated explicitly rather than statistically), the contribution of the stochastic forces to the change in a function over the time interval must be a small perturbation on the deterministic change, $|Dg'(x(t))| \ll |\dot{x}(t)g(x(t))|$. If this holds, then the difference between the two versions of the stochastic calculus is negligible. If it does not hold then one must use instead a time step smaller than the system relaxation time (because this reduces the stochastic influence, $\mathcal{O}(\Delta_t^2)$, more quickly than the adiabatic influence, $\mathcal{O}(\Delta_t)$).

One concludes from these arguments that the stochastic calculus of Itô and Stratinovich plays no role in the statistical mechanics of physical systems.

In some books and papers, the Fokker–Planck equation is derived using the stochastic calculus, and different versions result depending upon the choice. The differences depend upon the gradient of the diffusion constant with respect to the current trajectory variable, namely $\nabla \cdot D$ (Ermak and McCammon 1978, Gardiner 1983, Risken 1984, Tough *et al* 1986, Keizer 1987, Allen and Tildesley 1987, and Mazo 2008). These competing versions of the Fokker–Planck equation only occur for the standard 'no memory' Langevin equation; no such differences occur for the generalized Langevin equation that has memory.

It should be understood that the divergence of the transport coefficient in that analysis is with respect to the actual trajectory variable, which means that it is only non-zero if $D = D(x(t))$ in the present case. However since the transport coefficient is an average, as given for example by the time correlation function, or by Green–Kubo theory, in physical problems it is impossible for the transport coefficient to depend on the current point on the trajectory. (It may depend upon space and time, for example $D(\bar{x}(t), t)$, but not upon the system state $x$. Obviously, $\nabla \cdot D(\bar{x}(t), t) = 0$.) By the fluctuation–dissipation theorem, this independence holds equivalently for the variance of the stochastic forces. In a system without memory, the transport coefficient must also be independent of prior points on the trajectory. In the opinion of the present author, for systems without memory, $\nabla \cdot D = 0$, and there is no ambiguity in the Fokker–Planck equation or in the evolution of any physical variable.

There remains the case of non-Markov systems, which is to say systems with memory. For these the generalized Langevin equation is required,

$$\dot{x}(t) = \frac{1}{T}\int_0^t \mathrm{d}t'\, \ddot{q}_0^{\mathrm{s}}(t-t')F(x(t'), t') + \tilde{R}(t), \tag{4.60}$$

where $F(x, t)$ is an externally applied force. This and variants are derived in section 4.6 below. Detailed analysis (Attard 2012, chapters 10 and 11) shows that the different versions of the stochastic calculus have no influence on the Fokker–Planck equation or on any function of the trajectory for systems with memory.

## 4.4 Generalized equipartition theorem

For an equilibrium system with kinetic energy $\mathcal{K}(\mathbf{v}) = mv^2/2$, and potential energy $U(\mathbf{r})$, the entropy is $S_r(\mathbf{v}, \mathbf{r}) = -[\mathcal{K}(\mathbf{v}) + U(\mathbf{r})]/T$, and the probability is $\wp(\mathbf{v}, \mathbf{r}) = Z(T)^{-1}e^{S_r(\mathbf{v},\mathbf{r})/k_B}$. This is for a Brownian particle with no internal entropy so that the total entropy is the same as the reservoir entropy. Because of the Gaussian form of the velocity distribution, the variance of the velocity is

$$\langle \mathbf{v}\mathbf{v} \rangle = \frac{k_B T}{m} \mathrm{I}. \tag{4.61}$$

This is the equipartition theorem. It holds for any system with the entropy having the stated form.

For the case that there are $N$ particles, so that $\mathbf{v}$ is a $3N$-dimensional vector, then the trace of this is extensive, and fluctuations are relatively negligible. In this case the theorem applies locally,

$$\langle \mathbf{v} \cdot \mathbf{v} \rangle \approx \mathbf{v} \cdot \mathbf{v} = \frac{3Nk_B T}{m}. \tag{4.62}$$

Obviously one *could* choose sets of velocities where this fails badly. But if one chooses likely velocities, or if the velocities arise spontaneously in the system, then this is an increasingly good approximation as $N \to \infty$.

The equipartition theorem is a rather trivial consequence of the Gaussian nature of the probability. It is possible to obtain a powerful generalization of this theorem even when the probability is non-Gaussian. The derivation is quite simple and is based on the fact that the averand on the left-hand side can be considered to be in essence the dyadic product of the velocity gradient of the entropy, and the right-hand side can be considered to be the dyadic second velocity gradient of the entropy.

To this end, we represent the phase space of the sub-system of interest by $X$, and the reservoir entropy by $S_r(X, t)$. For greatest generality we consider a general non-equilibrium system dependent on time $t$. Explicit expressions for the non-equilibrium entropy are given in chapter 6 but will not be required here. The probability is as usual $\wp(X, t) = Z(t)^{-1}e^{S_r(X,t)/k_B}$. (It will be shown in section 5.2 below that phase space is the set of microstates, and that they have no internal entropy, $S_{total}(X, t) = S_r(X, t)$. When $X$ represents a macrostate with internal entropy, the total entropy $S_{total}(X, t) = S_s(X, t) + S_r(X, t)$ has to be used in place of the reservoir entropy.)

In view of the observation made in the penultimate paragraph above, the average of the dyadic product of the phase space gradient of the entropy is

$$\begin{aligned} k_B^{-2}\langle[\nabla S_r(X, t)][\nabla S_r(X, t)]\rangle &= k_B^{-2}\int \mathrm{d}X\, \wp(X, t)[\nabla S_r(X, t)][\nabla S_r(X, t)] \\ &= k_B^{-1}\int \mathrm{d}X\, [\nabla \wp(X, t)][\nabla S_r(X, t)] \\ &= -k_B^{-1}\int \mathrm{d}X\, \wp(X, t)\nabla\nabla S_r(X, t) \\ &= -k_B^{-1}\langle \nabla\nabla S_r(X, t)\rangle. \end{aligned} \tag{4.63}$$

The third equality follows after an integration by parts, the integrated portion vanishing because the probability vanishes on the boundaries of the integration region. This is the first generalized equipartition theorem. For a single component, say $x$, it reads

$$\left\langle \left( \frac{\partial S_r(X, t)}{k_B \partial x} \right)^2 \right\rangle + \left\langle \frac{\partial^2 S_r(X, t)}{k_B \partial x^2} \right\rangle = 0. \tag{4.64}$$

Taking $x = v$ and the reservoir entropy as above, one sees that this reduces to the usual equipartition theorem.

The first generalized equipartition theorem is equivalent to

$$\int \mathrm{d}X \ \nabla\nabla \wp(X, t) = 0. \tag{4.65}$$

This can be seen directly to be true by performing a single integration and noting that $\oint \mathrm{d}\mathbf{n}_X \cdot \nabla \wp(X, t) = 0$, since the probability and its gradient can be taken to vanish on the boundary. This is a global theorem, but one suspects that it may be a reasonable approximation applied point by point. That is

$$\begin{aligned} \nabla\nabla \wp(X, t) &\equiv [k_B^{-1} \nabla\nabla S_r(X, t) + k_B^{-2} \{\nabla S_r(X, t)\}\{\nabla S_r(X, t)\}] \, \wp(X, t) \\ &\approx 0. \end{aligned} \tag{4.66}$$

Of course there are values of $X$ when the term in brackets has a large magnitude, but one expects such $X$ to be rather improbable, $\wp(X, t) \approx 0$. Conversely probable values of $X$ correspond to the approximate cancelation of the two terms in the brackets. It is important to note that for this approximation to be true locally in phase space, both factors (the bracketed terms and the probability density) are required.

This local approximation can be made with even greater certainty by summing over components. For example, summing the diagonal elements of the dyadic matrix yields

$$[k_B^{-1} \nabla^2 S_r(X, t) + k_B^{-2} \{\nabla S_r(X, t)\} \cdot \{\nabla S_r(X, t)\}] \, \wp(X, t) \approx 0. \tag{4.67}$$

The reason that one can have confidence in this result is that the bracketed term is extensive, which is to say it scales with the size of the system. It will be recalled that fluctuations in extensive variables are relatively negligible. Hence in the thermodynamic limit one can write this as

$$k_B^{-1} \nabla^2 S_r(X, t) + k_B^{-2} \{\nabla S_r(X, t)\} \cdot \{\nabla S_r(X, t)\} = 0. \tag{4.68}$$

It is implicitly understood that the values of $X$ in this arise spontaneously in the system.

In the case of $N$ particles in three dimensions, for an equilibrium system with the kinetic energy given above, $S_r(X, t) = -mv^2/2T - U(\mathbf{r})/T$, using the velocity gradient in the last expression gives

$$\frac{-3Nm}{k_B T} + \frac{m^2}{k_B^2 T^2} v^2 = 0, \ \text{ or } v^2 = 3Nmk_B T. \tag{4.69}$$

One often sees the equipartition function written in this form. Again, the reason that one can expect it to apply locally and not just on average is that it is extensive. Provided that these are velocities that have arisen in the system, the fluctuations that would invalidate this result are unlikely to occur. Arguably it would be better to show the probability density explicitly here to ensure that unlikely fluctuations that violate the equipartition theorem have zero weight.

There is another result that can be called a generalized equipartition theorem. Consider the average of the dyadic,

$$\begin{aligned}\left\langle X\frac{\partial S_{\mathrm{r}}(X,\,t)}{\partial X}\right\rangle &= \frac{1}{Z}\int \mathrm{d}X\ e^{S_{\mathrm{r}}(X,t)/k_{\mathrm{B}}}X\frac{\partial S_{\mathrm{r}}(X,\,t)}{\partial X}\\ &= \frac{k_{\mathrm{B}}}{Z}\int \mathrm{d}X\ X\frac{\partial e^{S_{\mathrm{r}}(X,t)/k_{\mathrm{B}}}}{\partial X}\\ &= \frac{-k_{\mathrm{B}}}{Z}\int \mathrm{d}X\ e^{S_{\mathrm{r}}(X,t)/k_{\mathrm{B}}}\frac{\partial X}{\partial X}\\ &= -k_{\mathrm{B}}\,\mathrm{I}.\end{aligned} \tag{4.70}$$

The third equality follows from an integration by parts. This may be called the second generalized equipartition theorem.

Choosing $S_{\mathrm{r}}(X,\,t) = -\mathcal{H}(\mathbf{q},\,\mathbf{p})/T$, for a single momentum component this gives the conventional equipartition theorem,

$$\left\langle p_{j\alpha}^2\right\rangle_{N,V,T} = mk_{\mathrm{B}}T. \tag{4.71}$$

More generally it gives

$$\langle p_{j\alpha}p_{i\gamma}\rangle_{N,V,T} = \delta_{j,i}\delta_{\alpha,\gamma}\ mk_{\mathrm{B}}T, \tag{4.72}$$

$$\left\langle q_{j\alpha}\frac{\partial U(\mathbf{q})}{\partial q_{i\gamma}}\right\rangle_{N,V,T} = \delta_{j,i}\delta_{\alpha,\gamma}\ k_{\mathrm{B}}T, \tag{4.73}$$

and

$$\left\langle p_{j\alpha}\frac{\partial U(\mathbf{q})}{\partial q_{i\gamma}}\right\rangle_{N,V,T} = \langle q_{j\alpha}\ p_{i\gamma}\rangle_{N,V,T} = 0. \tag{4.74}$$

This last result exemplifies the principle that in an equilibrium system there can be no instantaneous correlation between variables of different time parity.

Finally, it should be pointed out that in general one should replace the reservoir entropy in the two generalized theorems everywhere above by the total entropy. In most case this makes no difference to the results, but for some choices of macrostate $X$ the sub-system entropy is non-constant and its inclusion does make a difference.

## 4.5 Liouville's approximation

The stochastic dissipative equation in the pure parity case, was given above as equation (4.31),

$$X(t+\tau) = X(t) + \tau \dot{X}^{\mathrm{det}}(t) + \tilde{R}(t). \tag{4.75}$$

Although derived with a Brownian particle in mind, the quantity $X$ can be interpreted quite generally. The deterministic velocity can be written as the sum of the adiabatic and the most likely reservoir-induced velocity, equation (4.39),

$$\dot{X}^{\mathrm{det}} = \dot{X}^0 + \dot{\overline{X}} = \dot{X}^0 + \frac{\hat{\tau}}{2}\Lambda\nabla S(X). \tag{4.76}$$

Here $S(X)$ is the total entropy, which for a Brownian particle and for phase space, is the same as the reservoir entropy. (For the mixed parity case, the transport matrix is instead $L(\hat{\tau})$, equation (3.136).) The stochastic contribution is Gaussian distributed, equation (4.32),

$$\wp(\tilde{R}) = \frac{1}{Z'} e^{-\Lambda^{-1}:\tilde{R}\tilde{R}/2k_{\mathrm{B}}|\tau|}, \quad Z' \equiv \{\mathrm{Det}\,[2\pi k_{\mathrm{B}}|\tau|\Lambda]\}^{1/2}. \tag{4.77}$$

The time development of a probability distribution due to this stochastic dissipative equation is given by the Fokker–Planck equation, derived above as equation (4.37). This can be rearranged as

$$\begin{aligned}\frac{\partial\wp(X,t)}{\partial t} &= -\nabla\cdot\dot{X}^{\mathrm{det}}\wp(X,t) - \dot{X}^{\mathrm{det}}\cdot\nabla\wp(X,t) + \frac{\hat{\tau}k_{\mathrm{B}}}{2}\Lambda:\nabla\nabla\wp(X,t)\\ &= -\nabla\cdot\{\dot{X}^{\mathrm{det}}\wp(X,t)\} + \frac{\hat{\tau}k_{\mathrm{B}}}{2}\nabla\cdot\{\Lambda\nabla\wp(X,t)\}.\end{aligned} \tag{4.78}$$

The final term can be written like this because the transport matrix $\Lambda$ is independent of $X$ (see the discussion of the stochastic calculus, section 4.3).

Written in this way, the right-hand side of the Fokker–Planck equation represents a conservation law for the probability, $\partial\wp(X,t)/\partial t = -\nabla\cdot J_{\wp}(X,t)$, with the probability flux being

$$J_{\wp}(X,t) = \dot{X}^{\mathrm{det}}\wp(X,t) + \frac{\hat{\tau}k_{\mathrm{B}}}{2}\Lambda\nabla\wp(X,t). \tag{4.79}$$

The first term on the right-hand side represents the flux of probability carried by the deterministic motion, and the second term is that carried by the stochastic part of the equations of motion.

Liouville's theorem for an open sub-system corresponds to the neglect of the stochastic contribution in the equations of motion and in the Fokker–Planck equation. In this case the total time derivative is

$$\begin{aligned}\frac{\mathrm{d}\wp(X,t)}{\mathrm{d}t} &= \frac{\partial\wp(X,t)}{\partial t} + \dot{X}^{\mathrm{det}}(X,t)\cdot\nabla\wp(X,t)\\ &= -\left[\nabla\cdot\dot{X}^{\mathrm{det}}(X,t)\right]\wp(X,t).\end{aligned} \tag{4.80}$$

The first line is exact for deterministic equations of motion, and the second line follows from the Fokker–Planck equation with the stochastic contribution neglected. This is called Liouville's theorem, and it was critically referred to at the start of this chapter, equation (4.2). This gives the probability flux as $J_{\wp}(X, t) = \dot{X}^{\rm det}(X, t)\wp(X, t)$.

It would be better to call this Liouville's approximation. As has been mentioned several times, when one projects the reservoir coordinates out of the problem, one must necessarily have both dissipative (most likely) and a stochastic (zero mean) forces in the equations of motion. When a reservoir is present, it is not possible to have a dissipative part without a fluctuation part. The only legitimate case with zero dissipative part and zero fluctuation part is the adiabatic equations of motion, which are valid for an isolated sub-system that does not interact with its environment or a reservoir. In this case, and only this case, Liouville's theorem is an exact theorem and it reads

$$\begin{aligned} \frac{{\rm d}\wp(X, t)}{{\rm d}t} &= -\left[\nabla \cdot \dot{X}^0(X, t)\right]\wp(X, t) \\ &= -\dot{X}^0(X, t) \cdot \nabla\wp(X, t). \end{aligned} \tag{4.81}$$

The second equality follows because adiabatic equations of motion are incompressible, $\nabla \cdot \dot{X}^0 = 0$, (assuming $X$ represents a point in phase space).

At the time that Liouville derived his theorem the only equations of motion that were used were Hamilton's. Over the years workers have developed ad hoc non-Hamiltonian equations of motion. One such example is the isokinetic equations of motion, equation (4.1), which were criticized in section 4.2.4. Unfortunately, some have naively assumed that Liouville's theorem holds also for these.

This raises three questions. Is it reasonable to neglect stochastic contributions to the evolution of the probability density? Is it a reasonable approximation to neglect stochastic contributions to the equations of motion themselves? And, is it reasonable to approximate the evolution of an arbitrary function as deterministic?

First, in equation (4.66) it was shown that the dyadic gradient of the probability density could be neglected not only globally, but also locally for extensive formulations such as the Laplacian. The relevance of this is that the stochastic contributions to the rate of change of the probability density, equation (4.78), involve the Laplacian of the probability density. If the local version of the generalized equipartition theorem, equation (4.66), is a reasonable approximation, then so also is Liouville's approximation for the evolution of the probability density. The double scaler product with the transport matrix makes the term extensive, and so local fluctuations from the generalized equipartition theorem are negligible in the thermodynamic limit. Arguably, Liouville's approximation gives the evolution of the probability density for the exact stochastic dissipative equations of motion, with the generalized equipartition theorem applied locally making the stochastic contribution negligible.

Second, deterministic, non-Hamiltonian equations of motion are invalid for a sub-system interacting with a reservoir or environment. Since the condition for the equilibrium probability density to be stationary is that the fluctuation–dissipation theorem be satisfied, and since the latter says that the variance of the fluctuations must be linearly proportional to the strength of the dissipation, it is not possible to have a dissipative deterministic non-Hamiltonian term in the equations of motion without having the corresponding stochastic term that is related by the fluctuation–dissipation theorem. We shall analyze non-equilibrium systems in chapter 6, but here one can already see that if purely deterministic non-Hamiltonian equations are forbidden for equilibrium systems, then they must necessarily be invalid for non-equilibrium systems. In general deterministic, non-Hamiltonian equations of motion are unphysical as they violate the equilibrium or non-equilibrium fluctuation dissipation theorem.

Deterministic, non-Hamiltonian equations of motion suffer from two additional errors. Not including the stochastic contribution to the equations of motion neglects the randomness that is an inevitable consequence of the projection from the total system to the sub-system. Because of this, purely deterministic non-Hamiltonian equations of motion cannot represent a real physical system. In addition, lacking a physical basis, the ad hoc non-Hamiltonian dissipative term has not only the wrong magnitude, but often also the wrong functional form. It is an illusion to imagine that the fitting parameters that this freedom from the laws of nature allow in any way compensates for the uncertainty that it introduces into the results.

Third, the evolution of an arbitrary function is dependent upon the chosen equations of motion. Since there is no dissipation without fluctuation, it is not reasonable to approximate such evolution as deterministic.

## 4.6 Generalized Langevin equation

### 4.6.1 Perturbation theory

Consider a Brownian particle at $\mathbf{r}$, acted upon by a time-varying external force $\mathbf{F}(\mathbf{r}, t) = -\nabla U(\mathbf{r}, t)$. The Brownian particle is the sub-system and the solvent is the thermal reservoir with temperature $T$. The aim here is to account for solvent memory effects when the external potential varies rapidly. To this end a perturbation or linear response theory for the particle motion is developed.

All of the change in external energy over a specified trajectory due to the movement of the Brownian particle comes from the solvent, and this changes the reservoir entropy from what it would be for the same trajectory in the absence of the external potential,

$$S_{\mathrm{r}}(t|[\mathbf{r}], [U]) = S_{\mathrm{r},0}(t|[\mathbf{r}]) - \frac{1}{T}\int_0^t \mathrm{d}t'\, \dot{\mathbf{r}}(t') \cdot \nabla U(\mathbf{r}(t'), t'). \tag{4.82}$$

The quantity $\dot{\mathbf{r}} \cdot \nabla U$ is the rate of change of external energy due to the particle motion, and, by energy conservation, this is equal and opposite to the rate of change of solvent energy additional to what would have occurred for the bare particle on the same trajectory.

The average velocity at time $t$ in the presence of the external potential is

$$\begin{aligned}
\langle \dot{\mathbf{r}}(t)\rangle_U &= \frac{\int \mathrm{d}[\dot{\mathbf{r}}]\, e^{S_r(t|[\mathbf{r}],[U])/k_B}\dot{\mathbf{r}}(t)}{\int \mathrm{d}[\dot{\mathbf{r}}]\, e^{S_r(t'|[\mathbf{r}],[U])/k_B}} \\
&= \frac{\int \mathrm{d}[\dot{\mathbf{r}}]\, e^{S_{r,0}(t|[\mathbf{r}])/k_B}\left[1-\beta\int_0^t \mathrm{d}t'\, \dot{\mathbf{r}}(t')\cdot\nabla U(\mathbf{r}(t'),t')\right]\dot{\mathbf{r}}(t)}{\int \mathrm{d}[\dot{\mathbf{r}}]\, e^{S_{r,0}(t|[\mathbf{r}])/k_B}\left[1-\beta\int_0^t \mathrm{d}t'\, \dot{\mathbf{r}}(t')\cdot\nabla U(\mathbf{r}(t'),t')\right]} \\
&= -\beta\int_0^t \mathrm{d}t'\langle \dot{\mathbf{r}}(t')\cdot\nabla U(\mathbf{r}(t'),t')\dot{\mathbf{r}}(t)\rangle_0 \\
&= -\beta\int_0^t \mathrm{d}t'\langle \dot{\mathbf{r}}(t)\dot{\mathbf{r}}(t')\rangle_0\cdot\nabla U(\mathbf{r}(t'),t') \\
&= \frac{1}{T}\int_0^t \mathrm{d}t'\, \ddot{q}_0(t-t')\mathbf{F}(\mathbf{r}(t'),t').
\end{aligned} \tag{4.83}$$

In the second equality the exponentials have been linearized with respect to the external potential. In the third equality the facts that the velocity in the bare system averages to zero, $\langle \dot{\mathbf{r}}(t')\rangle_0 = 0$, and that it is uncorrelated with the current position, $\langle \dot{\mathbf{r}}(t')\mathbf{r}(t')\rangle_0 = 0$, have been used. In the fourth equality the force has been taken outside of the average because in the bare system $\mathbf{r}(t')$ is uncorrelated with $\dot{\mathbf{r}}(t)\dot{\mathbf{r}}(t')$, $t > t'$. In the final equality, the bare velocity auto-correlation function has been defined as in equation (4.59),

$$\ddot{q}_0(\tau) = k_B^{-1}\langle \dot{\mathbf{r}}(t+\tau)\dot{\mathbf{r}}(t)\rangle_0. \tag{4.84}$$

This is a symmetric, even, matrix, $\ddot{q}_0(\tau) = \ddot{q}_0(-\tau) = \ddot{q}_0(\tau)^{\mathrm{T}}$.

The final result says that the velocity auto-correlation function of the bare Brownian particle is the response function that gives the current velocity due to the preceding values of the external force. This function contains the memory effects of the solvent. Because these are short-ranged, the result is independent of the lower limit of the integral.

In the event that the external potential is slowly varying, $\mathbf{F}(\mathbf{r}(t'), t') \approx \mathbf{F}(\mathbf{r}(t), t)$, this becomes

$$\begin{aligned}
\bar{\dot{\mathbf{r}}}(t) &= \langle \dot{\mathbf{r}}(t)\rangle_U \\
&\approx \frac{1}{T}\int_0^t \mathrm{d}t'\, \ddot{q}_0(t-t')\mathbf{F}(\mathbf{r}(t),t) \\
&= \beta\langle \dot{\mathbf{r}}(t)[\mathbf{r}(t)-\mathbf{r}(0)]\rangle_0\, \mathbf{F}(\mathbf{r}(t),t) \\
&= \beta D\mathbf{F}(\mathbf{r}(t),t).
\end{aligned} \tag{4.85}$$

The final equality uses equation (4.15) for the diffusion constant. This is known as the Smoluchowski equation and it is applicable in the large drag limit

(Smoluchowski 1906). It is equivalent to the over damped case in which the acceleration is set to zero in the Langevin equation (4.19), so that the drag force is always equal and opposite to the applied force, $\dot{\bar{\mathbf{r}}}(t) = \gamma^{-1}\mathbf{F}(\bar{\mathbf{r}}(t), t) = \beta D\mathbf{F}(\bar{\mathbf{r}}(t), t)$.

One can add the random force to the full result to obtain the Langevin equation with memory,

$$\dot{\mathbf{r}}(t) = \mathbf{R}_v(t) + \frac{1}{T}\int_{-\infty}^{t} \mathrm{d}t'\ \ddot{q}_0(t - t')\mathbf{F}(\mathbf{r}(t'), t'). \tag{4.86}$$

One can discretize this to generate the stochastic, dissipative trajectory of the Brownian particle driven by a rapidly varying external force. The variance of the random force (actually force divided by mass) can be obtained by setting the external force to zero, multiplying the respective sides of this by their value at $t = t'$, and taking the average,

$$\langle \mathbf{R}_v(t)\mathbf{R}_v(t')\rangle_0 = \langle \dot{\mathbf{r}}(t)\dot{\mathbf{r}}(t')\rangle_0 = k_{\mathrm{B}}\ddot{q}_0\big(t - t'\big). \tag{4.87}$$

When memory effects are important, this form of the fluctuation dissipation theorem tells how the random forces at different times are correlated.

The present perturbation theory has two advantages. First, an explicit formula is given for the memory function, namely it is the velocity auto-correlation function. Second, the difficult part of the calculation, the memory function, need be done for the bare system once only, and the same memory function can be used for arbitrary applied time-dependent forces.

### 4.6.2 Thermodynamic fluctuations with memory

We now analyze the effects of memory on the evolution of thermodynamic fluctuations. This has many similarities to the just treated case of a driven Brownian particle, with the main differences coming from the more general mixed parity system treated here. This gives rise to some terms that would otherwise be zero in the pure parity case of a Brownian particle.

What are here called thermodynamic variables can also be called phase functions or dynamical variables, and what is here called a perturbation approach can also be called linear response theory.

*Bare system*

Let $x$ denote a vector of fluctuating thermodynamic variables arranged so that they vanish on average $\langle x\rangle = 0$. Each variable has pure time parity, but the vector is of mixed parity, $x(\boldsymbol{\Gamma}^{\dagger}) = \varepsilon x(\boldsymbol{\Gamma})$, where, as in section 3.3, the parity matrix is diagonal with elements $\varepsilon_{jk} = [x_j(\boldsymbol{\Gamma}^{\dagger})/x_j(\boldsymbol{\Gamma})]\delta_{jk}$.

The velocity autocorrelation matrix is

$$\begin{aligned}\ddot{q}_0(\tau) &= k_{\mathrm{B}}^{-1}\,\langle \dot{x}(t+\tau)\dot{x}(t)\rangle_0\\ &= -k_{\mathrm{B}}^{-1}\,\langle \ddot{x}(t+\tau)x(t)\rangle_0.\end{aligned} \tag{4.88}$$

(This assumes that $x$ is real. For the case of Fourier transforms, this is defined with a complex conjugate.) From the parity rules one has $\ddot{q}_0(-\tau) = \varepsilon\ddot{q}_0(\tau)\varepsilon = \ddot{q}_0(\tau)^{\mathrm{T}}$. This is one difference from the pure parity case, equations (4.59) and (4.84).

The position velocity correlation matrix is

$$\begin{aligned}\dot{q}_0(\tau) &= k_{\mathrm{B}}^{-1}\,\langle x(t+\tau)\dot{x}(t)\rangle_0 \\ &= -k_{\mathrm{B}}^{-1}\,\langle \dot{x}(t+\tau)x(t)\rangle_0,\end{aligned} \tag{4.89}$$

with $\dot{q}_0(-\tau) = -\varepsilon\dot{q}_0(\tau)\varepsilon = -\dot{q}_0(\tau)^{\mathrm{T}}$. Note that $\dot{q}_0(\tau) \to 0,\ \tau \to \infty$. At $\tau = 0$ this is an antisymmetric matrix, $\dot{q}_0(0) = -\dot{q}_0(0)^{\mathrm{T}} \neq 0$. This is in contrast with Brownian motion above, where the pure parity position velocity correlation function vanishes at $\tau = 0$.

The position autocorrelation matrix is

$$q_0(\tau) = -k_{\mathrm{B}}^{-1}\,\langle x(t+\tau)x(t)\rangle_0, \tag{4.90}$$

with $q_0(\tau) \to 0,\ \tau \to \infty$. This differs from the symmetrized position autocorrelation function defined for a free Brownian particle, equation (4.58), because the present fluctuations are localized about the origin, unlike the free Brownian particle. The parity rule is $q_0(-\tau) = \varepsilon q_0(\tau)\varepsilon = q_0(\tau)^{\mathrm{T}}$. Note the negative sign in this definition of the position autocorrelation matrix, which is implied by the choice of a positive sign in the definition of the velocity autocorrelation matrix. With this convention, it is generally the case that the diagonal elements of the autocorrelation matrix are global maxima for the velocity and global minima for the position at $\tau = 0$. These definitions of the time correlation matrices may be confirmed by differentiation.

The preceding observation about global extrema is based on the generic properties of the time correlation function. In probability theory, the covariance inequality, which is a form of the Cauchy–Schwarz inequality, states that in the present case of real variables with zero mean,

$$\langle AB\rangle_0^2 \leqslant \langle A^2\rangle_0\,\langle B^2\rangle_0. \tag{4.91}$$

In the present context, one can evaluate $A$ at $t + \tau$ and $B$ at $t$, and each can be any of the components of $x$, $\dot{x}$, or $\ddot{x}$. For example, taking $A = x_j(t + \tau)$ and $B = x_k(t)$, this says

$$\left\{q_0(\tau)\right\}_{jk}^2 \leqslant \left\{q_0(0)\right\}_{jj}\left\{q_0(0)\right\}_{kk}. \tag{4.92}$$

Summing over both indices, this is $\mathrm{Tr}\,[q_0(\tau)q_0(-\tau)] \leqslant [\mathrm{Tr}\,q_0(0)]^2$. Taking $j = k$, this is $\{q_0(\tau)\}_{jj}^2 \leqslant \{q_0(0)\}_{jj}^2$, which shows that the magnitude of a diagonal element of the time correlation function decays from its magnitude at $\tau = 0$.

Consider a trajectory $x[t]$. This can be discretized into $n$ nodes of spacing $\tau$, and described by a vector consisting of the final position and the velocities leading up to that position, $\mathbf{X}^{(n+1)} = \{x(t), \dot{x}(t), \dot{x}(t-\tau), \ldots, \dot{x}(t-(n-1)\tau)\} = \{x(t), \dot{\mathbf{x}}^{(n)}\}$. The time step may be positive or negative. Each entry here is itself a vector of thermodynamic variables. Since we are dealing with fluctuations, we take the trajectory entropy to be a quadratic form,

$$\begin{aligned} S_0(\mathbf{X}^{(n+1)}) &= \frac{1}{2} S_0^{(n+1,n+1)} : \mathbf{X}^{(n+1)}\mathbf{X}^{(n+1)} \\ &= \frac{1}{2} S_{0;xx} : x^2 + S_{0;x\dot{x}}^{(n)} : \dot{\mathbf{x}}^{(n)} x + \frac{1}{2} S_{0;\dot{x}\dot{x}}^{(nn)} : \dot{\mathbf{x}}^{(n)}\dot{\mathbf{x}}^{(n)}, \end{aligned} \tag{4.93}$$

where $x = x(t)$ and the matrices have an implicit dependence on the time step $\tau$. The fluctuation matrix $S_0^{(n+1,n+1)}$ has been broken into four sub-matrices in an obvious notation.

The trajectory correlation matrix is

$$Q_0^{(n+1,n+1)} \equiv k_{\mathrm{B}}^{-1} \left\langle \mathbf{X}^{(n+1)}\mathbf{X}^{(n+1)} \right\rangle_0 . \tag{4.94}$$

This can also be decomposed into sub-matrices,

$$Q_{0;xx} \equiv k_{\mathrm{B}}^{-1} \langle x(t)x(t) \rangle_0 = -q_0(0), \tag{4.95}$$

$$Q_{0;\dot{x}\dot{x}}^{(nn)} \equiv k_{\mathrm{B}}^{-1} \left\langle \dot{\mathbf{x}}^{(n)}\dot{\mathbf{x}}^{(n)} \right\rangle_0 = \ddot{Q}_0^{(nn)}, \tag{4.96}$$

and

$$Q_{0;x\dot{x}}^{(n)} \equiv k_{\mathrm{B}}^{-1} \left\langle x(t)\dot{\mathbf{x}}^{(n)} \right\rangle_0 , \tag{4.97}$$

with $Q_{0;\dot{x}x}^{(n)} = [Q_{0;x\dot{x}}^{(n)}]^{\mathrm{T}}$.

The elements of the velocity autocorrelation matrix are

$$\ddot{Q}_{0;jk}^{(nn)} = k_{\mathrm{B}}^{-1} \langle \dot{x}(t_j)\dot{x}(t_k) \rangle_0 = \ddot{q}_0(t_j - t_k). \tag{4.98}$$

In the present mixed parity case $\ddot{Q}_0^{(nn)}$ is symmetric, $\ddot{Q}_{0;k\alpha,j\gamma}^{(nn)} = \ddot{Q}_{0;j\gamma,k\alpha}^{(nn)}$, where Roman letters range over the time nodes and Greek letters range over the thermodynamic components. Equivalently, $\ddot{q}_0(t) = \ddot{q}_0(-t)^{\mathrm{T}}$ or $\ddot{q}_{0;\alpha\gamma}(t) = \ddot{q}_{0;\gamma\alpha}(-t)$.

The position velocity time correlation $n$-vector has elements

$$Q_{0;x\dot{x};k}^{(n)} \equiv k_{\mathrm{B}}^{-1} \langle x(t)\dot{x}(t_k) \rangle_0 = \dot{q}_0(t - t_k). \tag{4.99}$$

Each of these elements is actually a matrix of the thermodynamic variables, $\{Q_{0;\dot{x}x;k}^{(n)}\}_{\alpha\beta} = \{Q_{0;x\dot{x};k}^{(n)}\}_{\beta\alpha} = \dot{q}_{0;\alpha\beta}(t - t_k) = -\dot{q}_{0;\beta\alpha}(t_k - t)$. Recall that $t_1 \equiv t$.

As usual, the fluctuation matrix is the negative inverse of the correlation matrix $Q_0^{(n+1,n+1)} S_0^{(n+1,n+1)} = -\mathrm{I}^{(n+1,n+1)}$. It is straightforward to show that the fluctuation sub-matrices are given by

$$S_{0;xx} = -\left[1 - \left[Q_{0;xx}\right]^{-1} Q_{0;x\dot{x}}^{(n)} \left[Q_{0;\dot{x}\dot{x}}^{(nn)}\right]^{-1} \left[Q_{0;x\dot{x}}^{(n)}\right]^{\mathrm{T}}\right]^{-1} \left[Q_{0;xx}\right]^{-1}, \tag{4.100}$$

$$S^{(nn)}_{0;\dot{x}\dot{x}} = -\left[\mathrm{I}^{(nn)} - \left[Q^{(nn)}_{0;\dot{x}\dot{x}}\right]^{-1} Q^{(n)}_{0;\dot{x}x}\left[Q_{0;xx}\right]^{-1} Q^{(n)}_{0;x\dot{x}}\right]^{-1}\left[Q^{(nn)}_{0;\dot{x}\dot{x}}\right]^{-1}, \tag{4.101}$$

and

$$S^{(n)}_{0;\dot{x}x} = \left[\mathrm{I}^{(nn)} - \left[Q^{(nn)}_{0;\dot{x}\dot{x}}\right]^{-1} Q^{(n)}_{0;\dot{x}x}\left[Q_{0;xx}\right]^{-1} Q^{(n)}_{0;x\dot{x}}\right]^{-1}\left[Q^{(nn)}_{0;\dot{x}\dot{x}}\right]^{-1} Q^{(n)}_{0;\dot{x}x}\left[Q_{0;xx}\right]^{-1}. \tag{4.102}$$

The last two give

$$-\left[S^{(nn)}_{0;\dot{x}\dot{x}}\right]^{-1} S^{(n)}_{0;\dot{x}x} = Q^{(n)}_{0;\dot{x}x}\left[Q_{0;xx}\right]^{-1}. \tag{4.103}$$

For the terminal node this is $Q^{(n)}_{0;\dot{x}x;1}[Q_{0;xx}]^{-1} = -\dot{q}_0(0)^{\mathrm{T}} q_0(0)^{-1}$, which will be used below.

The inverse of the velocity fluctuation matrix will turn out to be the memory matrix,

$$M^{(nn)} \equiv \left[S^{(nn)}_{0;\dot{x}\dot{x}}\right]^{-1} = -Q^{(nn)}_{0;\dot{x}\dot{x}} + Q^{(n)}_{0;\dot{x}x}\left[Q_{0;xx}\right]^{-1} Q^{(n)}_{0;x\dot{x}}. \tag{4.104}$$

The first term in essence is the unconditional contribution to the entropy from the direct correlation between the velocities at a pair of nodes. The second term is the indirect contribution conditional on the fixed terminal value of the fluctuation $x(t)$. For the case of a free Brownian particle, $Q_{0;xx} \to \infty$, leaving $M^{(nn)} = -Q^{(nn)}_{0;\dot{x}\dot{x}}$.

The components of the memory matrix are

$$M^{(nn)}_{jk} = -\ddot{q}_0(t_j - t_k) - \dot{q}_0(t - t_j)^{\mathrm{T}} q_0(0)^{-1} \dot{q}_0(t - t_k). \tag{4.105}$$

Recall that $t_1 \equiv t$. The memory matrix is in total symmetric, $M^{(nn)}_{j\alpha,k\gamma} = M^{(nn)}_{k\gamma,j\alpha}$.

The terminal case, $j = 1$, will be required below, and so with $t_1 \equiv t$ one can define the memory vector with components

$$M^{(n)}_{k} \equiv M^{(nn)}_{1k} = -\ddot{q}_0(t - t_k) - \dot{q}_0(0)^{\mathrm{T}} q_0(0)^{-1} \dot{q}_0(t - t_k). \tag{4.106}$$

In the continuum limit this is the memory function $M(t - t') \equiv -\ddot{q}_0(t - t') - \dot{q}_0(0)^{\mathrm{T}} q_0(0)^{-1} \dot{q}_0(t - t')$. The memory function is a matrix in the thermodynamic variables with symmetry $M_{\alpha\gamma}(t - t') = M_{\gamma\alpha}(t' - t)$. Because the correlation functions are short-ranged, so is the memory function, $M(t) \to 0$, $|t| \to \infty$. For the pure parity case, $\dot{q}_0(0) = 0$. Hence the pure parity memory function is just the velocity autocorrelation function alone.

*Regression of fluctuation*

The derivative of the trajectory entropy with respect to the velocities is

$$\frac{\partial S_0(\mathbf{X}^{(n+1)})}{\partial \dot{\mathbf{x}}^{(n)}} = S^{(n)}_{0;\dot{x}x} x(t) + S^{(nn)}_{0;\dot{x}\dot{x}} \dot{\mathbf{x}}^{(n)}. \tag{4.107}$$

Setting this to zero gives the most likely trajectory conditional on the terminal position $x_1 = x(t_1)$ as

$$\begin{aligned}\bar{\mathbf{x}}^{(n)} &= -\left[S^{(nn)}_{0;\dot{x}\dot{x}}\right]^{-1} S^{(n)\mathrm{T}}_{0;\dot{x}x} x_1 \\ &= Q^{(n)}_{0;\dot{x}x}\left[Q_{0;xx}\right]^{-1} x_1 \\ &= \langle \dot{\mathbf{x}}^{(n)} x_1 \rangle_0 \langle x_1 x_1 \rangle_0^{-1} x_1.\end{aligned} \tag{4.108}$$

The second equality follows from equation (4.103).

This must be the same as that given by the definitions of the time correlation functions, which may be confirmed by multiplying the $n$th element of this on the right by $x_1$ and taking the average. One obtains for the left-hand side

$$\mathrm{LHS} = \langle \bar{\dot{x}}_n x_1 \rangle_0 = \langle \dot{x}_n x_1 \rangle_0 = k_\mathrm{B} \dot{q}_0(t_1 - t_n)^\mathrm{T}, \tag{4.109}$$

and for the right-hand side

$$\mathrm{RHS} = -\dot{q}_0(t_1 - t_n)^\mathrm{T} q_0(0)^{-1} \langle x_1 x_1 \rangle_0 = k_\mathrm{B} \dot{q}_0(t_1 - t_n)^\mathrm{T}. \tag{4.110}$$

These are equal, which confirms the validity of the approach.

The element $\bar{\dot{x}}_n$, is the most likely velocity after an interval $t_n - t_1$ given that the system was (will be) at $x_1$ at $t_1$. This is explicitly

$$\begin{aligned}\bar{\dot{x}}(t_n | x_1, t_1) &= \langle \dot{x}(t_n) x_1 \rangle_0 \langle x_1 x_1 \rangle_0^{-1} x_1 \\ &= \dot{q}_0(t_n - t_1)\, S x_1.\end{aligned} \tag{4.111}$$

This uses the fluctuation form for the equilibrium entropy,

$$S(x) = \frac{1}{2} S'' : xx, \quad S'' = -k_\mathrm{B} \langle xx \rangle_0^{-1} = q_0(0)^{-1}. \tag{4.112}$$

The result for the most likely velocity is really just a form of Onsager's regression hypothesis, with $F = \partial S / \partial x = S'' x$ being the thermodynamic force, and $\dot{q}_0(t_n - t_1)$ being the mixed parity transport matrix for the given time interval.

This result for the regression of a fluctuation ignores the history of the system beyond $t_1$ (prior to $t_1$ if $t_n > t_1$; after $t_1$ if $t_1 > t_n$). This is equivalent to assuming that the system is dynamically disordered at $t_1$. This remark is of relevance in interpreting the perturbation theory that is now developed.

*Time varying perturbation force*

Now add to the bare system an external time varying force that acts on the fluctuation, $\mathbf{F}(x, t) = -\nabla_x U(x, t)$. For example, $x$ might include the electric or magnetic polarization of the system, and the external force would contain the corresponding applied field. We shall continue to regard $x$ as possibly a mixed parity vector, even if the external forces couple directly to only some of the variables.

As in section 4.6.1, the rate of change of entropy due to the motion of the thermodynamic variables is

$$\dot{S}(x, t) = \frac{-1}{T}\dot{x} \cdot \nabla_x U(x, t) = \frac{1}{T}\dot{x} \cdot \mathbf{F}(x, t). \tag{4.113}$$

The quantity $\dot{x} \cdot \nabla_x U = -\dot{x} \cdot \mathbf{F}(x, t)$ is the rate of change of external energy due to the particle motion. This energy comes from the solvent reservoir itself, which decreases the entropy of the reservoir accordingly; $T$ is the temperature of the reservoir.

As in equation (4.82), for a given trajectory the reservoir entropy is just the bare trajectory entropy plus the change in entropy over the trajectory due to the motion in the external field,

$$\begin{aligned} S(\mathbf{X}^{(n+1)}, [U]) &= S_0(\mathbf{X}^{(n+1)}) - \frac{1}{T}\int_0^t \mathrm{d}t' \ \dot{x}(t') \cdot \nabla_x U(x(t'), t') \\ &= \frac{1}{2}S_0^{(n+1,n+1)} : \mathbf{X}^{(n+1)}\mathbf{X}^{(n+1)} + \frac{\tau}{T}\dot{\mathbf{x}}^{(n)} \cdot \mathbf{F}^{(n)}. \end{aligned} \tag{4.114}$$

Differentiating this and using equation (4.107) gives

$$\frac{\partial S(\mathbf{X}^{(n+1)}, [U])}{\partial \dot{\mathbf{x}}^{(n)}} = S_{0;\dot{x}x}^{(n)} x(t) + S_{0;\dot{x}\dot{x}}^{(nn)} \dot{\mathbf{x}}^{(n)} + \frac{\tau}{T}\mathbf{F}^{(n)} + \mathcal{O}(F^2). \tag{4.115}$$

This neglects a contribution from the dependence of the force at time $t'$ on the position $x(t')$, which depends on the velocities $\dot{x}(t'')$, $t'' \in [t, t']$. This neglected term contains the product of $\dot{x}(t')$ and the derivative of the force, $\nabla_{x(t')}\mathbf{F}(x(t'), t')\partial x(t')/\partial \dot{x}(t'')$. Since the most likely value of the velocity is a linear function of the force and of the position $x(t)$, and since the most likely value of the latter is also linearly proportional to the force, the neglected term is quadratic in the force.

Setting the derivative to zero gives the most likely velocities and hence the trajectory as

$$\begin{aligned} \bar{\dot{\mathbf{x}}}^{(n)} &= -\left[S_{0;\dot{x}\dot{x}}^{(nn)}\right]^{-1} S_{0;\dot{x}x}^{(n)} x_1 - \frac{\tau}{T}\left[S_{0;\dot{x}\dot{x}}^{(n)}\right]^{-1} \mathbf{F}^{(n)} \\ &= Q_{0;\dot{x}x}^{(n)}\left[Q_{0;xx}\right]^{-1} x_1 - \frac{\tau}{T} M^{(n)}\mathbf{F}^{(n)}. \end{aligned} \tag{4.116}$$

This uses equation (4.103) and the memory matrix given in equation (4.104), $M^{(n)} \equiv [S_{0;\dot{x}\dot{x}}^{(n)}]^{-1}$. This confirms the above assertion that the most likely value of the velocity is a linear function of the force and of the position $x(t)$. This is essentially the same as the result for Brownian motion, section 4.6.1, with differences arising from terms that are non-zero in the present mixed parity case.

Evaluating this at the terminal node, $\bar{\dot{x}}(t_1)$, and writing $t_1 \equiv t$, and transforming back to the continuum this is

$$\bar{\dot{x}}(t) = -\,\dot{q}_0(0)^{\mathrm{T}} q_0(0)^{-1} x(t) - \frac{1}{T}\int_0^t \mathrm{d}t' \ M(t - t')\mathbf{F}(x(t'), t'). \tag{4.117}$$

The first term on the right-hand side is just the result for the (instantaneous) regression of a fluctuation in the absence of an external field, equation (4.111), $\bar{\dot{x}}(t_n|x_1, t_1) = \dot{q}_0(t_n - t_1)\, Sx_1$. The memory function is given by equation (4.106),

$$\begin{aligned} M(t - t') &\equiv -\ddot{q}_0\big(t - t'\big) - \dot{q}_0(0)^{\mathrm{T}} q_0(0)^{-1} \dot{q}_0(t - t') \\ &= -k_{\mathrm{B}}^{-1} \langle \dot{x}(t)\dot{x}(t')\rangle_0 + k_{\mathrm{B}}^{-1} \langle \dot{x}(t)x(t)\rangle_0 \, \langle x(t)x(t)\rangle_0^{-1} \, \langle x(t)\dot{x}(t')\rangle_0. \end{aligned} \tag{4.118}$$

The memory function is short-ranged and so the integral can be truncated after some fixed interval beyond the relaxation time.

One can perform a similar analysis to obtain the most likely velocity at $t$, given the position at some prior time $t'$ and the velocities leading up to $t$, $\dot{\mathbf{x}}$. In this case the memory function changes to reflect this different condition (Attard 2012, equation (10.161)).

*Internal thermodynamic force*

As already mentioned, the way in which the external force enters the above equations is that it gives the rate of entropy production on the trajectory, $\dot{S}(x, t) = \dot{x} \cdot \mathbf{F}(x, t)/T$. But as pointed out by Onsager (1931) in his regression hypothesis, the system behaves the same whether the current state is brought about by a previously applied external force or by a spontaneous fluctuation of the system itself. In the latter case the internal thermodynamic force is the gradient of the entropy, and the rate of change of system entropy is just $\dot{S}(x) = \dot{x} \cdot \nabla_x S(x)$. Hence for a 'bare' system one can write

$$\frac{\mathbf{F}(x, t)}{T} \Rightarrow \nabla_x S(x(t)) = S'' x(t), \tag{4.119}$$

the second equality holding in the linear regime. With this replacement the most likely velocity in the bare system without an applied force is

$$\begin{aligned} \bar{\dot{x}}(t) &= -\dot{q}_0(0)^{\mathrm{T}} q_0(0)^{-1} x(t) - \int_0^t \mathrm{d}t' \; M(t - t') \nabla_x S(x(t')) \\ &= \dot{q}_0(0) q_0(0)^{-1} x(t) \\ &\quad + \int_0^t \mathrm{d}t' \big\{\ddot{q}_0(t - t') - \dot{q}_0(0) q_0(0)^{-1} \dot{q}_0(t - t')\big\} S'' x(t'). \end{aligned} \tag{4.120}$$

This result gives $\bar{\dot{x}}(t|\mathbf{X})$, the most likely velocity conditional on the specified preceding trajectory $\mathbf{X} = \{x, \dot{\mathbf{x}}\}$, leading up to the current position $x = x(t)$. The result for the regression of a fluctuation, equation (4.111), namely $\bar{\dot{x}}(t'|x, t) = \dot{q}_0(t' - t)\, S'' x$, gives the most likely current velocity given the position at a different time and not specifying the trajectory leading up to that position. The present result is useful when memory effects are non-negligible and one has knowledge of how the system reached its current state.

One can form dyadic matrices by multiplying this on the right by $k_B^{-1}x(0)$. Taking the average gives

$$
\begin{aligned}
\dot{q}_0(t) &\equiv -k_B^{-1}\langle \dot{x}(t)x(0)\rangle_0 \\
&= -k_B^{-1}\langle \bar{\dot{x}}(t)x(0)\rangle_0 \\
&= -k_B^{-1}\dot{q}_0(0)q_0(0)^{-1}\langle x(t)x(0)\rangle_0 \\
&\quad - k_B^{-1}\int_0^t \mathrm{d}t'\left\{\ddot{q}_0(t-t') - \dot{q}_0(0)q_0(0)^{-1}\dot{q}_0(t-t')\right\}S''\langle x(t')x(0)\rangle_0 \\
&= \dot{q}_0(0)q_0(0)^{-1}q_0(t) \\
&\quad + \int_0^t \mathrm{d}t'\left\{\ddot{q}_0(t-t') - \dot{q}_0(0)q_0(0)^{-1}\dot{q}_0(t-t')\right\}q_0(0)^{-1}q_0(t').
\end{aligned}
\tag{4.121}
$$

This is, presumably, an identity.

*Generalized Langevin equation*

Adding a stochastic, zero mean term to the above result for the most likely velocity due to a perturbing time-dependent external force gives the generalized Langevin equation

$$\dot{x}(t) = R(t) + \dot{q}_0(0)q_0(0)^{-1}x(t) - \frac{1}{T}\int_0^t \mathrm{d}t'\, M(t-t')\mathbf{F}(x(t'), t'). \tag{4.122}$$

The covariance of the random forces is given by the memory matrix

$$\langle R(t)R(0)\rangle_0 = -k_B M(t), \tag{4.123}$$

where the memory matrix is given by equation (4.118).

This is readily checked for the discretized trajectory, in which case one has

$$
\begin{aligned}
\langle \mathbf{R}^{(n)}\mathbf{R}^{(n)}\rangle &= \langle[\dot{\mathbf{x}}^{(n)} - \bar{\dot{\mathbf{x}}}^{(n)}][\dot{\mathbf{x}}^{(n)} - \bar{\dot{\mathbf{x}}}^{(n)}]\rangle \\
&= \langle \dot{\mathbf{x}}^{(n)}\dot{\mathbf{x}}^{(n)}\rangle_0 \\
&= -k_B\left[S_{0;\dot{x}\dot{x}}^{(nn)}\right]^{-1} \\
&= -k_B M^{(nn)}.
\end{aligned}
\tag{4.124}
$$

The covariance of the fluctuation in velocity, the first equality, can be taken to be the same as in the bare system, the second equality. The fourth equality is the definition of the memory function as the velocity autocorrelation matrix, equation (4.104). The probability distribution of the last $n$ random forces is just the corresponding multivariate Gaussian.

The memory matrix satisfies $M(-t) = M(t)^{\mathrm{T}}$. The relationship between the covariance of the random 'force' and the memory function is the same as for Brownian motion, equation (4.87). That the memory function gives both the covariance of the random forces and the dissipative force is the fluctuation dissipation theorem for a system with memory. The generalized Langevin equation in this form has the advantage that the memory function is constructed from the bare time correlation functions and it can be used for arbitrary applied forces.

For the case of the internal thermodynamic force for a fluctuation in the bare system, $\mathbf{F}(x, t)/T \Rightarrow \nabla_x S(x(t)) = S''x(t)$, the generalized Langevin equation is

$$\dot{x}(t) = R(t) + \dot{q}_0(0)q_0(0)^{-1}x(t) - \int_0^t \mathrm{d}t' \; M(t - t')S''x(t'). \qquad (4.125)$$

The memory function, equation (4.118), and the covariance of the random force remain unchanged. It can be shown that these results are formally the same as the generalized Langevin equation given by the projector operator formalism of Zwanzig (1961) and Mori (1965) (see Attard 2012, section 10.7). The present approach gives the memory function explicitly in terms of the time correlation functions.

## Summary

- Brownian motion is the archetypical stochastic process in which hidden forces contribute as well as the extant adiabatic forces. These hidden forces arise from the projection of reservoir interactions, they must be treated statistically, and they obey certain thermodynamic and time symmetry rules.
- The Langevin equation and the fluctuation–dissipation theorem are consequences of the second entropy for transitions. The dissipative force drives the system up the reservoir entropy gradient, and the fluctuation force randomizes the transition, with a bias toward more probable sub-system macrostates. These compete with each other, reaching a balance in the equilibrium macrostate.
- The Fokker–Planck equation gives the evolution of the probability density under stochastic, dissipative equations of motion. The equilibrium probability density is stationary under the second entropy stochastic, dissipative equations of motion.
- The equation due to Liouville for the time derivative of the probability density is an exact theorem for adiabatic equations of motion. It is an approximation for deterministic, dissipative equations of motion that is valid when the generalized equipartition theorem can be applied locally.
- The first generalized equipartition theorem relates the dyadic of the gradient of the reservoir entropy to the dyadic gradient of the reservoir entropy. These two are equal on average, and they may be approximately equal locally on likely points in phase space. The second generalized equipartition theorem says that the average of dyad formed from the state coordinate and the coordinate gradient of the reservoir entropy is proportional to the identity matrix.
- The Langevin equation can be generalized to replace the friction coefficient by a memory function for a Brownian particle or thermodynamic variable that is driven by a time-dependent external potential. The formalism holds also for thermodynamic fluctuations.

## References


Allen M P and Tildesley D J 1987 *Computer Simulations of Liquids* (Oxford: Oxford University Press)

Attard P 2012 *Non-Equilibrium Thermodynamics and Statistical Mechanics: Foundations and Applications* (Oxford: Oxford University Press)

Boltzmann L 1871 Über das wärmegleichgewicht zwischen mehratomigen gasmolekule *Wien. Ber.* **63** 397–481

Brown R 1828 A brief account of microscopical observations made in the months of June, July, and August, 1827, on the particles contained in the pollen of plants; and on the general existence of active molecules in organic and inorganic bodies *Phil. Mag.* **4** 161

Einstein A 1905 Über die von der molekularkinetischen theorie der wärme geforderte bewegung von in ruhenden flüssigkeiten suspendierten teilchen *Ann. Phys.* **17** 549

Ermak D L and McCammon J A 1978 Brownian dynamics with hydrodynamic interactions *J. Chem. Phys.* **69** 1352

Fokker A 1914 Die mittlere energie rotierender elektrischer dipole im strahlungsfeld *Ann. Phys.* **43** 810

Gardiner C W 1983 *Handbook of Stochastic Methods* (New York: Springer)

Haw M D 2002 Colloidal suspensions, Brownian motion, molecular reality: a short history *J. Phys.: Condens. Matter* **14** 7769

Keizer J 1987 *Statistical Thermodynamics of Non-Equilibrium Processes* (New York: Springer)

Langevin P 1908 Sur la théorie du mouvement Brownien *C. R. Acad. Sci., Paris* **146** 530

Mazo R M 2008 *Brownian Motion: Fluctuations, Dynamics, and Applications* (Oxford: Oxford University Press)

Mori H 1965 Transport, collective motion, and Brownian motion *Prog. Theor. Phys.* **33** 423

Mori H 1965 A continued-fraction representation of the time-correlation functions *Prog. Theor. Phys.* **34** 399

Münster A 1969 *Statistical Thermodynamics* vol 1 (Berlin: Springer)

Onsager L 1931 Reciprocal relations in irreversible processes. I *Phys. Rev.* **37** 405

Onsager L 1931 Reciprocal relations in irreversible processes. II *Phys. Rev.* **38** 2265

Pathria R K 1972 *Statistical Mechanics* (Oxford: Pergamon Press)

Planck M 1916 Über einen satz der statistischen dynamik und seine erweiterung in der quanten theorie *Sitz.ber., Preuss. Akad. Wiss* **24** 324–41

Pope A 1734 *Essay on Man*

Risken H 1984 *The Fokker–Planck Equation* (Berlin: Springer)

Silver B L 1998 *The Ascent of Science* (Oxford: Oxford University Press)

Tough R J A, Pusey P N, Lekerkerker H N W and van den Broeck C 1986 Stochastic descriptions of the dynamics of interacting Brownian particles *Mol. Phys.* **59** 595

von Smoluchowski M 1906 Zur kinetischen theorie der Brownschen molekularbewegung und der suspensionen *Ann. Phys.* **21** 756

Weinert F 2005 *The Scientist as Philosopher: Philosophical Consequences of Great Scientific Discoveries* (Berlin: Springer) p 197

Zwanzig R 1961 *Lectures Theoretical Physics* ed W E Britton, B W Downs and J Downs vol III (New York: Wiley) p 135

# Chapter 5

## In phase with entropy

'Indeed it is clear that any individual uniform distribution, which might arise after a certain time from some particular initial state, is just as improbable as an individual non-uniform distribution... It is only because there are many more uniform distributions than non-uniform ones that the distribution of states will become uniform in the course of time.'

Boltzmann (1877)

'A new scientific truth does not triumph by convincing its opponents and making them see the light, but rather because its opponents eventually die, and a new generation grows up that is familiar with it.'

Planck (1949)

The main aim of this chapter is to formulate classical equilibrium statistical mechanics from first principles. The approach is to derive the Maxwell–Boltzmann distribution from Hamilton's classical equations of motion in phase space. The stochastic, dissipative equations of motion for phase space for an open sub-system are also derived. Two approximations for the information entropy are tested against exact results for the two-dimensional Ising model.

The Boltzmann distribution is often invoked in an equilibrium system,

$$\wp_\alpha = \frac{1}{Z(T)} e^{-U_\alpha/k_\mathrm{B}T}. \tag{5.1}$$

Here $\wp_\alpha$ is the probability of the state $\alpha$, $U_\alpha$ is its potential energy, and $T$ is the temperature. Despite its use being common in papers and textbooks, as a general

doi:10.1088/978-0-7503-1590-6ch5

result there is a quite serious problem with this that is discussed in section 5.3.3 below. The point of deriving the Maxwell–Boltzmann distribution from first principles is to explain why equilibrium statistical mechanics is formulated the way it is, and why the Boltzmann distribution in this form cannot be blindly applied to the states of an equilibrium system

## 5.1 Phase space and Hamilton's equations of motion

The preceding chapter dealt with the properties of Brownian motion, which is a generic model of the stochastic behavior of a macroscopic coordinate due to the hidden deterministic behavior of microscopic coordinates. Most of the results carry over directly to the stochastic behavior of microscopic coordinates of a sub-system due to the hidden deterministic behavior of a reservoir with which it interacts. Here, and in the remainder of this chapter, the focus is on the formulation of classical equilibrium statistical mechanics in terms of classical phase space, eventually leading to the Maxwell–Boltzmann probability distribution and to the stochastic, dissipative equation of motion in phase space.

### 5.1.1 Classical phase space

In this section points in phase space are identified as the microstates of a classical system. At the simplest level this is the space of positions and linear momenta of the particles in the sub-system. In the present case the sub-system is isolated from the rest of the Universe and is the same as the system. It is emphasized that this is the adiabatic case in which no interactions with a reservoir are permitted.

For $N$ particles in three dimensions phase space is $6N$-dimensional, with a phase space point being denoted $\mathbf{\Gamma} = \{\mathbf{q}, \mathbf{p}\}$, with the positions being

$$\mathbf{q} = \{\mathbf{q}_1, \mathbf{q}_2, \ldots, \mathbf{q}_N\}, \quad \mathbf{q}_j = \{q_{jx}, q_{jy}, q_{jz}\}, \tag{5.2}$$

and the momenta being

$$\mathbf{p} = \{\mathbf{p}_1, \mathbf{p}_2, \ldots, \mathbf{p}_N\}, \quad \mathbf{p}_j = \{p_{jx}, p_{jy}, p_{jz}\}. \tag{5.3}$$

At this level of description the particles are modeled as spherical. Although it is possible to augment phase space to include orientational degrees of freedom, the present choice is sufficiently realistic to derive the formalism of statistical mechanics without needlessly complicating the mathematical analysis.

The assertion that $6N$-dimensional classical phase space represents the microstates of the isolated sub-system is a statement about the level of description that is necessary and sufficient to describe the physical properties of the present sub-system with an acceptable level of accuracy. It is a statement that for this particular sub-system not only are orientational effects negligible, but so also are quantum effects, the electronic degrees of freedom, intra-nuclear configurations, etc. These effects may be subsumed grossly into an effective interaction potential. But in any case, having decided upon classical phase space, no finer level of description will be pursued.

Of course, this raises the question: if a point in classical phase space is an effective microstate, then might it not have some internal entropy, and might not this entropy vary from point to point? In other words, what is the entropy of a point in classical phase space?

Before addressing these questions we first give Hamilton's equations of motion for classic phase space. A derivation of the primacy of classical phase space and of Hamilton's classical equations of motion from the underlying quantum mechanical equations for the system is given in chapter 7.

### 5.1.2 Hamilton's equations of motion

Hamilton's equations of motion are

$$\dot{\mathbf{p}}^0 = \frac{-\partial \mathcal{H}(\mathbf{\Gamma})}{\partial \mathbf{q}}, \text{ and } \dot{\mathbf{q}}^0 = \frac{\partial \mathcal{H}(\mathbf{\Gamma})}{\partial \mathbf{p}}. \tag{5.4}$$

The Hamiltonian, $\mathcal{H}(\mathbf{\Gamma})$, is the total energy of the isolated sub-system, and it is the sum of the kinetic and potential energies, $\mathcal{H}(\mathbf{\Gamma}) = \mathcal{K}(\mathbf{p}) + U(\mathbf{q})$. Usually the kinetic energy is

$$\mathcal{K}(\mathbf{p}) = \frac{1}{2m} \sum_{j=1}^{N} \sum_{\alpha=x,y,z} p_{j\alpha}^2 = \frac{p^2}{2m}. \tag{5.5}$$

Here $m$ is the mass of the particles, assuming they are all identical. The momentum derivative of this yields the velocity, $\dot{\mathbf{q}}^0 = \partial \mathcal{K}(\mathbf{p})/\partial \mathbf{p} = \mathbf{p}/m$.

The superscript 0 here and throughout signifies the Hamiltonian or adiabatic evolution. In the present case the sub-system is isolated, and Hamilton's equations of motion are exact and entirely characterize the motion of a point in phase space. In later parts of this chapter and in most of this book the sub-system interacts with a reservoir, and Hamilton's equations of motion for the sub-system is just part of the total evolution. In such cases the superscript 0 is necessary to distinguish the adiabatic part internal to the sub-system from the dissipative and the stochastic parts due to the reservoir, and from the total evolution that is the sum of all three parts.

The energy is a constant of the motion of the isolated sub-system,

$$\begin{aligned}
\frac{\mathrm{d}\mathcal{H}(\mathbf{\Gamma})}{\mathrm{d}t} &= \dot{\mathbf{\Gamma}}^0 \cdot \nabla \mathcal{H}(\mathbf{\Gamma}) \\
&= \dot{\mathbf{q}}^0 \cdot \nabla_q \mathcal{H}(\mathbf{\Gamma}) + \dot{\mathbf{p}}^0 \cdot \nabla_p \mathcal{H}(\mathbf{\Gamma}) \\
&= \sum_{j=1}^{N} \sum_{\alpha=x,y,z} \left\{ \frac{\partial \mathcal{H}(\mathbf{\Gamma})}{\partial p_{j\alpha}} \frac{\partial \mathcal{H}(\mathbf{\Gamma})}{\partial q_{j\alpha}} - \frac{\partial \mathcal{H}(\mathbf{\Gamma})}{\partial q_{j\alpha}} \frac{\partial \mathcal{H}(\mathbf{\Gamma})}{\partial p_{j\alpha}} \right\} \\
&= 0.
\end{aligned} \tag{5.6}$$

In addition to the energy being a constant of the adiabatic motion, it is also approximately linear additive, which is to say it is extensive with the sub-system size. (It is exactly linear additive in the thermodynamic limit when interactions with and across boundaries are relatively negligible.) Because of these two facts energy plays a preeminent role in the analysis that follows.

There are six other constants of the adiabatic motion, namely the components of the linear and angular momentum. These tend to play a comparatively minor role in equilibrium thermodynamics and statistical mechanics. (They can be important in hydrodynamics and non-equilibrium thermodynamics.) The reason that momentum is usually ignored is possibly that the conservation law for them depends upon the system having linear and rotational symmetry. Since almost all systems have a fixed boundary, this symmetry is broken, and in practice linear and angular momentum are not conserved if the boundary interactions are accounted for.

In addition to conserving energy, Hamilton's equations of motion are incompressible. That is

$$\begin{aligned}\nabla \cdot \dot{\mathbf{\Gamma}}^0 &= \nabla_q \cdot \dot{\mathbf{q}}^0 + \nabla_p \cdot \dot{\mathbf{p}}^0 \\ &= \sum_{j=1}^{N} \sum_{\alpha=x,y,z} \left\{ \frac{\partial^2 \mathcal{H}(\mathbf{\Gamma})}{\partial q_{j\alpha} \partial p_{j\alpha}} - \frac{\partial^2 \mathcal{H}(\mathbf{\Gamma})}{\partial p_{j\alpha} q_{j\alpha}} \right\} \\ &= 0.\end{aligned} \tag{5.7}$$

This vanishing of the compressibility means that phase space volume is conserved during its adiabatic evolution.

To see the relationship between compressibility and volume change, let $\mathbf{\Gamma}_1$ be the initial phase space point, and let $\mathbf{\Gamma}_2 = \mathbf{\Gamma}_1 + \Delta_t \dot{\mathbf{\Gamma}}_1$ be the adiabatically evolved point in an infinitesimal time step $\Delta_t$. The relative change in volume element is

$$\begin{aligned}\left| \frac{\mathrm{d}\mathbf{\Gamma}_2}{\mathrm{d}\mathbf{\Gamma}_1} \right| &= \left| \mathrm{I} + \Delta_t \frac{\mathrm{d}\dot{\mathbf{\Gamma}}_1^0}{\mathrm{d}\mathbf{\Gamma}_1} \right| \\ &= 1 + \Delta_t \mathrm{Tr} \frac{\mathrm{d}\dot{\mathbf{\Gamma}}_1^0}{\mathrm{d}\mathbf{\Gamma}_1} + \mathcal{O}(\Delta_t^2) \\ &= 1 + \Delta_t \nabla \cdot \dot{\mathbf{\Gamma}}^0.\end{aligned} \tag{5.8}$$

The second equality follows because the determinant is the product of the eigenvalues, each one of which is of the form $1 + \Delta_t \lambda_k$, and $\prod_k (1 + \Delta_t \lambda_k) = 1 + \Delta_t \sum_k \lambda_k + \mathcal{O}(\Delta_t^2)$, where the sum of the eigenvalues is just the trace of the corresponding matrix.

The incompressibility of phase space during adiabatic evolution will be used to obtain an important result for the conservation of probability density on an adiabatic trajectory below.

*Uniqueness of the adiabatic trajectory*

The adiabatic trajectory passing through $\mathbf{\Gamma}_0$ at time $t_0$ is $\mathbf{\Gamma}^0(t|\mathbf{\Gamma}_0, t_0) = \mathbf{\Gamma}^0(t - t_0|\mathbf{\Gamma}_0)$. The equality holds because an equilibrium system is homogeneous in time. For Hamilton's equations of motion, for an infinitesimal time step $\Delta_t$, the evolution of the adiabatic trajectory is

$$\mathbf{\Gamma}_2 \equiv \mathbf{\Gamma}^0(t + \Delta_t|\mathbf{\Gamma}_1) = \mathbf{\Gamma}_1 + \Delta_t\dot{\mathbf{\Gamma}}^0(\mathbf{\Gamma}_1). \tag{5.9}$$

By the fundamental theorem of calculus, the full trajectory is obtained by integrating Hamilton's equations,

$$\mathbf{\Gamma}^0(t|\mathbf{\Gamma}_0) = \mathbf{\Gamma}_0 + \int_0^t \mathrm{d}t'\ \dot{\mathbf{\Gamma}}^0(\mathbf{\Gamma}^0(t'|\mathbf{\Gamma}_0)). \tag{5.10}$$

Since $t$ here may be taken to be positive or negative, Hamilton's equations specify a unique trajectory passing through each point in the isolated sub-system's phase space. Each phase space point has a unique destination in a transition forward in time, and a unique destination in a transition backward in time. This means that adiabatic trajectories can never cross or terminate.

This in turn means that a region of phase space adiabatically evolves in a unique and well-defined fashion: phase space points initially interior to the region can never cross the evolving boundaries as they themselves evolve. Conversely, points initially exterior to the specified region can never enter the region. The evolution of a region of phase space is sketched in figure 5.1.

Because adiabatic trajectories cannot cross the boundary of an evolving region of phase space, the number of trajectories inside an evolving region is a constant of the adiabatic motion. By the incompressibility of Hamilton's equations proved above, the volume of an evolving region is also a constant of the adiabatic motion. This is also sketched in figure 5.1, where the regions have different shapes but the same volume. These two constants of the motion can be combined to conclude that the number density of trajectories in phase space is a constant of the adiabatic motion. This result will be used in the following sub-section, section 5.2.1, to discuss a result due to Boltzmann, namely the uniformity of phase space weight on an adiabatic trajectory.

*Reversibility of Hamilton's equations*

Hamilton's equations of motion are reversible. To see what this means, define the conjugate of a phase space point of the isolated sub-system as the point with all the momenta reversed,

$$\mathbf{\Gamma} = \{\mathbf{q}, \mathbf{p}\} \Leftrightarrow \mathbf{\Gamma}^\dagger = \{\mathbf{q}, -\mathbf{p}\}. \tag{5.11}$$

The energy is insensitive to the direction of the particle's velocities,

$$\mathcal{H}(\mathbf{\Gamma}) = \mathcal{H}(\mathbf{\Gamma}^\dagger). \tag{5.12}$$

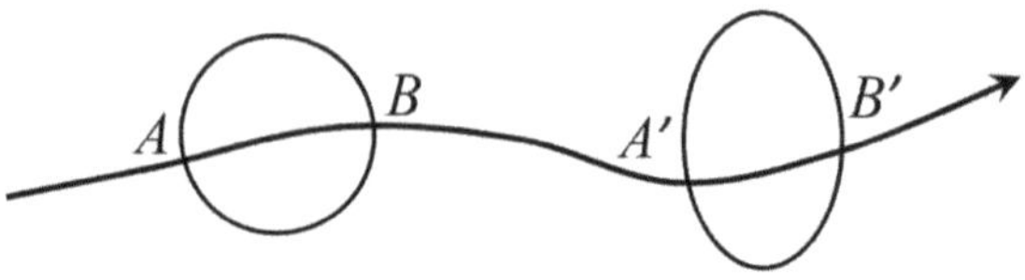


**Figure 5.1.** An adiabatic trajectory in phase space (curved arrow), and two regions of phase space (circle and ellipse) whose boundaries are related by adiabatic evolution. The labels refer to the times when the trajectory enters and exits the regions considered as fixed in phase space.

The adiabatic rate of change of a conjugate point is the negative conjugate of that of the original point,

$$\begin{aligned}\dot{\boldsymbol{\Gamma}}^0(\boldsymbol{\Gamma}^\dagger) &= \{\dot{\mathbf{q}}^0(\boldsymbol{\Gamma}^\dagger), \dot{\mathbf{p}}^0(\boldsymbol{\Gamma}^\dagger)\} \\ &= \left\{\frac{\partial H(\boldsymbol{\Gamma}^\dagger)}{\partial \mathbf{p}^\dagger}, \frac{-\partial H(\boldsymbol{\Gamma}^\dagger)}{\partial \mathbf{q}^\dagger}\right\} \\ &= \left\{\frac{-\partial H(\boldsymbol{\Gamma})}{\partial \mathbf{p}}, \frac{-\partial H(\boldsymbol{\Gamma})}{\partial \mathbf{q}}\right\} \\ &= -\dot{\boldsymbol{\Gamma}}^0(\boldsymbol{\Gamma})^\dagger.\end{aligned} \tag{5.13}$$

Consider an adiabatic transition over a single time step, $\boldsymbol{\Gamma}_2 = \boldsymbol{\Gamma}_1 + \Delta_t \dot{\boldsymbol{\Gamma}}^0(\boldsymbol{\Gamma}_1)$. To leading order it does not matter whether the velocity is evaluated at $\boldsymbol{\Gamma}_1$ or $\boldsymbol{\Gamma}_2$, $\dot{\boldsymbol{\Gamma}}^0(\boldsymbol{\Gamma}_1) = \dot{\boldsymbol{\Gamma}}^0(\boldsymbol{\Gamma}_2) + \mathcal{O}(\Delta_t)$. Hence, in view of the fact that a unique adiabatic trajectory passes through each point in phase space, and the above result for the velocity of the conjugate point, one has three symmetries

$$\begin{aligned}\boldsymbol{\Gamma}_2 = \boldsymbol{\Gamma}_1 + \Delta_t \dot{\boldsymbol{\Gamma}}^0(\boldsymbol{\Gamma}) &\Leftrightarrow \boldsymbol{\Gamma}_1 = \boldsymbol{\Gamma}_2 - \Delta_t \dot{\boldsymbol{\Gamma}}^0(\boldsymbol{\Gamma}), \text{ (temporal)} \\ &\Leftrightarrow \boldsymbol{\Gamma}_1^\dagger = \boldsymbol{\Gamma}_2^\dagger + \Delta_t \dot{\boldsymbol{\Gamma}}^0(\boldsymbol{\Gamma}^\dagger), \text{ (conjugate)} \\ &\Leftrightarrow \boldsymbol{\Gamma}_2^\dagger = \boldsymbol{\Gamma}_1^\dagger - \Delta_t \dot{\boldsymbol{\Gamma}}^0(\boldsymbol{\Gamma}^\dagger), \text{ (microscopic)}.\end{aligned} \tag{5.14}$$

Here $\boldsymbol{\Gamma}$ can be either $\boldsymbol{\Gamma}_1$ or $\boldsymbol{\Gamma}_2$. One sees that microscopic reversibility is a combination of temporal and conjugate reversibility.

Integrating these over time an adiabatic trajectory has symmetries

$$\begin{aligned}\boldsymbol{\Gamma}_2 = \boldsymbol{\Gamma}^0(\tau|\boldsymbol{\Gamma}_1) &\Leftrightarrow \boldsymbol{\Gamma}_1 = \boldsymbol{\Gamma}^0(-\tau|\boldsymbol{\Gamma}_2), \text{ (temporal)} \\ &\Leftrightarrow \boldsymbol{\Gamma}_1^\dagger = \boldsymbol{\Gamma}^0(\tau|\boldsymbol{\Gamma}_2^\dagger), \text{ (conjugate)} \\ &\Leftrightarrow \boldsymbol{\Gamma}_2^\dagger = \boldsymbol{\Gamma}^0(-\tau|\boldsymbol{\Gamma}_1^\dagger), \text{ (microscopic)}.\end{aligned} \tag{5.15}$$

The first of these can be called the temporal or time reversibility of Hamilton's equations of motion. Microscopic reversibility, the final one, says that if $\boldsymbol{\Gamma}_2$ is the adiabatic destination of $\boldsymbol{\Gamma}_1$ going forward in time, then $\boldsymbol{\Gamma}_2^\dagger$ is the adiabatic destination of $\boldsymbol{\Gamma}_1^\dagger$ going backward in time. This is sketched in figure 5.2. Note that time goes in the opposite direction on these trajectories. This is what is meant by the microscopic reversibility of Hamilton's equations of motion.

*Time-dependent potential*

At this point a small digression can be made to address the nature of microscopic reversibility in the case that the sub-system has applied to it an external potential that varies explicitly with time, $U(\boldsymbol{\Gamma}, t)$.

Consider the time interval $[t_1, t_2]$, and let the termini of the adiabatic trajectory satisfy $\boldsymbol{\Gamma}_2 = \boldsymbol{\Gamma}^0(t_2|\boldsymbol{\Gamma}_1, t_1)$. This of course can equally well be written $\boldsymbol{\Gamma}_1 = \boldsymbol{\Gamma}^0(t_1|\boldsymbol{\Gamma}_2, t_2)$. On this time interval one can also write $\boldsymbol{\Gamma}' = \boldsymbol{\Gamma}^0(t'|\boldsymbol{\Gamma}_1, t_1) = \boldsymbol{\Gamma}^0(t'|\boldsymbol{\Gamma}_2, t_2)$, $t' \in [t_1, t_2]$.

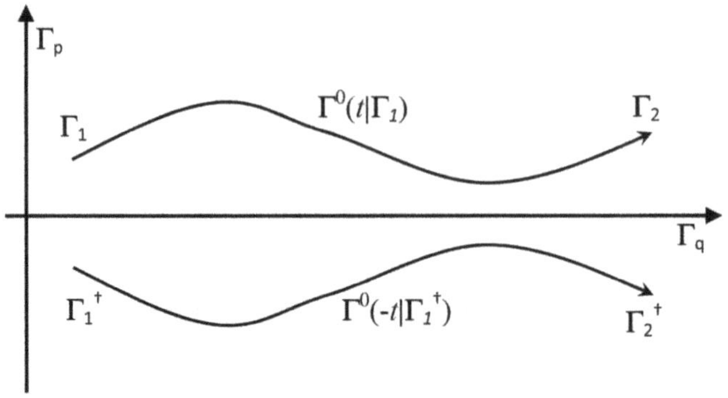


**Figure 5.2.** Microscopic reversibility for two adiabatic trajectories, from $\mathbf{\Gamma}_1$ (upper) and $\mathbf{\Gamma}_1^\dagger$ (lower). The arrows point in the direction of the transition; the two trajectories move in opposite directions in time.

Define the conjugate potential on this time interval as

$$\tilde{U}(\mathbf{\Gamma}, t) \equiv U(\mathbf{\Gamma}, t_1 + t_2 - t), \quad t \in [t_1, t_2]. \tag{5.16}$$

This is in effect the temporal mirror image of the original potential on the time interval. In so far as the applied potential arises from atoms and molecules external to the system, reversing its time dependence in this way as well as conjugating the momenta of the particles of the sub-system is equivalent to reversing the momenta of all the atoms and molecules in the Universe.

This conjugate potential is required for the reverse trajectory,

$$\mathbf{\Gamma}_2 = \mathbf{\Gamma}^0(t_2|\mathbf{\Gamma}_1, t_1; [U]) \Leftrightarrow \mathbf{\Gamma}_1^\dagger = \mathbf{\Gamma}^0(t_2|\mathbf{\Gamma}_2^\dagger, t_1; [\tilde{U}]), \tag{5.17}$$

and

$$\mathbf{\Gamma}' = \mathbf{\Gamma}^0(t'|\mathbf{\Gamma}_1, t_1; [U]) \Leftrightarrow \mathbf{\Gamma}'^\dagger = \mathbf{\Gamma}^0(t''|\mathbf{\Gamma}_2^\dagger, t_1; [\tilde{U}]), \quad t'' \equiv t_1 + t_2 - t'. \tag{5.18}$$

This is one form of microscopic reversibility for a time-dependent potential.

## 5.2 Entropy of a point in phase space

The aim in this section is to derive the entropy of a point in the phase space of the isolated system, $S(\mathbf{\Gamma})$. By the axioms of chapter 1, this is in essence equivalent to the weight density $w(\mathbf{\Gamma})$ and the probability density $\wp(\mathbf{\Gamma})$. This 'internal' entropy of a phase space point for the isolated system is a necessary prerequisite for the derivation of the Maxwell–Boltzmann probability for an open sub-system that can exchange energy with a heat reservoir (see section 5.3).

In the following sub-sections, three approaches are discussed, each with the same aim of proving that the entropy density is constant in phase space, which means that it can be set to zero, $S(\mathbf{\Gamma}) = 0$. Equivalently, the weight density and the probability density are constant, $w(\mathbf{\Gamma}) = \text{const.}$ and $\wp(\mathbf{\Gamma}) = \text{const.}$ The first approach, section 5.2.1, is the ergodic hypothesis, which essentially takes the result to be an axiom. The second approach, section 5.2.2, essentially proves that the probability density is uniform on an adiabatic trajectory, and assumes that this implies the final result. The third approach, section 5.2.3, derives the final result from time homogeneity.

### 5.2.1 Ergodic hypothesis

In most books on statistical mechanics the uniform weight density of phase space is but one of several axioms that together form what is called the ergodic hypothesis (Boltzmann 1871). This hypothesis is traditionally central to the formulation of statistical mechanics; Münster (1969), for example, discusses the role of the former in the axiomatic development of the latter.

The ergodic hypothesis comprises three points: For an isolated sub-system

1. all points in phase space with a given energy lie on a single trajectory
2. the probability density in phase space is uniform on an energy hypersurface
3. the probability density in phase space is uniform.

The second of these is sometimes called the equal *a priori* hypothesis (Pathria 1972, section 2.3).

The first of these assertions says that a trajectory is dense on the energy hypersurface that it is confined to. It further says that the hypersurface is not broken up into distinct regions each covered by its own trajectory that cannot communicate with the rest.

Since adiabatic trajectories do not terminate, they must be in the form of loops. Hence, since every point on the energy hypersurface belongs to an adiabatic trajectory, it is obvious that the entire energy hypersurface must be covered by one or more loops, each distinct and non-communicating. According to this first axiom, there is only one such loop for the whole energy hypersurface.

In the formal development of statistical mechanics this first axiom appears to be essential, because it is necessary to go from a property established for a single trajectory to a property of the entire energy hypersurface (and thence to a property of the entire phase space). In the following sub-subsection, following Boltzmann it is established that the probability density on a given trajectory has a uniform value. But it does not follow that this value must be the same on all distinct trajectories with the same energy. Hence without the first axiom of the ergodic hypothesis it is not possible to prove that the second axiom is a consequence of Boltzmann's result.

It strains credulity that an energy hypersurface with Avogadro's number of dimensions should be covered by one and only one trajectory loop. As a contrary example, if the isolated sub-system was itself composed of isolated sub-systems, then there would be at least one distinct adiabatic trajectory loop for each such sub-system, the starting position of each being arbitrary. Further, the total energy can be arbitrarily partitioned amongst the isolated sub-subsystems. Each combination of starting positions or energies corresponds to a different trajectory loop of the total sub-system. This example obviously violates the first axiom of the ergodic hypothesis.

Some texts emphasize the second assertion in the ergodic hypothesis, which is claimed to be self-evident: because there is nothing to choose between phase space points with the same energy, they must all have the same weight. (Against this one could point out that the speed of the trajectory varies on the energy hypersurface, and this could reasonably be expected to influence the weight.) Equivalently, on the basis of the first axiom, it is claimed that the system must pass through all possible points 'without fear or favor' (Pathria 1972, p 38).

Whether or not one finds the second assertion self-evident, it is the third assertion that is the one that is really needed to obtain the Maxwell–Boltzmann probability density. The point is that even if one accepts the second assertion, or the result derived next—that the probability density is uniform on a trajectory—one still has to find a way to prove the third assertion.

### 5.2.2 Constant probability on a trajectory

Consider the adiabatic trajectory $\mathbf{\Gamma}^0(t|\mathbf{\Gamma}_0)$, which passes through the point $\mathbf{\Gamma}_0$ at time $t = 0$. Consider also the possibly time-dependent phase space probability density, $\wp(\mathbf{\Gamma}, t)$ that has initial value $\wp(\mathbf{\Gamma}, 0) = \wp_0(\mathbf{\Gamma})$. Using the adiabatic equations of motion for the present isolated sub-system, the Fokker–Planck equation, equation (4.37), for the partial time derivative is

$$\begin{aligned}\frac{\partial \wp(\mathbf{\Gamma}, t)}{\partial t} &= -\nabla \cdot \dot{\mathbf{\Gamma}}^0 \wp(\mathbf{\Gamma}, t) - \dot{\mathbf{\Gamma}}^0 \cdot \nabla \wp(\mathbf{\Gamma}, t) \\ &= -\dot{\mathbf{\Gamma}}^0 \cdot \nabla \wp(\mathbf{\Gamma}, t).\end{aligned} \tag{5.19}$$

It is worth emphasizing that this result is a direct consequence of the conservation law for the transition probability, equation (3.17). The second equality follows because the adiabatic equations of motion are incompressible.

With this, the total rate of change of the probability density is

$$\begin{aligned}\frac{\mathrm{d}\wp(\mathbf{\Gamma}, t)}{\mathrm{d}t} &= \frac{\partial \wp(\mathbf{\Gamma}, t)}{\partial t} + \dot{\mathbf{\Gamma}}^0 \cdot \nabla \wp(\mathbf{\Gamma}, t) \\ &= -\dot{\mathbf{\Gamma}}^0 \cdot \nabla \wp(\mathbf{\Gamma}, t) + \dot{\mathbf{\Gamma}}^0 \cdot \nabla \wp(\mathbf{\Gamma}, t) \\ &= 0.\end{aligned} \tag{5.20}$$

The first equality is the definition of the total rate of change with time. Combining this with the Fokker–Planck equation gives the second equality. This is just Liouville's theorem, equation (4.81), which is exact for an adiabatic trajectory. This vanishing of the total time derivative says that the probability density in phase space is a constant of the adiabatic motion, $\wp(\mathbf{\Gamma}^0(t|\mathbf{\Gamma}_0), t) = \wp_0(\mathbf{\Gamma}_0)$.

One can give a straightforward physical interpretation to this result. As has been mentioned above in connection with figure 5.1, the adiabatic incompressibility of phase space means that volume is conserved during their evolution under Hamilton's equations of motion. Since adiabatic trajectories do not cross, trajectories inside the volume stay there for their entire evolution. Combined with the conservation law for weight during a transition, equation (3.17), this implies that the total weight inside the volume is a constant of the adiabatic motion. Since the volume itself is constant, this means that the weight density, and hence the probability density is constant along the length of a trajectory.

Boltzmann gave an equivalent argument that is also illuminating. In figure 5.1 two regions are shown, the second of which evolved from the first adiabatically.

An adiabatic trajectory is also shown, along with the times it entered, $t_A$ and $t_{A'}$, and exited, $t_B$ and $t_{B'}$, the respective regions (considered now as fixed in phase space). Since the evolution of the region is defined by the evolution of the boundary, the difference in these respective times must be equal, $t_{B'} - t_B = t_{A'} - t_A$. This means that the trajectory spent an equal amount of time in each region, $t_B - t_A = t_{B'} - t_{A'}$. Assuming that the weight of a region of phase space is proportional to the time spent in it, it follows that regions of phase space lying on a single trajectory have the same weight density.

Both arguments show that the probability density is a constant of the adiabatic motion $\wp(\mathbf{\Gamma}^0(t|\mathbf{\Gamma}_0), t) = \wp_0(\mathbf{\Gamma}_0)$.

Now for the present equilibrium system, the probability density must be independent of time,

$$\frac{\partial \wp(\mathbf{\Gamma}, t)}{\partial t} = 0. \tag{5.21}$$

Hence $\wp(\mathbf{\Gamma}, t) = \wp(\mathbf{\Gamma})$. These last two results imply that $\wp(\mathbf{\Gamma}^0(t|\mathbf{\Gamma}_0)) = \wp_0(\mathbf{\Gamma}_0)$. This says that the probability density is uniform along the length of an adiabatic trajectory. That is, if $\mathbf{\Gamma}_1$ and $\mathbf{\Gamma}_2$ lie on the same trajectory, $\mathbf{\Gamma}_2 = \mathbf{\Gamma}^0(t_2|\mathbf{\Gamma}_1, t_1)$, then they must have the same probability density, $\wp_0(\mathbf{\Gamma}_2) = \wp_0(\mathbf{\Gamma}_1)$.

With the first part of the ergodic hypothesis, this result becomes that all points that lie on the same energy hypersurface have equal probability density. As mentioned above, this is the equal *a priori* hypothesis (Pathria 1972, section 2.3). In equation form this says that given that the isolated sub-system has energy $E$, then the phase space probability density is

$$\wp(\mathbf{\Gamma}|E) = \frac{\delta(\mathcal{H}(\mathbf{\Gamma}) - E)}{\Omega(E)}. \tag{5.22}$$

Here the normalizing factor is essentially the number of phase space points in the energy hypersurface,

$$\Omega(E) = \int \mathrm{d}\mathbf{\Gamma}\ \delta(\mathcal{H}(\mathbf{\Gamma}) - E). \tag{5.23}$$

However, even if one accepts this result (notwithstanding the arguments given above against the first axiom), one still has to convert this result from a conditional probability density to an unconditional probability density. That is, it is the phase space weight $w(\mathbf{\Gamma})$ that is required.

Formally the unconditional probability density is

$$\begin{aligned} \wp(\mathbf{\Gamma}) &= \int \mathrm{d}E\ \wp(\mathbf{\Gamma}|E)\wp(E) \\ &= \int \mathrm{d}E \frac{\delta(\mathcal{H}(\mathbf{\Gamma}) - E)}{\Omega(E)} \frac{e^{S(E)/k_\mathrm{B}}}{W}. \end{aligned} \tag{5.24}$$

Again formally, the entropy of an energy macrostate is given by the weight of the macrostate, which is just the sum total of the phase space weight in the macrostate

$$e^{S(E)/k_{\rm B}} \equiv W(E) = \int \mathrm{d}\boldsymbol{\Gamma}\ \omega(\boldsymbol{\Gamma})\ \delta(\mathcal{H}(\boldsymbol{\Gamma}) - E). \tag{5.25}$$

With this the phase space probability density

$$\begin{aligned} \wp(\boldsymbol{\Gamma}) &= \int \mathrm{d}E \frac{\delta(\mathcal{H}(\boldsymbol{\Gamma}) - E)}{\Omega(E)} \frac{W(E)}{W} \\ &= \frac{W(\mathcal{H}(\boldsymbol{\Gamma}))}{\Omega(\mathcal{H}(\boldsymbol{\Gamma}))\ W}. \end{aligned} \tag{5.26}$$

In general, since neither $W(\mathcal{H}(\boldsymbol{\Gamma}))$ nor $\Omega(\mathcal{H}(\boldsymbol{\Gamma}))$ are known (contrast equation (5.23) with equation (5.25)), this is not a constant in phase space.

However, if one assumes uniform phase space weight, $w(\boldsymbol{\Gamma}) = 1$ (any set of weights can be re-scaled by a positive factor, and so if this is constant, then that constant may be set to unity), then one has

$$W(E) = \Omega(E) \text{ iff } \omega(\boldsymbol{\Gamma}) = 1, \tag{5.27}$$

and the phase space probability density becomes

$$\wp(\boldsymbol{\Gamma}) = \frac{1}{W}, \tag{5.28}$$

which is the desired constant in phase space. Regrettably this result begs the question: it shows that phase space has uniform weight if it is assumed that phase space has uniform weight.

### 5.2.3 Time and energy

Now the weight density of phase space is derived from a more satisfactory perspective that avoids invoking the three axioms of the ergodic hypothesis.

The adiabatic trajectory $\boldsymbol{\Gamma}^0(t|\boldsymbol{\Gamma}_0)$ that is determined by the Hamiltonian is confined to the energy hypersurface $\mathcal{H}(\boldsymbol{\Gamma}) = E$. This suggests proceeding in two stages to obtain the phase space weight density: first obtain the weight density on an energy hypersurface, and then obtain the density of energy hypersurfaces in phase space.

Define a coordinate system for the energy hypersurface, $\boldsymbol{\gamma}(\boldsymbol{\Gamma})$. The relationship is invertible, $\boldsymbol{\Gamma}(\gamma, E)$. The trajectory may be written as $\boldsymbol{\gamma}(t) = \boldsymbol{\gamma}(\boldsymbol{\Gamma}(t))$. It is axiomatic that an average of a phase space function is a simple time average over a trajectory, which is to say that time is homogeneous. In consequence of this, the weight density on the energy hypersurface must be inversely proportional to the speed of the trajectory at that point,

$$\begin{aligned} \omega(\gamma|E) &\propto \frac{1}{|\dot{\gamma}^0|} \\ &= \frac{1}{|\dot{\boldsymbol{\Gamma}}^0|}. \end{aligned} \tag{5.29}$$

The second equality follows because on a trajectory over a time step, $|\gamma_2 - \gamma_1| = |\boldsymbol{\Gamma}_2 - \boldsymbol{\Gamma}_1|$. Since a time average is a simple average, the weight is large in regions of slow speed because the systems spend more time there. This physical picture is incompatible with the equal *a priori* hypothesis. In the left of figure 5.3 an adiabatic trajectory of varying speed can be seen. Closely spaced marks signify slow speed, in which regions the trajectory spends more time, and which therefore have larger weight.

The proportionality constant neglected here is essentially the length of the time interval over which the weights are measured. It is important to note that this cannot depend upon the energy $E$, and that it must be the same constant on different energy hypersurfaces. The assertion that time is uniform, and that the weight of a state is proportional only to the time spent in the state, precludes any such energy dependence.

The weight density of phase space itself is related to this by the usual rule for the transformation of densities,

$$\begin{aligned}\omega(\mathbf{\Gamma}|E) &= \omega(\gamma|E)\frac{\mathrm{d}\gamma}{\mathrm{d}\mathbf{\Gamma}}, \quad |\mathcal{H}(\mathbf{\Gamma}) - E| < \mathrm{d}E \\ &= \frac{1}{|\dot{\mathbf{\Gamma}}^0|}\frac{|\nabla\mathcal{H}(\mathbf{\Gamma})|}{\mathrm{d}E}, \quad |\mathcal{H}(\mathbf{\Gamma}) - E| < \mathrm{d}E \\ &= \frac{|\nabla\mathcal{H}(\mathbf{\Gamma})|}{|\dot{\mathbf{\Gamma}}^0|}\delta(\mathcal{H}(\mathbf{\Gamma}) - E).\end{aligned} \tag{5.30}$$

The second factor from the transformation of coordinates, $|\nabla\mathcal{H}(\mathbf{\Gamma})|/\mathrm{d}E$, is essentially the number of energy hypersurfaces per unit phase space. One can perhaps see the need and form for this transformation more clearly by noting that the volume of the energy hypersurface can be equally written

$$\begin{aligned}A(E) &= \oint_E \mathrm{d}\gamma \\ &= \int \mathrm{d}\mathbf{\Gamma}\ |\nabla\mathcal{H}(\mathbf{\Gamma})|\ \delta(\mathcal{H}(\mathbf{\Gamma}) - E).\end{aligned} \tag{5.31}$$

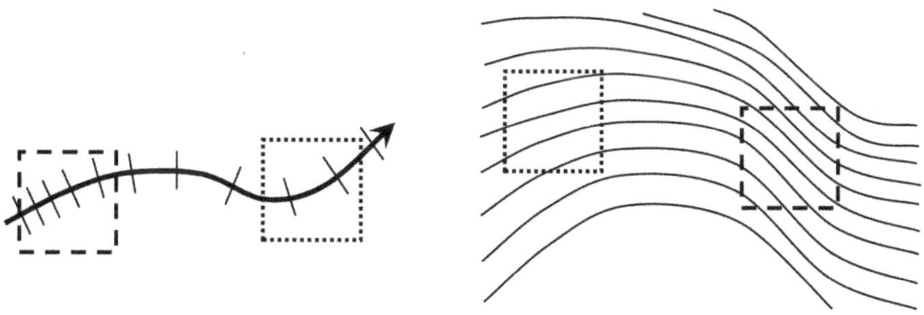

**Figure 5.3.** Left: an adiabatic trajectory in phase space marked in equal time increments. Right: contours of the energy hypersurface. The boxes signify regions of phase space, with the dashed region having greater weight than the dotted region in each case, all other things being equal.

With this transformation rule, the expression for the phase space weight gives the same total weight whether expressed as an integral over the energy hypersurface, or as an integral over phase space,

$$\begin{aligned} W(E) &= \oint_E \mathrm{d}\gamma \ \omega(\gamma|E) \\ &= \int \mathrm{d}\mathbf{\Gamma} \ \omega(\gamma|E) \, |\nabla\mathcal{H}(\mathbf{\Gamma})| \ \delta(\mathcal{H}(\mathbf{\Gamma}) - E) \\ &= \int \mathrm{d}\mathbf{\Gamma} \ \omega(\mathbf{\Gamma}|E). \end{aligned} \tag{5.32}$$

The result for $\omega(\mathbf{\Gamma}|E)$ says that the weight is larger for steep gradients because in such regions there are more energy hypersurfaces per unit phase space distance. This real physical effect is also inconsistent with the equal *a priori* hypothesis. In the right of figure 5.3 can be seen the contours of the energy hypersurface of an isolated system. The box crossed by many contour lines has greater weight than the box crossed by few. The former lies in a region with steeper energy gradient than the latter.

The function $\omega(\mathbf{\Gamma}|E)$ is the phase space weight density conditional on a given energy. The unconditional weight of an isolated system phase space point is

$$\begin{aligned} \omega(\mathbf{\Gamma}) &= \int \mathrm{d}E \ \omega(\mathbf{\Gamma}|E) \\ &= \frac{|\nabla\mathcal{H}(\mathbf{\Gamma})|}{|\dot{\mathbf{\Gamma}}^0|}. \end{aligned} \tag{5.33}$$

As was argued above, there cannot be an additional weight factor $\tilde{w}(E) \Rightarrow \tilde{w}(\mathcal{H}(\mathbf{\Gamma}))$ included in this final result because this would violate the principle of time homogeneity.

This has a straightforward physical interpretation. The weight of a phase space point is proportional to the time the system spends in the volume element about that point, which is to say that it is inversely proportional to the speed of the trajectory at that point. Since the system is confined to an energy hypersurface, the weight of a phase space point is also proportional to the number of energy hypersurfaces per unit volume, which is to say that it is proportional to the gradient of the energy at that point.

Both of these properties—the trajectory speed and the energy gradient—are local properties in phase space. The big advantage of this derivation of the weight density of phase space, $\omega(\mathbf{\Gamma})$, is that it relies only on the properties of the particular point, $\mathbf{\Gamma}$. This contrasts with the ergodic hypothesis, for example, where one has to postulate a global property, namely that a single trajectory passes infinitesimally close to every point on an energy hypersurface, and also that the weights on all energy hypersurfaces are equal.

In order to evaluate the energy gradient and the speed that appear here, position and momentum components have to be added together, and these have different dimensions. To get around this problem a metric element, the length scales $l_q$ and $l_p$, can be introduced into the definition of the scalar product and the gradient operator (Galea and Attard 2002). Scalar products have the form

$$\boldsymbol{\Gamma}_1 \cdot \boldsymbol{\Gamma}_2 = \frac{1}{l_q^2}\mathbf{q}_1 \cdot \mathbf{q}_2 + \frac{1}{l_p^2}\mathbf{p}_1 \cdot \mathbf{p}_2, \tag{5.34}$$

and gradients have the form

$$\nabla f(\boldsymbol{\Gamma}) = l_q^2\frac{\partial f(\boldsymbol{\Gamma})}{\partial \mathbf{q}} + l_p^2\frac{\partial f(\boldsymbol{\Gamma})}{\partial \mathbf{p}}. \tag{5.35}$$

The ratio of the particular values chosen for these length scales will change the value of the length of the gradient and of the speed of the trajectory, but it will make no difference to the final conclusion that the two cancel.

With these scale factors, the square of the magnitude of the gradient of the Hamiltonian is

$$\begin{aligned} |\nabla \mathcal{H}(\boldsymbol{\Gamma})|^2 &= \frac{l_q^4}{l_q^2}\frac{\partial \mathcal{H}}{\partial \mathbf{q}} \cdot \frac{\partial \mathcal{H}}{\partial \mathbf{q}} + \frac{l_p^4}{l_p^2}\frac{\partial \mathcal{H}}{\partial \mathbf{p}} \cdot \frac{\partial \mathcal{H}}{\partial \mathbf{p}} \\ &= l_q^2\dot{\mathbf{p}}^0 \cdot \dot{\mathbf{p}}^0 + l_p^2\dot{\mathbf{q}}^0 \cdot \dot{\mathbf{q}}^0 \\ &= l_q^2 l_p^2 \ |\dot{\boldsymbol{\Gamma}}^0|^2. \end{aligned} \tag{5.36}$$

The proportionality of these is a remarkable coincidence, and is a unique feature of Hamilton's equations of motion. The proportionality constant $l_q l_p$ has the dimensions of energy × time, which is the same dimensions as Planck's constant $h$. Since such positive multiplicative constants are immaterial, the phase space weight density of the isolated system is constant. Without loss of generality, it may be set equal to unity,

$$\omega(\boldsymbol{\Gamma}) = 1. \tag{5.37}$$

This is obviously a constant on an adiabatic trajectory. In this case this result is consistent with, but stronger than, the result derived in section 5.2.2 from the Fokker–Planck equation and Liouville's theorem: the probability density is uniform along the length of an adiabatic trajectory, $\wp(\boldsymbol{\Gamma}^0(t|\boldsymbol{\Gamma}_0)) = \wp_0(\boldsymbol{\Gamma}_0)$. Neither the trajectory speed nor the energy gradient are constants of the adiabatic motion, and hence it is essential that they cancel as here. This result says that the weight density is not just constant on a trajectory, but that it is a constant throughout phase space itself.

With this result, the total weight of an energy macrostate for the isolated system is just

$$W(E) = \int \mathrm{d}\boldsymbol{\Gamma}\ \delta(\mathcal{H}(\boldsymbol{\Gamma}) - E). \tag{5.38}$$

One can interpret this as the number of phase space points (microstates) in an energy macrostate of an isolated system. This is not the volume of the energy hypersurface, $A(E)$.

## 5.3 Canonical equilibrium system

### 5.3.1 Constrained entropy of a phase space point

The canonical equilibrium system consists of a sub-system that can exchange energy with a thermal reservoir. The Hamiltonian of the sub-system is independent of time, $\mathcal{H}(\mathbf{\Gamma})$. The microstates of the system, which are the smallest indivisible states that will be analyzed, will be taken to be the microstates of the sub-system, namely the points in the sub-system phase space $\mathbf{\Gamma}$.

Conversely, it could be argued that the sub-system phase space point $\mathbf{\Gamma}$ is actually a macrostate of the total system, since the microstates of the total system are in fact points in the total phase space of the sub-system and reservoir, $\mathbf{\Gamma}_{\text{total}} = \{\mathbf{\Gamma}, \mathbf{\Gamma}_{\text{r}}\}$. However, since the rationale of the reservoir formalism is to focus on the sub-system, I think it best to regard the sub-system phase space point $\mathbf{\Gamma}$ as a microstate of the total system. The projection of the reservoir phase space points $\mathbf{\Gamma}_{\text{r}}$ out of the formalism contributes to the internal entropy of the microstate $\mathbf{\Gamma}$. This argument over terminology is purely academic because the formalism for entropy is the same for microstates as for macrostates.

To obtain the probability of a microstate $\mathbf{\Gamma}$ we need its total entropy. The result has the same form as the total constrained entropy obtained in section 2.2.1. That analysis was also for the canonical equilibrium system, with the difference being that the sub-system energy macrostate $E$ was used. In the present case we essentially replace $E$ in equation (2.20) by the present sub-system microstate $\mathbf{\Gamma}$.

First we need the temperature of the reservoir. The entropy of an isolated system with energy $E$ is by definition

$$S(E) = k_{\text{B}} \ln W(E), \tag{5.39}$$

where the total weight is the number of phase space points in the energy hypersurface, as given by the immediately preceding analysis. Here the number and volume arguments have been suppressed.

By the definition (2.10), the temperature is given by

$$\frac{1}{T} \equiv \frac{\partial S(E)}{\partial E} = \frac{k_{\text{B}}}{W(E)} \frac{\partial W(E)}{\partial E}. \tag{5.40}$$

If the fixed total energy of the total isolated system is $E_{\text{total}}$, then the reservoir energy in the sub-system microstate $\mathbf{\Gamma}$ is $E_{\text{r}}(\mathbf{\Gamma}) = E_{\text{total}} - \mathcal{H}(\mathbf{\Gamma})$, and the reservoir entropy in that microstate is

$$\begin{aligned} S_{\text{r}}(E_{\text{r}}(\mathbf{\Gamma})) &\equiv S_{\text{r}}(E_{\text{total}}) - \mathcal{H}(\mathbf{\Gamma})\frac{\partial S_{\text{r}}(E_{\text{total}})}{\partial E_{\text{total}}} \\ &= \text{const.} - \frac{1}{T}\mathcal{H}(\mathbf{\Gamma}), \end{aligned} \tag{5.41}$$

where $T$ is the reservoir temperature. The constant part independent of the sub-system may be dropped.

As in equation (2.20), the constrained total entropy for the canonical equilibrium system in the microstate $\mathbf{\Gamma}$ is

$$\begin{aligned} S_{\text{total}}(\mathbf{\Gamma}|T) &= S_s(\mathbf{\Gamma}) + S_r(E_r(\mathbf{\Gamma})) \\ &= \frac{-1}{T}\mathcal{H}(\mathbf{\Gamma}), \end{aligned} \tag{5.42}$$

since there is no internal entropy for a point in the sub-system phase space, $S_s(\mathbf{\Gamma}) = k_B \ln \omega(\mathbf{\Gamma}) = 0$.

### 5.3.2 Maxwell–Boltzmann probability

With this result, the phase space probability distribution for the canonical equilibrium system is

$$\begin{aligned} \wp(\mathbf{\Gamma}|N, V, T) &= \frac{1}{Z'(T)} e^{S_r(\mathbf{\Gamma}|T)/k_B} \\ &= \frac{e^{-\beta\mathcal{H}(\mathbf{\Gamma})}}{N!h^{3N} Z(N, V, T)}. \end{aligned} \tag{5.43}$$

This is the Maxwell–Boltzmann distribution. Here and throughout the inverse temperature is $\beta \equiv 1/k_B T$. The pre-factor of $N!h^{3N}$ will be discussed shortly.

The normalizing partition function is

$$Z(N, V, T) = \frac{1}{N!h^{3N}} \int \mathrm{d}\mathbf{\Gamma} e^{-\beta\mathcal{H}(\mathbf{\Gamma})}. \tag{5.44}$$

This is the total weight of the system. As has been mentioned, weight is only defined up to an arbitrary positive constant, and so the prefactor $N!h^{3N}$ has relatively trivial effects.

Planck's constant $h = 6.626 \times 10^{-34}$J s has the dimensions of energy × time (equivalently, position × momentum), and including it makes the partition function dimensionless, which is often desirable even though it is not essential. This factor is often interpreted to mean that quantum mechanics gives a phase space volume per microstate of $h^{3N}$. This is a little loose, but not completely incorrect. Two points can be made: first, within classical mechanics there is nothing to give Planck's constant, and the above formalism would still be valid if it were replaced by something else with the same dimensions or even by nothing at all. Weight is only defined up to a positive scale factor. Second, in chapter 7 this result is derived as the leading order term in an expansion of quantum statistical mechanics, and in this exact expansion Planck's constant is mandated and appears exactly as here.

The factor of $N!$ was introduced by Gibbs as the resolution to the paradox of the entropy of mixing (Pathria 1972, section 1.5). The way to understand it is to note that the entropy or weight of a state should count only the weight of *distinct* configurations in the state. Since particles are indistinguishable, a permutation of the particle labels does not change the configuration. For example, for two particles, particle 1 at $\mathbf{r}$ and particle 2 at $\mathbf{s}$ is the same configuration as particle 1 at $\mathbf{s}$ and particle 2 at $\mathbf{r}$. The partition function above involves the integral over all of phase space, which means that each distinct configuration is counted $N!$ times. (In classical mechanics configurations with particles at the same phase space point form a set of measure zero and can be neglected.) This over counting is corrected by the pre-factor of $1/N!$ in the partition function. In chapter 7, this factor arises from wave function symmetrization in the zeroth order term in an expansion of quantum statistical mechanics.

These two factors together may be considered to be the constant 'internal' weight of a phase space of the isolated sub-system,

$$w(\boldsymbol{\Gamma}) = \frac{1}{N!h^{3N}}. \tag{5.45}$$

In this case the sub-system entropy of a phase space point is

$$S_s(\boldsymbol{\Gamma}) = k_B \ln w(\boldsymbol{\Gamma}) = Nk_B - Nk_B \ln(Nh^3), \tag{5.46}$$

where Stirling's approximation for $N!$ has been used. In general one can either show $w(\boldsymbol{\Gamma}) = 1/N!h^{3N}$ as an explicit prefactor for the phase space probability and use the reservoir entropy in the exponent, or else one can include it as the sub-system entropy and use the total entropy in the exponent.

The statistical mechanical Helmholtz free energy is essentially the logarithm of this partition function,

$$F(N, V, T) = -k_B T \ln Z(N, V, T). \tag{5.47}$$

This is a particular form of the generic analysis given in section 2.2.2 and 2.5.3.

### 5.3.3 Not Boltzmann distributed

At the beginning of this chapter, the widespread use of the Boltzmann distribution in the form of equation (5.1),

$$\wp_\alpha = \frac{1}{Z(T)} e^{-U_\alpha/k_B T}, \tag{5.48}$$

was described as questionable. Here $\wp_\alpha$ is meant to be the probability of the state $\alpha$, $U_\alpha$ is its potential energy, and $T$ is the temperature.

Comparison with the Maxwell–Boltzmann distribution, equation (5.43) shows that this neglects the kinetic energy. This is a relatively minor problem because often one is only interested in static properties and since the kinetic energy and the potential energy are separable, the velocity can be integrated out.

A much more serious problem is that this neglects any internal entropy associated with the state $\alpha$. The exponent that appears explicitly here is the entropy of the thermal reservoir alone. The point of section 5.2.3 was to rigorously justify that the internal entropy of a point in phase space could be neglected, $S_s(\boldsymbol{\Gamma}) = 0$. This was necessary for the Maxwell–Boltzmann phase space form. The problem with the above is that it neglects without justification $S_s(\alpha)$, the internal entropy of the sub-system macrostate $\alpha$. In general, for an arbitrary static property the entropy cannot simply be assumed constant or negligible. Instead one must use

$$\wp_\alpha = \frac{1}{Z(T)} e^{S_s(\alpha)/k_B} e^{-U_\alpha/k_B T} = \frac{1}{Z(T)} e^{-F_\alpha/k_B T}, \tag{5.49}$$

where $F_\alpha = U_\alpha - TS_s(\alpha)$ is the Helmholtz free energy of the macrostate.

For example, if the state $\alpha$ represents the conformation of a macromolecule in a solvent, in addition to the potential energy between the atoms of the macromolecule in this particular configuration, there is also the (weighted) number of possible arrangements of the solvent molecules, which varies with $\alpha$. As another example, if $\alpha$ represents the location of a defect in a crystal, in addition to the variation of the potential energy of the crystal with defect location, one may also need to account for the change in the spectrum of vibrations of the crystal with defect location, which is another form of entropy. Or if $\alpha$ represents a crystal type, one would need to include the entropy associated with the weighted displacements of the atoms about their mean locations that can occur without destroying the crystal structure.

It may well be that for the specific problem at hand one has reason to believe that any such internal entropy is either constant or its variation is relatively negligible. The important thing is to explicitly justify such an assumption in the given case. Conversely, it is poor practice and likely erroneous to invoke the Boltzmann distribution without explicit justification for neglecting the internal entropy of the state.

## 5.4 Stochastic, dissipative equations of motion

### 5.4.1 Equations of motion

We now turn to the equations of motion for the canonical equilibrium system. As above, $\mathbf{\Gamma} = \{\mathbf{q}, \mathbf{p}\}$ is a point in the phase space of the sub-system, $\mathcal{H}(\mathbf{\Gamma})$ is the sub-system energy, and $T$ is the temperature of the reservoir with which the sub-system can exchange energy.

The equations of motion are dominated by the adiabatic part that is internal to the sub-system. The additional external contribution due to the interactions with the reservoir must be a relatively small perturbation because they directly occur only in the region of the boundary of the sub-system, and this is much smaller than the sub-system itself.

These reservoir interactions are treated statistically rather than literally. The stochastic, dissipative terms in the equations of motion that they give rise to are applied equally throughout the sub-system rather than only in the boundary region. This is in the spirit of the reservoir formalism, where the focus is on the sub-system, and the reservoir only enters through certain generic parameters that abstract the microscopic details in any particular case. Of course the functional form and magnitude of the stochastic, dissipative terms are uniquely fixed by the thermodynamic properties of the reservoir.

The Langevin equation for Brownian motion was given in section 4.1.2, and its relationship to the second entropy form for the transition of a thermodynamic fluctuation, equation (3.86), was discussed. One can take the second entropy form of the Langevin equation for Brownian motion, equation (4.21), directly over to the present problem. The position and the velocity of the Brownian particle become the position and momentum coordinates in phase space, and the solvent entropy becomes the reservoir entropy. With these, the stochastic, dissipative equations of motion for the sub-system are

$$\begin{aligned} \mathbf{q}(t+\tau) &= \mathbf{q}(t) + \frac{\tau}{m}\mathbf{p}(t) \\ \mathbf{p}(t+\tau) &= \mathbf{p}(t) - \tau\nabla_q U(\mathbf{q}) + \frac{|\tau|}{2}\Lambda\nabla_p S_{\mathrm{r}}(\mathbf{\Gamma}) + \tilde{\mathbf{R}}. \end{aligned} \tag{5.50}$$

The adiabatic parts of these are

$$\mathbf{q}^0(\tau|\mathbf{\Gamma}) = \mathbf{q} + \tau\frac{\partial\mathcal{H}(\mathbf{\Gamma})}{\partial\mathbf{p}} = \mathbf{q} + \frac{\tau}{m}\mathbf{p}, \tag{5.51}$$

and

$$\mathbf{p}^0(\tau|\mathbf{\Gamma}) = \mathbf{p} - \tau\frac{\partial\mathcal{H}(\mathbf{\Gamma})}{\partial\mathbf{q}} = \mathbf{p} - \tau\nabla_q U(\mathbf{q}). \tag{5.52}$$

Since the reservoir entropy is $S_{\mathrm{r}}(\mathbf{\Gamma}) = -\mathcal{H}(\mathbf{\Gamma})/T$, the dissipative term in these equations of motion is

$$\frac{|\tau|}{2}\Lambda\nabla_p S_{\mathrm{r}}(\mathbf{\Gamma}) = \frac{-|\tau|}{2mT}\Lambda\mathbf{p}. \tag{5.53}$$

The $3N \times 3N$ transport matrix $\Lambda$ is symmetric, positive definite, and couples only the momentum components. The stochastic 'force' is Gaussian distributed, equation (3.88),

$$\wp(\tilde{\mathbf{R}}) = \frac{1}{[\mathrm{Det}\ 2\pi k_{\mathrm{B}}|\tau|\Lambda]^{1/2}}e^{-\Lambda^{-1}:\tilde{\mathbf{R}}\tilde{\mathbf{R}}/2k_{\mathrm{B}}|\tau|}. \tag{5.54}$$

That $\Lambda$ appears in these last two equations is the form that the fluctuation–dissipation theorem takes for phase space.

The equilibrium Maxwell–Boltzmann distribution is stationary under these equations of motion for any choice of the symmetric, positive definite, momentum transport matrix $\Lambda$. It is simplest, however, to assume that different components of the momenta are uncoupled, which means that $\Lambda$ is diagonal. In the simplest case that the particles are identical, it is proportional to the identity matrix, $\Lambda = \lambda\mathrm{I}$. Again, the Maxwell–Boltzmann distribution is stationary for any choice of the scalar $\lambda$. (The constant $\lambda/2T$ may be called the friction or drag coefficient.)

In this simplest case, the stochastic, dissipative equations of motion for the canonical equilibrium system are

$$\begin{aligned} \mathbf{q}(t+\tau) &= \mathbf{q}(t) + \frac{\tau}{m}\mathbf{p}(t) \\ \mathbf{p}(t+\tau) &= \mathbf{p}(t) - \tau\nabla_q U(\mathbf{q}) - \frac{|\tau|\lambda}{2mT}\mathbf{p} + \tilde{\mathbf{R}}, \end{aligned} \tag{5.55}$$

with

$$\wp(\tilde{\mathbf{R}}) = \frac{1}{[2\pi \lambda k_{\mathrm{B}}|\tau|]^{3N/2}} e^{-\tilde{\mathbf{R}}\cdot\tilde{\mathbf{R}}/2\lambda k_{\mathrm{B}}|\tau|}. \tag{5.56}$$

The physical interpretation of these stochastic, dissipative equations of motion is illuminating. The dissipative term, $\hat{\tau}\lambda\nabla_p S_{\mathrm{r}}$, drives the system up the reservoir entropy gradient, which is to say toward sub-system microstates of higher reservoir entropy. The irreversible factor $\hat{\tau}$ means that going forward in time, $\hat{\tau} = +1$, this term determines that the destination of the transition will have higher reservoir entropy than the present state. If one operates the equations of motion backward to predict where the system came from, $\hat{\tau} = -1$, this term determines that the previous state had higher reservoir entropy than the present state.

The stochastic term $\tilde{\mathbf{R}}$ randomizes the transition without regard to the reservoir entropy. It is however sensitive to the sub-system entropy in that there are more sub-system phase space points $\mathbf{\Gamma}$ at higher sub-system energy $\mathcal{H}(\mathbf{\Gamma})$ than at lower, because the sub-system entropy $S_{\mathrm{s}}(E, N, V)$ is a monotonic increasing function of energy. Since the stochastic term $\tilde{\mathbf{R}}$ chooses destination phase space points uniformly at random, it is more likely to increase the sub-system energy than to decrease it. This stochastic term on average acts like an irreversible deterministic driving force toward macrostates of higher sub-system energy.

There is a competition between the dissipative and the stochastic term. The former drives the sub-system to states of low sub-system energy, because these correspond to high reservoir entropy, $S_{\mathrm{r}}(\mathbf{\Gamma}) = -\mathcal{H}(\mathbf{\Gamma})/T$. The latter drives the sub-system to states of high sub-system energy, which is to say high sub-system entropy, $S_{\mathrm{s}}(\mathcal{H}(\mathbf{\Gamma}), N, V)$. These two competing forces balance when the current phase space point belongs to a sub-system energy macrostate corresponding to a sub-system temperature that equals that of the reservoir (see figure 5.4).

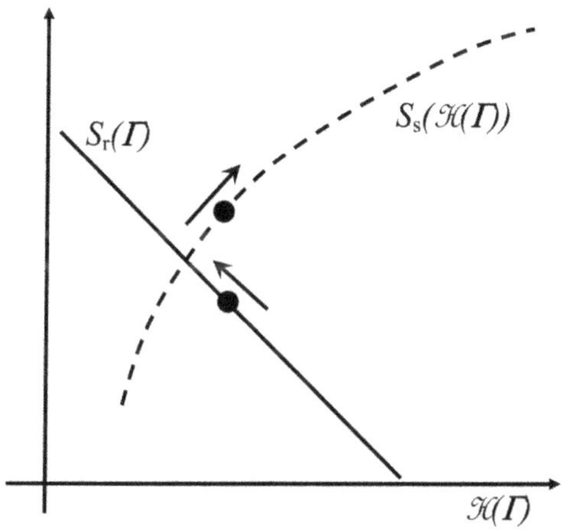


**Figure 5.4.** The reservoir entropy of a sub-system phase space point (full line) and the sub-system entropy of an energy macrostate (dashed curve) as a function of sub-system energy. The circles signify the point at which the tangents are equal and opposite.

It is worthwhile to reprise the justification for these stochastic, dissipative equations of motions in phase space. It is more or less the same as was given for the Langevin equation for Brownian motion, equation (4.21).

The position coordinate must evolve adiabatically over an infinitesimal time interval $|\tau|$. It is only over a sequence of consecutive intervals that forces from the sub-system particles and from the reservoir can indirectly effect the position evolution via their direct effect on the momentum evolution. This means that the position is a 'slave' to the momentum. It also means that only the momentum coordinate is to be treated statistically, which requires only the pure parity second entropy theory, equation (3.86).

For the present evolution of phase space over an infinitesimal time interval, the adiabatic evolution of the momentum appears explicitly. This is in contrast to the second entropy treatment of a fluctuation, equation (3.86). Although that case was for an isolated sub-system in which the evolution is purely adiabatic, because the time interval was long on molecular time scales, the adiabatic contributions to it had to be treated *in toto* rather than individually, which is to say statistically rather than explicitly. In the present case of phase space and an infinitesimal time interval, the forces due to the particles of the sub-system must appear explicitly in addition to the statistical forces from the reservoir.

The solvent forces acting on the Brownian particle in equation (4.21) are replaced here by the reservoir forces acting on the sub-system. The statistical treatment of the latter is identical to the pure parity second entropy analysis: the reservoir-induced transitions are treated as Gaussian fluctuations, with the coefficients being expanded for small time intervals in accord with general rules for the time symmetry of the transition in an equilibrium system. The result is that in the stochastic, dissipative equations of motion, the adiabatic forces are reversible, $\propto \tau$, whereas the dissipative and stochastic forces are irreversible, $\propto |\tau|$.

There are three reasons why it is valid to apply the second entropy analysis over infinitesimal time scales. First, the present momentum macrostate is the end point of a continuous evolution, and there is no need for an inertial period for it to become organized. Second, the relaxation time for individual particle momenta can be expected to be relatively short compared to that for a macroscopic fluctuation comprising Avogadro's number of particles. And third, any term linearly proportional to $|\tau_{\mathrm{macro}}|$ can be broken down into a series of infinitesimal transitions each proportional to $|\tau_{\mathrm{molec}}| \ll |\tau_{\mathrm{macro}}|$, and over the longer term these in series return the original expression.

It is important to note that these stochastic, dissipative equations of motions in phase space are justified by the second entropy analysis, and only by it. They cannot be justified by the Langevin equation for Brownian motion because the hydrodynamic drag term that is invoked therein has no justification on molecular length scales. The point is that the second entropy analysis shows that the dissipative force $\hat{\tau}\lambda\nabla_p S_{\mathrm{r}}$ is a direct consequence of the time symmetry of the second entropy. The hydrodynamic drag force is an effect, not a cause, of the second entropy dissipative force. It would be wrong to interpret the transport scalar $\lambda$ as the product of the radius of the molecule and the viscosity of the sub-system (i.e. Stoke's drag coefficient) since it is not determined by hydrodynamics. All that is essential is that the fluctuation–dissipation theorem be satisfied: the magnitude of the dissipation must be equal to the strength of the fluctuations. Since both are linearly proportional to $\lambda$, the actual value of $\lambda$ is immaterial (except, of course, that the reservoir forces should be negligible compared to the adiabatic forces.)

### 5.4.2 Reversibility of the stochastic, dissipative equations of motion

As was shown in section 5.1.2, the adiabatic equations of motion are time reversible,

$$\boldsymbol{\Gamma}_2 = \boldsymbol{\Gamma}^0(t|\boldsymbol{\Gamma}_1) \Leftrightarrow \boldsymbol{\Gamma}_1 = \boldsymbol{\Gamma}^0(-t|\boldsymbol{\Gamma}_2). \tag{5.57}$$

This may be called temporal reversibility. It follows because a unique adiabatic trajectory passes through each phase space point. The adiabatic equations of motion also display what might be called conjugate reversibility,

$$\boldsymbol{\Gamma}_2 = \boldsymbol{\Gamma}^0(t|\boldsymbol{\Gamma}_1) \Leftrightarrow \boldsymbol{\Gamma}_1^\dagger = \boldsymbol{\Gamma}^0(t|\boldsymbol{\Gamma}_2^\dagger). \tag{5.58}$$

Recall that each phase space point $\boldsymbol{\Gamma} = \{\mathbf{q}, \mathbf{p}\}$ has conjugate point with all the momenta reversed, $\boldsymbol{\Gamma}^\dagger = \{\mathbf{q}, -\mathbf{p}\}$. The combination of temporal reversibility and conjugate reversibility can be called microscopic reversibility,

$$\boldsymbol{\Gamma}_2 = \boldsymbol{\Gamma}^0(t|\boldsymbol{\Gamma}_1) \Leftrightarrow \boldsymbol{\Gamma}_2^\dagger = \boldsymbol{\Gamma}^0(-t|\boldsymbol{\Gamma}_1^\dagger). \tag{5.59}$$

The distinction between conjugate reversibility and microscopic reversibility is usually overlooked because in the adiabatic case temporal reversibility is both obvious and trivial. For stochastic dissipative equations it is necessary to distinguish the two.

Microscopic reversibility says that if both the velocities and the time interval are reversed, then the end point of this transition is the conjugate of the end point of the original transition. The physical origin of microscopic reversibility is that reversing the velocities causes the particles to retrace their previous paths, and hence reversing both the velocities and time is equivalent to the original transition. Mathematically, for a single time step,

$$\begin{aligned} \Delta_t\dot{\boldsymbol{\Gamma}}_q^0(\boldsymbol{\Gamma}^\dagger) &= \Delta_t\{\nabla_{p^\dagger}\mathcal{K}(\mathbf{p}^\dagger)\}_q = -\Delta_t\{\nabla_p\mathcal{K}(\mathbf{p})\}_q = -\Delta_t\{\nabla_p\mathcal{K}(\mathbf{p})\}_q^\dagger \\ \Delta_t\dot{\boldsymbol{\Gamma}}_p^0(\boldsymbol{\Gamma}^\dagger) &= \Delta_t\{\nabla_{q^\dagger}U(\mathbf{q}^\dagger)\}_p = \Delta_t\{\nabla_q U(\mathbf{q})\}_p = -\Delta_t\{\nabla_q U(\mathbf{q})\}_p^\dagger. \end{aligned} \tag{5.60}$$

These are a more explicit version of equation (5.13). One sees that the sign of the time step has been reversed in the conjugate system. Temporal, conjugate, and microscopic reversibility for an adiabatic transition are shown in table 5.1.

The stochastic part of the stochastic, dissipative equations of motion is dependent on the magnitude of the time interval, and is as likely positive as negative. Hence it displays each of temporal, conjugate, and microscopic reversibility.

**Table 5.1.** Time symmetries of equations of motion

| | Adiabatic | Dissipative |
|---|---|---|
| | $\Gamma_2 = \Gamma^0(\Delta_t\|\Gamma_1)$ | $\Gamma_2 = \bar{\Gamma}(\Delta_t\|\Gamma_1)$ |
| | $= \Gamma_1 + \Delta_t\dot{\Gamma}^0(\Gamma)$ | $= \Gamma_1 + \Delta_t\dot{\Gamma}^0(\Gamma) + \frac{\|\Delta_t\|}{2}\lambda\nabla_p S_r(\Gamma)$ |
| temporal | $\Gamma_1 = \Gamma^0(-\Delta_t\|\Gamma_2)$ | $\Gamma_3 = \bar{\Gamma}(-\Delta_t\|\Gamma_2)$ |
| | $= \Gamma_2 - \Delta_t\dot{\Gamma}^0(\Gamma)$ | $= \Gamma_2 - \Delta_t\dot{\Gamma}^0(\Gamma) + \frac{\|\Delta_t\|}{2}\lambda\nabla_p S_r(\Gamma)$ |
| conjugate | $\Gamma_1^\dagger = \Gamma^0(\Delta_t\|\Gamma_2^\dagger)$ | $\Gamma_4^\dagger = \bar{\Gamma}(\Delta_t\|\Gamma_2)^\dagger$ |
| | $= \Gamma_2^\dagger + \Delta_t\dot{\Gamma}^0(\Gamma^\dagger)$ | $= \Gamma_2^\dagger + \Delta_t\dot{\Gamma}^0(\Gamma^\dagger) + \frac{\|\Delta_t\|}{2}\lambda\nabla_p S_r(\Gamma^\dagger)$ |
| microscopic | $\Gamma_2^\dagger = \Gamma^0(-\Delta_t\|\Gamma_1^\dagger)$ | $\Gamma_2^\dagger = \bar{\Gamma}(-\Delta_t\|\Gamma_1)^\dagger$ |
| | $= \Gamma_1^\dagger - \Delta_t\dot{\Gamma}^0(\Gamma^\dagger)$ | $= \Gamma_1^\dagger - \Delta_t\dot{\Gamma}^0(\Gamma^\dagger) + \frac{\|\Delta_t\|}{2}\lambda\nabla_p S_r(\Gamma^\dagger)$ |

The dissipative part depends on the magnitude of the time interval, and hence it is irreversible in a temporal sense. The difference upon temporarily reversing a dissipative transition is

$$\begin{aligned}\bar{\Gamma}(-\Delta_t|\bar{\Gamma}(\Delta_t|\Gamma_1)) - \Gamma_1 &= \bar{\Gamma}(\Delta_t|\Gamma_1) - \Delta_t\dot{\Gamma}^0(\Gamma) + \frac{|\Delta_t|}{2}\lambda\nabla_p S_r(\Gamma) - \Gamma_1 \\ &= |\Delta_t|\lambda\nabla_p S_r(\Gamma) + \mathcal{O}(\Delta_t^2).\end{aligned} \tag{5.61}$$

The dissipative force increases the entropy at both stages of the cyclic transition, which causes the stochastic, dissipative equations of motion to be irreversible in a temporal sense.

Similarly, reversing the momenta at the end of the first transition and continuing forward another time step gives a difference of

$$\begin{aligned}\bar{\Gamma}(\Delta_t|\bar{\Gamma}(\Delta_t|\Gamma_1)^\dagger) - \Gamma_1^\dagger &= \bar{\Gamma}(\Delta_t|\Gamma_1)^\dagger + \Delta_t\dot{\Gamma}^0(\Gamma^\dagger) + \frac{|\Delta_t|}{2}\lambda\{\nabla_p S_r(\Gamma^\dagger)\}_p - \Gamma_1^\dagger \\ &= -|\Delta_t|\lambda\nabla_p S_r(\Gamma) + \mathcal{O}(\Delta_t^2).\end{aligned} \tag{5.62}$$

The second equality follows because for an equilibrium system $S_r(\Gamma)$ is an even function of the momenta, which means that the dissipative force, $\propto \nabla_p S_r(\Gamma)$, is an odd function of momenta. One sees that over this particular cycle the entropy has decreased, and that the dissipative equations of motion are irreversible in a conjugate sense.

Figure 5.5 shows several dissipative trajectories, both forward and backward in time from their initial point. It can be seen that a point in phase space does not have a unique trajectory through it. This is one reason for the temporal and conjugate irreversibility just mentioned. Although in figure 5.5 the square lies on the most likely backward trajectory from the circle, the circle does not lie on the most likely forward trajectory from the square. This is another reason for the temporal irreversibility of a dissipative trajectory. The discontinuity in the derivative of a trajectory at its initial point was discussed in connection with figure 4.2, where it was

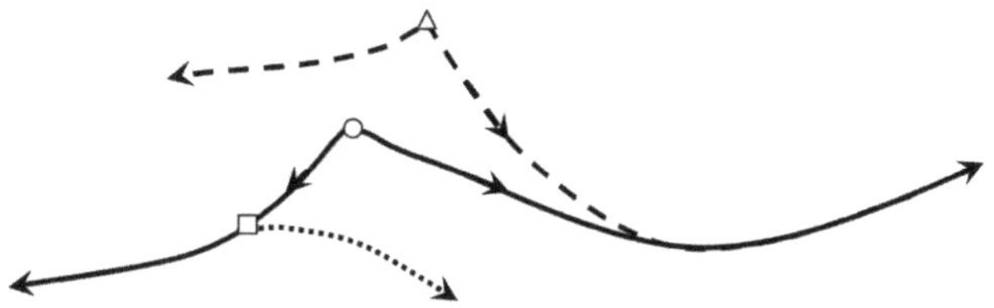

**Figure 5.5.** Six dissipative trajectories. The circle on the solid curve marks the initial point at time $t_1$ of a forward and a backward trajectory. The triangle on the dashed curve at $t_2$ marks the initial point of a forward and a backward trajectory. The square on the dotted curve at $t_3 < t_1$ marks the initial point of a forward and a backward (obscured) trajectory.

attributed to the fact that the dissipative contribution is proportional to the absolute value of the time interval. The essential physical reason for the kink in the trajectory at the initial point, and for the irreversibility, is that most likely the sub-system will head *toward* an equilibrium state, and most likely the sub-system came *from* an equilibrium state. The stochastic contributions will perturb the most likely trajectories shown in figure 5.5, increasingly so as time progresses or regresses from the initial time (see figure 4.2), but the qualitative picture remains as sketched.

As was mentioned in the discussion of figure 4.2 and equation (4.22), the discontinuity in the derivative of the trajectory means that the forward or backward time derivative of a function on the trajectory has to be specified. The irreversibility embodied by the dissipative and stochastic terms means that the forward and backward time derivatives are not equal to each other.

Microscopic reversibility for the dissipative transition is equivalent to

$$\begin{aligned}\bar{\mathbf{\Gamma}}(-\Delta_t|\mathbf{\Gamma}_1^\dagger) - \bar{\mathbf{\Gamma}}(\Delta_t|\mathbf{\Gamma}_1)^\dagger &= \mathbf{\Gamma}_1^\dagger - \Delta_t\dot{\mathbf{\Gamma}}^0(\mathbf{\Gamma}^\dagger) + \frac{|\Delta_t|}{2}\lambda\{\nabla_p S_{\rm r}(\mathbf{\Gamma}^\dagger)\}_p \\ &\quad - \mathbf{\Gamma}_1^\dagger - \Delta_t\dot{\mathbf{\Gamma}}^0(\mathbf{\Gamma})^\dagger - \frac{|\Delta_t|}{2}\lambda\{\nabla_p S_{\rm r}(\mathbf{\Gamma})\}_p^\dagger \\ &= \frac{-|\Delta_t|}{2}\lambda\{\nabla_p S_{\rm r}(\mathbf{\Gamma})\}_p + \frac{|\Delta_t|}{2}\lambda\{\nabla_p S_{\rm r}(\mathbf{\Gamma})\}_p \\ &\quad + \mathcal{O}(\Delta_t^2) \\ &= 0.\end{aligned} \tag{5.63}$$

This says that the original transition forward in time is equivalent to reversing the momenta, making a transition backward in time, and reversing the momenta again. Temporal and conjugate irreversibility, and microscopic reversibility for a dissipative transition are shown in table 5.1.

For the stochastic, dissipative equations of motion for the canonical equilibrium system equation (5.55), one can exhibit microscopic reversibility explicitly. The position evolution in a time step is purely adiabatic and hence microscopically reversible. The most likely momentum evolution is

$$\mathbf{p}_2 \equiv \bar{\mathbf{p}}(\Delta_t|\mathbf{\Gamma}_1) = \mathbf{p}_1 - \Delta_t\nabla_q U(\mathbf{q}_1) - \frac{|\Delta_t|\lambda}{2mT}\mathbf{p}_1. \tag{5.64}$$

Negating both sides this can be written as

$$\begin{aligned}\mathbf{p}_2^\dagger &= \mathbf{p}_1^\dagger + \Delta_t \nabla_q U\left(\mathbf{q}_1^\dagger\right) - \frac{|\Delta_t|\lambda}{2mT}\mathbf{p}_1^\dagger \\ &= \bar{\mathbf{p}}(-\Delta_t|\boldsymbol{\Gamma}_1^\dagger).\end{aligned} \tag{5.65}$$

This says explicitly that the most likely part of the stochastic, dissipative equations of motion for the momenta for the canonical equilibrium system are microscopically reversible. (The dissipative equations for the position are purely adiabatic, and so they are automatically microscopically reversible.)

Microscopic reversibility for the dissipative equations of motion may in general be written as

$$\boldsymbol{\Gamma}_2 = \bar{\boldsymbol{\Gamma}}(t|\boldsymbol{\Gamma}_1) \Leftrightarrow \boldsymbol{\Gamma}_2^\dagger = \bar{\boldsymbol{\Gamma}}(-t|\boldsymbol{\Gamma}_1^\dagger). \tag{5.66}$$

This is identical to microscopic reversibility for adiabatic equations of motion, which was sketched in figure 5.2. Since the stochastic part of the transition is insensitive to the direction of time, and is as likely positive as negative, $\wp(\tilde{\mathbf{R}}) = \wp(\tilde{\mathbf{R}}^\dagger)$, the stochastic, dissipative equations of motion for an equilibrium system themselves are microscopically reversible

$$\boldsymbol{\Gamma}_2 = \boldsymbol{\Gamma}(t|\boldsymbol{\Gamma}_1) \Leftrightarrow \boldsymbol{\Gamma}_2^\dagger = \boldsymbol{\Gamma}(-t|\boldsymbol{\Gamma}_2^\dagger). \tag{5.67}$$

This is to be understood in a probabilistic sense: the conditional probability of the transition $\boldsymbol{\Gamma}_1 \xrightarrow{t} \boldsymbol{\Gamma}_2$ is equal to that of $\boldsymbol{\Gamma}_1^\dagger \xrightarrow{-t} \boldsymbol{\Gamma}_2^\dagger$. Since for an equilibrium system the probability of a point in phase space is equal to that of the conjugate point, $\wp(\boldsymbol{\Gamma}_1) = \wp(\boldsymbol{\Gamma}_1^\dagger)$, the unconditional probability of the forward and the conjugate backward transition must also be equal.

### 5.4.3 Stationarity of the Maxwell–Boltzmann distribution

The Fokker–Planck equation for the evolution of an arbitrary phase space probability distribution under stochastic, dissipative equations of motion was derived above as equation (4.37). Replacing $X \Rightarrow \boldsymbol{\Gamma}$ it is

$$\begin{aligned}\frac{\partial\wp(\boldsymbol{\Gamma}, t)}{\partial t} &= -\left[\nabla \cdot \dot{\boldsymbol{\Gamma}}^{\text{det}}\right]\wp(\boldsymbol{\Gamma}, t) - \dot{\boldsymbol{\Gamma}}^{\text{det}} \cdot \nabla\wp(\boldsymbol{\Gamma}, t) + \frac{\hat{\tau}k_{\text{B}}}{2}\Lambda : \nabla\nabla\wp(\boldsymbol{\Gamma}, t) \\ &= \frac{3N\hat{\tau}\lambda}{2mT}\wp(\boldsymbol{\Gamma}, t) - \dot{\boldsymbol{\Gamma}}^{\text{det}} \cdot \nabla\wp(\boldsymbol{\Gamma}, t) + \frac{\hat{\tau}k_{\text{B}}}{2}\lambda\nabla_p^2\wp(\boldsymbol{\Gamma}, t).\end{aligned} \tag{5.68}$$

Since phase space is incompressible under the adiabatic equations of motion, $\nabla \cdot \dot{\boldsymbol{\Gamma}}^{\text{det}} = -\hat{\tau}\lambda\nabla_p \cdot \mathbf{p}/2mT$.

With the Maxwell–Boltzmann distribution, $\wp_{\text{MB}}(\boldsymbol{\Gamma}) = Z^{-1}e^{-\beta\mathcal{H}(\boldsymbol{\Gamma})}$, and the fact that energy is conserved on an adiabatic trajectory, $\dot{\boldsymbol{\Gamma}}^0 \cdot \nabla\mathcal{H}(\boldsymbol{\Gamma}) = 0$, the right-hand side of this is

$$\begin{aligned}
\text{RHS} &= \frac{3N\hat{\tau}\lambda}{2mT}\wp_{\text{MB}}(\boldsymbol{\Gamma}) + \frac{\hat{\tau}\lambda}{2mT}\mathbf{p}\cdot\nabla_p[-\beta\mathcal{H}(\boldsymbol{\Gamma})]\wp_{\text{MB}}(\boldsymbol{\Gamma}) \\
&\quad + \frac{\hat{\tau}k_{\text{B}}\lambda}{2}\Big[-\beta\nabla_p^2\mathcal{H}(\boldsymbol{\Gamma}) + \beta^2\big(\nabla_p\mathcal{H}(\boldsymbol{\Gamma})\big)^2\Big]\wp_{\text{MB}}(\boldsymbol{\Gamma}) \\
&= \left\{\frac{3N\hat{\tau}\lambda}{2mT} - \frac{\hat{\tau}\lambda}{2m^2k_{\text{B}}T^2}\mathbf{p}\cdot\mathbf{p} - \frac{\hat{\tau}3N\lambda}{2mT} + \frac{\hat{\tau}\lambda}{2m^2k_{\text{B}}T^2}\mathbf{p}\cdot\mathbf{p}\right\}\wp_{\text{MB}}(\boldsymbol{\Gamma}) \\
&= 0.
\end{aligned} \tag{5.69}$$

This shows that the Maxwell–Boltzmann distribution is stationary under the stochastic, dissipative equations of motion. Of course these equations of motion simply implement the second entropy transition probability for the canonical equilibrium system, and so they are consistent with the Maxwell–Boltzmann distribution by design.

### 5.4.4 Grand canonical system

A grand canonical system can exchange particles and energy with a reservoir of constant chemical potential $\mu$ and temperature $T$. As was discussed in section 2.3.2, the total constrained entropy for an energy-number macrostate is $S_{\text{total}}(N, E|\mu, V, T) = S_{\text{s}}(E, V, N) - E/T + \mu N/T$. The reservoir entropy for a point in the sub-system phase space $\boldsymbol{\Gamma}$ is therefore

$$S_{\text{r}}(\boldsymbol{\Gamma}, N|\mu, V, T) = \frac{-\mathcal{H}(\boldsymbol{\Gamma})}{T} + \frac{\mu}{T}N. \tag{5.70}$$

In the derivation of the Maxwell–Boltzmann distribution above, the sub-system entropy of a phase space point was given as equation (5.46), $S_{\text{s}}(\boldsymbol{\Gamma}) = Nk_{\text{B}} - Nk_{\text{B}}\ln(Nh^3)$. With this the total entropy for the microstate $\boldsymbol{\Gamma}$ is

$$\begin{aligned}
S_{\text{total}}(\boldsymbol{\Gamma}, N|\mu, V, T) &= S_{\text{s}}(\boldsymbol{\Gamma}) + S_{\text{r}}(\boldsymbol{\Gamma}, N|\mu, V, T) \\
&= Nk_{\text{B}} - Nk_{\text{B}}\ln(Nh^3) - \frac{\mathcal{H}(\boldsymbol{\Gamma})}{T} + \frac{\mu}{T}N.
\end{aligned} \tag{5.71}$$

Hence the probability density for the grand canonical system can be written explicitly either in terms of the reservoir entropy,

$$\wp(\boldsymbol{\Gamma}, N|\mu, V, T) = \frac{e^{S_{\text{r}}(\boldsymbol{\Gamma}, N|\mu, V, T)/k_{\text{B}}}}{N!h^{3N}\Xi(\mu, V, T)} = \frac{e^{-\beta\mathcal{H}(\boldsymbol{\Gamma})+\beta\mu N}}{N!h^{3N}\Xi(\mu, V, T)}, \tag{5.72}$$

or else in terms of the total entropy,

$$\wp(\boldsymbol{\Gamma}, N|\mu, V, T) = \frac{e^{S_{\text{total}}(\boldsymbol{\Gamma}, N|\mu, V, T)/k_{\text{B}}}}{\Xi(\mu, V, T)} = \frac{e^{N-N\ln(Nh^3)-\beta\mathcal{H}(\boldsymbol{\Gamma})+\beta\mu N}}{\Xi(\mu, V, T)}. \tag{5.73}$$

In the thermodynamic limit these are the same, although the former is the more conventional expression. In these the grand partition function is

$$\Xi(\mu, V, T) = \sum_{N=0}^{\infty} \frac{e^{\beta\mu N}}{N!h^{3N}} \int \mathrm{d}\boldsymbol{\Gamma} e^{-\beta\mathcal{H}(\boldsymbol{\Gamma})}. \tag{5.74}$$

Sometimes the fugacity, $z \equiv e^{\beta\mu}$, is used instead of the chemical potential as the independent variable of the grand canonical system.

*Grand canonical stochastic molecular dynamics*

Let the number of particles in the sub-system be described by a positive real number $\mathcal{N}$, with the integer part of this $N = \lfloor\mathcal{N}\rfloor$ being the number of fully coupled particles, and $\xi = \mathcal{N} - N$ being the extent of coupling of the $(N + 1)$ st particle. The Hamiltonian of the system is

$$\mathcal{H}(\boldsymbol{\Gamma}^{\mathcal{N}}) = \mathcal{H}(\boldsymbol{\Gamma}^{N}) + \xi\frac{p_{N+1}^{2}}{2m} + \xi U(\mathbf{q}^{N}, \mathbf{q}_{N+1}). \tag{5.75}$$

Here a linear coupling has been chosen for the partial particle, which is simplest but not essential. The reservoir entropy is

$$S_{\mathrm{r}}(\boldsymbol{\Gamma}^{\mathcal{N}}, \mathcal{N}|\mu, V, T) = \frac{-1}{T}\mathcal{H}(\boldsymbol{\Gamma}^{\mathcal{N}}) + \frac{\mu}{T}\mathcal{N}. \tag{5.76}$$

The internal entropy of a sub-system phase point is

$$S_{\mathrm{s}}(\boldsymbol{\Gamma}) = \mathcal{N}k_{\mathrm{B}} - \mathcal{N}k_{\mathrm{B}}\ln(\mathcal{N}h^{3}), \tag{5.77}$$

and so the total entropy is

$$\begin{aligned} S_{\mathrm{tot}}(\boldsymbol{\Gamma}, \mathcal{N}|\mu, V, T) &= S_{\mathrm{s}}(\boldsymbol{\Gamma}) + S_{\mathrm{r}}(\boldsymbol{\Gamma}, \mathcal{N}|\mu, V, T) \\ &= \mathcal{N}k_{\mathrm{B}} - \mathcal{N}k_{\mathrm{B}}\ln(\mathcal{N}h^{3}) - \frac{1}{T}\mathcal{H}(\boldsymbol{\Gamma}^{\mathcal{N}}) + \frac{\mu}{T}\mathcal{N}. \end{aligned} \tag{5.78}$$

The equations of motion for the particles are

$$\begin{aligned} \mathbf{q}^{N}(t + \Delta_t) &= \mathbf{q}^{N}(t) + \frac{\Delta_t}{m}\mathbf{p}^{N}(t) \\ \mathbf{q}_{N+1}(t + \Delta_t) &= \mathbf{q}_{N+1}(t) + \frac{\Delta_t\xi}{m}\mathbf{p}_{N+1}(t) \\ \mathbf{p}^{N+1}(t + \Delta_t) &= \mathbf{p}^{N+1}(t) - \Delta_t\frac{\partial\mathcal{H}(\boldsymbol{\Gamma}^{\mathcal{N}})}{\partial\mathbf{q}^{N+1}} + \frac{|\Delta_t|}{2}\lambda\frac{\partial S_{\mathrm{tot}}(\boldsymbol{\Gamma}^{\mathcal{N}})}{\partial\mathbf{p}^{N+1}} + \tilde{R}^{N+1}. \end{aligned} \tag{5.79}$$

Here $-\partial\mathcal{H}(\boldsymbol{\Gamma}^{\mathcal{N}})/\partial\mathbf{q}^{N+1}$ is the force for $N$ fully coupled particles and the $(N + 1)$ st partially coupled particle, and the gradient of the total entropy is

$$\frac{\partial S_{\mathrm{tot}}(\boldsymbol{\Gamma}^{\mathcal{N}}, \mathcal{N}|\mu, V, T)}{\partial\mathbf{p}^{N+1}} = \frac{-1}{mT}\{\mathbf{p}^{N}, \xi\mathbf{p}_{N+1}\}. \tag{5.80}$$

The transport matrix has been chosen to be diagonal with all elements equal to $\lambda$. The probability distribution for the stochastic force is

$$\wp(\tilde{R}^{N+1}) = \frac{1}{Z} e^{-\tilde{R}^{N+1} \cdot \tilde{R}^{N+1}/2\lambda k_{\rm B}|\Delta_t|}. \tag{5.81}$$

The equation of motion for the number is

$$\begin{aligned} \mathcal{N}(t + \Delta_t) &= \mathcal{N}(t) + \frac{|\Delta_t|}{2}\lambda' \frac{\partial S_{\rm tot}(\mathbf{\Gamma}^{\mathcal{N}}(t), \mathcal{N}|\mu, V, T)}{\partial \mathcal{N}} + \tilde{R}' \\ &= \mathcal{N}(t) - \frac{|\Delta_t|}{2T}\lambda' \left\{ \frac{p_{N+1}^2}{2m} + U\left(\mathbf{q}^N, \mathbf{q}_{N+1}\right) - \mu + k_{\rm B}T \ln(\mathcal{N}h^3) \right\} + \tilde{R}'. \end{aligned} \tag{5.82}$$

Notice how the sub-system entropy contributes to this. The probability distribution for the stochastic force in this case is

$$\wp(\tilde{R}') = \frac{1}{Z'} e^{-\tilde{R}'^2/2\lambda' k_{\rm B}|\Delta_t|}. \tag{5.83}$$

When $\xi(t + \Delta_t)$ exceeds 1, then the $(N + 1)$st particle becomes an ordinary particle in the sub-system, and a new partially coupled particle is created, coupled to the extent $\xi(t + \Delta_t) - 1$. This is randomly placed in the sub-system phase space according to the probability distribution $\propto \exp -\beta(\xi - 1)[p_{N+2}^2/2m + U(\mathbf{q}^{N+1}, \mathbf{q}_{N+2})]$. (The momentum can be chosen using the Gaussian; the position using several trial steps of the Metropolis algorithm, for example.) When $\xi(t + \Delta_t)$ drops below 0, the existing partially coupled particle is eliminated, and one of the remaining $N$ fully coupled particles is randomly chosen as the new partially coupled particle, and it is coupled to the extent $\xi(t + \Delta_t) + 1$.

### 5.4.5 Isobaric system

An isobaric system can exchange volume and energy with a reservoir of constant pressure $p$ and temperature $T$. As was discussed in section 2.3.2, the total constrained entropy for an energy-volume macrostate is $S_{\rm total}(E, V|N, p, T) = S_{\rm s}(E, V, N) - E/T - pV/T$. Hence the reservoir entropy for the sub-system phase space $\mathbf{\Gamma}$ is $S_{\rm r}(\mathbf{\Gamma}, V|N, p, T) = -\mathcal{H}(\mathbf{\Gamma})/T - pV/T$. The sub-system entropy for $\mathbf{\Gamma}$ is constant, $S_{\rm s}(\mathbf{\Gamma}) = Nk_{\rm B} - Nk_{\rm B}\ln(Nh^3)$. Hence the total entropy is

$$S_{\rm total}(\mathbf{\Gamma}, V|N, p, T) = Nk_{\rm B} - Nk_{\rm B}\ln(Nh^3) - \frac{\mathcal{H}(\mathbf{\Gamma})}{T} - \frac{p}{T}V. \tag{5.84}$$

This gives the probability density for the grand canonical system as

$$\wp(\mathbf{\Gamma}, V|N, p, T) = \frac{e^{S_{\rm r}(\mathbf{\Gamma}, V|N, p, T)/k_{\rm B}}}{N! h^{3N} \Delta_V Z(N, p, T)} = \frac{e^{-\beta\mathcal{H}(\mathbf{\Gamma}) - \beta p V}}{N! h^{3N} \Delta_V Z(N, p, T)}. \tag{5.85}$$

Here the isobaric partition function is

$$Z(N, p, T) = \frac{1}{N! h^{3N} \Delta_V} \int {\rm d}V\, e^{-\beta p V} \int {\rm d}\mathbf{\Gamma}\, e^{-\beta\mathcal{H}(\mathbf{\Gamma})}. \tag{5.86}$$

The volume width $\Delta_V$ is used to make the partition function dimensionless and its actual value has no physical consequences.

*Isobaric stochastic molecular dynamics*

Let the sub-system volume be cubic, $V = L^3$, and use the edge length to scale the positions, $\mathbf{q} = L\mathbf{x}$. The potential energy may be written $U(\mathbf{q}) = U(\mathbf{x}; L)$. The reservoir entropy is

$$S_r(\mathbf{p}, \mathbf{x}, L|N, p, T) = \frac{-1}{T}\mathcal{H}(\mathbf{p}, \mathbf{x}; L) - \frac{p}{T}L^3. \tag{5.87}$$

The internal entropy of a sub-system phase point is $S_s(\mathbf{\Gamma}) = Nk_B - Nk_B \ln(Nh^3)$, and so the total entropy is

$$\begin{aligned} S_{\text{tot}}(\mathbf{p}, \mathbf{x}, L|N, p, T) &= S_s(\mathbf{\Gamma}) + S_r(\mathbf{p}, \mathbf{x}, L|N, p, T) \\ &= Nk_B - Nk_B \ln(Nh^3) - \frac{1}{T}\mathcal{H}(\mathbf{p}, \mathbf{x}; L) - \frac{p}{T}L^3. \end{aligned} \tag{5.88}$$

The equations of motion for the particles are

$$\begin{aligned} \mathbf{x}(t + \Delta_t) &= \mathbf{x}(t) + \frac{\Delta_t}{mL}\mathbf{p}(t) \\ \mathbf{p}(t + \Delta_t) &= \mathbf{p}(t) - \Delta_t\frac{\partial U(\mathbf{q})}{\partial \mathbf{q}} + \frac{|\Delta_t|}{2}\lambda\frac{\partial S_{\text{tot}}(\mathbf{p}, \mathbf{x}, L|N, p, T)}{\partial \mathbf{p}} + \tilde{R}. \end{aligned} \tag{5.89}$$

Here the adiabatic force is $-\partial U(\mathbf{q})/\partial \mathbf{q} = -\partial U(\mathbf{x}; L)/L\partial \mathbf{x}$, and the gradient of the reservoir entropy is

$$\frac{\partial S_{\text{tot}}(\mathbf{p}, \mathbf{x}, L|N, p, T)}{\partial \mathbf{p}} = \frac{-1}{mT}\mathbf{p}. \tag{5.90}$$

The $N \times N$ transport matrix has been chosen to be diagonal with all elements equal to $\lambda$. The probability distribution for the stochastic force is

$$\wp(\tilde{R}) = \frac{1}{Z}e^{-\tilde{R}^2/2\lambda k_B|\Delta_t|}. \tag{5.91}$$

The equation of motion for the edge length is

$$\begin{aligned} L(t + \Delta_t) &= L(t) + \frac{|\Delta_t|\lambda'}{2}\frac{\partial S_{\text{tot}}(\mathbf{p}, \mathbf{x}, L|N, p, T)}{\partial L} + \tilde{R}' \\ &= L(t) - \frac{|\Delta_t|\lambda'}{2TL}\left\{\mathbf{q}\cdot\frac{\partial U(\mathbf{q})}{\partial \mathbf{q}} + 3pL^3\right\} + \tilde{R}'. \end{aligned} \tag{5.92}$$

Again the adiabatic force is $-\partial U(\mathbf{q})/\partial \mathbf{q} = -\partial U(\mathbf{x}; L)/L\partial \mathbf{x}$. The probability distribution for the stochastic force is

$$\wp(\tilde{R}') = \frac{1}{Z'}e^{-\tilde{R}'^2/2\lambda' k_B|\Delta_t|}. \tag{5.93}$$

## 5.5 Equilibrium phase space averages

### 5.5.1 Probability densities and averages

The stochastic, dissipative equations of motion derived above may be used to generate a trajectory through the sub-system phase space, $\mathbf{\Gamma}(t|\mathbf{\Gamma}_0)$. The canonical equilibrium statistical average of a phase function $f(\mathbf{\Gamma})$ can then be written as a time average,

$$\langle f \rangle_{N,V,T} = \frac{1}{t}\int_0^t \mathrm{d}t'\, f(\mathbf{\Gamma}(t'|\mathbf{\Gamma}_0)). \tag{5.94}$$

For a long enough trajectory, the results can be expected to be independent of the starting position $\mathbf{\Gamma}_0$.

The statistical average can also be written as an integral over phase space,

$$\langle f \rangle_{N,V,T} = \int \mathrm{d}\mathbf{\Gamma}\, \wp(\mathbf{\Gamma}|N, V, T) f(\mathbf{\Gamma}), \tag{5.95}$$

where the Maxwell–Boltzmann probability density is

$$\wp(\mathbf{\Gamma}|N, V, T) = \frac{e^{-\beta\mathcal{H}(\mathbf{\Gamma})}}{N!h^{3N}Z(N, V, T)}, \tag{5.96}$$

and the normalizing partition function is $Z(N, V, T) = (N!h^{3N})^{-1}\int \mathrm{d}\mathbf{\Gamma} e^{-\beta\mathcal{H}(\mathbf{\Gamma})}$. Where there is no ambiguity the Maxwell–Boltzmann probability density will be written simply as $\wp(\mathbf{\Gamma})$.

Results will be presented in this section for averages of various phase functions. We shall focus on the formula that gives these as an integral over phase space. In general the average is the same for all systems (canonical, grand canonical, isobaric, etc). Fluctuations about the average vary from system to system (e.g. the fluctuations in number are zero for a canonical system, but non-zero for a grand canonical system). For brevity most of the following results will be given for a canonical equilibrium system.

### 5.5.2 Helmholtz free energy and entropy

Since the partition function is the integral over phase space of the weight of phase space points, it is the total weight of the total system. Hence its logarithm is the total entropy of the total system,

$$\begin{aligned} S_{\text{total}}(N, V, T) &= k_\mathrm{B} \ln Z(N, V, T) \\ &\equiv - F(N, V, T)/T \\ &\approx - \overline{F}(N, V, T)/T \\ &\equiv - F(\overline{E}(N, V, T)|N, V, T)/T. \end{aligned} \tag{5.97}$$

Here $F(N, V, T)$ is the statistical mechanical and $\overline{F}(N, V, T)$ is the thermodynamic Helmholtz free energy (see chapter 2). Also, $\overline{E}(N, V, T)$ is the most likely sub-system energy, which is equal to the average sub-system energy.

Recall from chapter 2 that the constrained total entropy is $S_{\text{total}}(E|N, V, T) = S_s(E, N, V) - E/T$, and therefore the Helmholtz free energy is

$$F(\overline{E}|N, V, T) = \overline{E} - TS_s(\overline{E}, N, V). \tag{5.98}$$

Re-arranging these and equating the most likely energy to the average energy, the entropy of the sub-system may be written as

$$\begin{aligned} S_s(\overline{E}, N, V) &= \frac{\overline{E}}{T} - \frac{F(\overline{E}|N, V, T)}{T} \\ &= \frac{\overline{E}}{T} + S_{\text{total}}(N, V, T) \\ &= k_B \int d\Gamma\, \wp(\Gamma)\{\beta\mathcal{H}(\Gamma) + \ln Z(N, V, T)\} \\ &= -k_B \int d\Gamma\, \wp(\Gamma) \ln [h^{3N} N! \wp(\Gamma)] \\ &= -k_B \int d\Gamma\, \wp(\Gamma) \ln \wp(\Gamma). \end{aligned} \tag{5.99}$$

The third equality follows because the probability density is normalized to unity. The penultimate equality follows because $\wp(\Gamma) = e^{-\beta\mathcal{H}(\Gamma)}/h^{3N}N!Z(N, V, T)$. The final equality neglects an immaterial constant, $-k_B \ln h^{3N}N!$, which can be interpreted as the internal entropy of each phase space point. One sees from this that the so-called Gibbs or Shannon entropy is only the sub-system entropy, which is just part of the total entropy (see the information entropy, equation (1.1), and the discussion of it in reference to the full result equation (1.19)).

### 5.5.3 Energy and heat capacity

Multiplying both sides of the Helmholtz free energy, equation (5.98), by the inverse temperature and differentiating, yields the most likely energy of the sub-system,

$$\frac{\partial(\beta\overline{F})}{\partial\beta} = \overline{E}. \tag{5.100}$$

Here $\overline{E}(N, V, T)$ has been held constant during the differentiating because the constrained Helmholtz free energy, $F(E|N, V, T)$, is a variational principle for the energy. This was given in chapter 2 as equation (2.34). From the above, the left-hand side is the negative logarithmic derivative of the partition function,

$$\begin{aligned} \frac{-\partial \ln Z(N, V, T)}{\partial\beta} &= \frac{1}{N!h^{3N}Z} \int d\Gamma \frac{-\partial}{\partial\beta} e^{-\beta\mathcal{H}(\Gamma)} \\ &= \frac{1}{N!h^{3N}Z} \int d\Gamma e^{-\beta\mathcal{H}(\Gamma)} \mathcal{H}(\Gamma) \\ &= \langle\mathcal{H}\rangle_{N,V,T}. \end{aligned} \tag{5.101}$$

This is equation (2.135) with $x = 1$ and $X = E$.

The energy of the sub-system is extensive, and its fluctuations are Gaussian. Hence the average value equals the most likely value,

$$\langle \mathcal{H} \rangle_{N,V,T} = \overline{E}(N, V, T). \tag{5.102}$$

With this one sees that the thermodynamic derivative equals the statistical mechanical derivative, which shows the consistency of the two.

The second derivative of the Helmholtz free energy gives the heat capacity at constant volume, equation (2.108),

$$C_V = \frac{-1}{T^2}\left(\frac{\partial^2(\overline{F}/T)}{\partial(1/T)^2}\right)_{V,N} = -k_B\beta^2\frac{\partial}{\partial\beta}\frac{\partial(\beta\overline{F})}{\partial\beta}. \tag{5.103}$$

In terms of the partition function this is

$$\begin{aligned} C_V &= k_B\beta^2\frac{\partial}{\partial\beta}\frac{Z'(\beta)}{Z(\beta)} \\ &= k_B\beta^2\left\{\frac{Z''(\beta)}{Z(\beta)} - \left(\frac{Z'(\beta)}{Z(\beta)}\right)^2\right\} \\ &= k_B\beta^2\left\{\langle \mathcal{H}^2 \rangle_{N,V,T} - \langle \mathcal{H} \rangle^2_{N,V,T}\right\} \\ &= k_B\beta^2\left\langle [\mathcal{H} - \langle \mathcal{H} \rangle_{N,V,T}]^2\right\rangle_{N,V,T}. \end{aligned} \tag{5.104}$$

This is just equation (2.138) with $x = 1$ and $X = E$. This is evidently positive and extensive (because the free energy is extensive). The square root of the right-hand side tells how likely an individual measurement of the energy is to depart from the average energy due to statistical variations. The relative root mean square fluctuation, $\langle [\mathcal{H} - \langle \mathcal{H} \rangle]^2 \rangle^{1/2} / \langle \mathcal{H} \rangle \sim V^{-1/2}$, vanishes in the thermodynamic limit, which is to say that the statistical measurement error is relatively negligible.

### 5.5.4 Virial pressure

The volume derivative of the Helmholtz free energy gives the pressure, equation (2.31),

$$\left(\frac{\partial \overline{F}(N, V, T)}{\partial V}\right)_{T,N} = -\bar{p}. \tag{5.105}$$

Using $y = p$ and $Y = V$ in equation (2.136) gives the equivalent expression for the logarithmic derivative of the partition function,

$$\frac{\partial \ln Z(N, V, T)}{\partial V} = \left\langle \beta_s p \right\rangle_{N,V,T}. \tag{5.106}$$

Most commonly one regards the fluctuations in the sub-system temperature as negligible and one takes $\beta_s = \beta$ and takes it outside of the average.

That part of the partition function that depends upon the volume is called the configuration integral. It is

$$\begin{aligned} Q(N, V, T) &= \int_V \mathrm{d}\mathbf{q} e^{-\beta U(\mathbf{q})} \\ &= L^{3N} \int_0^1 \mathrm{d}\mathbf{x}\, e^{-\beta U(\mathbf{x};L)}. \end{aligned} \tag{5.107}$$

In the second equality $L$ has been used to scale all of the coordinates, $q_{j\alpha} = Lx_{j\alpha}$. Note that some authors use the opposite notation convention to here, denoting the partition function by $Q$ and the configuration integral by $Z$.

Note that the potential energy that appears here is solely that due to the intermolecular contributions. The external contributions, which provide the external forces that balance the internal pressure being obtained here, implicitly appear as the limit on the configuration integral.

The derivative of the potential energy is

$$\begin{aligned} \frac{\partial U(\mathbf{x};L)}{\partial L} &= \frac{\partial U(\mathbf{q})}{\partial \mathbf{q}} \cdot \frac{\partial \mathbf{q}}{\partial L} \\ &= L^{-1}\mathbf{q} \cdot \nabla U(\mathbf{q}) \\ &\equiv -L^{-1}\mathcal{V}. \end{aligned} \tag{5.108}$$

The final equality has been defined the virial of Clausius, $\mathcal{V}$. This is essentially the sum of the force acting on each particle due to the other particles. (We assume that only the internal potential energy between the particles alone appears here. Any external energy such as that due to the walls of the sub-system is incorporated into the limits on the integrals.) With this, the logarithmic derivative of the partition function is

$$\begin{aligned} \langle \beta_s p \rangle &= \frac{\partial \ln Z(N, V, T)}{\partial V} \\ &= \frac{1}{Q} \frac{\partial Q(N, V, T)}{3L^2 \partial L} \\ &= \frac{1}{3L^2 Q} \left[ 3NL^{3N-1} \int_0^1 \mathrm{d}\mathbf{x}\, e^{-\beta U(\mathbf{x};L)} - \beta L^{3N} \int_0^1 \mathrm{d}\mathbf{x}\, e^{-\beta U(\mathbf{x};L)} \frac{\partial U(\mathbf{x};L)}{\partial L} \right] \\ &= \frac{N}{V} + \frac{\beta \langle \mathcal{V} \rangle_{N,V,T}}{3V}. \end{aligned} \tag{5.109}$$

This is the virial equation for the pressure. The first term is the pressure of an ideal gas (see below), and the second is the average of the intermolecular forces. For a dilute real gas, this is dominated by the long ranged attractions between the particles, which makes the pressure less than that of the corresponding ideal gas. For solids and dense fluids, the short ranged core repulsion that defines the size of the particles typically dominates, which makes the pressure greater than that of the corresponding ideal gas. Because the core repulsion is short-ranged, it is steep, and therefore its contribution to the pressure increases rapidly with increasing density as the particles sample more of it. This means that dense fluids and solids have a pressure that increases more rapidly with density than an ideal gas, which means that they have a lower compressibility.

### 5.5.5 Ideal gas

The Hamiltonian separates into momentum and position parts, $\mathcal{H}(\mathbf{\Gamma}) = \mathcal{K}(\mathbf{p}) + U(\mathbf{q})$. For an equilibrium system one is generally interested in the average of a function of the velocity alone or of the position alone. Accordingly, the canonical equilibrium partition function can be written

$$\begin{aligned} Z(N, V, T) &= \frac{1}{N!h^{3N}} \int \mathrm{d}\mathbf{\Gamma}\, e^{-\beta\mathcal{H}(\mathbf{\Gamma})} \\ &= \frac{1}{N!h^{3N}} \int \mathrm{d}\mathbf{p}\, e^{-\beta p^2/2m} \int_V \mathrm{d}\mathbf{q}\, e^{-\beta U(\mathbf{q})} \\ &= \frac{\Lambda^{-3N}}{N!} \int_V \mathrm{d}\mathbf{q}\, e^{-\beta U(\mathbf{q})} \\ &= Z^{\mathrm{id}}(N, V, T) V^{-N} Q(N, V, T). \end{aligned} \tag{5.110}$$

The configuration integral was defined above as $Q(N, V, T) \equiv \int_V \mathrm{d}\mathbf{q}\, e^{-\beta U(\mathbf{q})}$.

The ideal gas has $U(\mathbf{q}) = 0$, and the corresponding partition function is

$$\begin{aligned} Z^{\mathrm{id}}(N, V, T) &= \frac{1}{N!h^{3N}} \int \mathrm{d}\mathbf{p}\, e^{-\beta p^2/2m} \int_V \mathrm{d}\mathbf{q} \\ &= \frac{1}{N!h^{3N}} [2\pi m k_{\mathrm{B}} T]^{3N/2} V^N \\ &\equiv \frac{V^N}{N!\Lambda^{3N}}. \end{aligned} \tag{5.111}$$

The thermal wave length here and below is

$$\Lambda \equiv \sqrt{\frac{2\pi\hbar^2}{mk_{\mathrm{B}}T}}, \tag{5.112}$$

where $\hbar \equiv h/2\pi$. This quantity will recur in the quantum analysis of chapter 7, where it will be shown that it is the lower limit of the length scales on which wave function symmetrization effects are negligible. One must glean from the context in each case whether $\Lambda$ means the thermal wave length or the transport coefficient matrix.

The average energy for the ideal gas is

$$\begin{aligned} \langle \mathcal{H}^{\mathrm{id}} \rangle_{N,V,T} &= \frac{-\partial \ln Z^{\mathrm{id}}(N, V, T)}{\partial \beta} \\ &= \frac{-\partial \ln \Lambda^{-3N}}{\partial \beta} \\ &= \frac{3N}{2} k_{\mathrm{B}} T. \end{aligned} \tag{5.113}$$

The total unconstrained entropy of the ideal gas for this canonical equilibrium system is just the logarithm of the partition function,

$$S^{\rm id}_{\rm tot}(N,\ V,\ T) = k_{\rm B} \ln Z^{\rm id}(N,\ V,\ T) = Nk_{\rm B}[1 - \ln \rho\Lambda^3], \tag{5.114}$$

where the density is $\rho = N/V$. The reservoir part of this is

$$S^{\rm id}_{\rm r}(N,\ V,\ T) = \frac{-\langle \mathcal{H}^{\rm id}\rangle_{N,V,T}}{T} = \frac{-3Nk_{\rm B}}{2}, \tag{5.115}$$

and the sub-system part is

$$S^{\rm id}_{\rm s}(N,\ V,\ T) = S^{\rm id}_{\rm tot}(N,\ V,\ T) - S^{\rm id}_{\rm r}(N,\ V,\ T) = Nk_{\rm B}\left[\frac{5}{2} - \ln \rho\Lambda^3\right]. \tag{5.116}$$

The ideal gas heat capacity at constant volume is

$$\begin{aligned} C^{\rm id}_V &= -\, k_{\rm B}\beta^2 \frac{\partial}{\partial\beta}\frac{\partial(\beta\overline{F}^{\rm id})}{\partial\beta} \\ &= -\, k_{\rm B}\beta^2 \frac{\partial\langle \mathcal{H}^{\rm id}\rangle_{N,V,T}}{\partial\beta} \\ &= \frac{3N}{2}k_{\rm B}. \end{aligned} \tag{5.117}$$

This is evidently independent of temperature.

The pressure of the ideal gas is given by

$$\left\langle \beta_{\rm s}\, p\right\rangle^{\rm id}_{N,V,T} = \frac{\partial \ln Z^{\rm id}(N,\ V,\ T)}{\partial V} = \frac{N}{V}. \tag{5.118}$$

Setting the sub-system temperature to that of the reservoir, $\beta_{\rm s} = \beta = 1/k_{\rm B}T$, and defining the number density as $\rho \equiv N/V$, this is $\bar{p}^{\rm id} = \rho k_{\rm B}T$, which is the well-known result. Rewriting equation (2.112) for a canonical equilibrium system, the ideal gas isothermal compressibility is

$$\chi^{\rm id}_T = \frac{-1}{V}\left(\frac{\partial \bar{p}^{\rm id}}{\partial V}\right)^{-1}_{T,N} = (\rho k_{\rm B}T)^{-1}. \tag{5.119}$$

This says that as the number density and temperature increase, the ideal gas becomes less compressible.

### 5.5.6 Particle densities and distributions

The one-particle density $\rho^{(1)}(\mathbf{r})$, usually denoted $\rho(\mathbf{r})$, or $\rho$ for a homogeneous system, gives the number of particles per unit volume at $\mathbf{r}$. For a canonical equilibrium system it is

$$\begin{aligned}\rho^{(1)}(\mathbf{r}) &= \left\langle \sum_{j=1}^{N} \delta(\mathbf{r} - \mathbf{q}_j) \right\rangle_{N,V,T} \\ &= \frac{1}{h^{3N} N! Z(N, V, T)} \int \mathrm{d}\mathbf{\Gamma}\, e^{-\beta\mathcal{H}(\mathbf{\Gamma})} \sum_{j=1}^{N} \delta(\mathbf{r} - \mathbf{q}_j) \\ &= \frac{1}{Q(N, V, T)} \int_V \mathrm{d}\mathbf{q}\, e^{-\beta U(\mathbf{q})} \sum_{j=1}^{N} \delta(\mathbf{r} - \mathbf{q}_j) \\ &= \frac{N}{Q(N, V, T)} \int_V \mathrm{d}\mathbf{q}\, e^{-\beta U(\mathbf{q})} \delta(\mathbf{r} - \mathbf{q}_N).\end{aligned} \tag{5.120}$$

The final equality follows because all the particles are identical. Equivalently,

$$\rho^{(1)}(\mathbf{q}_N) = \frac{N}{Q(N, V, T)} \int_V \mathrm{d}\mathbf{q}^{N-1}\, e^{-\beta U(\mathbf{q}^N)}, \tag{5.121}$$

where the integration is over the first $N - 1$ particles, and the $N$th particle is at $\mathbf{q}_N$.

The two-particle density is proportional to the probability of finding two different particles simultaneously at two positions. It is

$$\rho^{(2)}(\mathbf{r}, \mathbf{s}) = \left\langle \sum_{j=1}^{N} \sum_{k=1}^{N}{}^{(k \neq j)}\, \delta(\mathbf{r} - \mathbf{q}_j) \delta(\mathbf{s} - \mathbf{q}_k) \right\rangle_{N,V,T}, \tag{5.122}$$

or

$$\rho^{(2)}(\mathbf{q}_N, \mathbf{q}_{N-1}) = \frac{N(N-1)}{Q(N, V, T)} \int_V \mathrm{d}\mathbf{q}^{N-2}\, e^{-\beta U(\mathbf{q}^N)}. \tag{5.123}$$

Similarly, the $n$-particle density is

$$\rho^{(n)}(\mathbf{q}_N, \mathbf{q}_{N-1}, \ldots, \mathbf{q}_{N-n+1}) = \frac{N!/(N-n)!}{Q(N, V, T)} \int_V \mathrm{d}\mathbf{q}^{N-n}\, e^{-\beta U(\mathbf{q}^N)}. \tag{5.124}$$

Appending the subscript $N$ to make it clear that this is a closed or canonical density, this is evidently normalized such that

$$\int_V \mathrm{d}\mathbf{q}^n\, \rho_N^{(n)}(\mathbf{q}^n) = \frac{N!}{(N-n)!}. \tag{5.125}$$

For an open or grand canonical system, the normalization is

$$\int_V \mathrm{d}\mathbf{q}^n\, \rho_\mu^{(n)}(\mathbf{q}^n) = \left\langle \frac{N!}{(N-n)!} \right\rangle_{\mu,V,T}. \tag{5.126}$$

One can confirm that the fluctuations in particle number are given by

$$\begin{aligned}&\langle [N - \langle N \rangle^{\mu,V,T}]^2 \rangle_{\mu,V,T} \\ &= \int_V \mathrm{d}\mathbf{r}\, \mathrm{d}\mathbf{s} \left\{ \rho_\mu^{(2)}(\mathbf{r}, \mathbf{s}) - \rho_\mu^{(1)}(\mathbf{r}) \rho_\mu^{(1)}(\mathbf{s}) + \rho_\mu^{(1)}(\mathbf{r}) \delta(\mathbf{r} - \mathbf{s}) \right\}.\end{aligned} \tag{5.127}$$

This is also the second derivative of the grand potential, (see equation (2.138)), and it is related to the isothermal compressibility, (see equation (2.112)) as is shown in equation (5.144) below.

For an ideal gas, $Q^{\text{id}} = V^N$. Hence in the uniform case of no external one-body potential, the $n$-particle canonical density is

$$\rho_N^{(n),\text{id}}(\mathbf{q}^n) = \frac{N!V^{N-n}}{(N-n)!V^N} = \frac{N}{V}\frac{N-1}{V} \cdots \frac{N-n+1}{V}. \tag{5.128}$$

For $n \ll N$ this is just $\rho^n$, where the singlet density is $\rho = N/V$.

For the case of an ideal gas in an external field $U(\mathbf{r})$, this is

$$\rho_N^{(n),\text{id}}(\mathbf{q}^n) = \frac{N!}{(N-n)!\tilde{V}^n} \prod_{j=1}^{n} e^{-\beta U(\mathbf{q}_j)}, \quad \tilde{V} \equiv \int_V \mathrm{d}\mathbf{r}\; e^{-\beta U(\mathbf{r})}. \tag{5.129}$$

In general, particles are only correlated over short-ranges. At large separations between clusters of particles, the multi-particle density is a product form

$$\rho^{(m+n)}(\mathbf{r}^m, \mathbf{s}^n) \to \rho^{(m)}(\mathbf{r}^m)\rho^{(n)}(\mathbf{s}^n), \text{ all } |\mathbf{r}_j - \mathbf{s}_k| \to \infty. \tag{5.130}$$

In the opposite limit of small separations, real particles have finite size and a repulsive core that prevents overlap. Hence

$$\rho^{(n)}(\mathbf{r}^n) \to 0, \text{ any } |\mathbf{r}_j - \mathbf{r}_k| \to 0. \tag{5.131}$$

The $n$-particle distribution function is defined to be

$$g^{(n)}(\mathbf{r}^n) = \frac{\rho^{(n)}(\mathbf{r}^n)}{\prod_{j=1}^{n} \rho^{(1)}(\mathbf{r}_j)}. \tag{5.132}$$

The departure of this from unity is a measure of the correlation between the particles. Asymptotically,

$$g^{(n)}(\mathbf{r}^n) \to 1, \quad \text{all } r_{jk} \to \infty. \tag{5.133}$$

The most important distribution function is the pair one. For a uniform fluid (no one body potential) in which the particles interact with a radially symmetric pair potential, this is the radial distribution function,

$$g^{(2)}(\mathbf{r}_1, \mathbf{r}_2) = g(r_{12}). \tag{5.134}$$

Here $r_{12} = |\mathbf{r}_1 - \mathbf{r}_2|$ is the separation between the particles. One has

$$g(r) \to \begin{cases} 1, & r \to \infty, \\ 0, & r \to 0. \end{cases} \tag{5.135}$$

One also has

$$g(r) \to e^{-\beta u(r)}, \quad \rho \to 0. \tag{5.136}$$

Consider a uniform fluid, $U^{(1)}(\mathbf{r}) = 0$, with a radially symmetric pair potential,

$$U(\mathbf{q}) = \sum_{j=1}^{N} \sum_{k=j+1}^{N} u(q_{jk}) \equiv \sum_{k<j}^{N} u(q_{jk}). \tag{5.137}$$

The excess energy is the average potential energy, which is to say the energy without the kinetic energy. In this case for a canonical equilibrium system it is

$$\begin{aligned} E^{\text{ex}} &= \left\langle \sum_{k<j}^{N} u(q_{jk}) \right\rangle_{N,V,T} \\ &= \frac{N(N-1)}{2Q(N,\, V,\, T)} \int_V \mathrm{d}\mathbf{q}^N e^{-\beta U(\mathbf{q}^N)} u(q_{N,N-1}) \\ &= \frac{1}{2} \int_V \mathrm{d}\mathbf{q}_N \, \mathrm{d}\mathbf{q}_{N-1} \rho^{(2)}(\mathbf{q}_N,\, \mathbf{q}_{N-1}) u(q_{N,N-1}) \\ &= \frac{\rho^2 V}{2} \int_V \mathrm{d}\mathbf{r}\; g(r) u(r). \end{aligned} \tag{5.138}$$

The integral is convergent and independent of $V$ because generally $r^3 u(r) \to 0$, $r \to \infty$. (Coulomb and dipole interactions require special treatment.)

The virial of Clausius, equation (5.108), in this case is

$$\begin{aligned} \mathcal{V} &= -\mathbf{q} \cdot \nabla U(\mathbf{q}) \\ &= -\sum_{i=1}^{N} \mathbf{q}_i \cdot \sum_{j<k}^{N} \frac{\partial u(q_{jk})}{\partial \mathbf{q}_i} = -\sum_{i=1}^{N} \mathbf{q}_i \cdot \sum_{k=1}^{N} {}^{(k\neq i)} \frac{\partial u(q_{ik})}{\partial \mathbf{q}_i} \\ &= -\sum_{i\neq k}^{N} u'(q_{ik}) \mathbf{q}_i \cdot \frac{\mathbf{q}_{ik}}{q_{ik}} = -\sum_{i<k}^{N} u'(q_{ik}) \mathbf{q}_{ik} \cdot \frac{\mathbf{q}_{ik}}{q_{ik}} \\ &= \frac{-1}{2} \sum_{i\neq k}^{N} q_{ik} u'(q_{ik}). \end{aligned} \tag{5.139}$$

The canonical equilibrium average of this is

$$\begin{aligned} \langle \mathcal{V} \rangle_{N,V,T} &= \left\langle -\sum_{k<j}^{N} q_{kj} u'(q_{kj}) \right\rangle_{N,V,T} \\ &= \frac{-\rho^2 V}{2} \int_V \mathrm{d}\mathbf{r}\; g(r) r u'(r). \end{aligned} \tag{5.140}$$

The analysis is identical to that for the excess energy with the replacement $u(r) \Rightarrow -ru'(r)$. From the virial equation (5.109), the average pressure is

$$\begin{aligned} \langle \beta_{\mathrm{s}} p \rangle &= \frac{N}{V} + \frac{\beta \langle \mathcal{V} \rangle_{N,V,T}}{3V} \\ &= \rho - \frac{\beta \rho^2}{6} \int_V \mathrm{d}\mathbf{r}\; g(r) r u'(r). \end{aligned} \tag{5.141}$$

The first term is the ideal gas result. At low densities, $g(r) \sim 1$, the second term is dominated by the long range tail, which is generally attractive (i.e. a negative potential decreasing with increasing separation, which is to say a positive slope). Hence the second term is negative in this range, and so the pressure is less than that of an ideal gas at the same density and temperature.

In section 2.4.3, the isothermal compressibility was given as the derivative of the volume with pressure at constant number and temperature, equation (2.112),

$$\chi_T = \frac{-1}{\overline{V}}\left(\frac{\partial \overline{V}}{\partial p}\right)_{T,N} = \frac{1}{\rho}\left(\frac{\partial \rho}{\partial p}\right)_T. \tag{5.142}$$

In the second equality the number density $\rho \equiv N/V$ has been used. In section 2.4.3 this was written as the second pressure derivative of the Gibbs free energy.

This can also be written as the second chemical potential derivative of the grand potential as follows. Above the density was considered as a function of pressure and temperature, $\rho(p, T)$. But because of the uniqueness of the thermodynamic state, one can instead write $\rho(\mu, T)$ and $p(\mu, T)$, with which the compressibility becomes

$$\begin{aligned}\chi_T &= \frac{1}{\rho}\left(\frac{\partial \rho}{\partial \mu}\right)_T\left(\frac{\partial \mu}{\partial p}\right)_T \\ &= \frac{1}{\rho}\left(\frac{-\partial^2\Omega}{V\partial\mu^2}\right)_{T,V}\left(\frac{N^{-1}\partial G}{\partial p}\right)_{T,N} \\ &= \frac{-1}{\rho^2 V}\left(\frac{\partial^2\Omega}{\partial\mu^2}\right)_{T,V}.\end{aligned} \tag{5.143}$$

Writing $\Omega(\mu, V, T) = -k_{\mathrm{B}}T \ln \Xi(\mu, V, T)$, this is

$$\begin{aligned}\chi_T &= \frac{1}{k_{\mathrm{B}}T\rho^2 V}\frac{\partial^2 \ln \Xi(\mu, V, T)}{\partial(\beta\mu)^2} \\ &= \frac{1}{k_{\mathrm{B}}T\rho^2 V}\left\{\frac{\Xi''(\beta\mu)}{\Xi(\beta\mu)} - \left(\frac{\Xi'(\beta\mu)}{\Xi(\beta\mu)}\right)^2\right\} \\ &= \frac{1}{k_{\mathrm{B}}T\rho^2 V}\left\{\langle N^2\rangle_{\mu,V,T} - \langle N\rangle^2_{\mu,V,T}\right\} \\ &= \frac{1}{k_{\mathrm{B}}T\rho^2 V}\left\langle\left[N - \langle N\rangle_{\mu,V,T}\right]^2\right\rangle_{\mu,V,T}.\end{aligned} \tag{5.144}$$

This is just equation (2.138) with $x = -\mu$ and $X = N$.

In equation (5.127) above, the fluctuations in number were related to the integral over the pair density. For a homogeneous system, $\rho_\mu^{(1)}(\mathbf{r}) = \rho$, with spherically symmetric pair potential, $\rho_\mu^{(2)}(\mathbf{r}, \mathbf{s}) = \rho^2 g(|\mathbf{r} - \mathbf{s}|)$ this is

$$
\begin{aligned}
&\left\langle \left[ N - \langle N \rangle_{\mu,V,T} \right]^2 \right\rangle_{\mu,V,T} \\
&= \int_V \mathrm{d}\mathbf{r}\, \mathrm{d}\mathbf{s} \left\{ \rho_\mu^{(2)}(\mathbf{r}, \mathbf{s}) - \rho_\mu^{(1)}(\mathbf{r})\rho_\mu^{(1)}(\mathbf{s}) + \rho_\mu^{(1)}(\mathbf{r})\delta(\mathbf{r} - \mathbf{s}) \right\} \\
&= \int_V \mathrm{d}\mathbf{r}\, \mathrm{d}\mathbf{s} \{ \rho^2 g(|\mathbf{r} - \mathbf{s}|) - \rho^2 + \rho\delta(\mathbf{r} - \mathbf{s}) \} \\
&= \rho V + \rho^2 V \int \mathrm{d}\mathbf{r}\ \{ g(r) - 1 \}.
\end{aligned}
\tag{5.145}
$$

The quantity in the integrand is called the total correlation function, $h(r) \equiv g(r) - 1$. It evidently goes to zero at large separations.

The isothermal compressibility diverges at spinodal points where the system is infinitely compressible. The curve of spinodal points lies inside the coexistence curve and it marks the limit of absolute thermodynamic stability of a phase. The critical point is the extremum of the spinodal curve. It is also the point at which the spinodal and coexistence curves coincide. The divergence of the density fluctuations due to the infinitely compressibility gives rise to the phenomenon known as critical opalescence.

## 5.6 Sub-system entropy as a functional of particle densities

### 5.6.1 Gibbs–Shannon information entropy

As was discussed in section 2.3.2, the Gibbs–Shannon information entropy, $S_{\mathrm{GS}}[\wp_\alpha] = -k_{\mathrm{B}} \sum_\alpha \wp_\alpha \ln \wp_\alpha$, gives only part of the total entropy, because it lacks the entropy internal to each macrostate $\alpha$, $\sum_\alpha \wp_\alpha S_\alpha$. In thermodynamics and statistical mechanics, the 'external' entropy that is retained is the sub-system entropy, and the 'internal' entropy that is lacking is the reservoir entropy.

For example, for a canonical equilibrium system, the total entropy is

$$
\begin{aligned}
S_{\mathrm{tot}}(N, V, T) &= k_{\mathrm{B}} \ln Z(N, V, T) \\
&= S_{\mathrm{s}}(\overline{E}) + S_{\mathrm{r}}(\overline{E}),
\end{aligned}
\tag{5.146}
$$

the second equality holding for the most likely energy macrostate in the thermodynamic limit. The reservoir entropy is

$$
S_{\mathrm{r}}(\overline{E}) = \frac{-\overline{E}(N, V, T)}{T} = \frac{-1}{T} \langle \mathcal{H} \rangle_{N,V,T}, \tag{5.147}
$$

and, as shown in equation (5.99), the sub-system entropy is

$$
S_{\mathrm{s}}(\overline{E}, N, V) = -k_{\mathrm{B}} \int \mathrm{d}\boldsymbol{\Gamma}\ \wp(\boldsymbol{\Gamma}) \ln \wp(\boldsymbol{\Gamma}). \tag{5.148}
$$

### 5.6.2 Lattice models

We now give a rather useful approximation for calculating the sub-system entropy using the Gibbs–Shannon form and the multi-particle probability density. The scheme is most transparent for a lattice model.

Consider a lattice of $N$ sites labeled $j = 1, 2, \ldots, N$, with a discrete variable $\sigma_i$ at each site. This variable can represent whether or not a site is occupied by a particle, or the species of particle, or the magnetic spin state, etc. The microstate of the system is signified $\sigma^N$, and it is assumed that for an isolated system all microstates have equal weight. The energy of the system is $\mathcal{H}(\sigma^N)$. At this stage we do not need to specify this or the dimensionality of the system.

For the canonical equilibrium case, the probability of a configuration is

$$\wp(\sigma^N) = \frac{1}{Z(N, T)} e^{-\beta\mathcal{H}(\sigma^N)}, \tag{5.149}$$

the partition function being

$$Z(N, T) = \sum_{\sigma^N} e^{-\beta\mathcal{H}(\sigma^N)}. \tag{5.150}$$

There is no $N!$ here because the sites are regarded as distinguishable. The total unconstrained entropy is $S_{\text{tot}}(N, T) = k_{\text{B}} \ln Z(N, T)$. This is the sum of the reservoir entropy,

$$S_{\text{r}}(\overline{E}) = \frac{-\overline{E}}{T} = \frac{-1}{T} \sum_{\sigma^N} \wp(\sigma^N)\, \mathcal{H}(\sigma^N), \tag{5.151}$$

and the sub-system entropy, which is the Gibbs–Shannon information entropy,

$$S_{\text{s}}(\overline{E}) = S_{\text{GS}}[\wp] = -k_{\text{B}} \sum_{\sigma^N} \wp(\sigma^N) \ln \wp(\sigma^N). \tag{5.152}$$

A naive approximation to the sub-system entropy can be made based on its extensivity. The entropy per site may be expected to be independent of the size of the system provided that the system is not too small,

$$\begin{aligned} \frac{1}{N} S_{\text{GS}}[\wp] &= \frac{-k_{\text{B}}}{N} \sum_{\sigma^N} \wp(\sigma^N) \ln \wp(\sigma^N) \\ &\approx \frac{-k_{\text{B}}}{n} \sum_{\sigma^n} \wp(\sigma^n) \ln \wp(\sigma^n). \end{aligned} \tag{5.153}$$

(It will be shown that this is not a very accurate approximation for a highly coupled system.) Here $\wp(\sigma^n)$ is the $n$-site probability obtained in the full $N$-site system. For $n \ll N$, the sum over configurations $\sigma^n$ is much more tractable than the one over $\sigma^N$. Although this expression for the entropy is more accurate for large $n$, it is also more difficult to obtain $\wp(\sigma^n)$ when $n$ becomes larger.

The formal expression for the $n$-site probability is

$$\wp(\sigma'^n) = \sum_{\sigma^N} \wp(\sigma^N)\, \delta(\sigma^n - \sigma'^n). \tag{5.154}$$

In general terms it is relatively easy to numerically obtain accurate statistical results for the several site probability function, but this becomes more challenging as the order of the probability is increased. What is required is an approximation for the many-body probability in terms of the few body probability.

### 5.6.3 One dimension

It is most transparent to initially carry out the analysis for a one-dimensional crystal, and then to generalize it to two and three dimensions. We shall mainly consider consecutive sites, with the notation $\sigma_j^n$ denoting the configurations of the $n$ sites from site $j$, $\sigma_j^n = \{\sigma_j, \sigma_{j+1}, \ldots, \sigma_{j+n-1}\}$.

We shall write the $n$-site probability for $\sigma^n$ at arbitrary sites $j^n$ as

$$\wp^{(n)}(\sigma^n, j^n) = g^{(n)}(\sigma^n, j^n)\prod_{k=1}^{n} \wp^{(1)}(\sigma_k), \tag{5.155}$$

where the distribution function is unity for far separated sites. For $n$ consecutive sites beginning at $j$ we shall write

$$\wp^{(n)}\left(\sigma_j^n\right) = G^{(n)}\left(\sigma_j^n\right)\prod_{k=j}^{j+n-1} \wp^{(1)}(\sigma_k). \tag{5.156}$$

Assuming a homogeneous system the singlet site probability is $\wp^{(1)}(\sigma)$. The two site probability is $\wp^{(2)}(\sigma_1, \sigma_2)$. To leading order this can be taken to be the product of singlet probabilities,

$$\wp^{(2)}(\sigma_1, \sigma_2) = \wp^{(1)}(\sigma_1)\wp^{(1)}(\sigma_2)g^{(2)}(\sigma_1, \sigma_2) \approx \wp^{(1)}(\sigma_1)\wp^{(1)}(\sigma_2). \tag{5.157}$$

This is equivalent to setting the distribution function $g$ or $G$ to unity.

A very common approximation for the three consecutive site probability, $\wp^{(3)}(\sigma_1, \sigma_2, \sigma_3) = \wp^{(1)}(\sigma_1)\wp^{(1)}(\sigma_2)\wp^{(1)}(\sigma_3)G^{(3)}(\sigma_1, \sigma_2, \sigma_3)$, is the Kirkwood superposition approximation,

$$G^{(3)}(\sigma_1, \sigma_2, \sigma_3) \approx G^{(2)}(\sigma_1, \sigma_2)G^{(2)}(\sigma_2, \sigma_3)g^{(2)}(\sigma_1, \sigma_3; 1, 3). \tag{5.158}$$

The problem with this is that it double counts the correlation between sites 1 and 3, most of which is already counted in the product of the first two factors. It is far better to make a Markov approximation for this,

$$G^{(3)}(\sigma_1, \sigma_2, \sigma_3) \approx G^{(2)}(\sigma_1, \sigma_2)G^{(2)}(\sigma_2, \sigma_3). \tag{5.159}$$

One can formalize this by defining

$$G^{(3)}(\sigma_1, \sigma_2, \sigma_3) \equiv G^{(2)}(\sigma_1, \sigma_2)G^{(2)}(\sigma_2, \sigma_3)\Delta^{(3)}(\sigma_1, \sigma_2, \sigma_3). \tag{5.160}$$

The disparity $\Delta^{(3)}$ is what makes this exact. Setting the triplet disparity to unity yields the Markov approximation for the triplet distribution function.

Similarly, for four consecutive sites one defines

$$\begin{aligned} G^{(4)}(\sigma_1, \sigma_2, \sigma_3, \sigma_4) &\equiv \frac{G^{(3)}(\sigma_1, \sigma_2, \sigma_3)G^{(3)}(\sigma_2, \sigma_3, \sigma_4)}{G^{(2)}(\sigma_2, \sigma_3)}\Delta^{(4)}(\sigma_1, \sigma_2, \sigma_3, \sigma_4) \\ &= G^{(2)}(\sigma_1, \sigma_2)G^{(2)}(\sigma_2, \sigma_3)G^{(2)}(\sigma_3, \sigma_4) \\ &\quad \times \Delta^{(3)}(\sigma_1, \sigma_2, \sigma_3)\Delta^{(3)}(\sigma_2, \sigma_3, \sigma_4)\Delta^{(4)}(\sigma_1, \sigma_2, \sigma_3, \sigma_4). \end{aligned} \tag{5.161}$$

This again is exact. The $G^{(2)}$ in the denominator of the first equality accounts for double counting. Setting $\Delta^{(4)} = 1$ gives the triplet Markov approximation for the four particle distribution, and in addition setting $\Delta^{(3)} = 1$ gives the pair Markov approximation for the four particle distribution.

In general, one has the formally exact product form

$$\begin{aligned} G^{(N)}(\sigma_1, \ldots, \sigma_N) &\equiv \frac{G^{(N-1)}(\sigma_1, \ldots, \sigma_{N-1})G^{(N-1)}(\sigma_2, \ldots, \sigma_N)}{G^{(N-2)}(\sigma_2, \ldots, \sigma_{N-1})}\Delta^{(N)}(\sigma_1, \ldots, \sigma_N) \\ &= G^{(2)}(\sigma_1, \sigma_2)G^{(2)}(\sigma_2, \sigma_3)\ldots G^{(2)}(\sigma_{N-1}, \sigma_N) \\ &\quad \times \Delta^{(3)}(\sigma_1, \sigma_2, \sigma_3)\ldots\Delta^{(N)}(\sigma_1, \ldots, \sigma_N). \end{aligned} \tag{5.162}$$

Inserting this into the Gibbs–Shannon expression for the sub-system entropy, the logarithm of this product becomes a sum of terms, starting with the singlet, pair, triplet, etc. Summing over the redundant spin variables, the probability multiplying the logarithms is reduced to a singlet, pair, triplet, probability, etc,

$$\begin{aligned} S_s(N) &= -k_B \sum_{\sigma^N} \wp^{(N)}(\sigma^N) \ln\left\{ G^{(N)}(\sigma_1, \ldots, \sigma_N) \prod_{j=1}^{N} \wp^{(1)}(\sigma_j) \right\} \\ &= -k_B \sum_{j=1}^{N} \sum_{\sigma_j} \wp^{(1)}(\sigma_j) \ln \wp^{(1)}(\sigma_j) \\ &\quad - k_B \sum_{j=1}^{N-1} \sum_{\sigma_j, \sigma_{j+1}} \wp^{(2)}(\sigma_j, \sigma_{j+1}) \ln G^{(2)}(\sigma_j, \sigma_{j+1}) \\ &\quad - k_B \sum_{j=1}^{N-2} \sum_{\sigma_j, \sigma_{j+1}, \sigma_{j+2}} \wp^{(3)}(\sigma_j, \sigma_{j+1}, \sigma_{j+2}) \ln \Delta^{(3)}(\sigma_j, \sigma_{j+1}, \sigma_{j+2}) \\ &\quad - \cdots - k_B \sum_{\sigma^N} \wp^{(N)}(\sigma^N) \ln \Delta^{(N)}(\sigma^N). \end{aligned} \tag{5.163}$$

For a homogeneous system there are $N$ equivalent singlet terms, $N - 1$ equivalent pair terms, $N - 2$ equivalent triplet terms, etc. In the thermodynamic limit each of these may be replaced by $N$ and the expansion for the sub-system entropy per site is

$$\begin{aligned} S_s(N)/N &= -k_B \sum_{\sigma} \wp^{(1)}(\sigma) \ln \wp^{(1)}(\sigma) \\ &\quad - k_B \sum_{\sigma_1, \sigma_2} \wp^{(2)}(\sigma_1, \sigma_2) \ln G^{(2)}(\sigma_1, \sigma_2) \\ &\quad - k_B \sum_{\sigma_1, \sigma_2, \sigma_3} \wp^{(3)}(\sigma_1, \sigma_2, \sigma_3) \ln \Delta^{(3)}(\sigma_1, \sigma_2, \sigma_3) \\ &\quad - \cdots. \end{aligned} \tag{5.164}$$

If one has obtained the probability $\wp^{(n)}$, then one can evaluate this up to order $n$, since all the $\wp^{(j)}$ and $\Delta^{(j)}$, $j \leqslant n$ can be obtained by reducing $\wp^{(n)}$. This can be called the $n$th order Markov superposition approximation to the Gibbs–Shannon form of the sub-system entropy.

Terminating this expansion at some order $n$ is expected to be accurate because it is equivalent to invoking a Markov superposition approximation at that order. This is much more accurate than, for example, the Kirkwood superposition approximation, because it does not double count the correlations, which can be particularly problematic at high couplings. This $n$th order Markov superposition approximation is also more accurate than simply inserting $\wp^{(n)}$ directly into the Gibbs–Shannon form of the sub-system entropy,

$$S_{\mathrm{s}}^{(n)} \equiv -k_{\mathrm{B}} \sum_{\sigma^n} \wp^{(n)}(\sigma^n) \ln \wp^{(n)}(\sigma^n), \tag{5.165}$$

and taking $S_{\mathrm{s}}(N)/N \approx S_{\mathrm{s}}^{(n)}/n$.

There is another way to write the $n$th order Markov superposition approximation that is particularly easy to generalize to higher dimensions. In view of the definition of the disparity,

$$\Delta^{(n)}(\sigma_1, \ldots, \sigma_n) \equiv \frac{\wp^{(n)}(\sigma_1, \ldots, \sigma_n) \wp^{(n-2)}(\sigma_2, \ldots, \sigma_{n-1})}{\wp^{(n-1)}(\sigma_1, \ldots, \sigma_{n-1}) \wp^{(n-1)}(\sigma_2, \ldots, \sigma_n)}, \tag{5.166}$$

one can write the difference between successive approximations to the entropy as

$$\begin{aligned} S_{\mathrm{s}}^{(n)} - S_{\mathrm{s}}^{(n-1)} &= -k_{\mathrm{B}} \sum_{\sigma^n} \wp^{(n)}(\sigma^n) \ln \frac{\wp^{(n)}(\sigma^n)}{\wp^{(n-1)}(\sigma^{n-1})} \\ &= -k_{\mathrm{B}} \sum_{\sigma^n} \wp^{(n)}(\sigma^n) \ln \frac{\Delta^{(n)}(\sigma^n) \wp^{(n-1)}(\sigma^{n-1})}{\wp^{(n-2)}(\sigma^{n-2})} \\ &= S_{\mathrm{s}}^{(n-1)} - S_{\mathrm{s}}^{(n-2)} - k_{\mathrm{B}} \sum_{\sigma^n} \wp^{(n)}(\sigma^n) \ln \Delta^{(n)}(\sigma^n). \end{aligned} \tag{5.167}$$

Obviously one can repeat this down to $n = 3$ to obtain

$$\begin{aligned} S_{\mathrm{s}}^{(n)} - S_{\mathrm{s}}^{(n-1)} &= -k_{\mathrm{B}} \sum_{\sigma^n} \wp^{(n)}(\sigma^n) \ln \Delta^{(n)}(\sigma^n) \\ &\quad - k_{\mathrm{B}} \sum_{\sigma^{n-1}} \wp^{(n-1)}(\sigma^{n-1}) \ln \Delta^{(n-1)}(\sigma^{n-1}) \\ &\quad - \cdots - k_{\mathrm{B}} \sum_{\sigma^3} \wp^{(3)}(\sigma^3) \ln \Delta^{(3)}(\sigma^3) \\ &\quad - k_{\mathrm{B}} \sum_{\sigma^2} \wp^{(2)}(\sigma^2) \ln \wp^{(2)}(\sigma^2) + k_{\mathrm{B}} \sum_{\sigma} \wp^{(1)}(\sigma) \ln \wp^{(1)}(\sigma) \\ &= -k_{\mathrm{B}} \sum_{\sigma^n} \wp^{(n)}(\sigma^n) \ln \Delta^{(n)}(\sigma^n) - \cdots - k_{\mathrm{B}} \sum_{\sigma^3} \wp^{(3)}(\sigma^3) \ln \Delta^{(3)}(\sigma^3) \\ &\quad - k_{\mathrm{B}} \sum_{\sigma^2} \wp^{(2)}(\sigma^2) \ln G^{(2)}(\sigma^2) - k_{\mathrm{B}} \sum_{\sigma} \wp^{(1)}(\sigma) \ln \wp^{(1)}(\sigma). \end{aligned} \tag{5.168}$$

One sees that this is identical to the $n$th order Markov superposition approximation to the Gibbs–Shannon form of the sub-system entropy per site,

$$S_{\mathrm{s}}(N)/N \approx S_{\mathrm{s}}^{(n)} - S_{\mathrm{s}}^{(n-1)} \text{ (Markov)}. \tag{5.169}$$

The difference between the two terms on the right-hand side reflects the subtraction of the doubly counted correlations, which is a feature of the Markov approach. This approximation is equivalent to setting the higher order disparities to unity, $\Delta^{(m)} = 1$, $m \geqslant n + 1$. This expression is expected to be much more accurate than the simple minded $S_s(N)/N \approx S_s^{(n)}/n$, particularly at higher couplings between the sites.

### 5.6.4 Two and three dimensions

Consider a two-dimensional lattice. For simplicity take this to be a square lattice, but this is not essential. Let $\wp^{(n,m)}(\sigma^{nm})$ be the probability for the configuration $\sigma^{nm}$ on an $n \times m$ block of sites. The sub-system entropy of such a block is

$$S_s^{(n,m)} \equiv -k_B \sum_{\sigma^{nm}} \wp^{(n,m)}(\sigma^{nm}) \ln \wp^{(n,m)}(\sigma^{nm}). \tag{5.170}$$

The $n$, $m$th Markov superposition approximation to the Gibbs–Shannon form of the sub-system entropy in two dimensions is

$$S_s(N)/N \approx S_s^{(n,m)} - S_s^{(n-1,m)} - S_s^{(n,m-1)} + S_s^{(n-1,m-1)}. \tag{5.171}$$

This is a generalization of the one-dimensional result. It can be seen that the entropy of the common sites that is subtracted twice is added back by the final term. Again, since this correctly counts the correlated contributions, it can be expected to be much more accurate than the simple minded $S_s(N)/N \approx S_s^{(n,m)}/nm$, particularly at high couplings.

Analogously, the three-dimensional result is

$$\begin{aligned} S_s(N)/N \approx\ & S_s^{(n,m,k)} - S_s^{(n-1,m,k)} - S_s^{(n,m-1,k)} - S_s^{(n,m,k-1)} \\ & + S_s^{(n,m-1,k-1)} + S_s^{(n-1,m,k-1)} + S_s^{(n-1,m-1,k)} \\ & - S_s^{(n-1,m-1,k-1)}. \end{aligned} \tag{5.172}$$

Here the Gibbs–Shannon form of the sub-system entropy of a $n \times m \times k$ block of sites is

$$S_s^{(n,m,k)} \equiv -k_B \sum_{\sigma^{nmk}} \wp^{(n,m,k)}(\sigma^{nmk}) \ln \wp^{(n,m,k)}(\sigma^{nmk}). \tag{5.173}$$

*Ising model test*

These approximations are readily tested for the two-dimensional Ising model, for which the exact results are known (Baxter 1982). In this model the Hamiltonian is

$$\mathcal{H}(\sigma^N) = -J \sum_{j,k}^{(\mathrm{nn})} \sigma_j \sigma_k, \tag{5.174}$$

where nn denotes the nearest neighbors on the lattice, and the spin variable can be $\sigma = \pm 1$. The weight of each microstate for an isolated sub-system is unity, and hence the sub-system microstate entropy vanishes, $S_s(\sigma^N) = 0$.

From Onsager's exact solution, the unconstrained total entropy per site is (Wikipedia 2017)

$$\begin{aligned}\frac{S_{\text{total}}(N,\,T)}{N} &\equiv \frac{1}{N}\ln Z_N(T)\\ &= \frac{1}{2}\ln 2 + \frac{1}{2\pi}\int_0^{\pi} \mathrm{d}\theta\,\ln\left\{\cosh^2 2K + k^{-1}\sqrt{1 + k^2 - 2k\cos(2\theta)}\right\},\end{aligned} \tag{5.175}$$

where the dimensionless coupling parameter is $K \equiv J/k_{\text{B}}T$, and $k \equiv 1/\sinh^2 2K$. The average (equivalently, most likely) energy per site is (Wikipedia 2017)

$$\begin{aligned}&\frac{\beta}{N}\overline{E}(N,\,T)\\ &= -K\coth 2K\left\{1 + \frac{2}{\pi}[2\tanh^2 2K - 1]\int_0^{\pi/2}\frac{\mathrm{d}\theta}{\sqrt{1 - 4k(1+k)^{-2}\sin^2\theta}}\right\}.\end{aligned} \tag{5.176}$$

From these can be calculated the reservoir entropy per site

$$\frac{S_{\text{r}}(N,\,T)}{N} = \frac{-1}{NT}\overline{E}(N,\,T), \tag{5.177}$$

and the sub-system entropy per site

$$\frac{S_{\text{s}}(N,\,T)}{N} = \frac{S_{\text{tot}}(N,\,T)}{N} + \frac{1}{NT}\overline{E}(N,\,T). \tag{5.178}$$

Figure 5.6 shows the exact results for the entropy of the two-dimensional Ising model as a function of the coupling parameter. The reservoir entropy goes to zero as the coupling parameter goes to zero because in this regime the sub-system energy vanishes, and hence so does the reservoir entropy. The sub-system entropy decreases as the coupling parameter increases because the spins are increasingly correlated and aligned, which is an ordered arrangement.

The two approximations for the Gibbs–Shannon form of the sub-system entropy are tested, both based on the four spin probability $\wp^{(2,2)}(\sigma^4)$ on a $2 \times 2$ block. The probability was obtained by enumerating all configurations on a $5 \times 5$ lattice with periodic boundary conditions (Attard 1999). It can be seen that the Gibbs–Shannon entropy agrees with the exact sub-system entropy, which confirms that it is just part of the total entropy. The Markov superposition approximation, $S_{\text{s}}^{(2,2)} - 2S_{\text{s}}^{(2,1)} + S_{\text{s}}^{(1,1)}$, performs much better than the naive $S_{\text{s}}^{(2,2)}/4$ at higher couplings. It appears that the simple approach implicitly assumes a disordered state for spins beyond the domain upon which it is calculated. At high couplings when the correlation length extends beyond this domain, this underestimates the real degree of correlation beyond the domain, and hence overestimates the entropy of the system.

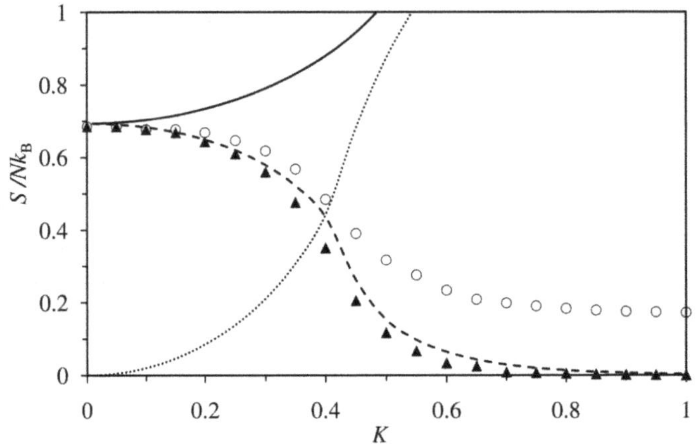


**Figure 5.6.** The entropy per site for the Ising model on a square lattice as a function of the coupling parameter for the canonical equilibrium system. The curves are the exact result for the total (solid), reservoir (dotted), and sub-system (dashed) entropy. The symbols are the Gibbs–Shannon information entropy using the four spin probability function on a 2 × 2 block, with the filled triangles being the Markov superposition formula, $S_s^{(2,2)} - 2S_s^{(2,1)} + S_s^{(1,1)}$, and the empty circles being $S_s^{(2,2)}/4$. Data from Attard (1999).

Evidently the Markov approach is able to correctly account for the correlations between the spins even at high couplings. It is surprising how well the Markov superposition approximation performs considering that only the four-spin distribution function was used.

### 5.6.5 Continuum disordered fluid

In this sub-section the focus is on the positional contribution to the sub-system entropy for particles in a continuum. The ideal gas contribution, which is to be added to this excess, was given above as equation (5.116).

The $n$-particle position probability, $\wp^{(n)}(\mathbf{r}_1, \mathbf{r}_2, \ldots, \mathbf{r}_n)$, is symmetric with respect to interchange of the arguments. In two and three dimensions, this appears to create difficulties for developing a Markov-type approximation for the entropy. Nevertheless, one can proceed at least formally.

Define the thee particle disparity as

$$g^{(3)}(\mathbf{r}_1, \mathbf{r}_2, \mathbf{r}_3) \equiv g^{(2)}(\mathbf{r}_1, \mathbf{r}_2) g^{(2)}(\mathbf{r}_2, \mathbf{r}_3) \Delta^{(3)}(\mathbf{r}_1, \mathbf{r}_2, \mathbf{r}_3). \tag{5.179}$$

Since the left-hand side is symmetric in its arguments, the disparity must be asymmetric in its arguments. One expects the disparity to be close to unity when $\mathbf{r}_2$ lies between the other two positions. One can define the higher order disparities also as in the one-dimensional spin lattice case.

Proceeding as in that case, the excess (i.e. without the ideal gas part) Gibbs–Shannon form of the sub-system entropy is

$$\begin{aligned} S_s^{ex}(N) &= -k_B \int_V d\mathbf{r}^N\ \wp^{(N)}(\mathbf{r}^N) \ln \{g^{(N)}(\mathbf{r}_1, \dots, \mathbf{r}_N) \prod_{j=1}^{N} \wp^{(1)}(\mathbf{r}_j)\} \\ &= -k_B \sum_{j=1}^{N} \int_V d\mathbf{r}_j\ \wp^{(1)}(\mathbf{r}_j) \ln \wp^{(1)}(\mathbf{r}_j) \\ &\quad - k_B \sum_{j=1}^{N-1} \int_V d\mathbf{r}_j\, d\mathbf{r}_{j+1}\ \wp^{(2)}(\mathbf{r}_j, \mathbf{r}_{j+1}) \ln g^{(2)}(\mathbf{r}_j, \mathbf{r}_{j+1}) \\ &\quad - k_B \sum_{j=1}^{N-2} \int_V d\mathbf{r}_j\, d\mathbf{r}_{j+1}\, d\mathbf{r}_{j+2}\ \wp^{(3)}(\mathbf{r}_j, \mathbf{r}_{j+1}, \mathbf{r}_{j+2}) \ln \Delta^{(3)}(\mathbf{r}_j, \mathbf{r}_{j+1}, \mathbf{r}_{j+2}) \\ &\quad - \cdots - k_B \int_V d\mathbf{r}^N\ \wp^{(N)}(\mathbf{r}^N) \ln \Delta^{(N)}(\mathbf{r}^N). \end{aligned} \tag{5.180}$$

For a homogeneous system each of these contributes essentially $N$ equivalent terms, and the expansion for the excess sub-system entropy per particle is

$$\begin{aligned} S_s^{ex}(N)/N &= -k_B \int_V d\mathbf{r}\ \wp^{(1)}(\mathbf{r}) \ln \wp^{(1)}(\mathbf{r}) \\ &\quad - k_B \int_V d\mathbf{r}\, d\mathbf{s}\ \wp^{(2)}(\mathbf{r}, \mathbf{s}) \ln g^{(2)}(\mathbf{r}, \mathbf{s}) \\ &\quad - k_B \int_V d\mathbf{r}\, d\mathbf{s}\, d\mathbf{t}\ \wp^{(3)}(\mathbf{r}, \mathbf{s}, \mathbf{t}) \ln \Delta^{(3)}(\mathbf{r}, \mathbf{s}, \mathbf{t}) - \cdots \end{aligned} \tag{5.181}$$

Recall that $\Delta^{(3)}(\mathbf{r}, \mathbf{s}, \mathbf{t})$ is symmetric with respect to the interchange of $\mathbf{r}$ and $\mathbf{t}$, but that it is asymmetric for any interchange with $\mathbf{s}$. This expansion is formally exact.

Define

$$S_s^{ex,(n)} = -k_B \int_V d\mathbf{r}^n\ \wp^{(n)}(\mathbf{r}^n) \ln \wp^{(n)}(\mathbf{r}^n). \tag{5.182}$$

One has

$$\begin{aligned} S_s^{ex,(n)} - S_s^{ex,(n-1)} &= -k_B \int_V d\mathbf{r}^n\ \wp^{(n)}(\mathbf{r}^n) \ln \frac{\wp^{(n)}(\mathbf{r}^n)}{\wp^{(n-1)}(\mathbf{r}^{n-1})} \\ &= -k_B \int_V d\mathbf{r}^n\ \wp^{(n)}(\mathbf{r}^n) \ln \frac{\Delta^{(n)}(\mathbf{r}^n) \wp^{(n-1)}(\mathbf{r}^{n-1})}{\wp^{(n-2)}(\mathbf{r}^{n-2})} \\ &= S_s^{ex,(n-1)} - S_s^{ex,(n-2)} - k_B \int_V d\mathbf{r}^n \wp^{(n)}(\mathbf{r}^n) \ln \Delta^{(n)}(\mathbf{r}^n), \end{aligned} \tag{5.183}$$

which telescopes to

$$\begin{aligned} S_s^{ex,(n)} - S_s^{ex,(n-1)} &= -k_B \int_V d\mathbf{r}^n\ \wp^{(n)}(\mathbf{r}^n) \ln \Delta^{(n)}(\mathbf{r}^n) \\ &\quad - \cdots - k_B \int_V d\mathbf{r}^3\ \wp^{(3)}(\mathbf{r}^3) \ln \Delta^{(3)}(\mathbf{r}^3) \\ &\quad - k_B \int_V d\mathbf{r}^2\ \wp^{(2)}(\mathbf{r}^2) \ln g^{(2)}(\mathbf{r}^2) - k_B \int_V d\mathbf{r}\ \wp^{(1)}(\mathbf{r}) \ln \wp^{(1)}(\mathbf{r}). \end{aligned} \tag{5.184}$$

This is just the Markov expression obtained above. Hence one can take the excess sub-system entropy per particle to be

$$S_s^{ex}(N)/N \approx S_s^{ex,(n)} - S_s^{ex,(n-1)}. \tag{5.185}$$

Whatever the general utility of the Markov expansion for the particle distribution functions in two and three dimensions, this is, presumably, the best approximation to the sub-system entropy that one can make.

## 5.7 Time correlation and van Hove functions

### 5.7.1 Time correlation function

Let $A(\mathbf{\Gamma})$ and $B(\mathbf{\Gamma})$ be functions of phase space. The time correlation function for them is usually defined in terms of the correlation between the fluctuations from their average value,

$$C_{AB}(\tau) = \langle [A(\tau) - \langle A \rangle] \, [B(0) - \langle B \rangle] \rangle_{N,V,T}. \tag{5.186}$$

In the long time limit the two variables must be uncorrelated and so one has

$$\begin{aligned} C_{AB}(\tau) &\to \langle [A(\tau) - \langle A \rangle] \rangle_{N,V,T} \, \langle [B(0) - \langle B \rangle] \rangle_{N,V,T}, \quad \tau \to \infty \\ &= 0. \end{aligned} \tag{5.187}$$

In terms of the stochastic, dissipative trajectory $\mathbf{\Gamma}(t|\mathbf{\Gamma}_0)$ this is

$$C_{AB}(\tau) = \frac{1}{t} \int_0^t \mathrm{d}t' [A(\mathbf{\Gamma}(t' + \tau|\mathbf{\Gamma}_0)) - \langle A \rangle] \, [B(\mathbf{\Gamma}(t'|\mathbf{\Gamma}_0)) - \langle B \rangle]. \tag{5.188}$$

For large $t$ this should be independent of the starting position $\mathbf{\Gamma}_0$. One often sees in the literature the stochastic, dissipative trajectory that appears here approximated by an adiabatic or even by a deterministic thermostatted trajectory.

Let $\varepsilon = \pm 1$ be the time parity such that

$$A(\mathbf{\Gamma}^\dagger) = \varepsilon_A A(\mathbf{\Gamma}) \text{ and } B(\mathbf{\Gamma}^\dagger) = \varepsilon_B B(\mathbf{\Gamma}). \tag{5.189}$$

Recall that the conjugate phase point has the same positions but reversed momenta. Note that $\langle A \rangle$ is non-zero only if $A$ has even parity. Hence one can always write $\langle A \rangle = \varepsilon_A \langle A \rangle$ and $\langle B \rangle = \varepsilon_B \langle B \rangle$.

As shown in section 5.4.2, the stochastic, dissipative equations of motion possess microscopic reversibility. Equation (5.67) reads

$$\mathbf{\Gamma}_2 = \mathbf{\Gamma}(t|\mathbf{\Gamma}_1) \Leftrightarrow \mathbf{\Gamma}_2^\dagger = \mathbf{\Gamma}(-t|\mathbf{\Gamma}_1^\dagger), \tag{5.190}$$

which is to be understood in a probabilistic sense. With these, and starting the trajectory from the equally likely conjugate phase space point $\mathbf{\Gamma}_0^\dagger$, the time correlation function has the symmetry

$$\begin{aligned} C_{AB}(\tau) &= \frac{1}{t}\int_0^t \mathrm{d}t'[A(\mathbf{\Gamma}(t' + \tau|\mathbf{\Gamma}_0^\dagger)) - \langle A\rangle]\,[B(\mathbf{\Gamma}(t'|\mathbf{\Gamma}_0^\dagger)) - \langle B\rangle] \\ &= \frac{1}{t}\int_0^t \mathrm{d}t'[A(\mathbf{\Gamma}(-t' - \tau|\mathbf{\Gamma}_0)^\dagger) - \langle A\rangle]\,[B(\mathbf{\Gamma}(-t'|\mathbf{\Gamma}_0)^\dagger) - \langle B\rangle] \\ &= \frac{\varepsilon_A\varepsilon_B}{t}\int_0^t \mathrm{d}t'[A(\mathbf{\Gamma}(-t' - \tau|\mathbf{\Gamma}_0)) - \langle A\rangle]\,[B(\mathbf{\Gamma}(-t'|\mathbf{\Gamma}_0)) - \langle B\rangle] \\ &= \varepsilon_A\varepsilon_B C_{AB}(-\tau). \end{aligned} \tag{5.191}$$

The final equality follows because of time homogeneity: only the relative times at which $A$ and $B$ are evaluated matters.

One can conclude from this that

$$C_{AB}(0) = 0, \quad \varepsilon_A \neq \varepsilon_B. \tag{5.192}$$

This says that there can be no instantaneous correlation between variables of opposite time parity.

### 5.7.2 van Hove function

The one-particle number density phase space function is defined as

$$\hat{\rho}^{(1)}(\mathbf{r}, t) = \sum_{j=1}^{N} \delta\big(\mathbf{r} - \mathbf{q}_j(t)\big), \tag{5.193}$$

where $\mathbf{q}_j$ is the position of molecule $j$. For a closed system, this integrates to the total number, $\int \mathrm{d}\mathbf{r}\,\hat{\rho}(\mathbf{r}, t) = N$.

The circumflex distinguishes this phase space function from the average particle density of the type discussed in section 5.5.6. The singlet density is

$$\langle\hat{\rho}^{(1)}(\mathbf{r}, t)\rangle = \rho^{(1)}(\mathbf{r}). \tag{5.194}$$

An equilibrium system is assumed here.

The density–density correlation function is a form of time correlation function. Taking $A \equiv \hat{\rho}^{(1)}(\mathbf{r}' + \mathbf{r}, t)$ and $B \equiv \hat{\rho}^{(1)}(\mathbf{r}', 0)$, this can be defined as

$$G^{(2)}(\mathbf{r}, \mathbf{r}', t) \equiv N^{-1}\langle\hat{\rho}^{(1)}(\mathbf{r}' + \mathbf{r}, t)\hat{\rho}^{(1)}(\mathbf{r}', 0)\rangle. \tag{5.195}$$

This gives the probability of a molecule being at $\mathbf{r}'$ and a molecule being located a distance $\mathbf{r}$ from $\mathbf{r}'$ at a time $t$ later. These may be the same molecule.

This density–density correlation function should not be confused with the pair distribution function $g^{(2)}$ introduced in section 5.5.6, or the consecutive site correlation function of equation (5.156). Asymptotically, $G^{(2)}(\mathbf{r}, \mathbf{r}', t) \sim \rho^{(1)}(\mathbf{r}' + \mathbf{r})\rho^{(1)}(\mathbf{r})/N$, as $r' \to \infty$ or $t \to \infty$. In contrast, $g^{(2)}(\mathbf{r}, \mathbf{r}') \sim 1$, as $|\mathbf{r} - \mathbf{r}'| \to \infty$.

For a homogenous system only the distance matters. Integrating the density–density correlation function over the volume gives the van Hove function (Hansen and McDonald 1990),

$$
\begin{aligned}
G(\mathbf{r}, t) &= \int \mathrm{d}\mathbf{r}' \, G(\mathbf{r}, \mathbf{r}', t) \\
&= \int \mathrm{d}\mathbf{r}' \, \frac{1}{N} \left\langle \sum_{i=1}^{N} \sum_{j=1}^{N} \delta\big(\mathbf{r}' + \mathbf{r} - \mathbf{q}_i(t)\big) \, \delta\big(\mathbf{r}' - \mathbf{q}_j(0)\big) \right\rangle \\
&= \frac{1}{N} \sum_{i=1}^{N} \sum_{j=1}^{N} \left\langle \delta\big(\mathbf{r} + \mathbf{q}_j(0) - \mathbf{q}_i(t)\big) \right\rangle .
\end{aligned}
\tag{5.196}
$$

This is equivalent to $G(\mathbf{r}, t) = VG(\mathbf{r}, \mathbf{0}, t) = \rho^{-1}\langle \hat{\rho}^{(1)}(\mathbf{r}, t)\hat{\rho}^{(1)}(\mathbf{0}, 0)\rangle$. The van Hove function can be measured by x-ray or neutron scattering techniques.

The self part of this gives the motion of an individual molecule over time,

$$
G_{\mathrm{s}}(\mathbf{r}, t) = \frac{1}{N} \sum_{j=1}^{N} \left\langle \delta\big(\mathbf{r} + \mathbf{q}_j(0) - \mathbf{q}_j(t)\big) \right\rangle . \tag{5.197}
$$

In essence, this is the probability that a molecule moves $\mathbf{r}$ in time $t$. The distinct part gives the correlation of different molecules

$$
G_{\mathrm{d}}(\mathbf{r}, t) = \frac{1}{N} \sum_{i=1}^{N} \sum_{j=1}^{N}{}^{(i \neq j)} \left\langle \delta\big(\mathbf{r} + \mathbf{q}_i(0) - \mathbf{q}_j(t)\big) \right\rangle . \tag{5.198}
$$

At large distances and at large times the molecules are uncorrelated, so that one has $\lim_{\mathbf{r}\to\infty} G_{\mathrm{d}}(\mathbf{r}, t) = \lim_{t\to\infty} G_{\mathrm{d}}(\mathbf{r}, t) = \rho$, assuming a homogeneous system.

## Summary

- Classical phase space forms the microstates for a classical system. Hamilton's equations of motion govern the adiabatic evolution of an isolated sub-system in phase space. For an isolated sub-system, classical phase space has uniform weight density and therefore zero internal entropy.
- The canonical equilibrium probability density in phase space is the Maxwell–Boltzmann distribution. In general, equilibrium probability distributions in phase space are proportional to the exponential of the reservoir entropy.
- Equilibrium averages can be given as either a time average over the stochastic, dissipative trajectory in phase space, or else as an integral over phase space weighted by the equilibrium probability distribution.
- The Gibbs–Shannon information entropy equals the sub-system entropy. Markov expansions for lattice and continuum systems in various dimensions facilitate its evaluation and are more accurate than direct approaches.
- The time correlation function for two phase functions can be obtained as a time average over the stochastic, dissipative trajectory.

## References

Attard P 1999 Markov superposition expansion for the entropy and correlation functions in two and three dimensions *Statistical Physics on the Eve of the Twenty-First Century* eds M T Batchelor and L T Wille (Singapore: World Scientific)

Baxter R J 1982 *Exactly Solved Models in Statistical Mechanics* (London: Academic)

Boltzmann L 1871 Über das wärmegleichgewicht zwischen mehratomigen gasmolekule *Wien. Ber.* **63** 397–481

Brush S G 1966 *Selected Readings in Physics* vol 2, pp 191–2

Galea T M and Attard P 2002 Constraint method for deriving non-equilibrium molecular dynamics equations of motion *Phys. Rev.* E **66** 041207

Hansen J-P and McDonald I R 1990 *Theory of Simple Liquids* 2nd edn (London: Academic)

Münster A 1969 *Statistical Thermodynamics* vol 1 (Berlin: Springer)

Pathria R K 1972 *Statistical Mechanics* (Oxford: Pergamon)

Planck M 1968 Scientific autobiography and other papers *Philos. Library* p 33 (English translation)

Wikipedia 2017 *Square-Lattice Ising Model* (Accessed: 8 April 2017)

**IOP** Publishing

Entropy Beyond the Second Law
Thermodynamics and statistical mechanics for equilibrium, non-equilibrium, classical, and quantum systems
**Phil Attard**

# Chapter 6

## Time for entropy

'Time that is moved by little fidget wheels
Is not my time, the flood that does not flow'

Slessor (1939)

'It can scarcely be denied that the supreme goal of all theory is to make the irreducible basic elements as simple and as few as possible without having to surrender the adequate representation of a single datum of experience'

Albert Einstein (1933)

'Medicine is a philosophy and this is not compatible with the renouncement of criticism of the leading authors'

al-Razi (Weinberg 2015)

The main aim of this chapter is to develop non-equilibrium classical statistical mechanics by deriving the general form for the probability density in phase space for a thermodynamic or a mechanical non-equilibrium system.

The need for such a first principles derivation is no better illustrated than by an expression often found in books on statistical mechanics that is meant to give the phase space probability density for a non-equilibrium system with Hamiltonian that varies with time,

$$\wp(\mathbf{\Gamma}, t) \propto e^{-\beta \mathcal{H}(\mathbf{\Gamma}, t)}. \tag{6.1}$$

Although this expression has the great virtue of being simple, it has the significant drawback of being qualitatively and quantitatively wrong. The chapter begins by showing why this expression is flawed.

doi:10.1088/978-0-7503-1590-6ch6

## 6.1 Reservoir entropy

There are two generic types of non-equilibriums systems, mechanical and thermodynamic. A mechanical non-equilibrium system has an explicitly time dependent Hamiltonian, $\mathcal{H}(\mathbf{\Gamma}, t)$. Typically the time variation is due to a time-varying external potential that is applied to the sub-system. Usually the sub-system is in thermal contact with a heat reservoir.

A thermodynamic non-equilibrium system is typically a sub-system across which there is a thermodynamic gradient applied by two spatially separated reservoirs with different values of certain field variables. In this case there is a steady flux of the conjugate exchangeable material from one reservoir to the other through the sub-system. Steady heat flow that was treated in chapter 3 is an example of thermodynamic non-equilibrium system.

Both of these generic non-equilibrium systems are sensitive to the direction of time, either because the external potential is explicitly time dependent, or else because of the steady induced flux that increases the entropy of the total system with time. This time sensitivity means that the probability distribution for the molecular configurations must be sensitive to the molecular velocities, since reversing time is equivalent to reversing all the velocities. Recall the definition of the conjugation operation, namely that each point in the phase space of the particles' positions and momenta, $\mathbf{\Gamma} \equiv \{\mathbf{q}, \mathbf{p}\}$, has a conjugate point with all the momenta reversed, $\mathbf{\Gamma}^{\dagger} \equiv \{\mathbf{q}, -\mathbf{p}\}$. The fact that by definition a non-equilibrium system must be sensitive to time means that the phase space point and its conjugate must have different probabilities,

$$\wp(\mathbf{\Gamma}, t) \neq \wp(\mathbf{\Gamma}^{\dagger}, t). \tag{6.2}$$

This is a formally exact requirement that must hold for both mechanical and thermodynamic non-equilibrium systems.

However, the kinetic energy in the time-dependent Hamiltonian is a quadratic function of velocities. Hence it is unchanged by their reversal,

$$\mathcal{H}(\mathbf{\Gamma}, t) = \mathcal{H}(\mathbf{\Gamma}^{\dagger}, t). \tag{6.3}$$

This means that the Maxwell–Boltzmann probability distribution that is the simplistic equation (6.1) cannot be the probability distribution for any non-equilibrium system.

### 6.1.1 Trajectory entropy

In chapter 5, it was shown that the weight density of a point in the phase space of an isolated system was uniform, which means that its entropy vanishes. (The constant contributions, Planck's constant $h^{3N}$ for the volume of a quantum microstate, and $N!$ for distinct microstates, can be defined as the phase point entropy, or else they can be shown explicitly, as will be done here.) This result was derived from the adiabatic equations of motion for an isolated system. Since the same Hamilton's equations hold for the isolated sub-system even when the Hamiltonian is time-dependent, one can assume that the same result holds in the non-equilibrium case. That is, the internal entropy of a point in the phase space of the isolated sub-system vanishes.

In view of this, for the case of a sub-system that can exchange with a reservoir, the entropy of a sub-system phase space point $\mathbf{\Gamma}$ is just the entropy of the reservoir that is associated with that point at that time, $S_r(\mathbf{\Gamma}, t)$. Hence the non-equilibrium phase space probability density is

$$\wp(\mathbf{\Gamma}, t) = \frac{e^{S_r(\mathbf{\Gamma}, t)/k_B}}{h^{3N} N! Z(t)}. \tag{6.4}$$

This result is formally exact but practically useless until an explicit expression for the reservoir entropy can be derived.

In order to be definite, and to give the clearest and most transparent derivation, we shall henceforth assume the canonical mechanical non-equilibrium system until it is stated otherwise. The sub-system Hamiltonian is $\mathcal{H}(\mathbf{\Gamma}, t)$, and the sub-system is assumed to be able to exchange energy with a heat reservoir of temperature $T$.

Consider a specific trajectory in the sub-system, $\mathbf{\Gamma}(t)$. Let us take this to be a forward trajectory from $t = 0$, which is convenient but not essential. This contains both adiabatic, superscript 0, and reservoir, $\mathbf{R}$, influences, as is sketched in figure 6.1. The rate of change of the trajectory is

$$\dot{\mathbf{\Gamma}}(t) = \dot{\mathbf{\Gamma}}^0(t) + \dot{\mathbf{R}}(t). \tag{6.5}$$

This is meant to be an actual trajectory, and the reservoir forces are the real forces that are necessary to give the specified departure from the adiabatic trajectory. For this reason we do not have to distinguish between forward and backward time derivatives. This is in contrast to the cases discussed with reference to equation (4.22) and to figure 5.5, where the statistical reservoir contributions to the trajectory are irreversible and create a discontinuity in the time derivative.

The rate of change of energy of the sub-system on the trajectory is

$$\begin{aligned} \frac{\mathrm{d}\mathcal{H}(\mathbf{\Gamma}(t), t)}{\mathrm{d}t} &= \frac{\partial \mathcal{H}(\mathbf{\Gamma}, t)}{\partial t} + \dot{\mathbf{\Gamma}}(t) \cdot \nabla \mathcal{H}(\mathbf{\Gamma}, t) \\ &= \frac{\partial \mathcal{H}(\mathbf{\Gamma}, t)}{\partial t} + \dot{\mathbf{R}}(t) \cdot \nabla \mathcal{H}(\mathbf{\Gamma}, t). \end{aligned} \tag{6.6}$$

The adiabatic velocity does not contribute to the change in energy. The final term is the reservoir-induced rate of change of energy. By energy conservation, this is equal and opposite to the rate of change of the reservoir energy. Hence the rate of change of the reservoir entropy on this particular trajectory is

**Figure 6.1.** Sketch of a trajectory in the sub-system phase space. The arrows represent the force at periodic intervals, with the solid tangential arrows being the total force, the dashed arrows being the adiabatic force, and the dotted arrows being the reservoir force.

$$\dot{S}_r(\mathbf{\Gamma}(t), t) = \frac{-1}{T}\dot{\mathbf{R}}(t) \cdot \nabla\mathcal{H}(\mathbf{\Gamma}(t), t). \tag{6.7}$$

Finally, the trajectory entropy, which is the change in reservoir entropy over the particular trajectory is

$$\begin{aligned} S_r([\mathbf{\Gamma}(t)], t) &= \frac{-1}{T}\int_0^t \mathrm{d}t'\, \dot{\mathbf{R}}(t') \cdot \nabla\mathcal{H}(\mathbf{\Gamma}(t'), t') \\ &= \frac{-1}{T}\int_0^t \mathrm{d}t'[\dot{\mathcal{H}}(\mathbf{\Gamma}(t'), t') - \dot{\mathcal{H}}^0(\mathbf{\Gamma}(t'), t')] \\ &= \frac{-1}{T}[\mathcal{H}(\mathbf{\Gamma}(t), t) - \mathcal{H}(\mathbf{\Gamma}(0), 0)] \\ &\quad + \frac{1}{T}\int_0^t \mathrm{d}t' \frac{\partial\mathcal{H}(\mathbf{\Gamma}(t'), t')}{\partial t'}. \end{aligned} \tag{6.8}$$

It is convenient, but not essential, to insist that this is a forward trajectory and that $t > 0$.

The first term on the right-hand side of the final equality is the instantaneous or static reservoir entropy at the end of the trajectory less that at the beginning. This static part of the entropy is the entropy based solely on the instantaneous structure of the sub-system, which is the entropy that would occur if the system were an equilibrium system. In the present mechanical case the static entropy is defined as

$$S_{r,st}(\mathbf{\Gamma}, t) = \frac{-1}{T}\mathcal{H}(\mathbf{\Gamma}, t). \tag{6.9}$$

The second term on the right-hand side may be called the dynamic part of the entropy. It is the correction to the static entropy that arises from the non-equilibrium nature of the system. It subtracts the sum total of the change of the static part of the reservoir entropy that is due to adiabatic processes,

$$\begin{aligned} S_{r,dyn}([\mathbf{\Gamma}(t)], t) &= \frac{1}{T}\int_0^t \mathrm{d}t' \frac{\partial\mathcal{H}(\mathbf{\Gamma}(t'), t')}{\partial t'} \\ &= \frac{1}{T}\int_0^t \mathrm{d}t'\, \dot{\mathcal{H}}^0(\mathbf{\Gamma}(t'), t') \\ &= -\int_0^t \mathrm{d}t' \dot{S}^0_{r,st}(\mathbf{\Gamma}(t'), t'). \end{aligned} \tag{6.10}$$

The integral of the first equality may be recognized as the work done by the external potential on the system over the particular trajectory. With these the trajectory entropy is

$$\begin{aligned} S_r([\mathbf{\Gamma}(t)], t) &= S_{r,st}(\mathbf{\Gamma}, t) + S_{r,dyn}([\mathbf{\Gamma}(t)], t) \\ &= S_{r,st}(\mathbf{\Gamma}, t) - \int_0^t \mathrm{d}t' \dot{S}^0_{r,st}(\mathbf{\Gamma}(t'), t'). \end{aligned} \tag{6.11}$$

This neglects the static entropy at the start of the trajectory, $S_{\mathrm{r,st}}(\mathbf{\Gamma}(0), 0)$, since for a long enough trajectory this is a constant that is uncorrelated with the end point.

Subtracting the second term, the dynamic part of the entropy, ensures that only reservoir-induced changes in energy contribute to the reservoir trajectory entropy. One virtue of writing the trajectory entropy this way is that the non-equilibrium part is explicit and can easily be treated as a perturbation about the equilibrium part, if desired.

A second virtue of writing the trajectory entropy this way is is that it holds formally unchanged for a thermodynamic non-equilibrium system. For example, in the case of steady heat that was treated in section 3.4, the static part of the entropy is just

$$S_{\mathrm{r,st}}(\mathbf{\Gamma}) = \frac{-E_0(\mathbf{\Gamma})}{T_0} - \frac{E_1(\mathbf{\Gamma})}{T_1}, \tag{6.12}$$

and its adiabatic rate of change is

$$\dot{S}^0_{\mathrm{r,st}}(\mathbf{\Gamma}) = \frac{-1}{T_1}\dot{E}_1^0(\mathbf{\Gamma}) = \frac{-1}{T_1}\dot{\mathbf{\Gamma}}^0 \cdot \nabla E_1(\mathbf{\Gamma}). \tag{6.13}$$

The larger point is that for any non-equilibrium system, it is trivial to write explicitly the expression for the static part of the reservoir entropy. It is defined as the instantaneous reservoir entropy that changes directly from exchanges between the sub-system and the reservoir.

From this point on, it is no longer assumed that the analysis is restricted to a mechanical non-equilibrium system.

### 6.1.2 Point entropy

What is required to formulate non-equilibrium statistical mechanics is the reservoir entropy associated with a phase point $\mathbf{\Gamma}$, whereas the above expression is for the reservoir entropy of a particular trajectory $\mathbf{\Gamma}(t)$. The obvious way to proceed is to neglect fluctuations and to focus on the most likely backward trajectory from the current point $\mathbf{\Gamma}$, namely $\overline{\mathbf{\Gamma}}(t'|\mathbf{\Gamma}, t)$, $t' \leqslant t$. (It is also possible to invoke instead the most likely forward trajectory from the current point.) With this the reservoir entropy for $\mathbf{\Gamma}$ for the non-equilibrium system is

$$S_{\mathrm{r}}(\mathbf{\Gamma}, t) = S_{\mathrm{r,st}}(\mathbf{\Gamma}, t) + S_{\mathrm{r,dyn}}(\mathbf{\Gamma}, t), \tag{6.14}$$

where the dynamic part of the entropy on the most likely backward trajectory is

$$\begin{aligned} S_{\mathrm{r,dyn}}(\mathbf{\Gamma}, t) &\equiv S_{\mathrm{r,dyn}}([\overline{\mathbf{\Gamma}}(t'|\mathbf{\Gamma}, t)], t) \\ &= -\int_0^t \mathrm{d}t' \dot{S}^0_{\mathrm{r,st}}(\overline{\mathbf{\Gamma}}(t'|\mathbf{\Gamma}, t), t'). \end{aligned} \tag{6.15}$$

Note that although the most likely trajectory has a discontinuous first time derivative, the adiabatic time derivative on this backward most likely trajectory is continuous, and so there is no distinction between the forward and backward adiabatic time derivatives.

This result formally gives the explicit expression for the phase space non-equilibrium probability density, equation (6.4). Details for the most likely trajectory in the non-equilibrium system are derived in section 6.2.2.

*Asymptote*
The integrand of the dynamic entropy has asymptote

$$\dot{S}^{0}_{\rm r,st}(\overline{\mathbf{\Gamma}}(t'|\mathbf{\Gamma}, t), t') \rightarrow \overline{\dot{S}^{0}_{\rm r,st}}(t'), \quad |t' - t| \rightarrow \infty. \tag{6.16}$$

This says that with overwhelming probability the system reverts back to the most likely rate of adiabatic entropy production no matter how far the current point $\mathbf{\Gamma}$ is from a likely point in phase space.

With this one can write

$$S_{\rm r,dyn}(\mathbf{\Gamma}, t) = \overline{S}_{\rm r,dyn}(t) - \int_0^t {\rm d}t' \left[ \dot{S}^{0}_{\rm r,st}(\overline{\mathbf{\Gamma}}(t'|\mathbf{\Gamma}, t), t') - \overline{\dot{S}^{0}_{\rm r,st}}(t') \right], \tag{6.17}$$

where the constant is $\overline{S}_{\rm r,dyn}(t) \equiv -\int_0^t {\rm d}t' \, \overline{\dot{S}^{0}_{\rm r,st}}(t')$. The integrand of the dynamic entropy is now short-ranged and so it is insensitive to the value chosen for the lower limit, which facilitates its numerical evaluation.

*Forward trajectory*
As was mentioned above, the dynamic part of the entropy can be written equally well as an integral over the forward trajectory as the backward one,

$$\begin{aligned} S_{\rm r,dyn}(\mathbf{\Gamma}, t) &= -\int_{-\infty}^{t} {\rm d}t' \dot{S}^{0}_{\rm r,st}(\overline{\mathbf{\Gamma}}(t'|\mathbf{\Gamma}, t), t') \\ &= -\int_{t}^{\infty} {\rm d}t' \dot{S}^{0}_{\rm r,st}(\overline{\mathbf{\Gamma}}(t'|\mathbf{\Gamma}, t), t'). \end{aligned} \tag{6.18}$$

These two expression may differ by an immaterial term that is constant in phase space.

### 6.1.3 Changes in entropy

*Reservoir-induced change in reservoir entropy*
In section 6.2.2 below we shall derive the stochastic, dissipative equations of motion for the non-equilibrium system. In view of the equilibrium result one can anticipate that they are of the form

$$\mathbf{\Gamma}_2 = \mathbf{\Gamma}_1 + \Delta_t \dot{\mathbf{\Gamma}}^0 + \mathbf{R}_p(\Delta_t, \mathbf{\Gamma}, t). \tag{6.19}$$

As usual, the superscript 0 denotes the adiabatic velocity and the reservoir contribution affects only the momenta over a single infinitesimal time step.

One can rewrite the result for the reservoir entropy directly in terms of its change over the most likely trajectory. With the initial most likely point being $\Gamma_0 \equiv \overline{\Gamma}(0|\Gamma, t)$, the change in reservoir entropy is

$$
\begin{aligned}
& S_r(\Gamma, t) - S_r(\Gamma_0, 0) \\
&= S_{r,st}(\Gamma, t) - S_{r,st}(\Gamma_0, 0) - \int_0^t dt' \dot{S}^0_{r,st}(\overline{\Gamma}(t'|\Gamma, t), t') \\
&= \int_0^t dt' \left\{ \dot{S}^-_{r,st}(\overline{\Gamma}(t'|\Gamma, t), t') - \dot{S}^0_{r,st}(\overline{\Gamma}(t'|\Gamma, t), t') \right\} \\
&= \int_0^t dt' \dot{\overline{\mathbf{R}}}^-_p(\overline{\Gamma}(t'|\Gamma, t), t') \cdot \nabla_p S_{r,st}(\overline{\Gamma}(t'|\Gamma, t), t').
\end{aligned}
\tag{6.20}
$$

On both sides of the first equality, the previously neglected constant static entropy at the start of the trajectory has been reinserted. Because the first time derivative is discontinuous on the most likely trajectory, it is necessary to specify that it is the backward derivative that appears here. One can see that the integrand is entirely the reservoir-induced change in reservoir entropy. That the reservoir force $\mathbf{R}_p$ couples to $\nabla_p S_{r,st}$ defines the static reservoir entropy as the instantaneous reservoir entropy that changes directly from exchanges between the sub-system and the reservoir.

One can also give an explicit expression for the change in reservoir entropy on the most likely backward trajectory, $\Delta_t < 0$. Since the integral for the dynamic entropy is precisely over this most likely backward trajectory, by the fundamental theorem of calculus one has

$$
\begin{aligned}
& S_r(\overline{\Gamma}(t + \Delta_t|\Gamma, t), t) - S_r(\Gamma, t) \\
&= \Delta_t \dot{S}^0_{r,st}(\Gamma, t) + \overline{\mathbf{R}}_p(\Gamma, t, \Delta_t) \cdot \nabla_p S_{r,st}(\Gamma, t) - \Delta_t \dot{S}^0_{r,st}(\Gamma, t) \\
&= \overline{\mathbf{R}}_p(\Gamma, t, \Delta_t) \cdot \nabla_p S_{r,st}(\Gamma, t), \quad \Delta_t < 0.
\end{aligned}
\tag{6.21}
$$

Notice that the adiabatic contributions cancel. Again given the definition of $\mathbf{R}_p$ as the reservoir forces, and the definition of the static entropy as the reservoir entropy instantaneously affected by exchange with the sub-system, this result is physically sensible.

Formally, the rate of change of the reservoir entropy on a trajectory is

$$
\begin{aligned}
\frac{dS_r(\Gamma, t)}{dt} &= \frac{\partial S_r(\Gamma, t)}{\partial t} + \dot{\Gamma}^0 \cdot \nabla S_r(\Gamma, t) + \frac{1}{\Delta_t} \mathbf{R}_p \cdot \nabla S_r(\Gamma, t) \\
&= \dot{S}^0_r(\Gamma, t) + \frac{1}{\Delta_t} \mathbf{R}_p \cdot \nabla_p S_r(\Gamma, t).
\end{aligned}
\tag{6.22}
$$

The first term on the right-hand side of the final equality is the adiabatic rate of change of the reservoir entropy. Evaluating the total rate of change on the most likely backward trajectory and using the above result the latter is

$$
\begin{aligned}
\dot{S}^0_r(\Gamma, t) &= \frac{1}{\Delta_t} \overline{\mathbf{R}}_p(\Gamma, t, \Delta_t) \cdot \nabla_p S_{r,st}(\Gamma, t) - \frac{1}{\Delta_t} \overline{\mathbf{R}}_p(\Gamma, t, \Delta_t) \cdot \nabla_p S_r(\Gamma, t) \\
&= \frac{-1}{\Delta_t} \overline{\mathbf{R}}_p(\Gamma, t, \Delta_t) \cdot \nabla_p S_{r,dyn}(\Gamma, t), \quad \Delta_t < 0.
\end{aligned}
\tag{6.23}
$$

The dynamic part of the reservoir entropy depends upon the most likely backward trajectory. In the next section, section 6.2, the stochastic dissipative equations that give this are derived. In section 6.3 the dynamic part of the entropy is reformulated in a more computationally efficient form, which includes utilizing adiabatic trajectories, and the connection of this to the Green–Kubo relations is elucidated.

*Rate of change of total entropy*

The partition function normalizes the probability distribution, equation (6.4),

$$Z(t) = \frac{1}{N!h^{3N}} \int \mathrm{d}\mathbf{\Gamma} e^{S_\mathrm{r}(\mathbf{\Gamma},t)/k_\mathrm{B}}. \tag{6.24}$$

Its logarithm gives the total unconstrained entropy of the total system,

$$S_\mathrm{tot}(t) = k_\mathrm{B} \ln Z(t). \tag{6.25}$$

The rate of change of the total entropy is

$$\begin{aligned} \dot{S}_\mathrm{tot}(t) &= \frac{k_\mathrm{B}\dot{Z}(t)}{Z(t)} \\ &= \frac{k_\mathrm{B}}{Z(t)} \frac{1}{N!h^{3N}} \int \mathrm{d}\mathbf{\Gamma} e^{S_\mathrm{r}(\mathbf{\Gamma},t)/k_\mathrm{B}} \frac{\partial S_\mathrm{r}(\mathbf{\Gamma}, t)}{k_\mathrm{B}\partial t} \\ &= \left\langle \frac{\partial S_\mathrm{r}(\mathbf{\Gamma}, t)}{\partial t} \right\rangle. \end{aligned} \tag{6.26}$$

Let $\mathbf{\Gamma}_2 = \mathbf{\Gamma}^0(t_2|\mathbf{\Gamma}_1, t_1) = \mathbf{\Gamma}_1 + t_{21}\dot{\mathbf{\Gamma}}_1^0$ be the adiabatic evolution of $\mathbf{\Gamma}_1$ over an infinitesimal time step. From the incompressibility of phase space under Hamilton's equations of motion one has

$$\begin{aligned} S_\mathrm{tot}(t_2) &= k_\mathrm{B} \ln \frac{1}{N!h^{3N}} \int \mathrm{d}\mathbf{\Gamma}_2 e^{S_\mathrm{r}(\mathbf{\Gamma}_2,t_2)/k_\mathrm{B}} \\ &= k_\mathrm{B} \ln \frac{1}{N!h^{3N}} \int \mathrm{d}\mathbf{\Gamma}_1 e^{\left\{S_\mathrm{r}(\mathbf{\Gamma}_1,t_1)+t_{21}\dot{S}_\mathrm{r}^0(\mathbf{\Gamma}_1,t_1)\right\}/k_\mathrm{B}} \\ &= k_\mathrm{B} \ln Z(t_1) \int \mathrm{d}\mathbf{\Gamma}_1\, \wp(\mathbf{\Gamma}_1, t_1)\left\{1 + t_{21}\dot{S}_\mathrm{r}^0(\mathbf{\Gamma}_1, t_1)/k_\mathrm{B}\right\} \\ &= S_\mathrm{tot}(t_1) + t_{21}\left\langle \dot{S}_\mathrm{r}^0(\mathbf{\Gamma}, t_1)\right\rangle. \end{aligned} \tag{6.27}$$

Hence the rate of change of the total entropy is also given by

$$\dot{S}_\mathrm{tot}(t) = \left\langle \dot{S}_\mathrm{r}^0(\mathbf{\Gamma}, t)\right\rangle. \tag{6.28}$$

Subtracting these two expressions for $\dot{S}_\mathrm{tot}(t)$ yields

$$\langle \dot{\mathbf{\Gamma}}^0 \cdot \nabla S_\mathrm{r}(\mathbf{\Gamma}, t)\rangle = 0. \tag{6.29}$$

Because these are the averages of an extensive variable, one can expect fluctuations to be relatively negligible. In this case these should hold locally, on points in phase space that are likely to occur. Hence to a good approximation,

$$\dot{S}_{\mathrm{r}}^{0}(\mathbf{\Gamma}, t) \approx \frac{\partial S_{\mathrm{r}}(\mathbf{\Gamma}, t)}{\partial t} \approx \dot{S}_{\mathrm{tot}}(t), \tag{6.30}$$

and

$$\dot{\mathbf{\Gamma}}^{0} \cdot \nabla S_{\mathrm{r}}(\mathbf{\Gamma}, t) \approx 0. \tag{6.31}$$

## 6.2 Stochastic, dissipative equations of motion

### 6.2.1 Foundations for the transition probability in the time varying case

Following Attard (2012, section 8.3.1), the set theoretic formulation of weight, entropy, and probability is now given in the case of time-varying systems. This extends the equilibrium formalities given in sections 1.1.2, 3.1.3, and 3.1.4. Although the transition analysis of section 3.1.4 was applied to the non-equilibrium system of steady heat flow, in a sense steady state systems are a special non-equilibrium system in which the sub-system does not change macroscopically with time. The analysis given here removes that restriction.

Suppose that the weights depend on time. These are microstate, $\omega(i, t)$, macrostate, $\omega(\alpha, t) = \sum_{i\in\alpha}\omega(i, t)$, and total $W(t) = \sum_{i}\omega(i, t) = \sum_{\alpha}\omega(\alpha, t)$. This is sketched in figure 6.2. The entropies are the logarithm of these: microstate, $S(i, t) = k_{\mathrm{B}} \ln \omega(i, t)$, macrostate, $S(\alpha, t) = k_{\mathrm{B}} \ln \omega(\alpha, t)$, and total $S(t) = k_{\mathrm{B}} \ln W(t)$. Again as usual, the probabilities are proportional to the weight, and hence to the exponential of the entropy, $\wp(i, t) = \omega(i, t)/W(t) = \exp[S(i, t)/k_{\mathrm{B}}]/W(t)$ and $\wp(\alpha, t) = \omega(\alpha, t)/W(t) = \exp[S(\alpha, t)/k_{\mathrm{B}}]/W(t)$.

The macrostate transition, $\{\alpha, t\} \rightarrow \{\gamma, t'\}$, has weight $\omega(\gamma, t'; \alpha, t)$. There is similarly a weight for microstate transitions, $\omega(j, t'; k, t)$. From statistical symmetry the unconditional weight is unchanged by swapping the order of the arguments, $\omega(\gamma, t'; \alpha, t) = \omega(\alpha, t; \gamma, t')$. This is the weight attached to the system being in the macrostate $\alpha$ at time $t$ *and* in the macrostate $\gamma$ at time $t'$.

For an equilibrium system two conservation laws for weight were established. The first was for the weight of a system being jointly in two macrostates, equation (3.9),

**Figure 6.2.** Sketch of the evolution of a collective of macrostates in a time varying system. The weight of a macrostate is proportional to the size of its cell.

$\sum_\gamma \omega(\alpha, \gamma) = \omega(\alpha)$. The second was for the system making a transition between two macrostates, equation (3.17), $\sum_\gamma \omega(\alpha, \gamma|\tau) = \omega(\alpha)$, and $\sum_\alpha \omega(\alpha, \gamma|\tau) = \omega(\gamma)$. These may also be called reduction conditions (of the first type), although they are slightly different to the reduction condition (of the second type) given in equation (3.38).

Fundamentally, the basis of both of these conservation laws is the fact that the macrostates of a given collective are disjoint and form a complete set. One therefore expects similar conservation conditions to hold in the present time varying case. One has to modify the conservation condition on the transition weight to account for the fact that since the total weight changes with time, $W(t)$, it cannot be conserved in a transition. For this reason perhaps it is better to call this a reduction condition of the first type, or else just a sum rule for the transition weight.

This variation in the total weight can be accounted for by including a time-dependent scale factor in the sum rule,

$$\sum_\gamma \omega(\gamma, t'; \alpha, t) = f(t', t)\omega(\alpha, t). \tag{6.32}$$

With this the total weight of the transition is

$$\sum_{\alpha,\gamma} \omega(\gamma, t'; \alpha, t) = f(t', t)W(t). \tag{6.33}$$

For this to be symmetric in both times the scale factor must be of the form

$$f(t', t) = W(t')^a W(t)^{a-1}. \tag{6.34}$$

This must give the equilibrium result in the time-independent case, namely $f(t', t) = 1$, which means that $a = 1/2$, and that $f(t', t) = \sqrt{W(t')/W(t)}$. That is, the sum rule for the transition weight in a time-varying system is

$$\sum_\gamma \omega(\gamma, t'; \alpha, t) = \sqrt{\frac{W(t')}{W(t)}}\,\omega(\alpha, t), \tag{6.35}$$

and the total weight of the transition is

$$\sum_{\alpha\gamma} \omega(\gamma, t'; \alpha, t) = \sqrt{W(t')W(t)}. \tag{6.36}$$

This is just the geometric mean of the total weights at the two termini of the transition, which has a pleasing symmetry.

In the time independent case of equation (3.17), the weight was conserved as it was distributed amongst the target states in a transition. In the present time dependent case, the weight of the initial state $\alpha$ is scaled and distributed amongst all the target states $\gamma$. This scaling is necessary whenever the weights of the states and the total weight change with time.

It is clear that a similar scaling has to occur for the sum over initial states,

$$\sum_{\alpha} \omega(\gamma, t'; \alpha, t) = \sqrt{\frac{W(t)}{W(t')}}\, \omega(\gamma, t'). \tag{6.37}$$

This says that the weight of the target macrostate $\gamma$ must have come from somewhere, and been scaled. The laws of probability require these two sum rules in the time varying case for its internal consistency.

The unconditional transition probability has the total transition weight as the normalizing factor,

$$\wp(\gamma, t'; \alpha, t) = \frac{\omega(\gamma, t'; \alpha, t)}{\sqrt{W(t')W(t)}}. \tag{6.38}$$

As usual, the conditional transition probability is

$$\begin{aligned}\wp(\gamma, t'|\alpha, t) &= \frac{\wp(\gamma, t'; \alpha, t)}{\wp(\alpha, t)} \\ &= \frac{\omega(\gamma, t'; \alpha, t)}{\omega(\alpha, t)} \sqrt{\frac{W(t)}{W(t')}} \\ &= \frac{\omega(\gamma, t'; \alpha, t)}{\sum_{\gamma} \omega(\gamma, t'; \alpha, t)}.\end{aligned} \tag{6.39}$$

The first equality is just Bayes' theorem, $\wp(\gamma, t'; \alpha, t) = \wp(\gamma, t'|\alpha, t)\wp(\alpha, t)$.

As in the equilibrium case, the second entropy is defined to be essentially the logarithm of the transition weight, $S^{(2)}(\gamma, t'; \alpha, t) = k_B \ln \omega(\gamma, t'; \alpha, t)$. In terms of it the transition probability is

$$\wp(\gamma, t'; \alpha, t) = \frac{e^{S^{(2)}(\gamma, t'; \alpha, t)/k_B}}{\sqrt{W(t')W(t)}}. \tag{6.40}$$

The conditional transition probability follows as

$$\wp(\gamma, t'|\alpha, t) = \sqrt{\frac{W(t)}{W(t')}}\, e^{[S^{(2)}(\gamma, t'; \alpha, t) - S(\alpha, t)]/k_B}. \tag{6.41}$$

This is evidently normalized to unity when summed over $\gamma$. The conditional probability refers to a forward transition if $t' > t$, which is to say that tells where the system will go to in the future given the current state. If $t' < t$ it refers to a backward transition, which tells where the system came from in the past to get to the current state.

The important equilibrium reduction condition of the second type, equation (3.38), also has to be modified slightly for the time dependent case. As previously, define the conditionally most likely target state $\bar{\gamma} \equiv \bar{\gamma}(t'|\alpha, t)$ as the state that maximizes the second entropy,

$$\left. \frac{\partial S^{(2)}(\gamma, t'; \alpha, t)}{\partial \gamma} \right|_{\gamma=\bar{\gamma}} = 0. \tag{6.42}$$

The reduction condition of the second type says that the maximal value of the conditional second entropy equals the entropy of the initial state $\alpha$. That is,

$$\begin{aligned} S(\alpha, t) &= k_B \ln \omega(\alpha, t) \\ &= k_B \ln \left[ \sqrt{\frac{W(t)}{W(t')}} \sum_{\gamma} \omega(\gamma, t'; \alpha, t) \right] \\ &\approx k_B \ln \left[ \sqrt{\frac{W(t)}{W(t')}} \omega(\bar{\gamma}, t'; \alpha, t) \right] \\ &= S^{(2)}(\bar{\gamma}, t'; \alpha, t) + \frac{1}{2}[S(t) - S(t')]. \end{aligned} \tag{6.43}$$

Since one expects the probability distribution to be sharply peaked in any macroscopic physical system, one can approximate the logarithm of a sum over states by the logarithm of the largest term in the sum, as in the third equality. This is equivalent to assuming that the fluctuations about the most likely state are relatively negligible. Here $S(t) = k_B \ln W(t)$ is the total unconstrained entropy. This second type of reduction condition can be rewritten as

$$S^{(2)}(\bar{\gamma}, t'; \alpha, t) = S(\alpha, t) + \frac{1}{2}[S(t') - S(t)]. \tag{6.44}$$

It is obvious that the reduction condition holds for both forward and backward transitions, and so one also has

$$S^{(2)}(\gamma, t'; \bar{\alpha}, t) = S(\gamma, t') + \frac{1}{2}[S(t) - S(t')], \tag{6.45}$$

where $\bar{\alpha} \equiv \bar{\alpha}(t|\gamma, t')$. Of course, the order of the arguments in the second entropy is irrelevant since the times at which each state occurs are stated explicitly. In the time independent case, equation (3.38), the maximal value of the second entropy equals the first entropy alone. In the present time varying non-equilibrium system, the maximal value of the second entropy equals the first entropy plus half the difference in the total entropy of the two states.

### 6.2.2 Second entropy and the transition probability

In previous work (Attard 2012, sections 8.2.3 and 8.3.2; Attard 2014), I derived the stochastic, dissipative equations of motion for a non-equilibrium system from a fluctuation form for the second entropy. The analysis that is now given improves upon this earlier work in three respects. First, the adiabatic contributions were originally treated as fluctuating quantities within the mixed parity second entropy formulation, whereas here the pure parity slave formulation used for Brownian motion, equation (4.21), and for a phase space trajectory in the canonical equilibrium system, equation (5.50), is invoked. Second, the original analysis was based on fluctuations about the most likely point in phase space, $\overline{\mathbf{\Gamma}}(t)$, aspects of which may be questioned. Fluctuations play a much smaller role in the present analysis, and a more direct physical interpretation of various quantities is given. And third, the present derivation is about one quarter the length of the original, which makes for greater simplicity and transparency, and which leaves less room for error. As the quote from Einstein given at the head of the chapter has been paraphrased: a theory should be as simple as possible, but no simpler.

We shall take the equations of motion in phase space over an infinitesimal time step to be of the form

$$\begin{aligned}\mathbf{\Gamma}_q(t+\Delta_t) &= \mathbf{\Gamma}_q(t) + \Delta_t\dot{\mathbf{\Gamma}}^0_q(\mathbf{\Gamma}, t),\\ \mathbf{\Gamma}_p(t+\Delta_t) &= \mathbf{\Gamma}_p(t) + \Delta_t\dot{\mathbf{\Gamma}}^0_p(\mathbf{\Gamma}, t) + \mathbf{R}_p(\mathbf{\Gamma}, t, \Delta_t).\end{aligned} \tag{6.46}$$

As usual, the superscript 0 represents the adiabatic velocity,

$$\dot{\mathbf{\Gamma}}^0_q(\mathbf{\Gamma}, t) = \frac{\partial\mathcal{H}(\mathbf{\Gamma}, t)}{\partial\mathbf{\Gamma}_p}, \text{ and } \dot{\mathbf{\Gamma}}^0_p(\mathbf{\Gamma}, t) = \frac{-\partial\mathcal{H}(\mathbf{\Gamma}, t)}{\partial\mathbf{\Gamma}_q}. \tag{6.47}$$

The force due to the reservoir $\mathbf{R}_p$ only effects the momenta directly. In this formulation the position coordinates are treated as slaves to the momenta, as was done in the case of the Langevin equation for Brownian motion, equation (4.21), and in the case of phase space for the canonical equilibrium system, equation (5.50). The reservoir force can be divided into a deterministic, dissipative part, $\overline{\mathbf{R}}_p$, and a stochastic part of zero mean $\tilde{\mathbf{R}}_p$,

$$\mathbf{R}_p = \overline{\mathbf{R}}_p + \tilde{\mathbf{R}}_p. \tag{6.48}$$

The explicit form for the former will shortly be determined from the second entropy form of the transition probability.

We assume that the probability distribution for the fluctuations in the present non-equilibrium system has the same time symmetry as for an equilibrium system, namely that the progression and regression of a fluctuation are equally likely. This means that the stochastic reservoir force is Gaussian distributed, with variance proportional to the magnitude of the time step,

$$\wp(\tilde{\mathbf{R}}_p) = \frac{e^{-\Lambda^{-1}:\tilde{\mathbf{R}}_p\tilde{\mathbf{R}}_p/2k_{\mathrm{B}}|\Delta_t|}}{\sqrt{\mathrm{Det}\{2\pi k_{\mathrm{B}}|\Delta_t|\Lambda\}}}. \tag{6.49}$$

Evidently the random force has zero mean, $\langle\tilde{\mathbf{R}}\rangle = 0$, and its variance is

$$\langle \tilde{\mathbf{R}}(t)\, \tilde{\mathbf{R}}(t)\rangle = |\Delta_t| k_{\rm B} \Lambda. \tag{6.50}$$

Random forces at different time steps are uncorrelated. It would be possible to have a time-dependent variance, $\Lambda(t)$, but this is not pursued here. Typically, and most simply, one may take the fluctuation matrix to be diagonal, $\Lambda = \lambda \mathrm{I}_{\rm pp}$. In some cases (e.g. boundary driven flow), there may be value in making the parameter position dependent.

In summary, the equations of motion for the transition $\mathbf{\Gamma}_1 \stackrel{\Delta_t}{\to} \mathbf{\Gamma}_2$ are

$$\mathbf{\Gamma}_2 = \mathbf{\Gamma}_1 + \Delta_t \dot{\mathbf{\Gamma}}^0 + \overline{\mathbf{R}}_p + \tilde{\mathbf{R}}_p. \tag{6.51}$$

The total reservoir force, $\mathbf{R}(t)$, for the present transition, $\mathbf{\Gamma}_1 \stackrel{\Delta_t}{\to} \mathbf{\Gamma}_2$, is equal and opposite to that for the opposite transition, $\mathbf{\Gamma}_2 \stackrel{-\Delta_t}{\longrightarrow} \mathbf{\Gamma}_1$. The same is true for the adiabatic force. The two individual components of the total reservoir force, the most likely and the stochastic, are not individually equal and opposite their respective counterpart.

The transition probability is just the exponential of the second entropy, equation (6.40),

$$\begin{aligned}\wp(\mathbf{\Gamma}_2, t_2; \mathbf{\Gamma}_1, t_1) &= \frac{e^{S^{(2)}(\mathbf{\Gamma}_2, t_2; \mathbf{\Gamma}_1, t_1)/k_{\rm B}}}{\sqrt{Z(t_2) Z(t_1)}} \\ &= \wp(\mathbf{\Gamma}_2, t_2 | \mathbf{\Gamma}_1, t_1)\, \wp(\mathbf{\Gamma}_1, t_1).\end{aligned} \tag{6.52}$$

The second equality writes this as the conditional probability times the probability of the initial state, which is Bayes' theorem, equation (6.39). The conditional transition probability is just the probability of the random force, $\wp(\mathbf{\Gamma}_2, t_2 | \mathbf{\Gamma}_1, t_1) = \wp(\tilde{\mathbf{R}}_p)$, since $\mathbf{\Gamma}_2$ is randomly distributed about $\overline{\mathbf{\Gamma}}(t_2 | \mathbf{\Gamma}_1, t_1)$. Hence the unconditional probability of the transition may also be written

$$\begin{aligned}\wp(\mathbf{\Gamma}_2, t_2; \mathbf{\Gamma}_1, t_1) &= \wp(\mathbf{\Gamma}_2, t_2 | \mathbf{\Gamma}_1, t_1)\, \wp(\mathbf{\Gamma}_1, t_1) \\ &= \frac{e^{-\Lambda^{-1}:\tilde{\mathbf{R}}_p \tilde{\mathbf{R}}_p / 2 k_{\rm B} |\Delta_t|}}{\sqrt{\mathrm{Det}\{2\pi k_{\rm B} |\Delta_t| \Lambda\}}} \frac{e^{S_{\rm r}(\mathbf{\Gamma}_1, t_1)/k_{\rm B}}}{Z(t_1)}.\end{aligned} \tag{6.53}$$

Writing the exponent in this form is equivalent to the reduction condition of the second type, equation (6.44).

Now we determine the most likely reservoir force. Unlike in chapter 3, we do not further use the reduction condition of the second type, equation (6.44), because this is for discrete states and it does not account for the compressibility of phase space. Instead we invoke the first type of reduction condition, equation (6.37), which involves an integral over the initial phase space transition point. This is equivalent to deriving the Fokker–Planck equation for the evolution of the probability density, as is discussed in section 6.2.4 below.

From the reduction condition of the first type, equation (6.37), one has

$$\begin{aligned}\wp(\Gamma_2, t_2) &= \int \mathrm{d}\Gamma_1 \wp(\Gamma_2, t_2; \Gamma_1, t_1)\\
&= \int \mathrm{d}\overline{\Gamma}_2 \left|\frac{\mathrm{d}\Gamma_1}{\mathrm{d}\overline{\Gamma}_2}\right| \frac{e^{-\Lambda^{-1}:\tilde{\mathbf{R}}_p\tilde{\mathbf{R}}_p/2k_B|\Delta_t|}}{\sqrt{\mathrm{Det}\{2\pi k_B|\Delta_t|\Lambda\}}} \frac{e^{S_r(\Gamma_1,t_1)/k_B}}{Z(t_1)}\\
&= \left|\frac{\mathrm{d}\Gamma_1}{\mathrm{d}\overline{\Gamma}_2}\right| \frac{e^{S_r(\Gamma_2,t_1)/k_B}}{Z(t_1)}\\
&\quad\times \int \mathrm{d}\overline{\Gamma}_2 \frac{e^{-\Lambda^{-1}:\tilde{\mathbf{R}}_p\tilde{\mathbf{R}}_p/2k_B|\Delta_t|}}{\sqrt{\mathrm{Det}\{2\pi k_B|\Delta_t|\Lambda\}}} \left\{1 + \frac{1}{k_B}[\Gamma_1 - \Gamma_2] \cdot \nabla S_r\right.\\
&\quad\left. + \frac{1}{2k_B}[\Gamma_1 - \Gamma_2]^2 : \nabla\nabla S_r + \frac{1}{2k_B^2}([\Gamma_1 - \Gamma_2] \cdot \nabla S_r)^2\right\}\\
&= \frac{e^{S_r(\Gamma_2,t_2)/k_B}}{Z(t_2)} \left\{1 - t_{21} \nabla \cdot \dot{\Gamma}^{\mathrm{det}}\right\} \left\{1 - t_{21} \frac{\partial S_r(\Gamma, t)}{k_B \partial t} + t_{21} \frac{\dot{Z}(t)}{Z(t)}\right\}\\
&\quad\times \left\{1 - \frac{t_{21}}{k_B} \dot{\Gamma}^{\mathrm{det}} \cdot \nabla S_r + \frac{1}{2k_B} k_B |t_{21}| \Lambda : \nabla_p \nabla_p S_r\right.\\
&\quad\left. + \frac{1}{2k_B^2} k_B |t_{21}| \Lambda : [\nabla_p S_r][\nabla_p S_r]\right\}.\end{aligned} \tag{6.54}$$

This uses $\tilde{\mathbf{R}}_p = \Gamma_2 - \overline{\Gamma}_2$, $\overline{\Gamma}_2 \equiv \overline{\Gamma}(t_2|\Gamma_1, t_1) = \Gamma_1 + t_{21}\dot{\Gamma}^{\mathrm{det}}$, $|\mathrm{d}\Gamma_1/\mathrm{d}\overline{\Gamma}_2| = \{1 - t_{21}\nabla \cdot \dot{\Gamma}^{\mathrm{det}}\}$, and $S_r(\Gamma_1, t_1) = S_r(\Gamma_2, t_1) + [\Gamma_1 - \Gamma_2] \cdot \nabla S_r + [\Gamma_1 - \Gamma_2]^2 : \nabla\nabla S_r/2$. Only terms to linear order in $t_{21}$ and in $\Lambda$ have been kept. This obviously places an upper limit on their magnitude in practical computations involving the stochastic, dissipative equations of motion.

Since the first factor on the right-hand side of the final equality is $\wp(\Gamma_2, t_2)$, the product of the three remaining factors must be unity to linear order in the time step. Hence the sum of the terms proportional to the time step must vanish,

$$\begin{aligned}0 &= -t_{21}\dot{S}_r^0(\Gamma, t) + t_{21}\dot{S}_{\mathrm{tot}}(t) - t_{21}k_B\nabla \cdot \dot{\Gamma}^{\mathrm{det}}\\
&\quad - \overline{\mathbf{R}}_p \cdot \nabla S_r + \frac{k_B|t_{21}|}{2}\Lambda : \nabla_p\nabla_p S_r + \frac{|t_{21}|}{2}\Lambda : [\nabla_p S_r][\nabla_p S_r],\end{aligned} \tag{6.55}$$

where $S_r \equiv S_r(\Gamma, t)$. Recall that the total unconstrained entropy is $S_{\mathrm{tot}}(t) = k_B \ln Z(t)$, and that its rate of change is $\dot{S}_{\mathrm{tot}}(t) = k_B\dot{Z}(t)/Z(t)$. This is equivalent to the Fokker–Planck equation, as is discussed in section 6.2.4 below.

The deterministic velocity is the sum of the adiabatic and the most likely reservoir-induced velocity,

$$\dot{\Gamma}^{\mathrm{det}}(\hat{t}_{21}, \Gamma, t) = \dot{\Gamma}^0(\Gamma, t) + \frac{1}{t_{21}}\overline{\mathbf{R}}_p(t_{21}, \Gamma, t). \tag{6.56}$$

Here and below, $\hat{t}_{21} \equiv \mathrm{sign}\, t_{21}$. The adiabatic compressibility vanishes, $\nabla \cdot \dot{\Gamma}^0 = 0$. Since the above equation contains both reversible, $\propto t_{21}$, and irreversible $\propto |t_{21}|$ terms, it is evident that the most likely reservoir force should be similarly decomposed,

$$\overline{\mathbf{R}}_p(t_{21}, \mathbf{\Gamma}, t) = |t_{21}|\mathbf{a}(\mathbf{\Gamma}, t) + t_{21}\mathbf{b}(\mathbf{\Gamma}, t). \tag{6.57}$$

This only has momentum components. According to the pure parity analysis, equations (3.86), (4.19) or (4.39), only the irreversible term survives, so that $\mathbf{b}(\mathbf{\Gamma}, t) = 0$. Strictly speaking, however, position coordinates are also present, which makes this a mixed parity case, and so the reversible term may be non-zero (Attard 2012, section 8.3.4). Using this and equating the irreversible terms to zero in the above yields

$$\begin{aligned} 0 = &- k_B\nabla_p \cdot \mathbf{a}(\mathbf{\Gamma}, t) - \mathbf{a}(\mathbf{\Gamma}, t) \cdot \nabla_p S_r(\mathbf{\Gamma}, t) \\ &+ \frac{k_B}{2}\Lambda : \nabla_p\nabla_p S_r(\mathbf{\Gamma}, t) + \frac{1}{2}\Lambda : \left[\nabla_p S_r(\mathbf{\Gamma}, t)\right]\left[\nabla_p S_r(\mathbf{\Gamma}, t)\right]. \end{aligned} \tag{6.58}$$

By inspection, this has the exact solution

$$\begin{aligned} \mathbf{a}(\mathbf{\Gamma}, t) &= \frac{1}{2}\Lambda\nabla_p S_r(\mathbf{\Gamma}, t) \\ &= \frac{1}{2}\Lambda\nabla_p S_{r,st}(\mathbf{\Gamma}, t) + \frac{1}{2}\Lambda\nabla_p S_{r,dyn}(\mathbf{\Gamma}, t). \end{aligned} \tag{6.59}$$

The reversible terms must satisfy

$$\dot{S}_r^0(\mathbf{\Gamma}, t) - \dot{S}_{tot}(t) = -k_B\nabla_p \cdot \mathbf{b}(\mathbf{\Gamma}, t) - \mathbf{b}(\mathbf{\Gamma}, t) \cdot \nabla_p S_r(\mathbf{\Gamma}, t). \tag{6.60}$$

By equation (6.30) (see also equation (6.115) below), the left-hand side is approximately zero, and so one expects $\mathbf{b}(\mathbf{\Gamma}, t)$ to be small or negligible.

The exact solution for $\mathbf{a}$ is difficult to work with, primarily because of the challenging nature of $S_{r,dyn}$. In contrast, $S_{r,st}$ is always relatively trivial to manipulate and to calculate. In the time honored tradition of physics, one simply discards such difficult terms, which gives the zeroth order approximation,

$$\mathbf{a}^{(0)}(\mathbf{\Gamma}, t) = \frac{1}{2}\Lambda\nabla_p S_{r,st}(\mathbf{\Gamma}, t), \quad \mathbf{b}^{(0)}(\mathbf{\Gamma}, t) = 0. \tag{6.61}$$

The first of these is equivalent to $\nabla_p S_{r,dyn}(\mathbf{\Gamma}, t) = 0$. Of course, in so far as the non-equilibrium contributions are expected to be a small perturbation on the instantaneous static structure of the system, $|S_{r,dyn}(\mathbf{\Gamma}, t)| \ll |S_{r,st}(\mathbf{\Gamma}, t)|$, one can expect this zeroth order approximation to be relatively accurate.

To do better than this, define the most likely configuration as the one that maximizes the reservoir entropy,

$$\nabla S_r(\mathbf{\Gamma}, t)|_{\mathbf{\Gamma}=\overline{\mathbf{\Gamma}}(t)} = \mathbf{0}. \tag{6.62}$$

Equivalently (see figure 6.3),

$$\nabla S_{r,st}(\overline{\mathbf{\Gamma}}(t), t) = -\nabla S_{r,dyn}(\overline{\mathbf{\Gamma}}(t), t). \tag{6.63}$$

The abbreviated notation here means that the gradient is evaluated at the most likely configuration. We shall not address issues such as the multiplicity of solutions to this, other than to say that $\overline{\mathbf{\Gamma}}(t)$ is to be interpreted as the zero of the reservoir entropy gradient that is closest to $\mathbf{\Gamma}$ at time $t$. With this, instead of discarding completely the dynamic part of the reservoir entropy, one can approximate it by its value at the most likely point,

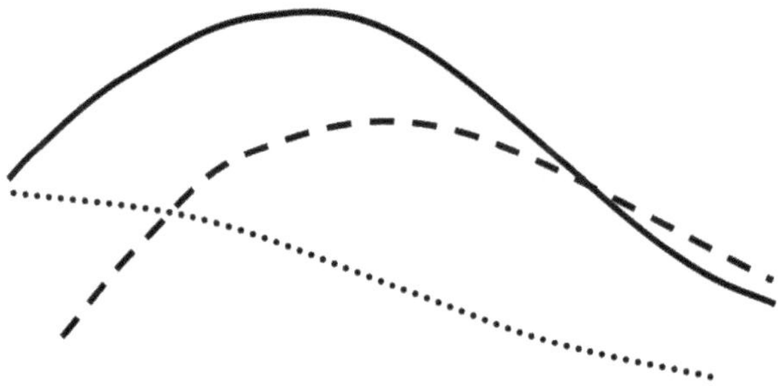

**Figure 6.3.** Sketch of the reservoir entropy (solid curve), and its static (dashed curve) and dynamic (dotted curve) parts in the vicinity of the most likely point in phase space at that time.

$$\nabla_p S_{\mathrm{r,dyn}}(\mathbf{\Gamma}, t) \approx \nabla_p S_{\mathrm{r,dyn}}(\overline{\mathbf{\Gamma}}(t), t) = -\nabla_p S_{\mathrm{r,st}}(\overline{\mathbf{\Gamma}}(t), t). \tag{6.64}$$

This approximation can be expected to be accurate in so far as departures from the most likely configuration are small. The utility of the most likely configuration is that it allows the dynamic entropy to be replaced by the static entropy, which is relatively easy to evaluate. It is not essential for what follows, but it may be reasonable to approximate this gradient at the most likely configuration by the most likely gradient, $\nabla_p S_{\mathrm{r,st}}(\overline{\mathbf{\Gamma}}(t), t) \approx \overline{\mathbf{S}'_{\mathrm{r,st}}}(t)$.

With this, the more refined expression for the irreversible contribution to the reservoir force is

$$\mathbf{a}(\mathbf{\Gamma}, t) = \frac{1}{2}\Lambda\nabla_p S_{\mathrm{r,st}}(\mathbf{\Gamma}, t) - \frac{1}{2}\Lambda\nabla_p S_{\mathrm{r,st}}(\overline{\mathbf{\Gamma}}(t), t). \tag{6.65}$$

The most likely configuration $\overline{\mathbf{\Gamma}}(t)$ evolves in time. This evolution can be expected to be given by the dissipative equations of motion, $\overline{\mathbf{\Gamma}}(t'|\overline{\mathbf{\Gamma}}(t), t) = \overline{\mathbf{\Gamma}}(t')$. Since this is assumed to be a single trajectory, this evolution must be reversible and given by the coefficient $\mathbf{b}(\mathbf{\Gamma}, t)$. Hence the more refined expression for the reversible contribution to the reservoir force is to set it equal to the zeroth approximation $\mathbf{a}^{(0)}$ evaluated at the most likely configuration,

$$\mathbf{b}(\mathbf{\Gamma}, t) = \mathbf{a}^{(0)}(\overline{\mathbf{\Gamma}}(t), t) = \frac{1}{2}\Lambda\nabla_p S_{\mathrm{r,st}}(\overline{\mathbf{\Gamma}}(t), t). \tag{6.66}$$

In the present approximation this is equivalent to $\mathbf{b}(\mathbf{\Gamma}, t) \approx -\Lambda\nabla_p S_{\mathrm{r,dyn}}(\mathbf{\Gamma}, t)/2$, which point is discussed in connection with equation (6.72) below. Since the dynamic part of the entropy is a small perturbation, this is consistent with the above observation that $\mathbf{b}(\mathbf{\Gamma}, t) \approx \mathbf{0}$.

With these, the most likely reservoir force is

$$\begin{aligned}\overline{\mathbf{R}}_p &= \frac{|\Delta_t|}{2}\Lambda\left[\nabla_p S_{\mathrm{r,st}}(\mathbf{\Gamma}, t) - \nabla_p S_{\mathrm{r,st}}(\overline{\mathbf{\Gamma}}(t), t)\right] + \frac{\Delta_t}{2}\Lambda\nabla_p S_{\mathrm{r,st}}(\overline{\mathbf{\Gamma}}(t), t) \\ &= \begin{cases} \dfrac{|\Delta_t|}{2}\Lambda\nabla_p S_{\mathrm{r,st}}(\mathbf{\Gamma}, t), & \Delta_t > 0, \\ \dfrac{|\Delta_t|}{2}\Lambda\nabla_p S_{\mathrm{r,st}}(\mathbf{\Gamma}, t) - |\Delta_t|\Lambda\nabla_p S_{\mathrm{r,st}}(\overline{\mathbf{\Gamma}}(t), t), & \Delta_t < 0. \end{cases}\end{aligned} \tag{6.67}$$

Notice that on a forward trajectory, the most likely configuration is not required. Further, one can see by inspection that on the most likely configuration, $\mathbf{\Gamma} = \overline{\mathbf{\Gamma}}(t)$, the most likely reservoir force is purely reversible, $\overline{\mathbf{R}}_p(\Delta_t, \overline{\mathbf{\Gamma}}(t), t) = \Delta_t \Lambda \nabla_p S_{\mathrm{r,st}}(\overline{\mathbf{\Gamma}}(t), t)/2$.

The extra term, $(\Delta_t - |\Delta_t|)\nabla_p S_{\mathrm{r,st}}(\overline{\mathbf{\Gamma}}(t), t)/2$, compared to the zeroth order approximation, (see equation (6.74 below), is a constant in phase space. Hence it can play no role in determining the relative probability of points in phase space. See equation (6.173) below for a discussion of this in the context of a detailed calculation of driven Brownian motion.

One can look at the consequences of the full result and check a number of consistency requirements. On the most likely backward trajectory, the change in reservoir entropy is given exactly by the reservoir force times the gradient in the static part of the reservoir entropy,

$$\begin{aligned}\Delta S_{\mathrm{r}} &= \overline{\mathbf{R}}_p(-|\Delta_t|, \mathbf{\Gamma}, t) \cdot \nabla_p S_{\mathrm{r,st}}(\mathbf{\Gamma}, t)\\ &= \left(\frac{|\Delta_t|}{2}\Lambda \nabla_p S_{\mathrm{r,st}}(\mathbf{\Gamma}, t) - |\Delta_t|\Lambda \nabla_p S_{\mathrm{r,st}}(\overline{\mathbf{\Gamma}}(t), t)\right) \cdot \nabla_p S_{\mathrm{r,st}}(\mathbf{\Gamma}, t).\end{aligned} \tag{6.68}$$

But formally one also has the Taylor expansion

$$\Delta S_{\mathrm{r}} = -|\Delta_t|\dot{S}_{\mathrm{r}}^0(\mathbf{\Gamma}, t) + \overline{\mathbf{R}}_p(-|\Delta_t|, \mathbf{\Gamma}, t) \cdot \nabla_p S_{\mathrm{r}}(\mathbf{\Gamma}, t). \tag{6.69}$$

Subtracting these and rearranging yields

$$\begin{aligned}\dot{S}_{\mathrm{r}}^0(\mathbf{\Gamma}, t) &= \frac{1}{|\Delta_t|}\overline{\mathbf{R}}_p(-|\Delta_t|, \mathbf{\Gamma}, t) \cdot \nabla_p S_{\mathrm{r,dyn}}(\mathbf{\Gamma}, t)\\ &= \left(\frac{1}{2}\Lambda \nabla_p S_{\mathrm{r,st}}(\mathbf{\Gamma}, t) - \Lambda \nabla_p S_{\mathrm{r,st}}(\overline{\mathbf{\Gamma}}(t), t)\right) \cdot \nabla_p S_{\mathrm{r,dyn}}(\mathbf{\Gamma}, t)\\ &\approx \frac{-1}{2}\Lambda : \left[\nabla_p S_{\mathrm{r,st}}(\overline{\mathbf{\Gamma}}(t), t)\right]\left[\nabla_p S_{\mathrm{r,dyn}}(\mathbf{\Gamma}, t)\right]\\ &\approx \frac{1}{2}\Lambda : \left[\nabla_p S_{\mathrm{r,dyn}}(\mathbf{\Gamma}, t)\right]\left[\nabla_p S_{\mathrm{r,dyn}}(\mathbf{\Gamma}, t)\right]\\ &\approx \frac{1}{2}\Lambda : \left[\nabla_p S_{\mathrm{r,st}}(\overline{\mathbf{\Gamma}}(t), t)\right]\left[\nabla_p S_{\mathrm{r,st}}(\overline{\mathbf{\Gamma}}(t), t)\right].\end{aligned} \tag{6.70}$$

The third equality approximates the gradient of the static part of the reservoir entropy at an arbitrary configuration by its value at the most likely configuration, $\nabla_p S_{\mathrm{r,st}}(\mathbf{\Gamma}, t) \approx \nabla_p S_{\mathrm{r,st}}(\overline{\mathbf{\Gamma}}(t), t)$, which is reasonable since one expects fluctuations to be small. The next two approximations invoke the above result for the dynamic part of the entropy, $\nabla_p S_{\mathrm{r,dyn}}(\mathbf{\Gamma}, t) \approx -\nabla_p S_{\mathrm{r,st}}(\overline{\mathbf{\Gamma}}(t), t)$.

Now equation (6.30) shows that the adiabatic rate of entropy production, the left-hand side, can be approximated by the time-dependent constant in phase space, namely the rate of total entropy production, $\dot{S}_{\mathrm{r}}^0(\mathbf{\Gamma}, t) \approx \dot{S}_{\mathrm{tot}}(t)$. (See also equation (6.115) below.) In view of the final equality, also a time-dependent constant in phase space, since $\Lambda$ is positive definite, one concludes that the rate of total entropy production is positive, as it must be. One sees therefore a certain consistency in the above expression for the most likely reservoir force.

Equation (6.60) gave the exact equation that the reversible coefficient had to satisfy. Inserting the present expression for $\mathbf{b}(\mathbf{\Gamma}, t)$ into this yields

$$\begin{aligned} \dot{S}_{\mathrm{r}}^{0}(\mathbf{\Gamma}, t) - \dot{S}_{\mathrm{tot}}(t) &= -k_{\mathrm{B}}\nabla_{p} \cdot \mathbf{b}(\mathbf{\Gamma}, t) - \mathbf{b}(\mathbf{\Gamma}, t) \cdot \nabla_{p} S_{\mathrm{r}}(\mathbf{\Gamma}, t) \\ &= \frac{-k_{\mathrm{B}}}{2}\nabla_{p} \cdot \left[\Lambda \nabla_{p} S_{\mathrm{r,st}}(\overline{\mathbf{\Gamma}}(t), t)\right] \\ &\quad - \frac{1}{2}\Lambda \nabla_{p} S_{\mathrm{r,st}}(\overline{\mathbf{\Gamma}}(t), t) \cdot \left[\nabla_{p} S_{\mathrm{r,st}}(\mathbf{\Gamma}, t) + \nabla_{p} S_{\mathrm{r,dyn}}(\mathbf{\Gamma}, t)\right] \\ &\approx \frac{-1}{2}\Lambda \nabla_{p} S_{\mathrm{r,st}}(\overline{\mathbf{\Gamma}}(t), t) \cdot \left[\nabla_{p} S_{\mathrm{r,st}}(\mathbf{\Gamma}, t) - \nabla_{p} S_{\mathrm{r,st}}(\overline{\mathbf{\Gamma}}(t), t)\right] \\ &\approx 0. \end{aligned} \tag{6.71}$$

The final two equalities invoke the most likely gradients, $\nabla_{p} S_{\mathrm{r,dyn}}(\mathbf{\Gamma}, t) \approx \nabla_{p} S_{\mathrm{r,dyn}}(\overline{\mathbf{\Gamma}}(t), t) = -\nabla_{p} S_{\mathrm{r,st}}(\overline{\mathbf{\Gamma}}(t), t)$, and $\nabla_{p} S_{\mathrm{r,st}}(\mathbf{\Gamma}, t) \approx \nabla_{p} S_{\mathrm{r,st}}(\overline{\mathbf{\Gamma}}(t), t)$. One sees again that this is consistent with equation (6.30), which is also based upon the concept that fluctuations are relatively negligible in the thermodynamic limit.

The above results were based on the approximation that the reversible coefficient was given by $\mathbf{b}(\mathbf{\Gamma}, t) = \Lambda \nabla_{p} S_{\mathrm{r,st}}(\overline{\mathbf{\Gamma}}(t), t)/2 \approx -\Lambda \nabla_{p} S_{\mathrm{r,dyn}}(\mathbf{\Gamma}, t)/2$. One might speculate that the final approximation is in fact the exact result. Combining this with the exact result for the irreversible coefficient, $\mathbf{a}(\mathbf{\Gamma}, t) = \Lambda \nabla_{p} S_{\mathrm{r}}(\mathbf{\Gamma}, t)/2$ this would mean the reservoir force is exactly given by

$$\begin{aligned} \overline{\mathbf{R}}_{p} &= \frac{|\Delta_{t}|}{2}\Lambda \nabla_{p} S_{\mathrm{r}}(\mathbf{\Gamma}, t) - \frac{\Delta_{t}}{2}\Lambda \nabla_{p} S_{\mathrm{r,dyn}}(\mathbf{\Gamma}, t) \\ &= \begin{cases} \dfrac{|\Delta_{t}|}{2}\Lambda \nabla_{p} S_{\mathrm{r,st}}(\mathbf{\Gamma}, t), & \Delta_{t} > 0, \\ \dfrac{|\Delta_{t}|}{2}\Lambda \nabla_{p} S_{\mathrm{r,st}}(\mathbf{\Gamma}, t) + |\Delta_{t}|\Lambda \nabla_{p} S_{\mathrm{r,dyn}}(\mathbf{\Gamma}, t), & \Delta_{t} < 0. \end{cases} \end{aligned} \tag{6.72}$$

This has a certain beauty to it, which may signify truth. Continuing with the hypothesis, the exact condition equation (6.60) becomes

$$\begin{aligned} &\dot{S}_{\mathrm{r}}^{0}(\mathbf{\Gamma}, t) - \dot{S}_{\mathrm{tot}}(t) \\ &= \frac{k_{\mathrm{B}}}{2}\Lambda : \nabla_{p}\nabla_{p} S_{\mathrm{r,dyn}}(\mathbf{\Gamma}, t) + \frac{1}{2}\Lambda : [\nabla_{p} S_{\mathrm{r,dyn}}(\mathbf{\Gamma}, t)]\,[\nabla_{p} S_{\mathrm{r}}(\mathbf{\Gamma}, t)]. \end{aligned} \tag{6.73}$$

The left-hand side of this equation is approximately zero, equation (6.30) (see also equation (6.115) below).

In any case, using either the approximate result or the putative exact result, the dynamic entropy gradient is a second order contribution that can be neglected. As such, the most likely reservoir force to leading order is

$$\overline{\mathbf{R}}_{p} \approx \frac{|\Delta_{t}|}{2}\Lambda \nabla_{p} S_{\mathrm{r,st}}(\mathbf{\Gamma}, t). \tag{6.74}$$

To this leading order the dissipative force is independent of the sign of the time step, $\overline{\mathbf{R}}_p^{21} = \overline{\mathbf{R}}_p^{12}$, where the superscript represents the direction of the transition. This makes it fully irreversible, just as in the equilibrium case. In fact it has the same functional form as in the equilibrium case, with the gradient of the static entropy providing the driving force toward the optimum non-equilibrium state. This zeroth order approximation is tested for driven Brownian motion in equation (6.173), and it is found to correctly give the relative entropy and hence relative probability of points in phase space.

The irreversible contributions break the symmetry of time reversibility, which is essential for consistency with the Second Law of Thermodynamics. The approximation to the gradient of the dynamic part of the entropy provides a correction to the zeroth order expression so that the reservoir force has both reversible and irreversible components. The fact that $\Lambda$ gives the magnitude of the dissipation and the variance of the stochastic force shows that the fluctuation dissipation theorem holds also for a non-equilibrium system.

Ultimately the approximations invoked in deriving these stochastic, dissipative equations of motion for the non-equilibrium system, should be judged by the results of their application to specific systems. It can be mentioned that they have been tested by computer simulation for both mechanical (Attard and Gray-Weale 2008, Attard 2009a, Attard 2009b) and thermodynamic (Attard 2006b, Attard 2009a) non-equilibrium systems. They are further tested analytically below with the derivation of the behavior of a driven Brownian particle, section 6.4.

### 6.2.3 Generalized equipartition theorem

It is of interest to discuss the result for the most likely reservoir force a little more generally in the context of the generalized equipartition theorem for a non-equilibrium system. Since the probability density and its gradient must vanish at the extremes of phase space, one must have that

$$\int \mathrm{d}\mathbf{\Gamma}\ \nabla\nabla\wp(\mathbf{\Gamma}, t) = \mathbf{0}. \tag{6.75}$$

This implies that

$$\langle \nabla\nabla S_\mathrm{r}(\mathbf{\Gamma}, t)\rangle = -k_\mathrm{B}^{-1}\langle[\nabla S_\mathrm{r}(\mathbf{\Gamma}, t)]\,[\nabla S_\mathrm{r}(\mathbf{\Gamma}, t)]\rangle. \tag{6.76}$$

This is a generalized equipartition theorem for a non-equilibrium system.

Taking the trace, or the double scalar product with a matrix, makes these the average of an extensive quantity. One can assume that in these cases they hold locally on the likely points in phase space. That is

$$\nabla^2 S_\mathrm{r}(\mathbf{\Gamma}, t) \approx -k_\mathrm{B}^{-1}[\nabla S_\mathrm{r}(\mathbf{\Gamma}, t)]\cdot[\nabla S_\mathrm{r}(\mathbf{\Gamma}, t)], \tag{6.77}$$

and

$$\Lambda : \nabla_p\nabla_p S_\mathrm{r}(\mathbf{\Gamma}, t) \approx -k_\mathrm{B}^{-1}\Lambda : \left[\nabla_p S_\mathrm{r}(\mathbf{\Gamma}, t)\right]\left[\nabla_p S_\mathrm{r}(\mathbf{\Gamma}, t)\right]. \tag{6.78}$$

In the thermodynamic limit these are likely exact.

Inserting the second of these into equation (6.58), one deduces that

$$0 \approx -k_B \nabla_p \cdot \mathbf{a}(\boldsymbol{\Gamma}, t) - \mathbf{a}(\boldsymbol{\Gamma}, t) \cdot \nabla_p S_r(\boldsymbol{\Gamma}, t). \tag{6.79}$$

This implies that $\mathbf{a}(\boldsymbol{\Gamma}, t)$ must be small, if not zero. This is consistent with the result invoked above, $\mathbf{a}(\boldsymbol{\Gamma}, t) = \Lambda[\nabla_p S_{r,st}(\boldsymbol{\Gamma}, t) - \nabla_p S_{r,st}(\overline{\boldsymbol{\Gamma}}(t), t)]/2$, equation (6.65), which is indeed zero on the most likely points in phase space.

### 6.2.4 The Fokker–Planck equation

From the formal expression for the phase space probability density, equation (6.4), the partial time derivative is

$$\begin{aligned}
&\frac{\partial^{\pm}\wp(\boldsymbol{\Gamma}, t)}{\partial t} \\
&\equiv \left\{ \frac{\partial S_r(\boldsymbol{\Gamma}, t)}{k_B \partial t} - \frac{\dot{Z}(t)}{Z(t)} \right\} \wp(\boldsymbol{\Gamma}, t) \\
&= \Bigg\{ -\nabla \cdot \dot{\boldsymbol{\Gamma}}^{det} - \dot{\boldsymbol{\Gamma}}^{det} \cdot \nabla S_r / k_B \\
&\quad + \frac{\hat{\Delta}_t}{2} \Lambda : \nabla_p \nabla_p S_r + \frac{\hat{\Delta}_t}{2k_B} \Lambda : \left[\nabla_p S_r\right]\left[\nabla_p S_r\right] \Bigg\} \wp(\boldsymbol{\Gamma}, t) \\
&= -[\nabla \cdot \dot{\boldsymbol{\Gamma}}^{det}]\wp(\boldsymbol{\Gamma}, t) - \dot{\boldsymbol{\Gamma}}^{det} \cdot \nabla \wp(\boldsymbol{\Gamma}, t) + \frac{\hat{\Delta}_t k_B}{2} \Lambda : \nabla_p \nabla_p \wp(\boldsymbol{\Gamma}, t).
\end{aligned} \tag{6.80}$$

The second equality is equation (6.55) rearranged. This is the Fokker–Planck equation. Since the stochastic, dissipative equations of motion for a non-equilibrium system have essentially the same functional form as those for an equilibrium system, it should not be surprising that this is identical in form to the equilibrium result, equation (5.68).

The superscript $\pm$ on the left-hand side is formally required because there are irreversible terms (those proportional to $\hat{\Delta}_t \equiv \text{sign}\, \Delta_t = \pm 1$, both explicit and implicit in $\dot{\boldsymbol{\Gamma}}^{det}$) on the right-hand side. However, these irreversible terms must cancel with each other because they are just the irreversible terms in equation (6.58). This makes the superscript $\pm$ on the left-hand side redundant. (It is redundant for the exact equations; the approximate ansatz will be discussed below.)

With the reversible part of the deterministic velocity that remains, $\dot{\boldsymbol{\Gamma}}^{det,rev} = \dot{\boldsymbol{\Gamma}}^0 + \mathbf{b}$, the Fokker–Planck equation becomes

$$\begin{aligned}
\frac{\partial \wp(\boldsymbol{\Gamma}, t)}{\partial t} &\equiv \left\{ \frac{\partial S_r(\boldsymbol{\Gamma}, t)}{k_B \partial t} - \frac{\dot{Z}(t)}{Z(t)} \right\} \wp(\boldsymbol{\Gamma}, t) \\
&= -[\nabla \cdot \mathbf{b}]\wp(\boldsymbol{\Gamma}, t) - [\dot{\boldsymbol{\Gamma}}^0 + \mathbf{b}] \cdot \nabla \wp(\boldsymbol{\Gamma}, t).
\end{aligned} \tag{6.81}$$

The formally exact expression for the reversible reservoir contribution, equation (6.60), makes this an identity.

Now let us analyze the Fokker–Planck equation for the approximation to the most likely reservoir force, equation (6.67). In this case the irreversible part of the equation is

$$
\begin{aligned}
&\frac{1}{2}\left\{\frac{\partial^+\wp(\mathbf{\Gamma}, t)}{\partial t} - \frac{\partial^-\wp(\mathbf{\Gamma}, t)}{\partial t}\right\} \\
&= -[\nabla \cdot \mathbf{a}(\mathbf{\Gamma}, t)]\wp(\mathbf{\Gamma}, t) - \mathbf{a}(\mathbf{\Gamma}, t) \cdot \nabla\wp(\mathbf{\Gamma}, t) + \frac{k_{\mathrm{B}}}{2}\Lambda : \nabla_p\nabla_p\wp(\mathbf{\Gamma}, t) \\
&= \left\{\frac{-1}{2}\Lambda : \nabla_p\nabla_p S_{\mathrm{r,st}}(\mathbf{\Gamma}, t)\right. \\
&\quad - \frac{1}{2k_{\mathrm{B}}}\Lambda : \left[\nabla_p S_{\mathrm{r,st}}(\mathbf{\Gamma}, t) - \nabla_p S_{\mathrm{r,st}}(\overline{\mathbf{\Gamma}}(t), t)\right]\left[\nabla_p S_{\mathrm{r}}(\mathbf{\Gamma}, t)\right] \\
&\quad \left.+ \frac{1}{2}\Lambda : \nabla_p\nabla_p S_{\mathrm{r}}(\mathbf{\Gamma}, t) + \frac{1}{2k_{\mathrm{B}}}\Lambda : \left[\nabla_p S_{\mathrm{r}}(\mathbf{\Gamma}, t)\right]\left[\nabla_p S_{\mathrm{r}}(\mathbf{\Gamma}, t)\right]\right\}\wp(\mathbf{\Gamma}, t) \\
&= \left\{\frac{1}{2}\Lambda : \nabla_p\nabla_p S_{\mathrm{r,dyn}}(\mathbf{\Gamma}, t)\right. \\
&\quad \left.+ \frac{k_{\mathrm{B}}^{-1}}{2}\Lambda : \left[\nabla_p S_{\mathrm{r,dyn}}(\mathbf{\Gamma}, t) - \nabla_p S_{\mathrm{r,dyn}}(\overline{\mathbf{\Gamma}}(t), t)\right]\left[\nabla_p S_{\mathrm{r}}(\mathbf{\Gamma}, t)\right]\right\}\wp(\mathbf{\Gamma}, t).
\end{aligned} \tag{6.82}
$$

This is evidently non-zero, although one might argue that it is small if not negligible. In this approximation the reversible part of the deterministic velocity is

$$
\dot{\mathbf{\Gamma}}^{\mathrm{det,rev}} = \dot{\mathbf{\Gamma}}^0 + \frac{1}{2}\Lambda\nabla_p S_{\mathrm{r,st}}(\overline{\mathbf{\Gamma}}(t), t). \tag{6.83}
$$

Since the reservoir part is a constant in phase space, one has that $\nabla \cdot \dot{\mathbf{\Gamma}}^{\mathrm{det,rev}} = 0$, and the reversible part of the Fokker–Planck equation in this case reduces to

$$
\begin{aligned}
\frac{\partial\wp(\mathbf{\Gamma}, t)}{\partial t} &= -\dot{\mathbf{\Gamma}}^{\mathrm{det,rev}} \cdot \nabla\wp(\mathbf{\Gamma}, t) \\
&= -\left[\dot{\mathbf{\Gamma}}^0 + \frac{1}{2}\Lambda\nabla_p S_{\mathrm{r,st}}(\overline{\mathbf{\Gamma}}(t), t)\right] \cdot \nabla\wp(\mathbf{\Gamma}, t).
\end{aligned} \tag{6.84}
$$

To the extent that equation (6.71) is satisfied, this is the expected result.

The hypothetical exact result for the most likely reservoir force, equation (6.72), gives the irreversible part of the Fokker–Planck equation as

$$
\begin{aligned}
&\frac{1}{2}\left\{\frac{\partial^+\wp(\mathbf{\Gamma}, t)}{\partial t} - \frac{\partial^-\wp(\mathbf{\Gamma}, t)}{\partial t}\right\} \\
&= -[\nabla \cdot \mathbf{a}(\mathbf{\Gamma}, t)]\wp(\mathbf{\Gamma}, t) - \mathbf{a}(\mathbf{\Gamma}, t) \cdot \nabla\wp(\mathbf{\Gamma}, t) + \frac{k_{\mathrm{B}}}{2}\Lambda : \nabla_p\nabla_p\wp(\mathbf{\Gamma}, t) \\
&= \left\{\frac{-1}{2}\Lambda : \nabla_p\nabla_p S_{\mathrm{r}}(\mathbf{\Gamma}, t)\left[\nabla_p S_{\mathrm{r}}(\mathbf{\Gamma}, t)\right] - \frac{1}{2k_{\mathrm{B}}}\Lambda : \left[\nabla_p S_{\mathrm{r}}(\mathbf{\Gamma}, t)\right]\left[\nabla_p S_{\mathrm{r}}(\mathbf{\Gamma}, t)\right]\right. \\
&\quad \left.+ \frac{1}{2}\Lambda : \nabla_p\nabla_p S_{\mathrm{r}}(\mathbf{\Gamma}, t) + \frac{1}{2k_{\mathrm{B}}}\Lambda : \left[\nabla_p S_{\mathrm{r}}(\mathbf{\Gamma}, t)\right]\left[\nabla_p S_{\mathrm{r}}(\mathbf{\Gamma}, t)\right]\right\}\wp(\mathbf{\Gamma}, t) \\
&= 0.
\end{aligned} \tag{6.85}
$$

This vanishes, which is not surprising because $\mathbf{a}(\mathbf{\Gamma}, t)$ is formally exact in this case. The reversible part is

$$\begin{aligned}\frac{\partial \wp(\mathbf{\Gamma}, t)}{\partial t} &= -\left[\nabla \cdot \mathbf{b}(\mathbf{\Gamma}, t)\right]\wp(\mathbf{\Gamma}, t) - \left[\dot{\mathbf{\Gamma}}^0 + \mathbf{b}(\mathbf{\Gamma}, t)\right] \cdot \nabla \wp(\mathbf{\Gamma}, t) \\ &= \left\{ \frac{1}{2}\Lambda : \nabla_p \nabla_p S_{\mathrm{r,dyn}}(\mathbf{\Gamma}, t) \right. \\ &\quad \left. - \left[\dot{\mathbf{\Gamma}}^0 - \frac{1}{2}\Lambda \nabla_p S_{\mathrm{r,dyn}}(\mathbf{\Gamma}, t)\right] \cdot \nabla_p S_{\mathrm{r}}(\mathbf{\Gamma}, t) \right\} \wp(\mathbf{\Gamma}, t).\end{aligned} \tag{6.86}$$

I have no independent evidence for the veracity or otherwise of this.

## 6.3 Odd projection of the dynamic entropy

This section is concerned in the first place with using only the odd time parity projection of the dynamic part of the reservoir entropy, and in the second place with replacing the backwards most likely trajectories that appear in the integrand with adiabatic trajectories.

There are several reasons for rewriting the dynamic part of the reservoir entropy in these two ways. First, it identifies and focusses on the dominant contributions to the reservoir entropy. Second, it allows certain analytic results for the rate of change of the entropy to be obtained. Third, it gives an explicit connection between the novel dynamic part of the reservoir entropy and the well-known Green–Kubo expressions for the transport coefficients of a steady state system, which provides additional confirmation of the validity of the former. And fourth it offers a different, possibly more efficient procedure for the numerical computation of the reservoir entropy associated with a point in the sub-system phase space.

### 6.3.1 Odd projection

It is now argued that the odd projection of the dynamic part of the reservoir entropy is dominant. Two definitions of the odd projection are given. The first has the virtue of being physically intuitive and it can be applied immediately to steady state thermodynamic systems. The second is more abstract and more general, and it can be applied both to steady state thermodynamic systems and to mechanical non-equilibrium systems.

*Simplest definition*

As was mentioned in section 6.1, the crucial distinction between an equilibrium and a non-equilibrium system is that the probability distribution for the latter depends upon the sign of the molecular velocities, $\wp(\mathbf{\Gamma}, t) \neq \wp(\mathbf{\Gamma}^\dagger, t)$. For a sub-system phase space point $\mathbf{\Gamma} = \{\mathbf{q}^N, \mathbf{p}^N\}$, the conjugate phase space point is the one with the velocities reversed, $\mathbf{\Gamma}^\dagger = \{\mathbf{q}^N, (-\mathbf{p})^N\}$. Since the probability density cannot have even parity, neither can its exponent, the reservoir entropy, $S_{\mathrm{r}}(\mathbf{\Gamma}, t) \neq S_{\mathrm{r}}(\mathbf{\Gamma}^\dagger, t)$. Since the static part of the reservoir entropy is a purely equilibrium quantity, it necessarily has even parity, $S_{\mathrm{r,st}}(\mathbf{\Gamma}, t) = S_{\mathrm{r,st}}(\mathbf{\Gamma}^\dagger, t)$. This means that the dynamic part of the reservoir entropy cannot be even, $S_{\mathrm{r,dyn}}(\mathbf{\Gamma}, t) \neq S_{\mathrm{r,dyn}}(\mathbf{\Gamma}^\dagger, t)$.

Further, because the non-equilibrium aspects of the system are a perturbation on the equilibrium aspects, one can neglect the even projection of $S_{r,dyn}(\mathbf{\Gamma}, t)$ in comparison with $S_{r,st}(\mathbf{\Gamma}, t)$ and so write

$$S_r(\mathbf{\Gamma}, t) \approx S_{r,st}(\mathbf{\Gamma}, t) + S_{r,dyn}^{odd}(\mathbf{\Gamma}, t) + \overline{S}_{r,dyn}(t). \tag{6.87}$$

First, the odd projection is discussed, and then the time-dependent constant that is retained here.

The simplest definition of the odd projection of the dynamic part of the reservoir entropy is

$$\begin{aligned} S_{r,dyn}^{odd}(\mathbf{\Gamma}, t) &\equiv \frac{1}{2}\left[S_{r,dyn}(\mathbf{\Gamma}, t) - S_{r,dyn}(\mathbf{\Gamma}^{\dagger}, t)\right] \\ &= \frac{-1}{2}\int_0^t \mathrm{d}t'\left[\dot{S}_{r,st}^{0}(\overline{\mathbf{\Gamma}}(t'|\mathbf{\Gamma}, t), t') - \dot{S}_{r,st}^{0}(\overline{\mathbf{\Gamma}}(t'|\mathbf{\Gamma}^{\dagger}, t), t')\right]. \end{aligned} \tag{6.88}$$

A more abstract form of the odd projection will be given shortly.

In addition to the odd projection, also retained in the above is the asymptotic contribution of the dynamic part of the reservoir entropy, which is constant in phase space and which therefore has even parity. On the most likely trajectory the asymptote of the integrand of the dynamic part of the reservoir entropy is

$$\dot{S}_{r,st}^{0}(\overline{\mathbf{\Gamma}}(t'|\mathbf{\Gamma}, t), t') \to \overline{\dot{S}_{r,st}^{0}}(t'), \quad |t' - t| \to \infty. \tag{6.89}$$

This asymptote arises from the fact that with overwhelming probability the system came from its most likely value in the past (and will return there in the future), independent of the current phase space point of the sub-system. As a consequence of this independence, the asymptote holds unchanged for the conjugate point, $\dot{S}_{r,st}^{0}(\overline{\mathbf{\Gamma}}(t'|\mathbf{\Gamma}^{\dagger}, t), t') \to \overline{\dot{S}_{r,st}^{0}}(t'), |t' - t| \to \infty$.

One can add and subtract this asymptote before taking the odd projection,

$$S_{r,dyn}(\mathbf{\Gamma}, t) = -\int_0^t \mathrm{d}t'\left[\dot{S}_{r,st}^{0}(\overline{\mathbf{\Gamma}}(t'|\mathbf{\Gamma}, t), t') - \overline{\dot{S}_{r,st}^{0}}(t')\right] + \overline{S}_{r,dyn}(t), \tag{6.90}$$

where

$$\overline{S}_{r,dyn}(t) = -\int_0^t \mathrm{d}t' \, \overline{\dot{S}_{r,st}^{0}}(t'). \tag{6.91}$$

(This result was given above as equation (6.17).) Since this is constant in phase space, it has even parity. Hence the asymptote does not contribute to the odd projection.

Although it could be neglected, it is arguably best to retain the even parity asymptotic contribution to the dynamic part of the entropy. This is a constant in phase space, so it has no effect on the relative probability of phase space points. It could simply be neglected, which effectively cancels it with the same contribution to the partition function. However, it does represent part of the total entropy produced as a function of time in the total system, which is a useful physical quantity. For this reason this even parity, constant term, has been retained explicitly.

*Abstract definition*
The argument for keeping the odd time projection, and only the odd projection, of the dynamic part of the reservoir entropy was that the even time projection was dominated by the static part, and that the non-equilibrium probability could not have even time parity if it is to satisfy the Second Law of Thermodynamics. Now a second definition of conjugation is given that has more general application than the physically appealing momentum reversal definition used above.

For each phase space point $\boldsymbol{\Gamma}$ denote a conjugate point as $\boldsymbol{\Gamma}^{\#}$. This new conjugation operation is meant to include velocity reversal, but it can also include other operations besides. (I shall be more specific in the treatment of mechanical non-equilibrium systems below.) The conjugation operation is defined to have two properties. First it leaves the static part of the reservoir entropy unchanged,

$$S_{\rm r,st}(\boldsymbol{\Gamma}^{\#}, t) = S_{\rm r,st}(\boldsymbol{\Gamma}, t). \tag{6.92}$$

This is exactly as in the case of velocity reversal alone, $S_{\rm r,st}(\boldsymbol{\Gamma}^{\dagger}, t) = S_{\rm r,st}(\boldsymbol{\Gamma}, t)$.

Second, it reverses the sign of the adiabatic rate of static entropy production,

$$\dot{S}^{0}_{\rm r,st}(\boldsymbol{\Gamma}^{\#}, t) = -\dot{S}^{0}_{\rm r,st}(\boldsymbol{\Gamma}, t). \tag{6.93}$$

For the case of a steady state system, which is not explicitly dependent on time, $\dot{S}^{0}_{\rm r,st}(\boldsymbol{\Gamma}) = \dot{\boldsymbol{\Gamma}}^{0} \cdot \nabla S_{\rm r,st}(\boldsymbol{\Gamma})$, this condition is also the same as for velocity reversal, $\dot{S}^{0}_{\rm r,st}(\boldsymbol{\Gamma}^{\dagger}) = -\dot{S}^{0}_{\rm r,st}(\boldsymbol{\Gamma})$. However, for a system explicitly dependent on time, the extra term in the adiabatic time derivative, $\partial S_{\rm r,st}(\boldsymbol{\Gamma}, t)/\partial t$, has even parity, and so $\dot{S}^{0}_{\rm r,st}(\boldsymbol{\Gamma}^{\dagger}, t) \neq -\dot{S}^{0}_{\rm r,st}(\boldsymbol{\Gamma}, t)$. Hence this second condition defines the more general conjugation operation # such that the adiabatic rate of static entropy production must reverse sign also for explicitly time dependent systems such as mechanical non-equilibrium systems.

Define the odd projection of the dynamic part of the reservoir entropy with respect to this conjugation operation as

$$\begin{aligned} S^{\rm odd}_{\rm r,dyn}(\boldsymbol{\Gamma}, t) &\equiv \frac{1}{2}\left[S_{\rm r,dyn}(\boldsymbol{\Gamma}, t) - S_{\rm r,dyn}(\boldsymbol{\Gamma}^{\#}, t)\right] \\ &= \frac{-1}{2}\int_0^t {\rm d}t'\left[\dot{S}^{0}_{\rm r,st}(\overline{\boldsymbol{\Gamma}}(t'|\boldsymbol{\Gamma}, t), t') - \dot{S}^{0}_{\rm r,st}(\overline{\boldsymbol{\Gamma}}(t'|\boldsymbol{\Gamma}^{\#}, t), t')\right]. \end{aligned} \tag{6.94}$$

As for the usual conjugation operation of velocity reversal, since the asymptote in independent of the starting point, one can add and subtract the asymptote before taking the odd projection.

The even projection is defined analogously and one has for the reservoir entropy

$$\begin{aligned} S_{\rm r}(\boldsymbol{\Gamma}, t) &= S_{\rm r,st}(\boldsymbol{\Gamma}, t) + S^{\rm odd}_{\rm r,dyn}(\boldsymbol{\Gamma}, t) + S^{\rm even}_{\rm r,dyn}(\boldsymbol{\Gamma}, t) \\ &\approx S_{\rm r,st}(\boldsymbol{\Gamma}, t) + S^{\rm odd}_{\rm r,dyn}(\boldsymbol{\Gamma}, t) + \overline{S_{\rm r,dyn}}(t), \end{aligned} \tag{6.95}$$

where as above, the asymptotic contribution is $\overline{S}_{\rm r,dyn}(t) = -\int_0^t {\rm d}t'\,\overline{\dot{S}^{0}_{\rm r,st}}(t')$. The justification for the second equality is as in the first definition of conjugation, namely that the static part of the reservoir entropy has even parity and it dominates the even projection of the dynamic part of the reservoir entropy. And since the Second Law of Thermodynamics mandates that a non-equilibrium system must have time asymmetry, it is necessary to retain the odd projection of the dynamic part of the reservoir entropy, since this is the only part of the reservoir entropy with such a time asymmetry. Since the Second Law of Thermodynamics refers to the change in entropy, it makes sense to take the definition of the conjugation operation to be that it reverses exactly the rate of entropy production, rather than the more obvious definition that it reverses the velocities, since there are some systems (i.e. mechanical non-equilibrium systems, which have a time-dependent external potential) where the two are not equivalent.

Retaining the even parity asymptotic contribution that is constant in phase space is not necessary for determining the relative probability of phase space points, but it does have a transparent physical interpretation and it contributes to the rate of dissipation of the total system, which make it useful to retain this constant term explicitly rather than incorporating it into the partition function.

### 6.3.2 Adiabatic odd projection, steady state system

In this subsection, consider a steady state, thermodynamic, non-equilibrium system. Examples include steady heat flow, shear flow, and diffusion. Since it is steady state, the static part of the reservoir entropy is independent of time, $S_{\mathrm{r,st}}(\mathbf{\Gamma}, t) = S_{\mathrm{r,st}}(\mathbf{\Gamma})$, and velocity reversal alone is sufficient to negate the adiabatic rate of static entropy production, $\dot{S}^0_{\mathrm{r,st}}(\mathbf{\Gamma}^\dagger) = \dot{\mathbf{\Gamma}}^0(\mathbf{\Gamma}^\dagger) \cdot \nabla^\dagger S_{\mathrm{r,st}}(\mathbf{\Gamma}^\dagger) = -\dot{S}^0_{\mathrm{r,st}}(\mathbf{\Gamma})$. Hence velocity reversal serves as the conjugation operation for this steady state system.

The behavior of the adiabatic rate of entropy production on various trajectories for a steady state thermodynamic system is sketched in figure 6.4. On the most likely trajectory the asymptotes are

$$\dot{S}^0_{\mathrm{r,st}}(\overline{\mathbf{\Gamma}}(t'|\mathbf{\Gamma}, t), t') \to \overline{\dot{S}^0_{\mathrm{r,st}}}(t'), \quad |t' - t| \to \infty. \tag{6.96}$$

This asymptote arises from the fact that with overwhelming probability the system came from its most likely value in the past (and will return there in the future), independent of the current phase space point of the sub-system. For the present steady state system, the asymptote is independent of $t'$ and $t$.

In contrast, the asymptotic behavior on the adiabatic trajectory is

$$\dot{S}^0_{\mathrm{r,st}}(\mathbf{\Gamma}^0(t'|\mathbf{\Gamma}, t), t') \sim \mathrm{sign}(t' - t)\, \overline{\dot{S}^0_{\mathrm{st}}}(t'). \tag{6.97}$$

This behavior can be seen in the case of steady heat flow in figure 3.2, where the tangent of the curve is proportional to the adiabatic rate of entropy production. The asymptotic behavior sets in for $|t' - t| \gtrsim \tau_{\mathrm{relax}}$, where $\tau_{\mathrm{relax}}$ is a relaxation time that is long enough for the system to reach its asymptote, but not so long that the structure has changed significantly, $|\tau_{\mathrm{relax}} \dot{S}^0_{\mathrm{r,st}}| \ll |S_{\mathrm{r,st}}|$. (One does not need to impose this condition for the dissipative trajectory because the interactions with the reservoir maintain the structure of the sub-system.) For an isolated system, the structure represents a fluctuation, and $\dot{S}^0_{\rm r,st}$ represents its regression, which must be an odd function of time, at least for a steady state system. For $t' > 0$, the adiabatic asymptote and the actual asymptote approximately coincide, which is just Onsager's regression hypothesis (Onsager 1931).

In view of the trajectories shown in figure 6.4 and the above discussion, the odd projection of the dynamic part of the reservoir entropy may be transformed from an integral over the most likely trajectory to an integral over the adiabatic trajectories. Successive transformations yield

$$
\begin{aligned}
&S^{\rm odd}_{\rm r,dyn}(\mathbf{\Gamma}, t) \\
&= \frac{-1}{2}\int_0^t {\rm d}t'\left[\dot{S}^0_{\rm r,st}(\overline{\mathbf{\Gamma}}(t'|\mathbf{\Gamma}, t), t') - \dot{S}^0_{\rm r,st}(\overline{\mathbf{\Gamma}}(t'|\mathbf{\Gamma}^\dagger, t), t')\right] \\
&\approx \frac{-1}{2}\int_t^{2t} {\rm d}t'\left[\dot{S}^0_{\rm r,st}(\overline{\mathbf{\Gamma}}(t'|\mathbf{\Gamma}, t), t') - \dot{S}^0_{\rm r,st}(\overline{\mathbf{\Gamma}}(t'|\mathbf{\Gamma}^\dagger, t), t')\right] \\
&\approx \frac{-1}{2}\int_t^{2t} {\rm d}t'\left[\dot{S}^0_{\rm r,st}(\mathbf{\Gamma}^0(t'|\mathbf{\Gamma}, t), t') - \dot{S}^0_{\rm r,st}(\mathbf{\Gamma}^0(t'|\mathbf{\Gamma}^\dagger, t), t')\right] \\
&= \frac{-1}{2}\int_0^t {\rm d}t'\left[\dot{S}^0_{\rm r,st}(\mathbf{\Gamma}^0(t'|\mathbf{\Gamma}^\dagger, t)^\dagger, t') - \dot{S}^0_{\rm r,st}(\mathbf{\Gamma}^0(t'|\mathbf{\Gamma}, t)^\dagger, t')\right] \\
&= \frac{-1}{2}\int_0^t {\rm d}t'\left[\dot{S}^0_{\rm r,st}(\mathbf{\Gamma}^0(t'|\mathbf{\Gamma}, t), t') - \dot{S}^0_{\rm r,st}(\mathbf{\Gamma}^0(t'|\mathbf{\Gamma}^\dagger, t), t')\right] \\
&\equiv S^{\rm odd;0}_{\rm r,dyn}(\mathbf{\Gamma}, t).
\end{aligned}
\tag{6.98}
$$

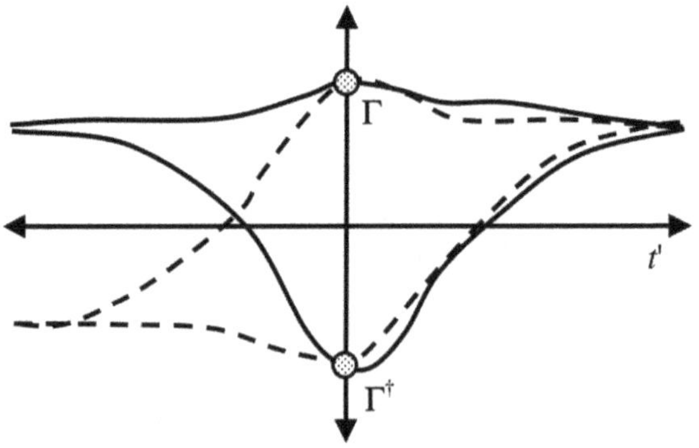


**Figure 6.4.** Sketch of the adiabatic rate of entropy production, $\dot{S}^0_{\rm r,st}$, for a steady state thermodynamic non-equilibrium system, in the future (right-hand side) and in the past (left-hand side), starting from a likely phase space point $\mathbf{\Gamma}$ (upper), and from its unlikely conjugate $\mathbf{\Gamma}^\dagger$ (lower). The solid curves are on most likely trajectories, $\overline{\mathbf{\Gamma}}(t'|\mathbf{\Gamma}, 0)$ and $\overline{\mathbf{\Gamma}}(t'|\mathbf{\Gamma}^\dagger, 0)$, and the short dashed curves are on adiabatic trajectories, $\mathbf{\Gamma}^0(t'|\mathbf{\Gamma}, 0)$ and $\mathbf{\Gamma}^0(t'|\mathbf{\Gamma}^\dagger, 0)$. After Attard (2009c).

The result says that for a steady state thermodynamic system, the odd projection of the dynamic part of the reservoir entropy can be well-approximated by adiabatic trajectories. The first equality is the area between the solid curves in the left half of the figure. The second equality is the area between the solid curves in the right half of the figure. This follows because the dissipation on the most likely trajectory is to a good approximation even in time. The third equality is the area between the short dashed curves in the right half of the figure. This follows from Onsager's regression hypothesis. The fourth equality is the area between the short dashed curves in the left half of the figure. This follows from the microscopic reversibility of Hamilton's equations of motion,

$$\boldsymbol{\Gamma}^0(t'|\boldsymbol{\Gamma}^\dagger, t) = \boldsymbol{\Gamma}^0(2t - t'|\boldsymbol{\Gamma}, t)^\dagger. \tag{6.99}$$

This holds for a steady state system in which the internal and any external potentials do not depend explicitly on time. The penultimate equality also holds for a steady state system, since in this case the adiabatic rate of change of the static part of the reservoir entropy is $\dot{S}^0_{\mathrm{r,st}}(\boldsymbol{\Gamma}) = \dot{\boldsymbol{\Gamma}}^0 \cdot \nabla S_{\mathrm{r,st}}(\boldsymbol{\Gamma})$, and this has odd parity, $\dot{S}^0_{\mathrm{r,st}}(\boldsymbol{\Gamma}^\dagger) = -\dot{S}^0_{\mathrm{r,st}}(\boldsymbol{\Gamma})$.

It is obvious from the figure that the integrand asymptotes to zero. This means that the lower limit of the integral can be replaced by $t - \tau$ for some convenient interval $\tau > 0$. Although the integrand is an exact differential, there is no point in analytically evaluating the integral because the actual value at the lower limit would be required, $S^{\mathrm{odd}}_{\mathrm{r,dyn}}(\boldsymbol{\Gamma}, t) = [S_{\mathrm{r,st}}(\boldsymbol{\Gamma}^0(t - \tau|\boldsymbol{\Gamma}, t), t - \tau) - S_{\mathrm{r,st}}(\boldsymbol{\Gamma}^0(t - \tau|\boldsymbol{\Gamma}^\dagger, t), t - \tau)]/2$. (Although $\dot{S}^0_{\mathrm{r,st}}$ has the same asymptote starting at $\boldsymbol{\Gamma}$ and at $\boldsymbol{\Gamma}^\dagger$, there is a finite difference between the respective asymptotes of $S_{\mathrm{r,st}}$ that corresponds to the area between the two curves in the left half of the figure.) It takes no more computational effort to perform the quadrature numerically than it does to calculate the adiabatic trajectories backward to their lower limit.

In summary, this section argues that in some circumstances the odd projection of the dynamic part of the reservoir entropy is either dominant or is all that is required. Further, it says that for a steady state thermodynamic system, the odd projection of the dynamic part of the reservoir entropy may be evaluated on the past adiabatic trajectories. With this result one does not need to evaluate the most likely backwards trajectory. (On the other hand, calculating the dissipative trajectory, or even the stochastic dissipative trajectory may be advantageous because it is thermostatted and therefore stable, in contrast to the adiabatic trajectory.) This adiabatic expression for the dynamic part of the reservoir entropy has been tested with computer simulations of non-equilibrium systems and found to be accurate (Attard 2006b). This adiabatic approximation for the trajectories of the dynamic part of the entropy, and the general formulation of the non-equilibrium probability, will be tested analytically by direct comparison with the Green–Kubo relations in section 6.3.6 below.

### 6.3.3 Adiabatic odd projection, mechanical system

In this sub-section, we consider a mechanical non-equilibrium system that has a time-dependent external potential. A specific example will be considered from which the general procedure can be extrapolated.

Consider a solvent in which there is a solute acted upon by a time varying potential, and also a thermal reservoir with which energy can be exchanged. Suppose that the sub-system Hamiltonian is of the form $\mathcal{H}(\mathbf{\Gamma}, t) = \mathcal{H}_0(\mathbf{\Gamma}) + U(\mathbf{x}, t)$. The time-independent Hamiltonian $\mathcal{H}_0(\mathbf{\Gamma})$ describes the interactions of the solvent particles amongst themselves and with the solute particle. The solute particle is acted upon by a time-dependent external potential that takes the form of a moving parabolic trap,

$$U(\mathbf{x}, t) = \frac{\kappa}{2}[\mathbf{x} - \mathbf{b}(t)] \cdot [\mathbf{x} - \mathbf{b}(t)]. \tag{6.100}$$

Here $\mathbf{x}$ is the position of the solute and $\mathbf{b}(t)$ is location of the trap. The sub-system phase space point $\mathbf{\Gamma}$ includes the solute as well as the solvent particles, $\mathbf{\Gamma} = \{\mathbf{\Gamma}_q, \mathbf{x}; \mathbf{\Gamma}_p, M\dot{\mathbf{x}}\}$, where $M$ is the solute mass. Velocity conjugation applies to the solvent and to the solute,

$$\mathbf{\Gamma}^\dagger = \left\{\mathbf{\Gamma}_q, \mathbf{x}; -\mathbf{\Gamma}_p, -M\dot{\mathbf{x}}\right\}. \tag{6.101}$$

The sub-system can exchange energy with a thermal reservoir of temperature $T$, and therefore the static part of the reservoir entropy is

$$S_{\mathrm{r,st}}(\mathbf{\Gamma}, t) = \frac{-1}{T}\mathcal{H}(\mathbf{\Gamma}, t). \tag{6.102}$$

Evidently this has even parity with respect to velocity reversal, $S_{\mathrm{r,st}}(\mathbf{\Gamma}^\dagger, t) = S_{\mathrm{r,st}}(\mathbf{\Gamma}, t)$. The adiabatic rate of change of the static entropy is

$$\dot{S}^0_{\mathrm{r,st}}(\mathbf{\Gamma}, t) = \frac{-1}{T}\left[\frac{\partial\mathcal{H}(\mathbf{\Gamma}, t)}{\partial t} + \dot{\mathbf{\Gamma}}^0 \cdot \nabla\mathcal{H}(\mathbf{\Gamma}, t)\right] = \frac{-1}{T}\frac{\partial U(\mathbf{x}, t)}{\partial t}. \tag{6.103}$$

This has even parity with respect to velocity reversal, $\dot{S}^0_{\mathrm{r,st}}(\mathbf{\Gamma}^\dagger, t) = \dot{S}^0_{\mathrm{r,st}}(\mathbf{\Gamma}, t)$, since the right-hand side is independent of the solvent and solute velocity.

In addition to velocity reversal symmetry, the Hamiltonian is symmetric with respect to reflection of the solute and all of the solvent particles about the trap minimum,

$$\mathbf{\Gamma}^\ddagger = \left\{2\mathbf{b}(t) - \mathbf{\Gamma}_q, 2\mathbf{b}(t) - \mathbf{x}; \mathbf{\Gamma}_p, M\dot{\mathbf{x}}\right\}. \tag{6.104}$$

Hence the static part of the reservoir entropy is unchanged by this reflection operation,

$$S_{\mathrm{r,st}}(\mathbf{\Gamma}^\ddagger, t) = S_{\mathrm{r,st}}(\mathbf{\Gamma}, t). \tag{6.105}$$

For this symmetry to hold, all the solvent particles have to be reflected at the same time as the solute particle. This assumes that the solvent and the solute are not chiral, and also that the sub-system is homogeneous apart from the moving trap potential.

This reflection symmetry as well as the velocity reversal symmetry is sketched in figure 6.5. The four configurations shown have the same energy and hence static part of the reservoir entropy,

$$S_{\rm r,st}(\mathbf{\Gamma}, t) = S_{\rm r,st}(\mathbf{\Gamma}^{\dagger}, t) = S_{\rm r,st}(\mathbf{\Gamma}^{\ddagger}, t) = S_{\rm r,st}(\mathbf{\Gamma}^{\dagger\ddagger}, t). \tag{6.106}$$

(Since this depends on the total energy of the solute and solvent, $\mathcal{H}(\mathbf{\Gamma}, t)$, one can see why the solvent particles also have to be reflected in the trap minimum, since this is the only way that their interaction potential with the solute is unchanged.)

As mentioned, the adiabatic rate of change of the static part of the reservoir entropy is

$$\dot{S}^{0}_{\rm r,st}(\mathbf{\Gamma}, t) = \frac{-1}{T}\frac{\partial U(\mathbf{x}, t)}{\partial t} = \frac{\kappa}{T}\dot{\mathbf{b}}(t) \cdot [\mathbf{x} - \mathbf{b}(t)]. \tag{6.107}$$

By inspection one sees that this is independent of solvent and solute velocity, $\dot{S}^{0}_{\rm r,st}(\mathbf{\Gamma}, t) = \dot{S}^{0}_{\rm r,st}(\mathbf{\Gamma}^{\dagger}, t)$, but that it changes sign upon position reflection,

$$\dot{S}^{0}_{\rm r,st}(\mathbf{\Gamma}, t) = -\dot{S}^{0}_{\rm r,st}(\mathbf{\Gamma}^{\ddagger}, t), \text{ and also } \dot{S}^{0}_{\rm r,st}(\mathbf{\Gamma}, t) = -\dot{S}^{0}_{\rm r,st}(\mathbf{\Gamma}^{\dagger\ddagger}, t). \tag{6.108}$$

In view of the definition of the generalized conjugation operation, equation (6.93), one can choose either $\# \equiv \ddagger$, or else $\# \equiv \dagger\ddagger$.

In previous work it has been argued that the combined symmetry operation $\dagger\ddagger$ should be used to define the odd projection (Attard 2009c), and this has been successfully used in computer simulations of a driven Brownian particle (Attard 2009b). A possible reason for needing to combine both is that non-equilibrium systems are necessarily sensitive to both velocity reversal and the rate of entropy production. The combined symmetry operation is discussed for an explicit calculation of driven Brownian motion in equation (6.174) below. Following that earlier

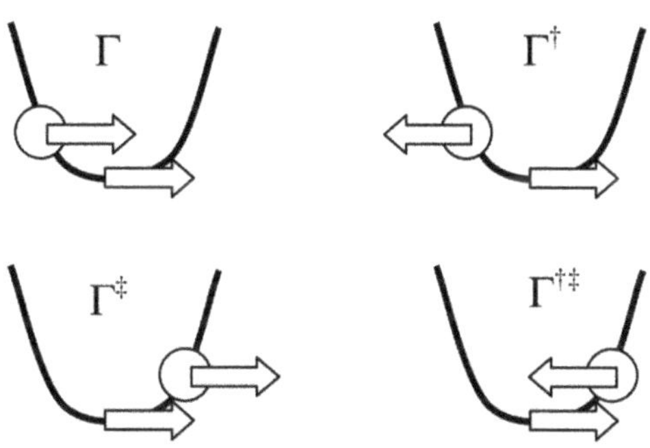


**Figure 6.5.** Four configurations of a Brownian particle (circle) in a solvent (not shown) in a moving potential trap (curve). The arrows indicate the velocity of the particle and of the trap. The phase space point $\mathbf{\Gamma}$ is a likely point. The dagger signifies velocity reversal, and the double dagger signifies reflection in the trap minimum. These operations apply to the solute and the solvent particles. All four configurations have the same Hamiltonian energy.

work, the odd projection of the dynamic part of the reservoir entropy is also defined here as

$$\begin{aligned} S^{\rm odd}_{\rm r,dyn}(\mathbf{\Gamma}, t) &\\ = \frac{-1}{2}\int_0^t {\rm d}t' \Big[ \dot{S}^0_{\rm r,st}(\overline{\mathbf{\Gamma}}(t'|\mathbf{\Gamma}, t), t') - \dot{S}^0_{\rm r,st}(\overline{\mathbf{\Gamma}}(t'|\mathbf{\Gamma}^{\dagger\ddagger}, t), t') \Big] &\\ \approx \frac{-1}{2}\int_0^t {\rm d}t' \Big[ \dot{S}^0_{\rm r,st}(\mathbf{\Gamma}^0(t'|\mathbf{\Gamma}, t), t') - \dot{S}^0_{\rm r,st}(\mathbf{\Gamma}^0(t'|\mathbf{\Gamma}^{\dagger\ddagger}, t), t') \Big] &\\ \equiv S^{\rm odd;0}_{\rm r,dyn}(\mathbf{\Gamma}, t). & \end{aligned} \tag{6.109}$$

In the second equality the most likely trajectories have been replaced by adiabatic trajectories. This result is formally identical to equation (6.98) for a steady state thermodynamic system (but with the present definition of conjugation).

Figure 6.6 sketches the backwards evolution from a likely point $\mathbf{\Gamma}_1$ and from its unlikely conjugate $\mathbf{\Gamma}_1^{\dagger\ddagger}$. As in the preceding derivation, it is assumed that the most likely trajectory is approximately the same as the adiabatic trajectory in each case. This is a reasonable assumption in view of the fact that the solute is surrounded by solvent that is included in the adiabatic evolution. In this case over not too long time periods, the reservoir is redundant, since the solvent itself acts like a heat sink. Hence the dissipative trajectories of the system can be replaced by adiabatic trajectories of the sub-system, provided that the time period concerned is short enough that any perturbation by the reservoir is insignificant.

The two different starting configurations at $t_1$ in figure 6.6 both most likely end up at the same destination at $t_2 < t_1$. This means that the integrand for $S^{\rm odd;0}_{\rm r,dyn}(\mathbf{\Gamma}, t)$ goes to zero, and that the lower limit on the integral can be replaced by $t - \tau$ for some convenient time interval.

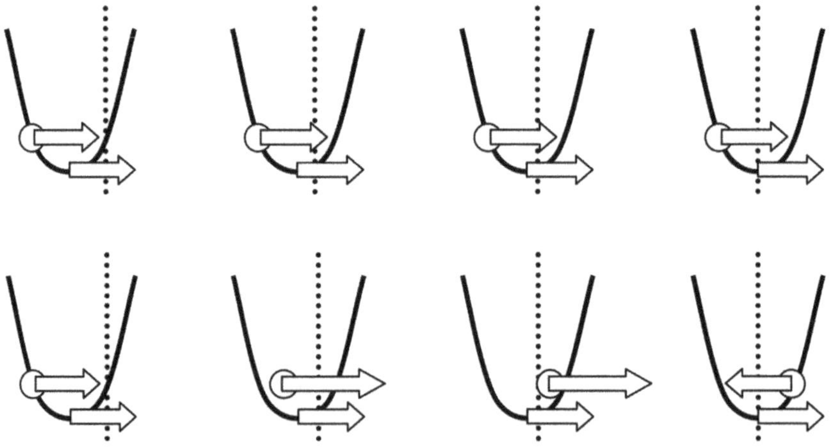

**Figure 6.6.** A Brownian particle in a solvent in a moving potential trap. Time increases from left to right, with the current time $t_1$ being on the far right. The first line shows a backward trajectory from the current point $\mathbf{\Gamma}_1$, $\overline{\mathbf{\Gamma}}(t'|\mathbf{\Gamma}_1, t_1) \approx \mathbf{\Gamma}^0(t'|\mathbf{\Gamma}_1, t_1)$, and the second line shows a trajectory from the unlikely conjugate point, $\overline{\mathbf{\Gamma}}(t'|\mathbf{\Gamma}_1^{\dagger\ddagger}, t_1) \approx \mathbf{\Gamma}^0(t'|\mathbf{\Gamma}_1^{\dagger\ddagger}, t_1)$. The dotted line is fixed in space as a guide to the eye.

It is quite common in the treatment of a driven Brownian particle to make the particle the entire sub-system and to include all of the solvent as the reservoir; such a model is explored in section 6.4 below. In this approach solvent effects are subsumed entirely into the diffusion constant (or the related memory function for rapidly varying applied potentials). The adiabatic approximation would be quite inappropriate in such a picture, since one can only justify replacing the most likely dissipative trajectories by adiabatic trajectories if the latter include the solvent to provide a mechanism and sink for dissipation.

### 6.3.4 Adiabatic time derivative of the entropy

In this section the adiabatic time derivative of the odd projection of the dynamic part of the reservoir entropy is sought. It is actually much simpler to evaluate the adiabatic derivative first, and then take the odd projection. However, the operations of trajectory derivative and odd projection do not commute. In using the result now derived, one implicitly assumes that the non-commuting part is negligible compared to the commuting part.

The odd projection of the adiabatic derivative of the dynamic part of the reservoir entropy may be obtained by approximating the dissipative trajectories by adiabatic ones. This is more valid for a mechanical non-equilibrium system than for a steady state thermodynamic system. One has

$$\begin{aligned}\frac{\mathrm{d}^0 S_{\mathrm{r,dyn}}(\mathbf{\Gamma}, t)}{\mathrm{d}t} &\approx \frac{\mathrm{d}^0 S^0_{\mathrm{r,dyn}}(\mathbf{\Gamma}, t)}{\mathrm{d}t} \\ &= \frac{S^0_{\mathrm{r,dyn}}(\mathbf{\Gamma} + \Delta_t \dot{\mathbf{\Gamma}}^0, t + \Delta_t) - S^0_{\mathrm{r,dyn}}(\mathbf{\Gamma}, t)}{\Delta_t} \\ &= \frac{-1}{\Delta_t}\left\{\Delta_t \dot{S}^0_{\mathrm{r,st}}(\mathbf{\Gamma}^0(t|\mathbf{\Gamma}, t), t) - \int_0^t \mathrm{d}t' \Big[\dot{S}^0_{\mathrm{r,st}}(\mathbf{\Gamma}^0(t'|\mathbf{\Gamma}, t), t') \right. \\ &\qquad \left. - \dot{S}^0_{\mathrm{r,st}}(\mathbf{\Gamma}^0(t'|\{\mathbf{\Gamma} + \Delta_t \dot{\mathbf{\Gamma}}^0\}, t + \Delta_t), t')\Big]\right\} \\ &= -\dot{S}^0_{\mathrm{r,st}}(\mathbf{\Gamma}, t).\end{aligned} \tag{6.110}$$

The surviving term is the integrand evaluated at the upper limit of the integral (Leibnitz rule). The remaining terms cancel because a point on an adiabatic trajectory at time $t'$ is unchanged by specifying any point and time that lies on it, $\mathbf{\Gamma}^0(t'|\mathbf{\Gamma} + \Delta_t \dot{\mathbf{\Gamma}}^0, t + \Delta_t) = \mathbf{\Gamma}^0(t'|\mathbf{\Gamma}, t)$. With this the odd projection of the adiabatic derivative of the dynamic part of the reservoir entropy is

$$\begin{aligned}\dot{S}^{0;\mathrm{odd}}_{\mathrm{r,dyn}}(\mathbf{\Gamma}, t) &\equiv \frac{1}{2}\left\{\frac{\mathrm{d}^0 S_{\mathrm{r,dyn}}(\mathbf{\Gamma}, t)}{\mathrm{d}t} - \frac{\mathrm{d}^0 S_{\mathrm{r,dyn}}(\mathbf{\Gamma}^{\#}, t)}{\mathrm{d}t}\right\} \\ &\approx -\dot{S}^0_{\mathrm{r,st}}(\mathbf{\Gamma}, t).\end{aligned} \tag{6.111}$$

Since by definition, $\dot{S}^0_{\mathrm{r,st}}(\mathbf{\Gamma}^{\#}, t) = -\dot{S}^0_{\mathrm{r,st}}(\mathbf{\Gamma}, t)$. The accuracy or utility of this result is unclear.

This result should not be confused with the adiabatic derivative of the odd projection of the dynamic part of the reservoir entropy. The latter is not so easy to obtain explicitly because the conjugate starting points of the adiabatic trajectories do not lie on the same trajectory,

$$\begin{aligned}\mathbf{\Gamma}^0(t'|\{\mathbf{\Gamma} + \Delta_t\dot{\mathbf{\Gamma}}^0\}^\dagger, t + \Delta_t) &= \mathbf{\Gamma}^0(t'|\mathbf{\Gamma}^\dagger + \Delta_t\{\dot{\mathbf{\Gamma}}^0\}^\dagger - \Delta_t\dot{\mathbf{\Gamma}}^0(\mathbf{\Gamma}^\dagger), t) \\ &= \mathbf{\Gamma}^0(t'|\mathbf{\Gamma}^\dagger + \Delta_t\{\dot{\mathbf{\Gamma}}^0\}^\dagger + \Delta_t\{\dot{\mathbf{\Gamma}}^0\}^\dagger), t) \\ &= \mathbf{\Gamma}^0(t'|\mathbf{\Gamma}^\dagger + 2\Delta_t\{\dot{\mathbf{\Gamma}}^0\}^\dagger, t).\end{aligned} \tag{6.112}$$

The even projection of the dynamic part of the reservoir entropy can be approximated by the asymptote of the integrand. Hence the adiabatic derivative of the even projection of the dynamic part of the reservoir entropy is approximately

$$\begin{aligned}\frac{\mathrm{d}^0 S_{\mathrm{r,dyn}}^{\mathrm{even}}(\mathbf{\Gamma}, t)}{\mathrm{d}t} &\approx \frac{\mathrm{d}^0 \overline{S}_{\mathrm{r,dyn}}(t)}{\mathrm{d}t} \\ &= \frac{\partial \overline{S}_{\mathrm{r,dyn}}(t)}{\partial t} \\ &= - \overline{\dot{S}_{\mathrm{r,st}}^0}(t).\end{aligned} \tag{6.113}$$

With this the adiabatic total time derivative of the dynamic part of the reservoir entropy may be taken to be

$$\dot{S}_{\mathrm{r,dyn}}^0(\mathbf{\Gamma}, t) \equiv \frac{\mathrm{d}^0 S_{\mathrm{r,dyn}}(\mathbf{\Gamma}, t)}{\mathrm{d}t} \approx -\dot{S}_{\mathrm{r,st}}^0(\mathbf{\Gamma}, t) - \overline{\dot{S}_{\mathrm{r,st}}^0}(t). \tag{6.114}$$

As mentioned above, the form of the first term on the right-hand side assumes that the commuting part of the adiabatic derivative of the odd projection dominates the non-commuting part.

It follows that the adiabatic total time derivative of the reservoir entropy is

$$\begin{aligned}\dot{S}_{\mathrm{r}}^0(\mathbf{\Gamma}, t) &\equiv \frac{\mathrm{d}^0 S_{\mathrm{r}}(\mathbf{\Gamma}, t)}{\mathrm{d}t} \\ &\approx - \overline{\dot{S}_{\mathrm{r,st}}^0}(t).\end{aligned} \tag{6.115}$$

This is evidently a constant in phase space. This holds for both steady state thermodynamic systems and for mechanical non-equilibrium systems (subject to the validity of the various approximations invoked above).

This is consistent with equation (6.30) above, which assumed that the exact result for the average adiabatic time derivative of the reservoir entropy held locally. That result is

$$\dot{S}_{\mathrm{r}}^0(\mathbf{\Gamma}, t) = \frac{k_{\mathrm{B}}\dot{Z}(t)}{Z(t)} = \dot{S}_{\mathrm{tot}}(t). \tag{6.116}$$

One can therefore conclude that the rate of change of the total entropy is equal and opposite to the most likely adiabatic rate of change of the static part of the reservoir entropy,

$$\dot{S}_{\text{tot}}(t) = -\overline{\dot{S}^{0}_{\text{r,st}}(t)}. \tag{6.117}$$

### 6.3.5 Steady state system

This general result applied to a general steady state system is

$$\dot{S}_{\text{tot}}(x_{\text{r}}, \nabla x_{\text{r}}, t) = \dot{S}_{\text{r}}(x_{\text{r}}, \nabla x_{\text{r}}, t) = -\overline{\dot{S}^{0}_{\text{r,st}}}. \tag{6.118}$$

Here $x_{\text{r}}$ is the reservoir field variable or variables. Because this is a steady state system, the structure and hence the sub-system entropy do not change with time, $\dot{S}_{\text{s}}(t) = 0$, and so only the reservoir entropy contributes to the rate of change of the total entropy. For a steady state system, $\overline{\dot{S}^{0}_{\text{r,st}}}$ is independent of time. The physical interpretation of this result is that in a steady state, the internal change in the structure of the sub-system is canceled by exchange with the reservoir, which means that the internal change in entropy is equal and opposite to the change in reservoir entropy.

Since the material flux is constant, so is the rate of change of reservoir entropy, and so integrating the latter gives the reservoir entropy as a linear function of time, $S_{\text{r}}(x_{\text{r}}, \nabla x_{\text{r}}, t) = S_{\text{r}}(x_{\text{r}}, \nabla x_{\text{r}}, 0) - t\,\overline{\dot{S}^{0}_{\text{r,st}}}$. The partition function is therefore of the form

$$Z(x_{\text{r}}, \nabla x_{\text{r}}, t) = Z(x_{\text{r}}, \nabla x_{\text{r}})e^{-t\,\overline{\dot{S}^{0}_{\text{r,st}}}/k_{\text{B}}}, \tag{6.119}$$

which gives the total unconstrained entropy as

$$S_{\text{tot}}(x_{\text{r}}, \nabla x_{\text{r}}, t) = S_{\text{tot}}(x_{\text{r}}, \nabla x_{\text{r}}) - t\,\overline{\dot{S}^{0}_{\text{r,st}}}. \tag{6.120}$$

A steady state system is homogeneous in time,

$$\overline{\mathbf{\Gamma}}(t|\mathbf{\Gamma}_0, t_0) = \overline{\mathbf{\Gamma}}(t - t_0|\mathbf{\Gamma}_0). \tag{6.121}$$

Hence the dynamic part of the reservoir entropy may be written

$$\begin{aligned} S_{\text{r,dyn}}(\mathbf{\Gamma}|\nabla x_{\text{r}}, t) &= -\int_0^t \mathrm{d}t' \dot{S}^{0}_{\text{r,st}}(\overline{\mathbf{\Gamma}}(t'|\mathbf{\Gamma}, t)) \\ &= -\int_{-t}^0 \mathrm{d}t'' \dot{S}^{0}_{\text{r,st}}(\overline{\mathbf{\Gamma}}(t''|\mathbf{\Gamma})), \quad t'' = t' - t. \end{aligned} \tag{6.122}$$

The partial time derivative of this is

$$\begin{aligned} \frac{\partial S_{\text{r,dyn}}(\mathbf{\Gamma}|\nabla x_{\text{r}}, t)}{\partial t} &= -\dot{S}^{0}_{\text{r,st}}(\overline{\mathbf{\Gamma}}(-t|\mathbf{\Gamma}, 0)) \\ &= -\overline{\dot{S}^{0}_{\text{r,st}}}. \end{aligned} \tag{6.123}$$

This is evidently a constant in phase space. The second equality holds because the dynamic part of the entropy is formulated such that $t$ is assumed large enough for the system to have settled into a steady state and the integrand to have reached its asymptotic limit.

For a steady state system the macroscopic structure of the sub-system is fixed and the static part of the reservoir entropy is not explicitly dependent on time, $S_{\rm r,st}(\mathbf{\Gamma})$. Hence in this case the partial time derivative of the reservoir entropy is just that of the dynamic part alone,

$$\frac{\partial S_{\rm r}(\mathbf{\Gamma}|x_{\rm r}, \nabla x_{\rm r}, t)}{\partial t} = \frac{\partial S_{\rm r,dyn}(\mathbf{\Gamma}|\nabla x_{\rm r}, t)}{\partial t} = -\overline{\dot{S}^0_{\rm r,st}}. \tag{6.124}$$

In view of equation (6.115), $\dot{S}^0_{\rm r}(\mathbf{\Gamma}, t) = -\overline{\dot{S}^0_{\rm r,st}}(t)$, this implies that

$$\dot{\mathbf{\Gamma}}^0 \cdot \nabla S_{\rm r}(\mathbf{\Gamma}, t) = 0. \tag{6.125}$$

This agrees with equation (6.31).

With this result for the partial time derivative of the dynamic entropy for a steady state system, one can write

$$S_{\rm r,dyn}(\mathbf{\Gamma}|x_{\rm r}, \nabla x_{\rm r}, t) = S_{\rm r,dyn}(\mathbf{\Gamma}|\nabla x_{\rm r}) - t\,\overline{\dot{S}^0_{\rm r,st}} \tag{6.126}$$

with

$$S_{\rm r,dyn}(\mathbf{\Gamma}|\nabla x_{\rm r}) = -\int_{-\infty}^{0} {\rm d}t'' \left\{ \dot{S}^0_{\rm r,st}(\overline{\mathbf{\Gamma}}(t''|\mathbf{\Gamma}, 0)) - \overline{\dot{S}^0_{\rm r,st}} \right\}. \tag{6.127}$$

The probability density is

$$\begin{aligned} \wp(\mathbf{\Gamma}|x_{\rm r}, \nabla x_{\rm r}) &= \frac{e^{S_{\rm r,st}(\mathbf{\Gamma}|x_{\rm r},\nabla x_{\rm r})/k_{\rm B}} e^{S_{\rm r,dyn}(\mathbf{\Gamma}|\nabla x_{\rm r},t)/k_{\rm B}}}{h^{3N} N! Z(x_{\rm r}, \nabla x_{\rm r}, t)} \\ &= \frac{e^{S_{\rm r,st}(\mathbf{\Gamma}|x_{\rm r},\nabla x_{\rm r})/k_{\rm B}} e^{S_{\rm r,dyn}(\mathbf{\Gamma}|\nabla x_{\rm r})/k_{\rm B}}}{h^{3N} N! Z(x_{\rm r}, \nabla x_{\rm r})}. \end{aligned} \tag{6.128}$$

The passage from the first to the second equality confirms the first three equations in this subsection, equation (6.118) *et seq.*

The constant part of the partition function is just

$$\begin{aligned} Z(x_{\rm r}, \nabla x_{\rm r}) &= \frac{1}{h^{3N} N!} \int {\rm d}\mathbf{\Gamma}\, e^{S_{\rm r,st}(\mathbf{\Gamma}|x_{\rm r},\nabla x_{\rm r})/k_{\rm B}} e^{S_{\rm r,dyn}(\mathbf{\Gamma}|\nabla x_{\rm r})/k_{\rm B}} \\ &= e^{S_{\rm s}\left(\overline{X_{\rm s}}, \overline{\dot{X}^0_{\rm s}}|x_{\rm r},\nabla x_{\rm r}\right)/k_{\rm B}} e^{S_{\rm r,st}(\overline{X_{\rm s}}|x_{\rm r},\nabla x_{\rm r})/k_{\rm B}}. \end{aligned} \tag{6.129}$$

The second equality is exact when fluctuations are negligible, which they are in the thermodynamic limit. In terms of the total unconstrained entropy this is

$$S_{\rm tot}(x_{\rm r}, \nabla x_{\rm r}) = S_{\rm s}\left(\overline{X_{\rm s}}, \overline{\dot{X}^0_{\rm s}}|x_{\rm r}, \nabla x_{\rm r}\right) + S_{\rm r,st}(\overline{X_{\rm s}}|x_{\rm r}, \nabla x_{\rm r}). \tag{6.130}$$

The sub-system entropy in the most likely macrostate, $\overline{S}_s$, contains the effects of the dynamic order due to the induced energy flux, which is embodied in the dynamic part of the entropy.

*Steady heat flow*

For steady heat flow, the static part of the reservoir entropy is

$$S_{r,st}(\mathbf{\Gamma}) = \frac{-E_0(\mathbf{\Gamma})}{T_0} - \frac{E_1(\mathbf{\Gamma})}{T_1}. \tag{6.131}$$

The most likely heat flux is

$$\begin{aligned} \overline{J}_E^0 &= \frac{1}{V}\overline{\dot{E}_1^0} \\ &= \frac{\hat{\tau}}{2V}\Lambda\frac{1}{T_1}. \end{aligned} \tag{6.132}$$

Since the present case is that of an applied temperature gradient rather than a spontaneous fluctuation, one has $\hat{\tau} = +1$ here.

The adiabatic rate of change of the static part of the reservoir entropy is

$$\overline{\dot{S}_{r,st}^0} = \frac{-1}{T_1}\overline{\dot{E}_1^0}. \tag{6.133}$$

Since the flux is constant, the reservoir entropy increases linearly with time, and it may be written as $S_r(t|T_0, T_1) = S_r(0|T_0, T_1) + t\overline{\dot{E}_1^0}/T_1$, with the (immaterial) value at $t = 0$ taken to be

$$S_r(0|T_0, T_1) = \overline{S}_{r,st}(T_0, T_1) = \frac{-\overline{E}_0(\mathbf{\Gamma})}{T_0} - \frac{\overline{E}_1(\mathbf{\Gamma})}{T_1}. \tag{6.134}$$

The partition function is therefore of the form

$$Z(T_0, T_1, t) = Z(T_0, T_1)e^{-t\,\overline{\dot{S}_{r,st}^0}/k_B}, \tag{6.135}$$

which gives the total unconstrained entropy as

$$S_{tot}(T_0, T_1, t) = S_{tot}(T_0, T_1) - t\,\overline{\dot{S}_{r,st}^0}. \tag{6.136}$$

The partial time derivative of the dynamic part of the reservoir entropy is

$$\frac{\partial S_{r,dyn}(\mathbf{\Gamma}, t|T_1)}{\partial t} = -\overline{\dot{S}_{r,st}^0} = \frac{1}{T_1}\overline{\dot{E}_1^0}. \tag{6.137}$$

The probability density is

$$\begin{aligned} \wp(\mathbf{\Gamma}|T_0, T_1) &= \frac{e^{S_{r,st}(\mathbf{\Gamma}|T_0,T_1)/k_B}e^{S_{r,dyn}(\mathbf{\Gamma}|T_1,t)/k_B}}{h^{3N}N!Z(T_0, T_1, t)} \\ &= \frac{e^{S_{r,st}(\mathbf{\Gamma}|T_0,T_1)/k_B}e^{S_{r,dyn}(\mathbf{\Gamma}|T_1)/k_B}}{h^{3N}N!Z(T_0, T_1)}. \end{aligned} \tag{6.138}$$

The time-independent part of the partition function is just

$$\begin{aligned} Z(T_0, T_1) &= \frac{1}{h^{3N} N!} \int \mathrm{d}\mathbf{\Gamma} e^{S_{\mathrm{r,st}}(\mathbf{\Gamma}|T_0, T_1)/k_{\mathrm{B}}} e^{S_{\mathrm{r,dyn}}(\mathbf{\Gamma}|T_1)/k_{\mathrm{B}}} \\ &= e^{S_{\mathrm{s}}\left(\overline{E}_0, \overline{E}_1, \overline{\dot{E}_1^0}\middle| T_0, T_1\right)/k_{\mathrm{B}}} e^{S_{\mathrm{r,st}}(\overline{E}_0, \overline{E}_1| T_0, T_1)/k_{\mathrm{B}}}. \end{aligned} \tag{6.139}$$

The second equality is exact when fluctuations are negligible, which they are in the thermodynamic limit. In terms of the total unconstrained entropy this is

$$\begin{aligned} S_{\mathrm{tot}}(T_0, T_1) &= S_{\mathrm{s}}\left(\overline{E}_0, \overline{E}_1, \overline{\dot{E}_1^0} | T_0, T_1\right) + S_{\mathrm{r,st}}(\overline{E}_0, \overline{E}_1 | T_0, T_1) \\ &= S_{\mathrm{s}}\left(\overline{E}_0, \overline{E}_1, \overline{\dot{E}_1^0} | T_0, T_1\right) - \frac{\overline{E}_0}{T_0} - \frac{\overline{E}_1}{T_1}. \end{aligned} \tag{6.140}$$

The sub-system entropy in the most likely macrostate, $S_{\mathrm{s}}(\overline{E}_0, \overline{E}_1, \overline{\dot{E}_1^0} | T_0, T_1)$, contains the effects of the dynamic order due to the induced energy flux, which is embodied in the dynamic part of the entropy.

### 6.3.6 Green–Kubo relations

The validity and utility of the expression for the non-equilibrium probability will now be illustrated with a simple derivation of the Green–Kubo relations (Onsager 1931, Green 1954, Kubo 1966). These relate the hydrodynamic transport coefficients to the equilibrium time correlation functions of the fluxes.

For the particular case of heat flow, the static part of the reservoir entropy is (see chapter 3)

$$S_{\mathrm{r,st}}(\mathbf{\Gamma}) = \frac{-E_0(\mathbf{\Gamma})}{T_0} - \frac{E_1(\mathbf{\Gamma})}{T_1}, \tag{6.141}$$

where the $n$th energy moment in the $z$-direction is $E_n(\mathbf{\Gamma}) \equiv \int \mathrm{d}\mathbf{r}\, \epsilon(\mathbf{r}; \mathbf{\Gamma}) z^n$, with $\epsilon(\mathbf{r}; \mathbf{\Gamma})$ being the energy density at $\mathbf{r}$. Also the zeroth temperature is the mid-temperature of the two reservoirs, $T_0^{-1} \equiv [T_+^{-1} + T_-^{-1}]/2 = T^{-1} + \mathcal{O}(\nabla T)^2$, and the first temperature is essentially the temperature gradient imposed by them, $T_1^{-1} \equiv [T_+^{-1} - T_-^{-1}]/L_z = -T^{-2}\nabla T + \mathcal{O}(\nabla T)^2$.

The instantaneous heat flux, a phase function of the isolated sub-system, is essentially the adiabatic rate of change of the first energy moment

$$J_{\mathrm{E}}(\mathbf{\Gamma}) \equiv \dot{E}_1^0(\mathbf{\Gamma})/V, \tag{6.142}$$

where $V$ is the volume of the sub-system. Due to energy conservation of the isolated system, $\dot{E}_0^0(\mathbf{\Gamma}) = 0$ and $\dot{S}_{\mathrm{r,st}}^0(\mathbf{\Gamma}) = -\dot{E}_1^0(\mathbf{\Gamma})/T_1$.

Fourier's law gives the heat flow in the presence of an applied thermal gradient, and it is (Kubo *et al* 1978, Zwanzig 2001, Bellac *et al* 2004, Pottier 2010, Attard 2012)

$$\overline{J}_{\mathrm{E}} = -\lambda \nabla T, \tag{6.143}$$

where $\lambda$ is the thermal conductivity. The left-hand side is the most likely heat flux, which equals the average heat flux. This law of course holds to linear order in the temperature gradient.

The average heat flux given by the present non-equilibrium theory is

$$\begin{aligned}
\langle J_{\mathrm{E}}\rangle_{\text{non-equil}} &= \frac{1}{V}\int \mathrm{d}\mathbf{\Gamma}\,\wp(\mathbf{\Gamma}|T_0,\,T_1)\dot{E}_1^0(\mathbf{\Gamma}) \\
&= \frac{1}{V}\frac{\int \mathrm{d}\mathbf{\Gamma}\,e^{[S_{\mathrm{r,st}}(\mathbf{\Gamma})+S_{\mathrm{r,dyn}}(\mathbf{\Gamma})]/k_{\mathrm{B}}}\dot{E}_1^0(\mathbf{\Gamma})}{\int \mathrm{d}\mathbf{\Gamma}\,e^{[S_{\mathrm{r,st}}(\mathbf{\Gamma})+S_{\mathrm{r,dyn}}(\mathbf{\Gamma})]/k_{\mathrm{B}}}} \\
&= \frac{1}{Vk_{\mathrm{B}}}\frac{\int \mathrm{d}\mathbf{\Gamma}\,e^{-E_0(\mathbf{\Gamma})/k_{\mathrm{B}}T_0}\dot{E}_1^0(\mathbf{\Gamma})S_{\mathrm{r,dyn}}^{\mathrm{odd;0}}(\mathbf{\Gamma})}{\int \mathrm{d}\mathbf{\Gamma}\,e^{-E_0(\mathbf{\Gamma})/k_{\mathrm{B}}T_0}} + \mathcal{O}(\nabla T)^2 \\
&= \frac{1}{2Vk_{\mathrm{B}}T_1}\int_{-\tau}^{0}\mathrm{d}t'\left\langle \dot{E}_1^0(\mathbf{\Gamma})\left[\dot{E}_1^0(\mathbf{\Gamma}^0(t'|\mathbf{\Gamma},0)) - \dot{E}_1^0(\mathbf{\Gamma}^0(t'|\mathbf{\Gamma}^\dagger,0))\right]\right\rangle_{\text{equil}} \\
&= \frac{-\nabla T}{2Vk_{\mathrm{B}}T_0^2}\int_{-\tau}^{\tau}\mathrm{d}t'\left\langle \dot{E}_1^0(\mathbf{\Gamma})\dot{E}_1^0(\mathbf{\Gamma}^0(t'|\mathbf{\Gamma},0))\right\rangle_{\text{equil}}.
\end{aligned} \tag{6.144}$$

In the third equality the exponentials have been expanded in powers of the temperature gradient and second order terms have been neglected. As well, terms that are the product of an even parity function, $S_{\mathrm{r,st}}(\mathbf{\Gamma})$ or $S_{\mathrm{r,dyn}}^{\mathrm{even}}(\mathbf{\Gamma})$, and an odd parity function, $\dot{S}_{\mathrm{r,st}}^0(\mathbf{\Gamma})$ or $\dot{E}_1^0(\mathbf{\Gamma})$, vanish upon integration over phase space. In addition, the most likely trajectory has been replaced by the adiabatic trajectory, equation (6.98). The equilibrium average arises because $Z(T_0)^{-1}e^{-E_0(\mathbf{\Gamma})/k_{\mathrm{B}}T_0}$ is the Maxwell–Boltzmann distribution. Comparing this to Fourier's law, one can identify the thermal conductivity as

$$\begin{aligned}
\lambda &= \frac{1}{2Vk_{\mathrm{B}}T_0^2}\int_{-\tau}^{\tau}\mathrm{d}t'\left\langle E_1^0(\mathbf{\Gamma})\dot{E}_1^0(\mathbf{\Gamma}^0(t'|\mathbf{\Gamma},0))\right\rangle_{\text{equil}} \\
&= \frac{1}{2Vk_{\mathrm{B}}T_0^2}\left\langle \dot{E}_1^0(t)[E_1(t+\tau) - E_1(t-\tau)]\right\rangle_{\text{equil}}.
\end{aligned} \tag{6.145}$$

The right-hand side is independent of $\tau$ for $\tau \gtrsim \tau_{\text{relax}}$. This can be written in a number of different ways, but all involve the equilibrium time correlation function of the heat flux or an integral thereof. This is a typical example of a Green–Kubo relation (Onsager 1931, Green 1954, Kubo 1966). It is to be noted that the time correlation function in any Green–Kubo relation *always* invokes adiabatic trajectories (Kubo *et al* 1978, Zwanzig 2001, Bellac *et al* 2004, Pottier 2010, Attard 2012).

From this analysis one sees that the general formula for obtaining the Green–Kubo relations for a steady state thermodynamic system is

$$
\begin{aligned}
&\left\langle \dot{S}^{0}_{\mathrm{r,st}}(\mathbf{\Gamma}) \right\rangle_{\mathrm{non-equil}} \\
&= \int \mathrm{d}\mathbf{\Gamma}\ \wp(\mathbf{\Gamma}|x_{\mathrm{r}}, \nabla x_{\mathrm{r}}) \dot{S}^{0}_{\mathrm{r,st}}(\mathbf{\Gamma}) \\
&= \frac{\int \mathrm{d}\mathbf{\Gamma} e^{\left[S_{\mathrm{r,st}}(\mathbf{\Gamma}|x_{\mathrm{r}}, \nabla x_{\mathrm{r}}) + S_{\mathrm{r,dyn}}(\mathbf{\Gamma}|\nabla x_{\mathrm{r}})\right]/k_{\mathrm{B}}} \dot{S}^{0}_{\mathrm{r,st}}(\mathbf{\Gamma})}{\int \mathrm{d}\mathbf{\Gamma} e^{\left[S_{\mathrm{r,st}}(\mathbf{\Gamma}|x_{\mathrm{r}}, \nabla x_{\mathrm{r}}) + S_{\mathrm{r,dyn}}(\mathbf{\Gamma}|\nabla x_{\mathrm{r}})\right]/k_{\mathrm{B}}}} \\
&= k_{\mathrm{B}}^{-1} \frac{\int \mathrm{d}\mathbf{\Gamma} e^{S_{\mathrm{r,st}}(\mathbf{\Gamma}|x_{\mathrm{r}})/k_{\mathrm{B}}} S_{\mathrm{r,dyn}}(\mathbf{\Gamma}|\nabla x_{\mathrm{r}}) \dot{S}^{0}_{\mathrm{r,st}}(\mathbf{\Gamma})}{\int \mathrm{d}\mathbf{\Gamma} e^{S_{\mathrm{r,st}}(\mathbf{\Gamma}|x_{\mathrm{r}})/k_{\mathrm{B}}}} + \mathcal{O}(\nabla x_{\mathrm{r}})^2 \\
&= k_{\mathrm{B}}^{-1} \left\langle S^{\mathrm{odd}}_{\mathrm{r,dyn}}(\mathbf{\Gamma}|\nabla x_{\mathrm{r}}) \dot{S}^{0}_{\mathrm{r,st}}(\mathbf{\Gamma}) \right\rangle_{\mathrm{equil}} \\
&\approx k_{\mathrm{B}}^{-1} \left\langle S^{\mathrm{odd;0}}_{\mathrm{r,dyn}}(\mathbf{\Gamma}|\nabla x_{\mathrm{r}}) \dot{S}^{0}_{\mathrm{r,st}}(\mathbf{\Gamma}) \right\rangle_{\mathrm{equil}}.
\end{aligned}
\tag{6.146}
$$

In the third equality the exponentials have been expanded in powers of the gradient in the reservoir field variable, and second order terms have been neglected. In the denominator the term linear in the gradient vanishes because it has odd spatial symmetry. In the fourth equality the odd temporal parity projection of the dynamic part of the reservoir entropy is the only non-zero contribution because the adiabatic flux has odd temporal parity. In the fifth equality, the most likely trajectory has been replaced by the adiabatic trajectory, equation (6.98). The equilibrium average arises because $S_{\mathrm{r,st}}(\mathbf{\Gamma}|x_{\mathrm{r}}, \nabla x_{\mathrm{r}})$ is the equilibrium reservoir entropy for the point in phase space. The left-hand side is the flux times the volume averaged in the non-equilibrium system with applied reservoir gradient. The right-hand side is the equilibrium average of that flux multiplied by the temporal integral that is here identified with the dynamic part of the reservoir entropy.

The fact that the theory gives the Green–Kubo relations should give one confidence both in the adiabatic transformation of the dynamic part of the reservoir entropy, equation (6.98), and in the general expression for the phase space probability for non-equilibrium systems, equations (6.4) and (6.14). Of course, the Green–Kubo relations are a linear theory, whereas the full expression for the phase space probability for non-equilibrium systems, equations (6.4) and (6.14), applies in all circumstances, linear and non-linear.

## 6.4 Driven Brownian motion

Consider a mechanical non-equilibrium system consisting of a solute that is acted upon by a time-varying potential,

$$
U(\mathbf{x}, t) = \frac{\kappa}{2}[\mathbf{x} - \mathbf{b}(t)] \cdot [\mathbf{x} - \mathbf{b}(t)]. \tag{6.147}
$$

Here $\mathbf{x}$ is the position of the solute. The potential is in the form of a trap, with $\kappa > 0$. We shall assume that the location of the trap is in uniform motion,

$$\mathbf{b}(t) = \dot{\mathbf{b}}t. \tag{6.148}$$

All of the solvent is subsumed into the reservoir of temperature $T$, and so the Hamiltonian of the sub-system is

$$\mathcal{H}(\dot{\mathbf{x}}, \mathbf{x}; t) = \frac{m}{2}\dot{x}^2 + U(\mathbf{x}, t), \tag{6.149}$$

where $m$ is the mass of the solute. The static part of the reservoir entropy is

$$\begin{aligned} S_{\mathrm{r,st}}(\dot{\mathbf{x}}, \mathbf{x}; t) &= \frac{-\mathcal{H}(\dot{\mathbf{x}}, \mathbf{x}; t)}{T} \\ &= \frac{-m\dot{x}^2}{2T} - \frac{\kappa}{2T}[\mathbf{x} - \mathbf{b}(t)] \cdot [\mathbf{x} - \mathbf{b}(t)]. \end{aligned} \tag{6.150}$$

From physical considerations, for uniform trap velocity, it is most likely that the particle will move with the same velocity as the trap, and that it will lag the trap location such that the mechanical force is equal and opposite to the drag force (see figure 6.7). That is, the most likely configuration can be expected to be

$$\overline{\mathbf{X}}(t) \equiv \{\overline{\mathbf{x}}(t), \dot{\overline{\mathbf{x}}}(t)\} = \{\mathbf{b}(t) - \gamma\dot{\mathbf{b}}/\kappa, \dot{\mathbf{b}}\}. \tag{6.151}$$

This maximizes the reservoir entropy,

$$\nabla S_{\mathrm{r}}(\mathbf{X}, t)|_{\overline{\mathbf{X}}(t)} = \mathbf{0}. \tag{6.152}$$

This result will be confirmed by explicit calculation toward the end of the following analysis.

The adiabatic rate of static entropy production is

$$\begin{aligned} \dot{S}^0_{\mathrm{r,st}}(\mathbf{X}, t) &= \frac{\partial S_{\mathrm{r,st}}(\mathbf{X}, t)}{\partial t} \\ &= \frac{\kappa}{T}\dot{\mathbf{b}} \cdot (\mathbf{x} - \mathbf{b}(t)). \end{aligned} \tag{6.153}$$

According to equation (6.117), we should expect that the rate of total entropy production is equal and opposite of the most likely value of this

$$\dot{S}_{\mathrm{tot}}(t) = -\overline{\dot{S}^0_{\mathrm{r,st}}}(t) = -\dot{S}^0_{\mathrm{r,st}}(\overline{\mathbf{X}}(t), t) = \frac{\gamma}{T}\dot{b}^2. \tag{6.154}$$

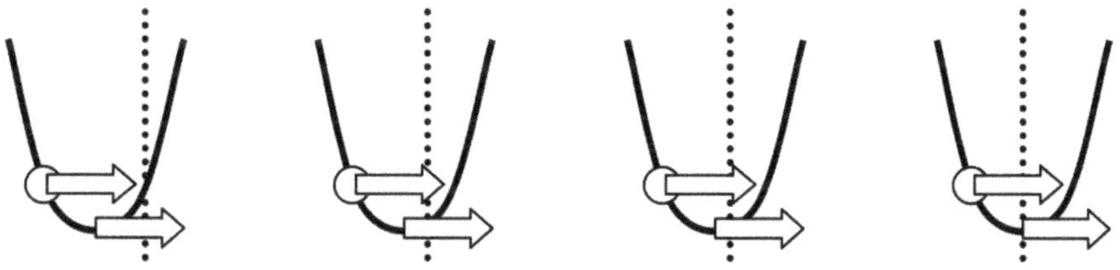

**Figure 6.7.** The most likely configuration of a Brownian particle in a solvent in a steadily moving potential trap. Time increases from left to right, and the dotted line is fixed in space as a guide to the eye.

From the results of section 6.2.2, the stochastic dissipative equations of motion over an infinitesimal time step are

$$\begin{aligned} \mathbf{x}(t+\Delta_t) &= \mathbf{x}(t) + \Delta_t \dot{\mathbf{x}}(t), \\ \dot{\mathbf{x}}(t+\Delta_t) &= \dot{\mathbf{x}}(t) - \frac{\Delta_t}{m}\frac{\partial \mathcal{H}(\dot{\mathbf{x}}, \mathbf{x}; t)}{\partial \mathbf{x}} + \mathbf{R}(\dot{\mathbf{x}}, \mathbf{x}; t, \Delta_t) \\ &= \dot{\mathbf{x}}(t) - \frac{\Delta_t \kappa}{m}[\mathbf{x}(t) - \mathbf{b}(t)] + \overline{\mathbf{R}} + \tilde{\mathbf{R}}. \end{aligned} \tag{6.155}$$

Note that the reservoir 'force' is here in velocity space rather than the more conventional momentum space. Hence it is the actual reservoir force divided by the particle mass.

In the present model all of the solvent is subsumed into the reservoir, and the subsystem consists of a single Brownian particle. In this case it would be an exceedingly poor approximation to replace the dissipative equations of motion by adiabatic equations of motion.

Making contact with Langevin's equation (4.19), specifically in the second entropy form, equation (4.21), the transport matrix can be written in terms of the the drag coefficient, $\Lambda = 2m^{-2}T\gamma \mathrm{I}$. Hence the probability distribution of the stochastic part of the force is

$$\wp(\tilde{\mathbf{R}}|\Delta_t) = \frac{e^{-m^2\tilde{R}^2/4\gamma|\Delta_t|k_{\mathrm{B}}T}}{(4\pi m^{-2}\gamma|\Delta_t|k_{\mathrm{B}}T)^{3/2}}. \tag{6.156}$$

The diffusion constant $D$ and the drag coefficient $\gamma$ are related as $D = k_{\mathrm{B}}T/\gamma$.

According to equation (6.67), the dissipative force is

$$\begin{aligned} \overline{\mathbf{R}}_\nu &= \frac{|\Delta_t|}{2}\Lambda\nabla_\nu S_{\mathrm{r,st}}(\mathbf{X}(t), t) - \frac{|\Delta_t| - \Delta_t}{2}\Lambda\nabla_\nu S_{\mathrm{r,st}}(\overline{\mathbf{X}}(t), t) \\ &= \frac{-|\Delta_t|\gamma}{m}\dot{\mathbf{x}}(t) + \frac{(|\Delta_t| - \Delta_t)\gamma}{m}\overline{\dot{\mathbf{x}}}(t) \\ &= \frac{-|\Delta_t|\gamma}{m}[\dot{\mathbf{x}}(t) - \overline{\dot{\mathbf{x}}}(t)] - \frac{\Delta_t\gamma}{m}\overline{\dot{\mathbf{x}}}(t). \end{aligned} \tag{6.157}$$

The usual drag force going forward in time is $\mathbf{F}_{\mathrm{d}} = -\gamma\dot{\mathbf{x}}$, which can be identified in this.

For future comparison, the zeroth order approximation, equation (6.74), which is here $\overline{\mathbf{R}}_\nu = |\Delta_t|\Lambda\nabla_\nu S_{\mathrm{r,st}}(\mathbf{\Gamma}, t)/2$, is equivalent to setting the second term of the first and second equalities to zero.

Therefore, the stochastic, dissipative equations of motion for the uniformly driven Brownian particle are explicitly

$$\begin{aligned} \mathbf{x}(t+\Delta_t) &= \mathbf{x}(t) + \Delta_t\dot{\mathbf{x}}(t), \\ \dot{\mathbf{x}}(t+\Delta_t) &= \dot{\mathbf{x}}(t) - \frac{\Delta_t\kappa}{m}[\mathbf{x}(t) - \mathbf{b}(t)] - \frac{|\Delta_t|\gamma}{m}\dot{\mathbf{x}} + \frac{(|\Delta_t| - \Delta_t)\gamma}{m}\dot{\mathbf{b}} + \tilde{\mathbf{R}}_\nu. \end{aligned} \tag{6.158}$$

Setting $\tilde{\mathbf{R}}_v = \mathbf{0}$, and taking the limit $\Delta_t \to 0$, these dissipative equations may be re-written in the form

$$\left.\frac{\mathrm{d}\bar{\mathbf{x}}(t'|\mathbf{X}(t), t)}{\mathrm{d}t'}\right|_{t'=t^\pm} = \dot{\mathbf{x}}(t),$$
$$\left.\frac{\mathrm{d}\bar{\dot{\mathbf{x}}}(t'|\mathbf{X}(t), t)}{\mathrm{d}t'}\right|_{t'=t^\pm} = \frac{-\kappa}{m}[\mathbf{x}(t) - \mathbf{b}(t)] - \frac{\hat{\Delta}_t\gamma}{m}\dot{\mathbf{x}}(t) + \frac{(\hat{\Delta}_t - 1)\gamma}{m}\dot{\mathbf{b}}, \tag{6.159}$$

where $\hat{\Delta}_t = \text{sign}\, \Delta_t$.

Since $\mathbf{b}(t) = \dot{\mathbf{b}}t$, in vector form this is

$$\dot{\bar{\mathbf{X}}}(t|\mathbf{X}(t), t) = \mathbf{A}\mathbf{X}(t) + \mathbf{B}t + \mathbf{C}, \tag{6.160}$$

with

$$\mathbf{A} \equiv \begin{pmatrix} \mathbf{0} & \mathbf{I} \\ (-\kappa/m)\mathbf{I} & (-\hat{\Delta}_t\gamma/m)\mathbf{I} \end{pmatrix},$$
$$\mathbf{B} \equiv \{\mathbf{0}, (\kappa/m)\dot{\mathbf{b}}\},$$
$$\text{and } \mathbf{C} \equiv \{\mathbf{0}, ((\hat{\Delta}_t - 1)\gamma/m)\dot{\mathbf{b}}\}. \tag{6.161}$$

In these matrices and vectors, the three components of direction are independent and can be considered separately. On a backward trajectory, $\hat{\Delta}_t = -1$, and $\mathbf{C} = \{\mathbf{0}, (-2\gamma/m)\dot{\mathbf{b}}\}$.

Again for future comparison, the zeroth order approximation, equation (6.74), takes $\mathbf{C} = \mathbf{0}$.

As can be confirmed by direct substitution, the solution to the differential equation is

$$\bar{\mathbf{X}}(t'|\mathbf{X}, t) = e^{(t'-t)\mathbf{A}}[\mathbf{X} + t\mathbf{A}^{-1}\mathbf{B} + \mathbf{A}^{-1}\mathbf{C} + \mathbf{A}^{-2}\mathbf{B}] - t'\mathbf{A}^{-1}\mathbf{B} - \mathbf{A}^{-1}\mathbf{C} - \mathbf{A}^{-2}\mathbf{B}. \tag{6.162}$$

As aforementioned, the three components of direction are independent from each other.

For future reference, the eigenvalues of $\mathbf{A}$ are given by the characteristic equation

$$\lambda_\pm\left[\lambda_\pm + \frac{\hat{\Delta}_t\gamma}{m}\right] + \frac{\kappa}{m} = 0, \tag{6.163}$$

or

$$\lambda_\pm = \frac{-\hat{\Delta}_t\gamma}{2m}\left\{1 \pm \sqrt{1 - 4\kappa m/\gamma^2}\right\} \sim \left\{\frac{-\hat{\Delta}_t\gamma}{m}, \frac{-\hat{\Delta}_t\kappa}{\gamma}\right\}, \kappa \to 0. \tag{6.164}$$

On a backwards trajectory, $\hat{\Delta}_t = -1$, both eigenvalues are positive (assuming $4\kappa m/\gamma^2 < 1$).

Using the adiabatic rate of static entropy production, $\dot{S}^0_{\mathrm{r,st}}(\mathbf{X}, t) = \kappa\dot{\mathbf{b}} \cdot [\mathbf{x}(t) - \mathbf{b}(t)]/T$, and the full expression for the most likely backward trajectory, the dynamic part of the reservoir entropy is

$$\begin{aligned} S_{\mathrm{r,dyn}}&(\mathbf{X}, t) \\ &= -\int_0^t \mathrm{d}t' \dot{S}^0_{\mathrm{r,st}}(\overline{\mathbf{X}}(t'|\mathbf{X}, t), t') \\ &= \frac{-\kappa}{T}\dot{\mathbf{b}} \cdot \int_0^t \mathrm{d}t'[\bar{\mathbf{x}}(t'|\mathbf{X}, t) - \dot{\mathbf{b}}t'] \\ &= \frac{\kappa\dot{b}^2 t^2}{2T} - \frac{\kappa}{T}\dot{\mathbf{b}} \cdot \{\mathbf{A}^{-1}[\mathbf{I} - e^{-t\mathbf{A}}][\mathbf{X} + t\mathbf{A}^{-1}\mathbf{B} + \mathbf{A}^{-1}\mathbf{C} + \mathbf{A}^{-2}\mathbf{B}] \\ &\quad - \frac{t^2}{2}\mathbf{A}^{-1}\mathbf{B} - t\mathbf{A}^{-1}\mathbf{C} - t\mathbf{A}^{-2}\mathbf{B}\Big\}_{\mathbf{x}}. \end{aligned} \tag{6.165}$$

Note that because this is the most likely backward trajectory in the integrand, $\hat{\Delta}_t = -1$ in $\mathbf{A}$ and $\mathbf{C}$. Assuming both eigenvalues are positive, since $t > 0$ the exponential term may be neglected and this becomes

$$\begin{aligned} S_{\mathrm{r,dyn}}&(\mathbf{X}, t) \\ &= \frac{\kappa\dot{b}^2 t^2}{2T} - \frac{\kappa}{T}\dot{\mathbf{b}} \cdot \left\{\mathbf{A}^{-1}\mathbf{X} + \mathbf{A}^{-2}\mathbf{C} + \mathbf{A}^{-3}\mathbf{B} - \frac{t^2}{2}\mathbf{A}^{-1}\mathbf{B} - t\mathbf{A}^{-1}\mathbf{C}\right\}_{\mathbf{x}}. \end{aligned} \tag{6.166}$$

As has been mentioned, the zeroth order approximation, equation (6.74), takes $\mathbf{C} = \mathbf{0}$.

The term quadratic in $t$ in the braces is

$$\begin{aligned} \{\mathbf{A}^{-1}\mathbf{B}\}_{\mathbf{x}} &= \left\{\begin{pmatrix} \mathbf{0} & \mathbf{I} \\ (-\kappa/m)\mathbf{I} & (\gamma/m)\mathbf{I} \end{pmatrix}^{-1} \begin{pmatrix} \mathbf{0} \\ \kappa\dot{\mathbf{b}}/m \end{pmatrix}\right\}_{\mathbf{x}} \\ &= \frac{m}{\kappa}\left\{\begin{pmatrix} (\gamma/m)\mathbf{I} & -\mathbf{I} \\ (\kappa/m)\mathbf{I} & \mathbf{0} \end{pmatrix} \begin{pmatrix} \mathbf{0} \\ \kappa\dot{\mathbf{b}}/m \end{pmatrix}\right\}_{\mathbf{x}} \\ &= -\dot{\mathbf{b}}. \end{aligned} \tag{6.167}$$

This term is equal and opposite the term outside the braces and it therefore cancels with it.

The term linear in time is

$$\begin{aligned} \{\mathbf{A}^{-1}\mathbf{C}\}_{\mathbf{x}} &= \frac{m}{\kappa}\left\{\begin{pmatrix} (\gamma/m)\mathbf{I} & -\mathbf{I} \\ (\kappa/m)\mathbf{I} & \mathbf{0} \end{pmatrix} \begin{pmatrix} \mathbf{0} \\ -2\gamma\dot{\mathbf{b}}/m \end{pmatrix}\right\}_{\mathbf{x}} \\ &= \frac{2\gamma}{\kappa}\dot{\mathbf{b}}. \end{aligned} \tag{6.168}$$

One of the terms independent of time and of **X** is

$$\begin{aligned}\{\mathbf{A}^{-3}\mathbf{B}\}_{\mathrm{x}} &= \frac{m^3}{\kappa^3}\left\{\begin{pmatrix}(\gamma/m)\mathbf{I} & -\mathbf{I}\\(\kappa/m)\mathbf{I} & \mathbf{0}\end{pmatrix}^3\begin{pmatrix}\mathbf{0}\\ \kappa\dot{\mathbf{b}}/m\end{pmatrix}\right\}_{\mathrm{x}}\\
&= \frac{m^3}{\kappa^3}\left\{\begin{pmatrix}(\gamma/m)\mathbf{I} & -\mathbf{I}\\(\kappa/m)\mathbf{I} & \mathbf{0}\end{pmatrix}^2\begin{pmatrix}-\kappa\dot{\mathbf{b}}/m\\ \mathbf{0}\end{pmatrix}\right\}_{\mathrm{x}}\\
&= \frac{m^3}{\kappa^3}\left\{\begin{pmatrix}(\gamma/m)\mathbf{I} & -\mathbf{I}\\(\kappa/m)\mathbf{I} & \mathbf{0}\end{pmatrix}\begin{pmatrix}-\gamma\kappa\dot{\mathbf{b}}/m^2\\ -\kappa^2\dot{\mathbf{b}}/m^2\end{pmatrix}\right\}_{\mathrm{x}}\\
&= \frac{m^3}{\kappa^3}\left[\frac{-\gamma^2\kappa}{m^3} + \frac{\kappa^2}{m^2}\right]\dot{\mathbf{b}}\\
&= \left[\frac{m}{\kappa} - \frac{\gamma^2}{\kappa^2}\right]\dot{\mathbf{b}}.\end{aligned} \tag{6.169}$$

The other term independent of time and of **X** is

$$\begin{aligned}\{\mathbf{A}^{-2}\mathbf{C}\}_{\mathrm{x}} &= \frac{m^2}{\kappa^2}\left\{\begin{pmatrix}(\gamma/m)\mathbf{I} & -\mathbf{I}\\(\kappa/m)\mathbf{I} & \mathbf{0}\end{pmatrix}^2\begin{pmatrix}\mathbf{0}\\ -2\gamma\dot{\mathbf{b}}/m\end{pmatrix}\right\}_{\mathrm{x}}\\
&= \frac{m^2}{\kappa^2}\left\{\begin{pmatrix}(\gamma/m)\mathbf{I} & -\mathbf{I}\\(\kappa/m)\mathbf{I} & \mathbf{0}\end{pmatrix}\begin{pmatrix}2\gamma\dot{\mathbf{b}}/m\\ \mathbf{0}\end{pmatrix}\right\}_{\mathrm{x}}\\
&= \frac{2\gamma^2}{\kappa^2}\dot{\mathbf{b}}.\end{aligned} \tag{6.170}$$

The term dependent on **X** is

$$\begin{aligned}\{\mathbf{A}^{-1}\mathbf{X}\}_{\mathrm{x}} &= \frac{m}{\kappa}\left\{\begin{pmatrix}(\gamma/m)\mathbf{I} & -\mathbf{I}\\(\kappa/m)\mathbf{I} & \mathbf{0}\end{pmatrix}\begin{pmatrix}\mathbf{x}\\ \dot{\mathbf{x}}\end{pmatrix}\right\}_{\mathrm{x}}\\
&= \frac{\gamma}{\kappa}\mathbf{x} - \frac{m}{\kappa}\dot{\mathbf{x}}.\end{aligned} \tag{6.171}$$

Putting these together the dynamic part of the reservoir entropy is

$$\begin{aligned}&S_{\mathrm{r,dyn}}(\mathbf{X}, t)\\
&= \frac{\kappa\dot{b}^2 t^2}{2T} - \frac{\kappa}{T}\dot{\mathbf{b}}\cdot\left\{\mathbf{A}^{-1}\mathbf{X} + \mathbf{A}^{-2}\mathbf{C} + \mathbf{A}^{-3}\mathbf{B} - \frac{t^2}{2}\mathbf{A}^{-1}\mathbf{B} - t\mathbf{A}^{-1}\mathbf{C}\right\}_{\mathrm{x}}\\
&= \left[\frac{\gamma^2}{\kappa T} - \frac{m}{T}\right]\dot{b}^2 - \frac{\gamma}{T}\dot{\mathbf{b}}\cdot\mathbf{x} + \frac{m}{T}\dot{\mathbf{b}}\cdot\dot{\mathbf{x}} - \frac{2\gamma^2}{\kappa T}\dot{b}^2 + t\frac{2\gamma}{T}\dot{b}^2\\
&= \frac{-\gamma}{T}\dot{\mathbf{b}}\cdot\left[\mathbf{x} - \mathbf{b}(t) + \frac{\gamma}{\kappa}\dot{\mathbf{b}}\right] + \frac{m}{T}\dot{\mathbf{b}}\cdot[\dot{\mathbf{x}} - \dot{\mathbf{b}}] + t\frac{\gamma}{T}\dot{b}^2.\end{aligned} \tag{6.172}$$

The final equality uses the fact that $\mathbf{b}(t) = \dot{\mathbf{b}}t$. The gradient of this is constant in phase space, $\nabla_v S_{\mathrm{r,dyn}}(\mathbf{X}, t) = m\dot{\mathbf{b}}/T = -\nabla_v S_{\mathrm{r,st}}(\overline{\mathbf{X}}, t)$. This is consistent with the approximation equation (6.64), which underpins the expression for the most likely reservoir force, equation (6.67). The final constant linear in time is the reservoir entropy produced to date, equation (6.154), as will be confirmed shortly.

The zeroth order approximation, equation (6.74), sets $\mathbf{C} = \mathbf{0}$, in which case this reduces to

$$\begin{aligned} S_{\mathrm{r,dyn}}(\mathbf{X}, t) &= \frac{\kappa \dot{b}^2 t^2}{2T} - \frac{\kappa}{T}\dot{\mathbf{b}} \cdot \left\{ \mathbf{A}^{-1}\mathbf{X} + \mathbf{A}^{-3}\mathbf{B} - \frac{t^2}{2}\mathbf{A}^{-1}\mathbf{B} \right\}_{\mathbf{x}} \\ &= \left[ \frac{\gamma^2}{\kappa T} - \frac{m}{T} \right] \dot{b}^2 - \frac{\gamma}{T}\dot{\mathbf{b}} \cdot \mathbf{x} + \frac{m}{T}\dot{\mathbf{b}} \cdot \dot{\mathbf{x}} \\ &= \frac{-\gamma}{T}\dot{\mathbf{b}} \cdot \left[ \mathbf{x} - \mathbf{b}(t) - \frac{\gamma}{\kappa}\dot{\mathbf{b}} \right] + \frac{m}{T}\dot{\mathbf{b}} \cdot [\dot{\mathbf{x}} - \dot{\mathbf{b}}]. \end{aligned} \tag{6.173}$$

Compared to the full result, this has lost the term linear in time, which represents the important physical quantity of the entropy produced to date, and it has changed the first term in brackets. Both of these changes are constants in phase space, so in fact the zeroth order approximation, equation (6.74), gives the same relative probability for points in phase space as the full expression, equation (6.67).

The full result for the dynamic part of the reservoir entropy is *not* odd with respect to velocity reversal, nor with respect to reflection in the trap minimum. The even terms that would be projected out are significant in these two individual projections. It is not *exactly* odd with respect to both operations simultaneously, but the even part in this case is immaterial. That is, the odd projection of this with respect to both operations is

$$\begin{aligned} S^{\mathrm{odd}}_{\mathrm{r,dyn}}(\mathbf{X}, t) &\equiv \frac{1}{2}\left[ S_{\mathrm{r,dyn}}(\mathbf{X}, t) - S_{\mathrm{r,dyn}}(\mathbf{X}^{\dagger\ddagger}, t) \right] \\ &= -\frac{\gamma}{T}\dot{\mathbf{b}} \cdot [\mathbf{x} - \mathbf{b}(t)] + \frac{m}{T}\dot{\mathbf{b}} \cdot \dot{\mathbf{x}}. \end{aligned} \tag{6.174}$$

This differs from the full expression by constant terms (one of which is the even asymptotic contribution). One can see that it is necessary to perform both operations simultaneously otherwise one or other of $\mathbf{x}$ or $\dot{\mathbf{x}}$ is projected out by the operation. As far as determining the relative probability of phase space points is concerned, this odd projection with respect to both operations is equally as good as the full expression.

The static part of the reservoir entropy,

$$S_{\mathrm{r,st}}(\mathbf{X}, t) = \frac{-m}{2T}\dot{\mathbf{x}} \cdot \dot{\mathbf{x}} - \frac{\kappa}{2T}[\mathbf{x} - \mathbf{b}(t)] \cdot [\mathbf{x} - \mathbf{b}(t)], \tag{6.175}$$

added to the dynamic part of the reservoir entropy gives the reservoir entropy,

$$
\begin{aligned}
S_r(\mathbf{X}, t) &= \frac{-m}{2T}\dot{\mathbf{x}} \cdot \dot{\mathbf{x}} - \frac{\kappa}{2T}[\mathbf{x} - \mathbf{b}(t)] \cdot [\mathbf{x} - \mathbf{b}(t)] \\
&\quad - \frac{\gamma}{T}\dot{\mathbf{b}} \cdot \left[\mathbf{x} - \mathbf{b}(t) + \frac{\gamma}{\kappa}\dot{\mathbf{b}}\right] + \frac{m}{T}\dot{\mathbf{b}} \cdot [\dot{\mathbf{x}} - \dot{\mathbf{b}}] + t\frac{\gamma}{T}\dot{b}^2 \\
&= \frac{-m}{2T}[\dot{\mathbf{x}} - \dot{\mathbf{b}}] \cdot [\dot{\mathbf{x}} - \dot{\mathbf{b}}] \\
&\quad - \frac{\kappa}{2T}\left[\mathbf{x} - \mathbf{b}(t) + \frac{\gamma}{\kappa}\dot{\mathbf{b}}\right] \cdot \left[\mathbf{x} - \mathbf{b}(t) + \frac{\gamma}{\kappa}\dot{\mathbf{b}}\right] \\
&\quad + t\frac{\gamma}{T}\dot{b}^2 - \frac{m}{2T}\dot{b}^2 - \frac{\kappa}{2T}\frac{\gamma^2}{\kappa^2}\dot{b}^2.
\end{aligned}
\tag{6.176}
$$

By inspection, one can see that this is a maximum with respect to $\mathbf{x}$ when $\bar{\mathbf{x}}(t) = \mathbf{b}(t) - \gamma\dot{\mathbf{b}}/\kappa$, and it is a maximum with respect to $\dot{\mathbf{x}}$ when $\bar{\dot{\mathbf{x}}}(t) = \dot{\mathbf{b}}$. This confirms the result deduced on physical grounds in equation (6.151).

The partial time derivative of this in the most likely state is

$$
\left.\frac{\partial S_r(\mathbf{X}, t)}{\partial t}\right|_{\mathbf{X}=\bar{\mathbf{X}}(t)} = \frac{\gamma}{T}\dot{b}^2. \tag{6.177}
$$

This is equal to the rate of total entropy production, and is equal and opposite to the adiabatic rate of static entropy production, equation (6.154).

The $\mathbf{X}$-dependent part of the reservoir entropy is in the form of a fluctuation entropy about the most likely state. It would give a Gaussian probability distribution. The fluctuation coefficients (i.e. $-m/2T$ for the velocity, $-\kappa/2T$ for the position) come from the second derivatives of the static entropy, and they would be the same in the corresponding equilibrium system.

## 6.5 Path entropy and trajectory probability

### 6.5.1 Path entropy

For a transition $\{\mathbf{\Gamma}_1, t_1\} \to \{\mathbf{\Gamma}_2, t_2\}$, the actual change in reservoir entropy is

$$
\begin{aligned}
\overrightarrow{\Delta} S_r(\mathbf{\Gamma}_2, t_2; \mathbf{\Gamma}_1, t_1) &\equiv S_r(\mathbf{\Gamma}_2, t_2|\mathbf{\Gamma}_1, t_1) - S_r(\mathbf{\Gamma}_1, t_1) \\
&= \mathbf{R}_p \cdot \nabla_p S_{r,st}(\mathbf{\Gamma}, t) \\
&= [\mathbf{\Gamma}_2 - \mathbf{\Gamma}_1 - t_{21}\dot{\mathbf{\Gamma}}^0] \cdot \nabla S_{r,st}(\mathbf{\Gamma}, t) \\
&= S_{r,st}(\mathbf{\Gamma}_2, t_2) - S_{r,st}(\mathbf{\Gamma}_1, t_1) - t_{21}\dot{S}^0_{r,st}(\mathbf{\Gamma}, t).
\end{aligned}
\tag{6.178}
$$

Here $S_r(\mathbf{\Gamma}_2, t_2|\mathbf{\Gamma}_1, t_1)$ is the conditional reservoir entropy, which is the entropy for the point $\{\mathbf{\Gamma}_2, t_2\}$ given that the system had previously been at $\{\mathbf{\Gamma}_1, t_1\}$. Here and in what follows the forward trajectory is considered, $t_2 > t_1$. Also $\mathbf{R}$ is the actual force from the reservoir, and so by the conservation laws, $\mathbf{R} \cdot \nabla S_{r,st}(\mathbf{\Gamma}, t)$ is the actual change in reservoir entropy. We always work to linear order in the time step and so $\{\mathbf{\Gamma}, t\}$ can be anywhere on the interval. The final two equalities are simply rearrangements that make the adiabatic contributions explicit.

Note that this, $\overrightarrow{\Delta} S_{\rm r}$ the actual change in entropy on an actual transition, is not the same as the difference in entropy, $\Delta S_{\rm r} \equiv S_{\rm r}(\mathbf{\Gamma}_2, t_2) - S_{\rm r}(\mathbf{\Gamma}_1, t_1)$. Both quantities in the latter are calculated as if the phase space point had been reached via the most likely trajectory (see figure 6.8).

From equation (6.67), the most likely reservoir force for this transition is

$$\begin{aligned}\overline{\mathbf{R}}_p &= \frac{|t_{21}|}{2}\Lambda\left[\nabla_p S_{\rm r,st}(\mathbf{\Gamma}, t) - \nabla_p S_{\rm r,st}(\overline{\mathbf{\Gamma}}(t), t)\right] + \frac{t_{21}}{2}\Lambda\nabla_p S_{\rm r,st}(\overline{\mathbf{\Gamma}}(t), t) \\ &= \frac{t_{21}}{2}\Lambda\nabla_p S_{\rm r,st}(\mathbf{\Gamma}, t),\end{aligned} \tag{6.179}$$

since $t_{21} > 0$.

In view of equations (6.52) and (6.53), the second entropy for the transition may be written

$$S^{(2)}(\mathbf{\Gamma}_2, t_2; \mathbf{\Gamma}_1, t_1) = \frac{-\Lambda^{-1}: \tilde{\mathbf{R}}_p\tilde{\mathbf{R}}_p}{2|t_{21}|} + S_{\rm r}(\mathbf{\Gamma}_1, t_1) + \frac{1}{2}[S_{\rm tot}(t_2) - S_{\rm tot}(t_1)]. \tag{6.180}$$

This neglects the (constant) contribution from the fluctuations, the normalizing factor for the Gaussian. Writing the random part of the reservoir force as $\tilde{\mathbf{R}}_p = \mathbf{R}_p - \overline{\mathbf{R}}_p$, the conditional second entropy can be defined and rewritten as

$$\begin{aligned} & S^{(2)}(\mathbf{\Gamma}_2, t_2|\mathbf{\Gamma}_1, t_1) \\ &\equiv S^{(2)}(\mathbf{\Gamma}_2, t_2; \mathbf{\Gamma}_1, t_1) - S_{\rm r}(\mathbf{\Gamma}_1, t_1) - \frac{1}{2}[S_{\rm tot}(t_2) - S_{\rm tot}(t_1)] \\ &= \frac{-1}{2|t_{21}|}\Lambda^{-1}: \tilde{\mathbf{R}}_p\tilde{\mathbf{R}}_p \\ &= \frac{-1}{2|t_{21}|}\Lambda^{-1}: \left[\mathbf{R}_p\mathbf{R}_p + \overline{\mathbf{R}}_p\overline{\mathbf{R}}_p\right] + \frac{1}{|t_{21}|}\Lambda^{-1}: \mathbf{R}_p\overline{\mathbf{R}}_p \\ &= \frac{-1}{2|t_{21}|}\Lambda^{-1}: \left[\mathbf{R}_p\mathbf{R}_p + \overline{\mathbf{R}}_p\overline{\mathbf{R}}_p\right] + \frac{1}{2}\mathbf{R}_p \cdot \nabla_p S_{\rm r,st}(\mathbf{\Gamma}, t) \\ &= \frac{-1}{2|t_{21}|}\Lambda^{-1}: \left[\mathbf{R}_p\mathbf{R}_p + \overline{\mathbf{R}}_p\overline{\mathbf{R}}_p\right] + \frac{1}{2}\overrightarrow{\Delta} S_{\rm r}(\mathbf{\Gamma}_2, t_2; \mathbf{\Gamma}_1, t_1).\end{aligned} \tag{6.181}$$

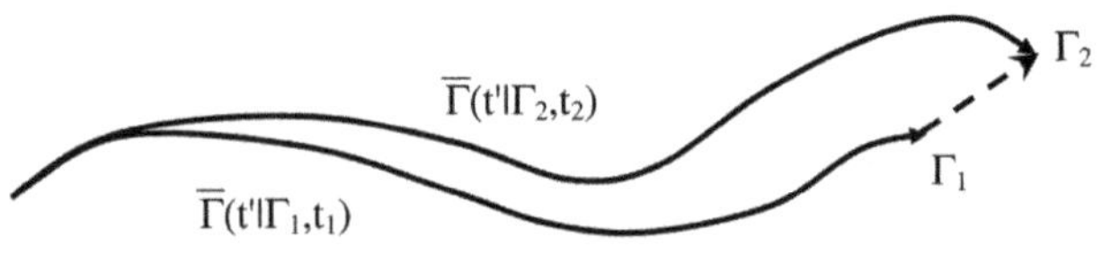


**Figure 6.8.** The actual transition between two points in phase space (dashed arrow), and the two most likely trajectories leading to the points (solid arrows). The change in entropy, $\overrightarrow{\Delta} S_{\rm r}$, is calculated on the former, whereas the difference in entropy, $\Delta S_{\rm r}$, is calculated using the latter.

Since $\overline{\mathbf{R}}_p$ and $\mathbf{R}_p$ scale with the size of the time step, the conditional second entropy itself also scales with the size of the time step, as can also be seen directly from the definition.

This result for a single transition is now extended to a trajectory, which is a series of such transitions. Denote the trajectory by $\underline{\mathbf{\Gamma}}$. Divide the interval $[t_0, t_f]$ of the trajectory into uniformly spaced nodes, $t_n = t_0 + n\Delta_t, n = 0, 1, \ldots, f$, with $\mathbf{\Gamma}_n = \mathbf{\Gamma}(t_n)$ and the time step being $\Delta_t = [t_f - t_0]/f > 0$. We assume that the change in entropy for each nodal transition is independent of the prior trajectory of the system. With these, the actual change in entropy for this trajectory is

$$\begin{aligned}
\overrightarrow{\Delta} S_{\rm r}(\underline{\mathbf{\Gamma}}) &= \sum_{n=0}^{f-1} \overrightarrow{\Delta} S_{\rm r}(\mathbf{\Gamma}_{n+1}, t_{n+1}; \mathbf{\Gamma}_n, t_n) \\
&= \sum_{n=0}^{f-1} \mathbf{R}_n \cdot \nabla_p S_{\rm r,st}(\mathbf{\Gamma}_n, t_n) \\
&= \sum_{n=0}^{f-1} \left\{ S_{\rm r,st}(\mathbf{\Gamma}_{n+1}, t_{n+1}) - S_{\rm r,st}(\mathbf{\Gamma}_n, t_n) - \Delta_t \dot{S}^0_{\rm r,st}(\mathbf{\Gamma}_n, t_n) \right\} \\
&= S_{\rm r,st}\left(\mathbf{\Gamma}_f, t_f\right) - S_{\rm r,st}(\mathbf{\Gamma}_0, t_0) - \int_{t_0}^{t_f} \mathrm{d}t' \dot{S}^0_{\rm r,st}(\mathbf{\Gamma}(t'), t') \\
&\equiv S_{\rm r,st}\left(\mathbf{\Gamma}_f, t_f\right) - S_{\rm r,st}(\mathbf{\Gamma}_0, t_0) + S_{\rm r,dyn}(\underline{\mathbf{\Gamma}}).
\end{aligned} \tag{6.182}$$

Here $\mathbf{R}_n$ is the actual reservoir force applied at the $n$th node, and the final equality defines the dynamic part of the trajectory entropy. This is sketched in figure 6.9.

It is a significant assumption that the change in entropy for each nodal interval is independent of the preceding trajectory. This assumes that on the present infinitesimal time steps, Hamilton's equations apply to both the sub-system and the reservoir, and these are strictly Markovian. Of course a certain statistical re-summation of the reservoir perturbation has taken place in the formulations of $\overline{\mathbf{R}}$ and of $\tilde{\mathbf{R}}$. But this is not same, as for example, the integration of fast modes in Brownian dynamics, where memory effects can be important.

In view of the Markov assumption, the trajectory entropy is just the sum of the second entropy for the transition between each node,

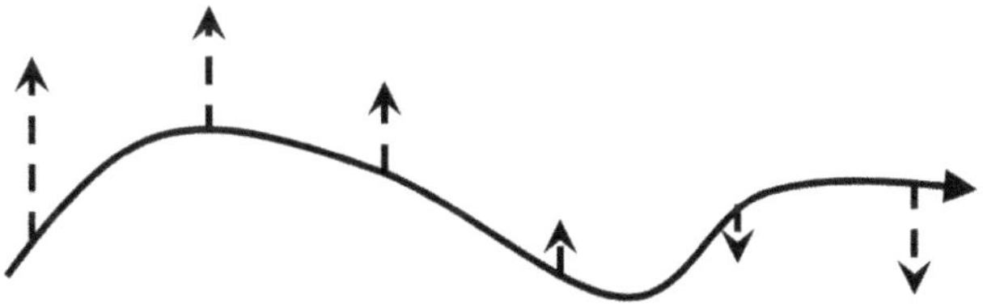

**Figure 6.9.** Sketch of a trajectory in the sub-system phase space (solid arrow) showing the adiabatic rate of entropy production on it at regular intervals (dashed arrows).

$$\begin{aligned} S_{\mathrm{r}}(\underline{\Gamma}|\Gamma_0, t_0) &= \sum_{n=0}^{f-1} S^{(2)}(\Gamma_{n+1}, t_{n+1}|\Gamma_n, t_n) \\ &= \sum_{n=0}^{f-1} \frac{-1}{2|\Delta_t|}\Lambda^{-1}: \tilde{\mathbf{R}}_n\tilde{\mathbf{R}}_n \\ &= \sum_{n=0}^{f-1}\left\{\frac{-\Lambda^{-1}:}{2|\Delta_t|}[\mathbf{R}_n\mathbf{R}_n + \overline{\mathbf{R}}_n\overline{\mathbf{R}}_n] + \frac{1}{2}\vec{\Delta} S_{\mathrm{r}}(\Gamma_{n+1}, t_{n+1}; \Gamma_n, t_n)\right\} \\ &= \frac{1}{2}\vec{\Delta} S_{\mathrm{r}}(\underline{\Gamma}) - \frac{1}{2|\Delta_t|}\sum_{n=0}^{f-1} \Lambda^{-1}: [\mathbf{R}_n\mathbf{R}_n + \overline{\mathbf{R}}_n\overline{\mathbf{R}}_n]. \end{aligned} \tag{6.183}$$

Here the total reservoir force at the $n$th node is $\mathbf{R}_n = \Gamma_{n+1} - \Gamma_n - \Delta_t\dot{\Gamma}^0_n$.

The probability of the trajectory conditioned on the starting point is just the exponential of this trajectory entropy. This is usefully written as the product of Gaussians, the second equality above,

$$\begin{aligned} \wp(\underline{\Gamma}|\Gamma_0, t_0) &= \prod_{n=0}^{f-1} \wp(\Gamma_{n+1}, t_{n+1}|\Gamma_n, t_n) \\ &= \prod_{n=0}^{f-1} \frac{e^{S^{(2)}(\Gamma_{n+1}, t_{n+1}|\Gamma_n, t_n)/k_{\mathrm{B}}}}{Z(t_{n+1}, t_n)} \\ &= \prod_{n=0}^{f-1} \frac{\Theta_\Lambda(\tilde{\mathbf{R}}_n)}{[\mathrm{Det}\, 2\pi|\Delta_t|k_{\mathrm{B}}\Lambda]^{1/2}}. \end{aligned} \tag{6.184}$$

Here and below the un-normalized Gaussian distribution is

$$\Theta_\Lambda(\mathbf{R}_p) \equiv e^{-\Lambda_{pp}^{-1}:\mathbf{R}_p\mathbf{R}_p/2k_{\mathrm{B}}|\Delta_t|}. \tag{6.185}$$

By inspection, one sees that this expression for the conditional trajectory probability is correctly normalized.

This result for the conditional trajectory probability is based on the second equality for the trajectory entropy. The conditional trajectory probability can also be written in terms of the change in entropy over the trajectory, which occurs in the final equality for the trajectory entropy,

$$\begin{aligned} \wp(\underline{\Gamma}|\Gamma_0, t_0) &= \frac{1}{Z(\underline{t})} e^{S_{\mathrm{r}}(\underline{\Gamma}|\Gamma, t_0)/k_{\mathrm{B}}} \\ &= e^{\left[S_{\mathrm{r,st}}(\Gamma_f, t_f) - S_{\mathrm{r,st}}(\Gamma_0, t_0)\right]/2k_{\mathrm{B}}} e^{S_{\mathrm{r,dyn}}(\underline{\Gamma})/2k_{\mathrm{B}}} \\ &\quad\times \prod_{n=0}^{f-1} \frac{\Theta_\Lambda(\mathbf{R}_n)\Theta_\Lambda(\overline{\mathbf{R}}_n)}{[\mathrm{Det}\, 2\pi|\Delta_t|k_{\mathrm{B}}\Lambda]^{1/2}}. \end{aligned} \tag{6.186}$$

Because the exponent here is equal to the previous Gaussian form, this must also be correctly normalized.

It is usually the case that the initial probability distribution is that of the non-equilibrium system at time $t_0$,

$$\wp(\mathbf{\Gamma}, t_0) = \frac{1}{Z(t_0)} e^{S_r(\mathbf{\Gamma}, t_0)/k_B} = \frac{1}{Z(t_0)} e^{\left[S_{r,st}(\mathbf{\Gamma}, t_0) + S_{r,dyn}(\mathbf{\Gamma}, t_0)\right]/k_B}. \tag{6.187}$$

For convenience, sometimes one instead assumes a dynamically disordered system, namely one that is initially distributed according to the static part of the reservoir entropy alone,

$$\wp_{st}(\mathbf{\Gamma}, t_0) = \frac{1}{Z_{st}(t_0)} e^{S_{r,st}(\mathbf{\Gamma}, t_0)/k_B}. \tag{6.188}$$

With this latter assumption the unconditional trajectory probability in the second form is

$$\begin{aligned}\wp(\underline{\mathbf{\Gamma}}) &= \wp(\underline{\mathbf{\Gamma}}|\mathbf{\Gamma}_0, t_0)\wp_{st}(\mathbf{\Gamma}_0, t_0) \\ &= \frac{1}{Z_{st}(t_0)} e^{\left[S_{r,st}(\mathbf{\Gamma}_f, t_f) + S_{r,st}(\mathbf{\Gamma}_0, t_0)\right]/2k_B} e^{S_{r,dyn}(\underline{\mathbf{\Gamma}})/2k_B} \\ &\times \prod_{n=0}^{f-1} \frac{\Theta_\Lambda(\mathbf{R}_n)\Theta_\Lambda(\overline{\mathbf{R}}_n)}{[\mathrm{Det}\, 2\pi|\Delta_t|k_B\Lambda]^{1/2}}.\end{aligned} \tag{6.189}$$

For a long trajectory, the actual initial distribution of states makes little difference to the unconditional trajectory probability.

*Conjugate trajectory*

We now wish to analyse the properties of the conjugate trajectory upon which the velocities are reversed. To this end, in the event that the Hamiltonian is explicitly time dependent, we define the conjugate Hamiltonian on the trajectory time interval as $\tilde{\mathcal{H}}(\mathbf{\Gamma}, t) = \mathcal{H}(\mathbf{\Gamma}, t_0 + t_f - t)$. This is the same conjugation operation as was defined in equation (5.16) for the discussion of microscopic reversibility for a time-dependent potential. With $\dagger$ representing velocity reversal, the phase space point $\mathbf{\Gamma} = \{\mathbf{q}, \mathbf{p}\}$ has conjugate $\mathbf{\Gamma}^\dagger = \{\mathbf{q}, -\mathbf{p}\}$. The trajectory in the conjugate system, $\tilde{\underline{\mathbf{\Gamma}}}$, has the property that $\tilde{\mathbf{\Gamma}}(t) = \mathbf{\Gamma}(t_0 + t_f - t)^\dagger$. Equivalently, $\tilde{\mathbf{\Gamma}}_n = \mathbf{\Gamma}^\dagger_{f-n}$. Time proceeds in the positive direction on the conjugate trajectory, with the initial point being the conjugate of the original final point, $\tilde{\mathbf{\Gamma}}_0 = \mathbf{\Gamma}^\dagger_f$.

The static entropy is the instantaneous equilibrium entropy, and as such it has even parity, $S_{r,st}(\mathbf{\Gamma}^\dagger, t) = S_{r,st}(\mathbf{\Gamma}, t)$. Hence $\tilde{S}_{r,st}(\tilde{\mathbf{\Gamma}}(t), t) = S_{r,st}(\mathbf{\Gamma}(t_0 + t_f - t), t_0 + t_f - t)$. In the conjugate system this is a function of the conjugate Hamiltonian, $\tilde{\mathcal{H}}(\mathbf{\Gamma}, t)$. The adiabatic derivatives are equal and opposite in the two systems,

$$\dot{\tilde{S}}^0_{r,st}(\tilde{\mathbf{\Gamma}}(t), t) = -\dot{S}^0_{r,st}\big(\mathbf{\Gamma}(t_0 + t_f - t), t_0 + t_f - t\big). \tag{6.190}$$

This result only holds for the conjugate Hamiltonian as defined here. Of course there are some systems where the conjugate Hamiltonian is the same as the original, such as equilibrium and thermodynamic steady state systems, where the Hamiltonian does not depend on time. Also harmonic systems with time dependence of the form $\cos(2\pi l t/(t_0 + t_f))$, with $l$ an integer.

The conditional trajectory probability in the conjugate system is

$$\begin{aligned}
\wp(\underline{\tilde{\boldsymbol{\Gamma}}}|\tilde{\boldsymbol{\Gamma}}_0;\tilde{\mathcal{H}}) &= e^{[\tilde{S}_{\mathrm{r,st}}(\tilde{\boldsymbol{\Gamma}}_f,\tilde{t}_f)-\tilde{S}_{\mathrm{r,st}}(\tilde{\boldsymbol{\Gamma}}_0,\tilde{t}_0)]/2k_{\mathrm{B}}} e^{\tilde{S}_{\mathrm{r,dyn}}(\underline{\tilde{\boldsymbol{\Gamma}}})/2k_{\mathrm{B}}} \\
&\quad \times \prod_{n=0}^{f-1} \frac{\Theta_\Lambda(\mathbf{R}_n^{\mathrm{conj}})\Theta_\Lambda(\overline{\mathbf{R}}_n^{\mathrm{conj}})}{[\mathrm{Det}\, 2\pi|\Delta_t|k_{\mathrm{B}}\Lambda]^{1/2}} \\
&= e^{\left[S_{\mathrm{r,st}}(\boldsymbol{\Gamma}_0,t_0)-S_{\mathrm{r,st}}\left(\boldsymbol{\Gamma}_f,t_f\right)\right]/2k_{\mathrm{B}}} e^{-S_{\mathrm{r,dyn}}(\underline{\boldsymbol{\Gamma}})/2k_{\mathrm{B}}} \\
&\quad \times \prod_{n=0}^{f-1} \frac{\Theta_\Lambda(\mathbf{R}_n)\Theta_\Lambda(\overline{\mathbf{R}}_n)}{[\mathrm{Det}\, 2\pi|\Delta_t|k_{\mathrm{B}}\Lambda]^{1/2}}.
\end{aligned} \tag{6.191}$$

The second equality follows from three facts. First, the dynamic trajectory entropy is negated on the conjugate trajectory,

$$\begin{aligned}
\tilde{S}_{\mathrm{r,dyn}}(\underline{\tilde{\boldsymbol{\Gamma}}}) &= -\int_{t_0}^{t_f} \mathrm{d}t\; \dot{\tilde{S}}^0_{\mathrm{r,st}}(\tilde{\boldsymbol{\Gamma}}(t), t) \\
&= \int_{t_0}^{t_f} \mathrm{d}t\; \dot{S}^0_{\mathrm{r,st}}\left(\boldsymbol{\Gamma}(t_0 + t_f - t), t_0 + t_f - t\right) \\
&= \int_{t_0}^{t_f} \mathrm{d}t' \dot{S}^0_{\mathrm{r,st}}(\boldsymbol{\Gamma}(t'), t') \\
&= -S_{\mathrm{r,dyn}}(\underline{\boldsymbol{\Gamma}}).
\end{aligned} \tag{6.192}$$

Second, the reservoir force on the conjugate trajectory is the negative conjugate of that on the original trajectory,

$$\begin{aligned}
\mathbf{R}_n^{\mathrm{conj}} &= \tilde{\boldsymbol{\Gamma}}_{n+1} - \tilde{\boldsymbol{\Gamma}}_n - \Delta_t \dot{\tilde{\boldsymbol{\Gamma}}}^0_n \\
&= \boldsymbol{\Gamma}^\dagger_{f-n-1} - \boldsymbol{\Gamma}^\dagger_{f-n} - \Delta_t \dot{\boldsymbol{\Gamma}}^0(\boldsymbol{\Gamma}^\dagger_{f-n-1}, t_{f-n-1}) \\
&= \boldsymbol{\Gamma}^\dagger_{f-n-1} - \boldsymbol{\Gamma}^\dagger_{f-n} + \Delta_t \dot{\boldsymbol{\Gamma}}^0(\boldsymbol{\Gamma}_{f-n-1}, t_{f-n-1})^\dagger \\
&= -\mathbf{R}^\dagger_{f-n-1}.
\end{aligned} \tag{6.193}$$

Hence the Gaussian exponent is unchanged, $\Lambda_{pp} : [-\mathbf{R}^\dagger_{f-n-1}]\,[-\mathbf{R}^\dagger_{f-n-1}] = \Lambda_{pp} : \mathbf{R}_{f-n-1}\mathbf{R}_{f-n-1}$. The sum of the Gaussian exponents in both cases adds together all the reservoir forces. Third, the most likely reservoir also simply changes sign,

$$\begin{aligned}
\overline{\mathbf{R}}_n^{\mathrm{conj}} &= \frac{|\Delta_t|}{2}\Lambda_{pp}\tilde{\nabla}_p\tilde{S}_{\mathrm{r,st}}(\tilde{\boldsymbol{\Gamma}}(t_n), t_n) \\
&= \frac{-|\Delta_t|}{2}\Lambda_{pp}\nabla_p S_{\mathrm{r,st}}\left(\boldsymbol{\Gamma}_{f-n}, t_{f-n}\right).
\end{aligned} \tag{6.194}$$

Recall that the time step is positive on both trajectories, and so the most likely force is proportional to the gradient of the static part of the reservoir entropy alone. And so again the Gaussian factor of these is unchanged.

The ratio of the conditional trajectory probabilities for the two systems is

$$
\begin{aligned}
\frac{\wp(\underline{\Gamma}|\Gamma_0)}{\wp(\tilde{\underline{\Gamma}}|\tilde{\Gamma}_0;\tilde{\mathcal{H}})} &= e^{[S_{\mathrm{r,st}}(\Gamma_f,t_f)-S_{\mathrm{r,st}}(\Gamma_0,t_0)]/k_{\mathrm{B}}} e^{S_{\mathrm{r,dyn}}(\underline{\Gamma})/k_{\mathrm{B}}} \\
&= e^{\vec{\nabla} S_{\mathrm{r}}(\underline{\Gamma})/k_{\mathrm{B}}} \\
&\approx e^{S_{\mathrm{r,dyn}}(\underline{\Gamma})/k_{\mathrm{B}}}.
\end{aligned} \tag{6.195}
$$

The time interval is implicit in the trajectory, and for simplicity it is not shown explicitly as a conditioning argument. The exponent on the right-hand side of the second equality is the actual change in reservoir entropy over the original trajectory. This obviously scales with the length of the trajectory, and so for a long enough trajectory the initial conditions can be neglected. This means that this result also holds approximately for the ratio of unconditional probabilities.

Alternatively, for the unconditional probability, one assumption that one can make is that the system is initially and finally in the non-equilibrium state given by the usual expression for the reservoir entropy. In this case the ratio of the unconditional probabilities is

$$
\begin{aligned}
\frac{\wp(\underline{\Gamma}|\Gamma_0)\wp(\Gamma_0,t_0)}{\wp(\tilde{\underline{\Gamma}}|\tilde{\Gamma}_0;\tilde{\mathcal{H}})\wp(\tilde{\Gamma}_0,t_0;\tilde{\mathcal{H}})} &= \frac{Z(t_0;\tilde{\mathcal{H}})}{Z(t_0)} \frac{e^{S_{\mathrm{r,dyn}}(\Gamma_0,t_0)/k_{\mathrm{B}}}}{e^{S_{\mathrm{r,dyn}}(\tilde{\Gamma}_0,t_0;\tilde{\mathcal{H}})/k_{\mathrm{B}}}} e^{S_{\mathrm{r,dyn}}(\underline{\Gamma})/k_{\mathrm{B}}} \\
&\approx e^{S_{\mathrm{r,dyn}}(\underline{\Gamma})/k_{\mathrm{B}}}.
\end{aligned} \tag{6.196}
$$

The final approximation assumes that the total unconstrained entropy does not increase monotonically over time.

A different set of initial conditions is for the system to be in the static state initially and finally. Now the ratio of the unconditional probabilities is

$$
\begin{aligned}
\frac{\wp(\underline{\Gamma}|\Gamma_0)\wp_{\mathrm{st}}(\Gamma_0,t_0)}{\wp(\tilde{\underline{\Gamma}}|\tilde{\Gamma}_0;\tilde{\mathcal{H}})\wp_{\mathrm{st}}(\tilde{\Gamma}_0,t_0;\tilde{\mathcal{H}})} &= \frac{Z_{\mathrm{st}}(t_f)}{Z_{\mathrm{st}}(t_0)} e^{S_{\mathrm{r,dyn}}(\underline{\Gamma})/k_{\mathrm{B}}} \\
&\approx e^{S_{\mathrm{r,dyn}}(\underline{\Gamma})/k_{\mathrm{B}}}.
\end{aligned} \tag{6.197}
$$

In general, the logarithm of the partition function is the total unconstrained entropy, which when multiplied by the negative of the temperature gives a free energy. Hence the first factor in the first equality is essentially the exponential of the difference in the static total entropy or static free energy. For a cyclic trajectory, $\mathcal{H}(\Gamma, t_f) = \mathcal{H}(\Gamma, t_0)$, the final approximation is exact as $Z_{\mathrm{st}}(t_f) = Z_{\mathrm{st}}(t_0)$.

In any case, these examples confirm the general conclusion that for a large enough time interval one can ignore the state of the system beyond it. The ratio of probabilities, conditional or unconditional, of the original and conjugate trajectories is the exponential of the change in the reservoir entropy over the interval.

### 6.5.2 Fluctuation and work theorems

Although the Second Law of Thermodynamics asserts that the entropy increases with time, from the discussion in the preceding sections of this book, specifically sections 1.2.4, 1.3, and 3.1.5, it is clear that this is meant in a probabilistic sense only. Hence the question arises: how much more probable is the entropy to increase than to decrease?

The expressions given above for the trajectory probability give a quantitative answer to this question. The probability of observing the reservoir entropy change by $\vec{\Delta}S_{\rm r}$ over the given time interval can be related to the probability of observing the opposite change by elementary manipulations of the above results. For a specified time interval one has

$$\begin{aligned} \wp(\vec{\Delta}S_{\rm r}|t_{21}) &= \int \mathrm{d}\underline{\Gamma}\; \delta(\vec{\Delta}S_{\rm r} - \vec{\Delta}S_{\rm r}(\underline{\Gamma}))\wp(\underline{\Gamma}) \\ &= \int \mathrm{d}\underline{\tilde{\Gamma}}\; \delta(\vec{\Delta}S_{\rm r} - \vec{\Delta}S_{\rm r}(\underline{\Gamma}))\wp(\underline{\tilde{\Gamma}}|\tilde{\mathcal{H}})e^{\vec{\Delta}S_{\rm r}(\underline{\Gamma})/k_{\rm B}} \\ &= e^{\vec{\Delta}S_{\rm r}/k_{\rm B}} \int \mathrm{d}\underline{\tilde{\Gamma}}\; \delta(\vec{\Delta}S_{\rm r} + \vec{\Delta}S_{\rm r}(\underline{\tilde{\Gamma}}))\wp(\underline{\tilde{\Gamma}}|\tilde{\mathcal{H}}) \\ &= e^{\vec{\Delta}S_{\rm r}/k_{\rm B}}\wp(-\vec{\Delta}S_{\rm r}|t_{21}, \tilde{\mathcal{H}}). \end{aligned} \tag{6.198}$$

Here $\vec{\Delta}S_{\rm r}$ is just a number, the specified change in entropy over the given time interval. All possible trajectories are integrated over in obtaining this result. Again the time interval is implicit in the trajectory, and for simplicity it is not shown explicitly in the trajectory probability.

This result relies upon the conjugate system as defined above. As has been mentioned, in some cases the conjugate system is the same as the original system. Examples include equilibrium systems, steady state thermodynamic systems, and mechanical non-equilibrium systems with harmonic Hamiltonian.

This result is exact for the conditional trajectory probability, which is to say that the initial and final points of the trajectory are specified. In any case, for a sufficiently long trajectory end effects can be neglected and it also holds approximately for the unconditional probability.

The result says that the probability of a particular entropy increase is exponentially more likely than the same decrease. This shows why the Second Law of Thermodynamics can be considered exact for a macroscopic system. In essence this result is the fluctuation theorem that was first derived by Bochkov and Kuzovlev (1981). Different versions and applications of the theorem have subsequently been presented by others (Evans *et al* 1993, Jarzynski 1997, Evans 2003, Attard 2006b). A feature of the present derivation is that it includes the stochastic, dissipative forces from the reservoir (i.e. it is not restricted to adiabatic trajectories), and it applies also to non-equilibrium thermodynamic systems (i.e. it is not restricted to mechanical non-equilibrium systems).

The work theorem is closely related to the fluctuation theorem. This is derived from the average over trajectories of the exponential of the negative of the entropy change on each trajectory. For a specified time interval one has

$$\begin{aligned}\left\langle e^{-\tilde{\Delta} S_{\mathrm{r}}(\underline{\Gamma})/k_{\mathrm{B}}}\right\rangle &= \int \mathrm{d}\underline{\Gamma}\ \wp(\underline{\Gamma}) e^{-\tilde{\Delta} S_{\mathrm{r}}(\underline{\Gamma})/k_{\mathrm{B}}} \\ &= \int \mathrm{d}\underline{\tilde{\Gamma}}\ \wp(\underline{\tilde{\Gamma}}|\tilde{\mathcal{H}}) \\ &= 1.\end{aligned} \tag{6.199}$$

On the left-hand side the negative of the change in reservoir entropy over the trajectory appears. End effects have again been neglected, which is valid for long enough trajectories. What is interesting about this result is that the exponent on the left-hand side scales linearly with the length of the time interval, but the average itself is independent of the time interval.

As an illustration of this, consider the average of the exponential of the negative of the dynamic part of the reservoir entropy. This is just the thermodynamic work. Assuming static equilibrium initially and finally, this is

$$\begin{aligned}&\left\langle e^{-S_{\mathrm{r,dyn}}(\underline{\Gamma})/k_{\mathrm{B}}}\right\rangle \\ &= \int \mathrm{d}\underline{\Gamma}\ \wp(\underline{\Gamma}|\Gamma_0)\wp_{\mathrm{st}}(\Gamma_0, t_0) e^{-S_{\mathrm{r,dyn}}(\underline{\Gamma})/k_{\mathrm{B}}} \\ &= \int \mathrm{d}\underline{\tilde{\Gamma}}\ \wp(\underline{\tilde{\Gamma}}|\tilde{\Gamma}_0; \tilde{\mathcal{H}})\wp_{\mathrm{st}}(\tilde{\Gamma}_0, t_0; \tilde{\mathcal{H}}) \frac{\wp_{\mathrm{st}}(\Gamma_0, t_0)}{\wp_{\mathrm{st}}(\tilde{\Gamma}_0, t_0; \tilde{\mathcal{H}})} \\ &\quad \times e^{-S_{\mathrm{r,st}}(\Gamma_0, t_0)/k_{\mathrm{B}}} e^{S_{\mathrm{r,st}}(\Gamma_f, t_f)/k_{\mathrm{B}}} \\ &= \frac{Z_{\mathrm{st}}(t_f)}{Z_{\mathrm{st}}(t_0)} \\ &= e^{-\beta \Delta F_{\mathrm{st}}}.\end{aligned} \tag{6.200}$$

This uses the fact that $Z_{\mathrm{st}}(t_0; \tilde{\mathcal{H}}) = Z_{\mathrm{st}}(t_f)$, and also the fact that the free energy is $-k_{\mathrm{B}}T$ times the logarithm of the partition function. The dynamic trajectory entropy, equation (6.182), is in essence the work done on the sub-system (at least it is for a mechanical non-equilibrium system; the present definition also applies to a thermodynamic non-equilibrium system), and so one sees that the left-hand side is the exponential of the negative of the work done. On the right-hand side, the exponent is the negative of the difference in the static free energy divided by $k_{\mathrm{B}}T$.

The work theorem was first given by Bochkov and Kuzovlev (1981), who assumed a long cyclic trajectory. A similar result was also given by Jarzynski (1997), who also treated a mechanical non-equilibrium system and assumed a Boltzmann distribution at the beginning and end of the trajectory. The present result is valid both for applied thermodynamic gradients and for external time dependent mechanical fields.

## 6.6 Path entropy for mechanical work

### 6.6.1 Stochastic dissipative equations of motion

We now analyze a mechanical non-equilibrium system, which is a sub-system with a time varying Hamiltonian, $\mathcal{H}(\Gamma, t) = \mathcal{K}(\mathbf{p}) + U(\mathbf{q}, t)$. Energy can be exchanged with a thermal reservoir of temperature $T$. The kinetic energy may be taken to be

$\mathcal{K}(\mathbf{p}) = \sum_{j\alpha} p_{j\alpha}^2/2m$, where $j = 1, 2\ldots, N$ labels the atoms and $\alpha = x, y, z$ labels the axes. The time dependence of the potential is generally contained in an externally applied potential, and the initial aim is to calculate the mechanical work that this does on the sub-system over time.

The static part of the reservoir entropy is the usual equilibrium expression, $S_{\mathrm{r,st}}(\mathbf{\Gamma}, t) = -\mathcal{H}(\mathbf{\Gamma}, t)/T$. Hence the change in entropy for a transition, $\mathbf{\Gamma}_2 \stackrel{\Delta_t}{\rightarrow} \mathbf{\Gamma}_1$, $\Delta_t \equiv t_2 - t_1$, is explicitly

$$
\begin{aligned}
\vec{\Delta} S_{\mathrm{r}}(\mathbf{\Gamma}_2, t_2; \mathbf{\Gamma}_1, t_1) &= \mathbf{R} \cdot \nabla_{\mathrm{p}} S_{\mathrm{r,st}}(\mathbf{\Gamma}, t) \\
&= \frac{-1}{mT} \sum_{j\alpha} R_{j\alpha}\, p_{j\alpha}.
\end{aligned}
\tag{6.201}
$$

The total reservoir force is the force required beyond the adiabatic force to give the designated transition, $\mathbf{R} = \mathbf{\Gamma}_2 - \mathbf{\Gamma}_1 - \Delta_t \dot{\mathbf{\Gamma}}^0$. As usual, it comprises a dissipative part and a stochastic part, $\mathbf{R} = \overline{\mathbf{R}} + \tilde{\mathbf{R}}$.

We shall take the second entropy fluctuation matrix to be proportional to the identity matrix in momentum space, $\Lambda = \lambda \mathrm{I}_{\mathrm{pp}}$. In this case the variance of the stochastic part of the force is

$$
\langle \tilde{\mathbf{R}}(t)\, \tilde{\mathbf{R}}(t) \rangle = k_{\mathrm{B}} |\Delta_t| \lambda\, \mathrm{I}_{\mathrm{pp}},
\tag{6.202}
$$

and the most likely reservoir force is

$$
\begin{aligned}
\overline{\mathbf{R}}(\Delta_t, \mathbf{\Gamma}, t) &= \frac{|\Delta_t|\lambda}{2}\left[\nabla_{\mathrm{p}} S_{\mathrm{r,st}}(\mathbf{\Gamma}, t) + [\hat{\Delta}_t - 1]\nabla_p S_{\mathrm{r,st}}(\overline{\mathbf{\Gamma}}(t), t)\right] \\
&= \frac{-|\Delta_t|\lambda}{2mT}\left[\mathbf{\Gamma}_{\mathrm{p}} + [\hat{\Delta}_t - 1]\overline{\mathbf{\Gamma}}_{\mathrm{p}}(t)\right].
\end{aligned}
\tag{6.203}
$$

Here and throughout, $\hat{\Delta}_t \equiv \mathrm{sign}\, \Delta_t$.

The scalar $\lambda$ represents the influence of the reservoir on the sub-system. In practical computations, the magnitude of this is to a large extent arbitrary. (One does have to consider the twin effects of the length of the time step and the variance of the stochastic force in solving the equations of motion by simple time stepping. These are not independent, and the statistical averages must be insensitive to the chosen values.) Of course, the same constant $\lambda$ must be used for the variance of the stochastic force as for the magnitude of the dissipative force, which is just the fluctuation–dissipation theorem. For a positive time step, $\Delta_t > 0$, which is usually the case in the results analyzed below, the most likely momenta are not required because $\hat{\Delta}_t - 1 = 0$.

The stochastic, dissipative equations of motion over a single time step for the present mechanical non-equilibrium mechanical system are explicitly

$$
\begin{aligned}
q_{j\alpha}(t + \Delta_t) &= q_{j\alpha}(t) + \Delta_t \dot{q}^0_{j\alpha}(t), \\
p_{j\alpha}(t + \Delta_t) &= p_{j\alpha}(t) + \Delta_t \dot{p}^0_{j\alpha}(t) - \frac{|\Delta_t|\lambda}{2\, mT}\left[p_{j\alpha} + [\hat{\Delta}_t - 1]\bar{p}_{j\alpha}(t)\right] + \tilde{R}_{j\alpha}.
\end{aligned}
\tag{6.204}
$$

As usual, the adiabatic rates of change are

$$\dot{q}^0_{j\alpha}(t) = \frac{\partial \mathcal{H}(\mathbf{\Gamma}, t)}{\partial p_{j\alpha}} = \frac{p_{j\alpha}}{m} \text{ and } \dot{p}^0_{j\alpha}(t) = \frac{-\partial \mathcal{H}(\mathbf{\Gamma}, t)}{\partial q_{j\alpha}}. \tag{6.205}$$

In these equations of motion, the stochastic forces at each time step are uncorrelated. The dissipative force, $-(|\Delta_t|\lambda/2mT)p_{j\alpha}$, has the form of a drag force. The final term that appears here, $-(|\Delta_t|\lambda/2mT)[\hat{\Delta}_t - 1]\bar{p}_{j\alpha}(t)$, is only non-zero for a backward time step. As mentioned, that equality of the drag coefficient and the variance of the stochastic force is known as the fluctuation dissipation theorem. These stochastic, dissipative equations of motion have the same form as the Langevin equation for Brownian motion, section 4.1.2, but of course the justification for the present molecular application is statistical rather than hydrodynamic.

The conditional second entropy for the transition is

$$\begin{aligned} S^{(2)}(\mathbf{\Gamma}_2, t_2|\mathbf{\Gamma}_1, t_1) &= \frac{-1}{2|\Delta_t|\lambda}\tilde{\mathbf{R}}\cdot\tilde{\mathbf{R}} \\ &= \frac{-1}{2|\Delta_t|\lambda}[\mathbf{R}\cdot\mathbf{R} + \overline{\mathbf{R}}\cdot\overline{\mathbf{R}}] + \frac{1}{2}\vec{\Delta}S_{\mathrm{r}}(\mathbf{\Gamma}_2, t_2; \mathbf{\Gamma}_1, t_1) \\ &= \frac{-1}{2|\Delta_t|\lambda}\sum_{j\alpha}\left[R_{j\alpha}^2 + \frac{|\Delta_t|^2\lambda^2}{4m^2T^2}p_{j\alpha}^2\right] - \frac{1}{2mT}\sum_{j\alpha}R_{j\alpha}\,p_{j\alpha},\quad \Delta_t > 0. \end{aligned} \tag{6.206}$$

With this, the conditional transition probability is

$$\begin{aligned} \wp(\mathbf{\Gamma}_2, t_2|\mathbf{\Gamma}_1, t_1) &= \frac{1}{Z(t_2, t_1)}e^{S^{(2)}(\mathbf{\Gamma}_2, t_2|\mathbf{\Gamma}_1, t_1)/k_{\mathrm{B}}} \\ &= \frac{1}{Z(t_2, t_1)}e^{-\tilde{\mathbf{R}}\cdot\tilde{\mathbf{R}}/2k_{\mathrm{B}}|\Delta_t|\lambda}\delta\left(\mathbf{\Gamma}_{\mathrm{q}2} - \mathbf{\Gamma}_{\mathrm{q}1} - \Delta_t\dot{\mathbf{\Gamma}}^0_{\mathrm{q}}\right) \\ &= \frac{1}{Z(t_2, t_1)}\delta\left(\mathbf{\Gamma}_{\mathrm{q}2} - \mathbf{\Gamma}_{\mathrm{q}1} - \Delta_t\dot{\mathbf{\Gamma}}^0_{\mathrm{q}}\right) \\ &\quad\times\prod_{j\alpha} e^{-[R_{j\alpha}^2 + |\Delta_t|^2\lambda^2 p_{j\alpha}^2/4m^2T^2]/2k_{\mathrm{B}}|\Delta_t|\lambda}\prod_{j\alpha} e^{-\beta R_{j\alpha}\,p_{j\alpha}/2m}. \end{aligned} \tag{6.207}$$

The normalization factor is $Z(t_2, t_1) = (2\pi k_{\mathrm{B}}|\Delta_t|\lambda)^{3N/2}$, and $\beta \equiv 1/k_{\mathrm{B}}T$.

The final exponent, $-\beta\sum_{j\alpha}R_{j\alpha}\,p_{j\alpha}/2m = -\beta\mathbf{R}\cdot\nabla_{\mathrm{p}}\mathcal{H}(\mathbf{\Gamma}, t)/2$, is essentially half the change in entropy for the transition. It is interesting to note that the present second entropy analysis yields the same factor of $\beta/2$ as in Glauber and Kawasaki dynamics, which are stochastic approaches to dynamical systems that neglect adiabatic evolution (Glauber 1963, Kawasaki 1966, Langer 1969, Metiu *et al* 1975). The present result for the conditional stochastic transition probability satisfies the detailed balance for an equilibrium Boltzmann distribution. It has been applied with a stochastic molecular dynamics algorithm for equilibrium systems (Attard 2002, Boinepalli and Attard 2003). It has also been used as the basis for a non-equilibrium molecular dynamics algorithm for steady heat flow and for driven Brownian motion (Attard 2006b, 2009a).

### 6.6.2 Work done and free energy difference

As in section 6.5.1, denote a trajectory by the vector $\underline{\Gamma}$, with elements $\Gamma_n = \Gamma(t_n)$, and nodes $t_n = t_0 + n\Delta_t$, $n = 0, 1, \ldots, f$, the time step being $\Delta_t = [t_f - t_0]/f > 0$. The change in reservoir entropy for a mechanical non-equilibrium system is

$$\begin{aligned}\vec{\Delta} S_{\rm r}(\underline{\Gamma}) &= S_{\rm r,st}(\Gamma_f, t_f) - S_{\rm r,st}(\Gamma_0, t_0) - \int_{t_0}^{t_f} {\rm d}t' \dot{S}^0_{\rm r,st}(\Gamma(t'), t') \\ &= \frac{1}{T}\left[\mathcal{H}(\Gamma_0, t_0) - \mathcal{H}(\Gamma_f, t_f)\right] + \frac{1}{T}\int_{t_0}^{t_f} {\rm d}t' \frac{\partial U(\Gamma', t')}{\partial t'},\end{aligned} \tag{6.208}$$

where $\Gamma' \equiv \Gamma(t')$. The dynamic part of the reservoir entropy, the final term, is the work done over the actual trajectory divided by temperature, $S_{\rm r,dyn}(\underline{\Gamma}) = W(\underline{\Gamma})/T$, with the work done being

$$\begin{aligned}W(\underline{\Gamma}) &= \int_{t_0}^{t_f} {\rm d}t' \frac{\partial U(\Gamma', t')}{\partial t'} \\ &= \int_{t_0}^{t_f} {\rm d}t' \, \dot{\mathcal{H}}^0(\Gamma(t'), t') \\ &= \Delta_t \sum_{n=0}^{f-1} \dot{\mathcal{H}}^0(\Gamma_n, t_n).\end{aligned} \tag{6.209}$$

Figure 6.9, shown earlier, also serves to illustrate this. The adiabatic rate of change of the energy is just the partial time derivative of the potential, $\dot{\mathcal{H}}^0(\Gamma, t) = \partial U(\Gamma, t)/\partial t$. Other recipes for discretizing the integral give the same result to leading order in the time step. As has been mentioned in section 6.5.1, the work term dominates the change in entropy for a sufficiently long trajectory and the contributions of the initial and final states may be neglected. The exception to this is if the time dependent part of the potential grossly changes in proportion to the length of the time interval.

The conditional trajectory entropy is

$$\begin{aligned}S(\underline{\Gamma}|\Gamma_0, t_0) &= \frac{1}{2}\vec{\Delta} S_{\rm r}(\underline{\Gamma}) - \frac{1}{2|\Delta_t|\lambda}\sum_{n=0}^{f-1}\sum_{j\alpha}\left[R_{n;j\alpha}^2 + \frac{|\Delta_t|^2\lambda^2}{4m^2T^2}p_{n;j\alpha}^2\right] \\ &= \frac{1}{2T}\left[\mathcal{H}(\Gamma_0, t_0) - \mathcal{H}(\Gamma_f, t_f)\right] + \frac{W(\underline{\Gamma})}{2T} \\ &\quad - \frac{1}{2|\Delta_t|\lambda}\sum_{n=0}^{f-1}\sum_{j\alpha}\left[R_{n;j\alpha}^2 + \frac{|\Delta_t|^2\lambda^2}{4m^2T^2}p_{n;j\alpha}^2\right].\end{aligned} \tag{6.210}$$

The total reservoir force at the $n$th node is $\mathbf{R}_n \equiv \Gamma_{{\rm p},n+1} - \Gamma_{{\rm p},n} - \Delta_t \dot{\Gamma}^0_{{\rm p},n}$, with the adiabatic velocity being evaluated anywhere on the $n$th interval.

The conditional trajectory probability is the exponential of this divided by Boltzmann's constant,

$$\begin{aligned}\wp(\underline{\Gamma}|\Gamma_0, t_0) &= \prod_{n=0}^{f-1} \frac{\Theta_\lambda(\tilde{\mathbf{R}}_n)}{(2\pi k_B|\Delta_t|\lambda)^{3N/2}} \\ &= e^{\bar{\Delta}S_r(\underline{\Gamma})/2k_B} \prod_{n=0}^{f-1} \frac{\Theta_\lambda(\mathbf{R}_n)\Theta_\lambda(\overline{\mathbf{R}}_n)}{(2\pi k_B|\Delta_t|\lambda)^{3N/2}} \\ &= e^{-\beta[\mathcal{H}(\Gamma_f, t_f)-\mathcal{H}(\Gamma_0, t_0)]/2} e^{\beta W(\underline{\Gamma})/2} \prod_{n=0}^{f-1} \frac{\Theta_\lambda(\mathbf{R}_n)\Theta_\lambda(\overline{\mathbf{R}}_n)}{(2\pi k_B|\Delta_t|\lambda)^{3N/2}},\end{aligned} \tag{6.211}$$

where the un-normalized Gaussian probability distribution is

$$\Theta_\lambda(\mathbf{R}_n) \equiv e^{-\mathbf{R}_n\cdot\mathbf{R}_n/2k_B|\Delta_t|\lambda}. \tag{6.212}$$

As in section 6.5.1, define a conjugate system that on the interval $t \in [t_0, t_f]$ has Hamiltonian $\tilde{\mathcal{H}}(\Gamma, t) = \mathcal{H}(\Gamma, t_f + t_0 - t)$. The conjugate trajectory satisfies $\tilde{\Gamma}(t) = \Gamma(t_f + t_0 - t)^\dagger$, where the dagger denotes velocity reversal. The respective conditional probabilities are in the ratio

$$\frac{\wp(\underline{\Gamma}|\Gamma_0)}{\wp(\tilde{\underline{\Gamma}}|\tilde{\Gamma}_0; \tilde{\mathcal{H}})} = e^{\bar{\Delta}S_r(\underline{\Gamma})/k_B} = e^{-\beta[\mathcal{H}(\Gamma_f, t_f)-\mathcal{H}(\Gamma_0, t_0)]} e^{\beta W(\underline{\Gamma})}. \tag{6.213}$$

As in the general case, the Gaussians cancel in this ratio.

Assuming that the system was in the non-equilibrium state at the start of the original trajectory, $\wp(\Gamma_0, t_0) = Z(t_0)^{-1} e^{S_r(\Gamma_0, t_0)/k_B}$, and also of the conjugate trajectory, $\wp(\tilde{\Gamma}_0, t_0; \tilde{\mathcal{H}}) = Z(\tilde{t}_0)^{-1} e^{S_r(\tilde{\Gamma}_0, t_0; \tilde{\mathcal{H}})/k_B}$, with $S_{r,st}(\tilde{\Gamma}_0, t_0; \tilde{\mathcal{H}}) = S_{r,st}(\Gamma_f, t_f)$, the ratio of the unconditional trajectory probabilities is

$$\frac{\wp(\underline{\Gamma}|\Gamma_0)\wp(\Gamma_0, t_0)}{\wp(\tilde{\underline{\Gamma}}|\tilde{\Gamma}_0; \tilde{\mathcal{H}})\wp(\tilde{\Gamma}_0, t_0; \tilde{\mathcal{H}})} = \frac{Z(t_0; \tilde{\mathcal{H}})}{Z(t_0)} \frac{e^{\beta\overline{W}(\Gamma_0, t_0)}}{e^{\beta\overline{W}(\tilde{\Gamma}_0, t_0; \tilde{\mathcal{H}})}} e^{\beta W(\underline{\Gamma})}. \tag{6.214}$$

This follows because the dynamic part of the reservoir entropy is essentially the most likely work done,

$$\begin{aligned}\overline{W}(\Gamma, t) &\equiv TS_{r,dyn}(\Gamma, t) \\ &= -T\int_0^t dt'\, \dot{S}_r^0(\overline{\Gamma}(t'|\Gamma, t), t') \\ &= \int_0^t dt' \frac{U(\Gamma', t')}{\partial t'}, \quad \Gamma' \equiv \overline{\Gamma}(t'|\Gamma, t).\end{aligned} \tag{6.215}$$

These require the definition of the Hamiltonian to be extended beyond the time interval. For a cyclic system, $\mathcal{H}(\Gamma, t_0) = \mathcal{H}(\Gamma, t_f)$, with extended Hamiltonian $\tilde{\mathcal{H}}(\Gamma, t) = \mathcal{H}(\Gamma, t)$, $t \leqslant t_0$, the dynamic part of the entropies are equal, $S_{r,dyn}(\Gamma, t; \tilde{\mathcal{H}}) = S_{r,dyn}(\Gamma, t)$. In this case $Z(t_0; \tilde{\mathcal{H}}) = Z(t_0)$ and

$$\frac{\wp(\underline{\Gamma}|\Gamma_0)\wp(\Gamma_0, t_0)}{\wp(\underline{\tilde{\Gamma}}|\tilde{\Gamma}_0; \tilde{\mathcal{H}})\wp(\tilde{\Gamma}_0, t_0; \tilde{\mathcal{H}})} = e^{\beta W(\underline{\Gamma})}. \tag{6.216}$$

This result also holds for a system in which the pre-factors do not change grossly over the time interval, provided that the time interval is long enough.

A similar result holds for a dynamically disordered system prior to and after the trajectory. This corresponds to the instantaneous static distribution which is just the Maxwell–Boltzmann distribution,

$$\wp_{\mathrm{st}}(\Gamma, t) \equiv \frac{1}{Z_{\mathrm{st}}(t)} e^{-\beta \mathcal{H}(\Gamma, t)}, \quad t = t_0 \text{ or } t_f. \tag{6.217}$$

In this case the unconditional probability ratio is given exactly by,

$$\begin{aligned} \frac{\wp(\underline{\Gamma}|\Gamma_0)\wp_{\mathrm{st}}(\Gamma_0, t_0)}{\wp(\underline{\tilde{\Gamma}}|\tilde{\Gamma}_0; \tilde{\mathcal{H}})\wp_{\mathrm{st}}(\tilde{\Gamma}_0, t_0; \tilde{\mathcal{H}})} &= \frac{Z_{\mathrm{st}}(t_0; \tilde{\mathcal{H}})}{Z_{\mathrm{st}}(t_0)} e^{\beta W(\underline{\Gamma})} \\ &= e^{-\beta\left[F_{\mathrm{st}}(t_f) - F_{\mathrm{st}}(t_0)\right]} e^{\beta W(\underline{\Gamma})}. \end{aligned} \tag{6.218}$$

Here the static or instantaneous Helmholtz free energy is the logarithm of the total entropy, which is just the partition function,

$$F_{\mathrm{st}}(t) = -k_{\mathrm{B}}T \ln Z_{\mathrm{st}}(t) = -k_{\mathrm{B}}T \ln \frac{1}{h^{3N}N!} \int \mathrm{d}\Gamma e^{-\beta \mathcal{H}(\Gamma, t)}. \tag{6.219}$$

For a cyclic system, $\mathcal{H}(\Gamma, t_0) = \mathcal{H}(\Gamma, t_f)$, the static partition functions are equal, and the Helmholtz free energy is the same at the start as at the end of the time interval.

The fluctuation and work theorems were given above for the general non-equilibrium case, equations (6.198) and (6.199). Those results obviously can be applied to the present case of mechanical work. To be explicit, equation (6.200) is the average Boltzmann factor of the work done. Assuming thermal equilibrium at the beginning and at the end of the interval, this is

$$\begin{aligned} \left\langle e^{-\beta W(\underline{\Gamma})} \right\rangle &= \int \mathrm{d}\underline{\Gamma}\, \wp(\underline{\Gamma}|\Gamma_0)\wp_{\mathrm{MB}}(\Gamma_0, t_0) e^{-\beta W(\underline{\Gamma})} \\ &= \int \mathrm{d}\underline{\tilde{\Gamma}} \tilde{\wp}(\underline{\tilde{\Gamma}}|\tilde{\Gamma}_0)\tilde{\wp}_{\mathrm{MB}}(\tilde{\Gamma}_0, t_0) e^{-\beta\left[F_{\mathrm{st}}(t_f) - F_{\mathrm{st}}(t_0)\right]} \\ &= e^{-\beta\left[F_{\mathrm{st}}(t_f) - F_{\mathrm{st}}(t_0)\right]}. \end{aligned} \tag{6.220}$$

The left-hand side is the average of the exponential of the negative of the mechanical work actually done over the time interval, divided by $k_{\mathrm{B}}T$. This average is over all possible trajectories. The exponent on the right-hand side of the final equality is the negative of the difference in the Helmholtz free energy divided by $k_{\mathrm{B}}T$. The assumption that the system is in an equilibrium state at the termini of the trajectory is valid if the time dependence of the Hamiltonian is negligible near the termini of the trajectory: $\mathcal{H}(\Gamma, t) \approx \mathcal{H}(\Gamma, t_0)$, for $t \in [t_0, t_0 - \tau]$, $\mathcal{H}(\Gamma, t) \approx \mathcal{H}(\Gamma, t_f)$, for $t \in [t_f, t_f + \tau]$, where $\tau$ is some relaxation time.

For example, if

$$\mathcal{H}(\mathbf{\Gamma}, t) \rightarrow \begin{cases} \mathcal{H}_1(\mathbf{\Gamma}), t \lesssim t_0 \\ \mathcal{H}_2(\mathbf{\Gamma}), t \gtrsim t_f, \end{cases} \tag{6.221}$$

then the exponent on the right-hand side is the difference in Helmholtz free energy for the two equilibrium systems with the respective Hamiltonians. Most commonly, $\mathcal{H}(\mathbf{\Gamma}, t)$ is chosen to smoothly transform from one to the other. Also common is for the trajectories to be calculated adiabatically, and the average of the work done over them is taken with respect to a Maxwell–Boltzmann distribution of the initial point and initial Hamiltonian. The present formulation allows the average to be more correctly taken over stochastic dissipative trajectories with the correct Gaussian weight, which is the right-hand side of the first equality above.

As has been mentioned, this mechanical work theorem was first given by Bochkov and Kuzovlev (1981) (for the case of a long cyclic trajectory), and was later given by Jarzynski (1997). The advantage of the present derivation is that it explicitly accounts for the exchange of heat between the sub-system and the reservoir during the performance of the work.

## Summary

- The non-equilibrium probability in the sub-system phase space is the exponential of the reservoir entropy, which is the sum of a static or instantaneous part, and a dynamic part. The latter is the time integral over the most likely backward trajectory of the adiabatic rate of static reservoir entropy production. In the mechanical case it may be interpreted as the negative of the most likely work done by the external potential on the sub-system.
- The stochastic, dissipative equations of motion for a non-equilibrium system can be derived from the second entropy for the transition probability. The stochastic reservoir force is Gaussian distributed, and the dissipative force is dominated by an irreversible term proportional to the gradient of the static part of the reservoir entropy. That a single transport matrix gives the variance of the former and the strength of the latter is a non-equilibrium version of the fluctuation–dissipation theorem in phase space.
- The dynamic part of the reservoir entropy can be approximated by its odd projection, which in turn can often be approximated by an integral over adiabatic trajectories.
- The Green–Kubo relations for the transport coefficients can be derived from the average flux by linearizing the non-equilibrium probability with respect to the dynamic part of the reservoir entropy.
- A Brownian particle in a potential trap in uniform motion serves to illustrate results for a mechanical non-equilibrium system simply and explicitly. From the stochastic, dissipative equations of motion, which have Langevin form, the dynamic part of the reservoir entropy may be obtained. Maximizing the full reservoir entropy of the non-equilibrium system shows that the most

likely configuration has the particle moving with the same velocity as the trap, and lagging the trap such that the drag force is equal and opposite to the mechanical restoring force.

- Analysis of the path entropy in general gives the probability of a trajectory, from which the average of various functions of the change in entropy may be deduced. These give the fluctuation and work theorems for both mechanical and thermodynamic non-equilibrium systems.

## References


Attard P 2002 Stochastic molecular dynamics: a combined Monte Carlo and molecular dnamics technique for isothermal simulations *J. Chem. Phys.* **116** 9616

Attard P 2006a Statistical mechanical theory for steady state systems. IV. Transition probability and simulation algorithm demonstrated for heat flow *J. Chem. Phys.* **124** 024109

Attard P 2006b Statistical mechanical theory for steady state systems. V. Non-equilibrium probability density *J. Chem. Phys.* **124** 224103

Attard P and Gray-Weale A 2008 Statistical mechanical theory for steady state systems. VIII. General theory for a Brownian particle driven by a time- and space-varying force *J. Chem. Phys.* **128** 114509

Attard P 2009a Statistical mechanical theory for non-equilibrium systems. IX. Stochastic molecular dynamics *J. Chem. Phys.* **130** 194113

Attard P 2009b Non-equilibrium Monte Carlo simulation for a driven Brownian particle *Phys. Rev.* E **80** 041126

Attard P 2009c The second entropy: A general theory for non-equilibrium thermodynamics and statistical mechanics *Annu. Rep. Prog. Chem.* **105** 63

Attard P 2012 *Non-Equilibrium Thermodynamics and Statistical Mechanics: Foundations and Applications* (Oxford: Oxford University Press)

Attard P 2014 Simplified derivation of the non-equilibrium probability distribution arXiv:1405.1469v1

Bochkov G N and Kuzovlev Yu E 1981 Non-linear fluctuation–dissipation relations and stochastic models in non-equilibrium thermodynamics *Physica* **106** 443

Boinepalli S and Attard P 2003 Grand canonical molecular dynamics *J. Chem. Phys.* **119** 12769

Evans D J, Cohen E G D and Morriss G P 1993 Probability of second law violations in shearing steady states *Phys. Rev. Lett.* **71** 2401

Evans D J 2003 A non-equilibrium free energy theorem for deterministic systems *Mol. Phys.* **101** 1551

Einstein A 1933 *On the Method of Theoretical Physics* (The Herbert Spencer Lecture, delivered at Oxford, 10 June)

Glauber R J 1963 Time-dependent statistics of the Ising model *J. Math. Phys.* **4** 294

Green M S 1954 Markoff random processes and the statistical mechanics of time-dependent phenomena. II. Irreversible processes in fluids *J. Chem. Phys.* **22** 398

Jarzynski C 1997 Non-equilibrium equality for free energy differences *Phys. Rev. Lett.* **78** 2690

Kawasaki K 1966 Diffusion constants near the critical point for time-dependent Ising models *I Phys. Rev.* **145** 244

Kubo R 1966 The fluctuation–dissipation theorem *Rep. Prog. Phys.* **29** 255

Kubo R, Toda M and Hashitsume N 1978 *Statistical Physics II. Non-Equilibrium Statistical Mechanics* (Berlin: Springer)

Langer J S 1969 Statistical theory of the decay of metastable states *Ann. Phys.* **54** 258
Le Bellac M, Mortessagne F and Batrouni G G 2004 *Equilibrium and Non-equilibrium Statistical Thermodynamics* (Cambridge: Cambridge University Press)
Metiu H, Kitahara K and Ross J 1975 Derivation of stochastic equations for nonequilibrium Ising mean field model *J. Chem. Phys.* **63** 5116
Onsager L 1931 Reciprocal relations in irreversible processes I *Phys. Rev.* **37** 405
Onsager L 1931 Reciprocal relations in irreversible processes II *Phys. Rev.* **38** 2265
Pottier N 2010 *Non-Equilibrium Statistical Physics: Linear Irreversible Processes* (Oxford: Oxford University Press)
Slessor K 1939 *Five bells: XX Poems* (Sydney: F.C.Johnson)
Weinberg S 2015 *To Explain the World: The Discovery of Modern Science* (London: Penguin) p 111
Zwanzig R 2001 *Non-Equilibrium Statistical Mechanics* (Oxford: Oxford University Press)

**IOP** Publishing

Entropy Beyond the Second Law
Thermodynamics and statistical mechanics for equilibrium, non-equilibrium, classical, and quantum systems
**Phil Attard**

# Chapter 7

## Entropy collapses

> 'Inspiration and aesthetic judgement are important in the development of scientific theories, but the verification of these theories relies finally on impartial experimental tests of their predictions. Though mathematics is used in the formulation of physical theories and in working out their consequences, science is not a branch of mathematics, and scientific theories cannot be deduced by purely mathematical reasoning'
>
> Weinberg (2015)

> 'Let me tell you the secret that has led me to my goal. My only strength lies in my tenacity'
>
> Louis Pasteur (Dubos 1950)

The main aim of this chapter is to formulate quantum statistical mechanics. Specifically, the von Neumann trace (i.e. mixture of states) for the partition function and statistical averages derived from the postulates of quantum mechanics. A dual expansion for these that accounts for non-commutativity and particle symmetry is developed, with the leading order term giving classical statistical mechanics, and hence classical mechanics.

Perhaps the key equation in quantum statistical mechanics is the trace over the density operator form of the partition function,

$$Z = \mathrm{TR}\,\hat{\rho} = \sum_n e^{-\beta \mathcal{H}_n}. \tag{7.1}$$

This is due to von Neumann (1932) and it is said to give the canonical partition function of a quantum system as the trace of the density operator, which in turn is said to equal the Maxwell–Boltzmann weighted sum over all the energy eigenstates.

doi:10.1088/978-0-7503-1590-6ch7

The challenge is to actually derive both of these from first principles, and to give their correct form and interpretation.

## 7.1 Conventional quantum statistical average

### 7.1.1 Expectation value

We begin by establishing the basic notation and by setting out the conventional approach to quantum statistical mechanics.

The state of a quantum system may be specified by a particular wave function $\psi$. A wave function will usually be denoted in the position representation, $\psi(\mathbf{r})$, or else in the bra-ket notation, $|\psi\rangle$ and $\langle\psi|$. The squared norm (or squared modulus) of the wave function is

$$N(\psi) \equiv \langle\psi|\psi\rangle = \int \mathrm{d}\mathbf{r}\ \psi(\mathbf{r})^* \psi(\mathbf{r}), \tag{7.2}$$

where the asterisk denotes the complex conjugate. It is usually the case that the wave function is normalized, $N(\psi) = 1$.

The expectation value of an operator $\hat{O}$ when the system is in the wave state $\psi$ is (Messiah 1961, Merzbacher 1970)

$$O(\psi) = \frac{1}{N(\psi)}\langle\psi|\hat{O}|\psi\rangle = \frac{1}{N(\psi)}\int \mathrm{d}\mathbf{r}\ \psi(\mathbf{r})^* \hat{O}(\mathbf{r})\psi(\mathbf{r}). \tag{7.3}$$

The system can also be described as being in a particular quantum state $n = 0, 1, 2, \ldots$, which are the eigenstates of some operator. The corresponding eigenfunctions $\zeta_n$ are orthonormal, $\langle\zeta_n|\zeta_m\rangle = \delta_{n,m}$, and they form a complete basis for the Hilbert space. Hence the wave function can be written

$$\psi(\mathbf{r}) = \sum_n \psi_n \zeta_n(\mathbf{r}), \quad \psi_n = \langle\zeta_n|\psi\rangle. \tag{7.4}$$

An arbitrary operator can also be represented in this basis, $O_{mn} = \langle\zeta_m|\hat{O}|\zeta_n\rangle$, and its expectation value then takes the form (Messiah 1961, Merzbacher 1970)

$$O(\psi) = \frac{1}{N(\psi)}\sum_{m,n}\psi_m^* O_{mn}\,\psi_n. \tag{7.5}$$

Suppose $\zeta_n^0(\mathbf{r})$ is a normalized eigenfunction of the operator with eigenvalue $O_n$, $\hat{O}\zeta_n^O = O_n\zeta_n^O$. The eigenstates may possibly be degenerate, but nevertheless the eigenfunctions can be organized to form a complete set, $\psi = \sum_n \psi_n^O \zeta_n^O$. In the eigenfunction basis the operator is diagonal, $O_{mn}^O = O_n\delta_{mn}$, and the expectation value in the wave state $\psi$ is

$$O(\psi) = \frac{1}{N(\psi)}\sum_n \psi_n^{O*}\psi_n^O O_n. \tag{7.6}$$

The quantity $\psi_n^{O*}\psi_n^O/N(\psi)$ has the interpretation as the probability of the system being in the quantum state $n$ (given that it is in the wave state $\psi$).

In this form, the eigenvalues $O_n$ are the possible outcomes of a measurement with the operator $\hat{O}$, and the square of the amplitude $\psi_n^{O*}\psi_n^O$ is proportional to the probability that that particular outcome will occur. In the more general form for the expectation value as the double sum in an arbitrary basis, equation (7.5), the complex numbers $\psi_m^*\psi_n$ cannot be interpreted in terms of classical probability theory because, as discussed in section 1.1.2, the latter is predicated upon real, non-negative weights. The general form exhibits the phenomenon of quantum superposition and interference: the quantum system can be in more than one state at a time, and the different states affect each other.

The wave function collapses upon measurement, $\psi \Rightarrow \zeta_n^O$, which is to say that the system no longer exists in a superposition of states but rather becomes a pure state. The wave function could be said to be decoherent in that there is no interference between states. In general, a decoherent system is a mixture of pure states, and its expectation value is formally identical to a classical average. The specifically quantum aspects of the system (interference, superposition) are no longer present.

One may use the bra-ket notation for the wave function to construct the (single wave function) density operator,

$$\hat{\rho} \equiv \frac{1}{N(\psi)}|\psi\rangle\,\langle\psi|. \tag{7.7}$$

With this the expectation value may be written in the form of the trace of the operator product,

$$O(\psi) = \mathrm{TR}\,\hat{\rho}\;\hat{O} = \frac{1}{N(\psi)}\sum_{m,n}\psi_n\psi_m^* O_{mn}. \tag{7.8}$$

Because the density operator constructed from a single wave function has a dyadic form, it cannot be diagonalized. To see this, consider some representation in which it is diagonal. If $\rho_{n_1n_1} = |\psi_{n_1}|^2$ and $\rho_{n_2n_2} = |\psi_{n_2}|^2$ are two distinct non-zero diagonal elements, then the corresponding off-diagonal elements, $\rho_{n_1n_2} = \rho_{n_2n_1}^* = \psi_{n_1}^*\psi_{n_2}$, would also be non-zero. This contradiction proves that the single wave function density operator cannot be diagonal in any representation. The exception to this is if it has precisely one non-zero element, as only occurs upon collapse due to measurement.

### 7.1.2 Statistical average

In the context of the statistical average in quantum statistical mechanics, the non-diagonal nature of the single wave function density operator is directly relevant. This is because conventionally the statistical average is expressed as a trace over the density operator (von Neumann 1932). We shall return to this matter shortly, but for the present let us assume the existence of a probability operator, $\hat{\wp}$. Let us further assume that the statistical average can be expressed as the trace of it and the observable operator

$$\begin{aligned}\langle \hat{O} \rangle_{\text{stat}} &= \text{TR}\hat{\wp}\,\hat{O} \\ &= \sum_{m,n} \wp_{mn}\, O_{nm} \\ &= \sum_{n} \wp^{O}_{nn}\, O^{O}_{nn} \\ &= \sum_{n} \wp^{S}_{nn}\, O^{S}_{nn}. \end{aligned} \tag{7.9}$$

In the second last equality the eigenstates of the observable operator have been used. This form only requires the diagonal elements of the probability operator, whether or not the latter is diagonalizable. This is identical to the classical probability form for a statistical average, namely the sum over states of the operator eigenvalue, $O^{O}_{nn} = O_n$, times the probability of the state, $\wp^{O}_{nn}$. From the point of view of classical probability this is rather appealing, but it does raise the question of how it arises from the quantum expectation value form that reflects a superposition of states and interference between states.

The final equality for the statistical average is also problematic. Here the eigenstates of the probability operator are invoked. Although not essential for the present discussion, and anticipating somewhat results to come, these can be assumed to be entropy eigenstates and denoted by the superscript $S$, $\hat{\wp}|\zeta^{S}_{n}\rangle = \wp_n|\zeta^{S}_{n}\rangle$, and $\wp^{S}_{mn} = \wp_n\delta_{m,n}$. Again this is apparently dissonant with the expectation value for a wave state, where the density operator cannot be diagonalized. This demands a more careful examination of the origin of the trace form for the statistical average and of the properties of the probability operator. One can already see that any such diagonalizable probability operator cannot be a single wave function density operator.

The challenge is to get from the expectation value form to the statistical average form. In the former, the density operator always includes off-diagonal contributions, (except in the observable operator representation itself). In the latter, the off-diagonal contributions vanish in both the entropy and the observable operator representations. The qualitative difference between these two expressions can be seen in figure 7.1.

The conventional view of quantum statistical mechanics is in terms of the density operator rather than the probability operator (von Neumann 1932, Messiah 1961, Merzbacher 1970). Because the density operator of a single-wave function cannot be diagonalized, most workers instead imagine an ensemble of systems, labeled $a = 1, 2, \ldots, M$, each with its own wave function $\psi_a$. The ensemble average of the single-wave function density operators defines the many wave function density operator,

$$\hat{\bar{\rho}} \equiv \frac{1}{M} \sum_{a=1}^{M} \frac{|\psi_a\rangle\,\langle\psi_a|}{\langle\psi_a|\psi_a\rangle}. \tag{7.10}$$

(Almost always the wave functions are taken to be normalized, $\langle\psi_a|\psi_a\rangle = 1$, and the denominator is redundant.) This can also be called the averaged density operator.

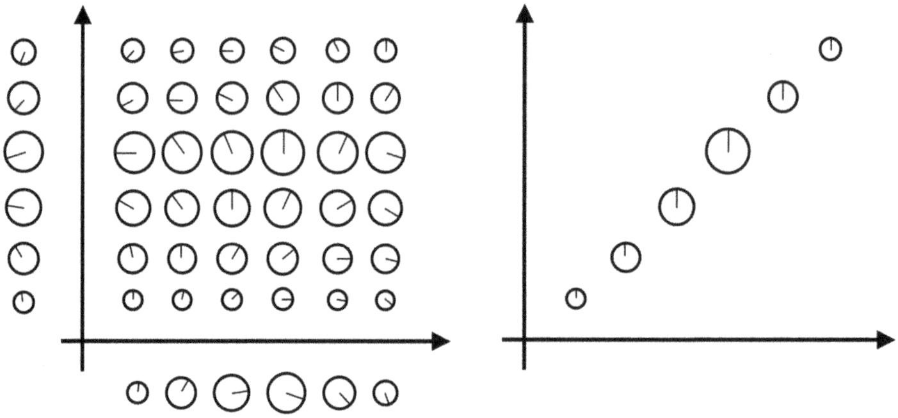

**Figure 7.1.** Density operator for a superposition of states (left) and for a mixture of pure states (right). The amplitude and phase of the wave function is represented on the axes.

This does not factorize into a dyadic product, which is the crucial difference from the single-wave function density operator.

There is no reason to suppose that this cannot be diagonalized. Hence the many-wave function density operator can be equated to the probability operator introduced above. In the representation in which it is diagonal (presumably the entropy representation), the entries equal the state probabilities,

$$\bar{\rho}^{\mathrm{S}}_{mn} = \frac{1}{M}\sum_{a=1}^{M}\frac{\psi^{S*}_{a,m}\psi^{S}_{a,n}}{\langle\psi_a|\psi_a\rangle} = \wp_n\delta_{m,n}. \tag{7.11}$$

One can see that the off-diagonal terms average to zero if the phase angles of the coefficients are uncorrelated (decoherent). The diagonal terms all have zero phase angle exactly and so they cannot cancel with each other. This shows that the crucial difference between the one- and many-wave function density operators is this averaging process.

In view of this, in so far as one or other exists, the averaged density operator is equal to the probability operator, $\hat{\bar{\rho}} = \hat{\wp}$. This operator relationship holds in any representation, not just the entropy representation, in which both were said to be diagonal. If the probability operator is an ordinary operator, then it can only be diagonal in the basis formed from the eigenfunctions of the entropy operator (or of any operator that commutes with the entropy operator). This would mean that the off-diagonal phase cancelation just discussed for the averaged density operator only occurred in the entropy basis and not in any other basis. It remains to prove this, either for the averaged density operator in the ensemble picture, or else for the probability operator directly.

With this definition of the many-wave function density operator, and its equality with the probability operator, the statistical average can be written as the trace of it and the observable operator,

$$\langle\hat{O}\rangle_{\mathrm{stat}} = \mathrm{TR}\,\hat{\bar{\rho}}\,\hat{O} = \frac{1}{M}\sum_{a=1}^{M}\frac{\langle\psi_a|\hat{O}|\psi_a\rangle}{\langle\psi_a|\psi_a\rangle}. \tag{7.12}$$

The conventional presentation of the ensemble picture is that the wave functions $\psi_a$ are each a pure quantum state, in which case the ensemble comprises a mixture of pure quantum states. This means that each system in the ensemble has a fully collapsed wave function. The above discussion departs slightly from this in that it is sufficient for equation (7.11) to hold, which is to say that the phases of the entropy states are decoherent. In other words we need not insist that each member of the ensemble be in a pure quantum state.

In the presentation of classical statistical mechanics, the ensemble picture was criticized, sections 1.4.1 and 1.4.2. Since the many-wave particle density operator is a direct manifestation of the ensemble, it is similarly unsatisfactory. The problem is that such a collection of independent replica systems has no physical reality. The fact of the matter is that an actual measurement is made on a single system. This is not to say that the trace form of the quantum statistical average (with a probability operator equivalent to an averaged density operator) is incorrect; in fact this trace form has long been successfully used. Rather it says that the explanation and justification for it does not lie in the ensemble picture. And of course, without the ensemble picture there is no need or basis for the many-wave function density operator.

Instead of an ensemble average one might consider that the decoherent system arises from a time average for the density operator. Let us imagine that a measurement extends over a brief time interval, $\tau$. In this period the wave function evolves, $\psi(t)$, and one can define the averaged density operator as

$$\hat{\rho} \equiv \frac{1}{\tau N_0} \int_0^\tau \mathrm{d}t \; |\psi(t)\rangle \, \langle \psi(t)|, \tag{7.13}$$

assuming unitary evolution of the wave function, $N(\psi(t)) \equiv \langle \psi(t)|\psi(t)\rangle = N_0$. Again assuming that this is the same as the probability operator, $\hat{\rho} = \hat{\wp}$, it must be diagonal in the entropy representation,

$$\bar{\rho}_{ln}^S = \frac{1}{\tau N_0} \int_0^\tau \mathrm{d}t \; \psi_l^{S*}(t)\psi_n^S(t) = \wp_n \delta_{l,n}. \tag{7.14}$$

Presumably, the phase factors of the entropy modes evolve independently, which makes them effectively random. This means that the off-diagonal elements average to zero over the measurement time.

It is not clear that this putative temporal averaging over the measurement time is any more realistic than the ensemble averaging. Both explanations have the appearance of *post facto* rationalization. Under the circumstances it would perhaps be more honest not to pretend to justify equation (7.9), but rather simply to assert it as the definition of the statistical average, taking the existence of the probability operator as a given. In my opinion it is better not to introduce or to invoke the averaged density operator as it has no physical reality.

In the following two sections, a derivation from first principles is given of the von Neumann trace form for the statistical average equation (7.9). The derivation gives the explicit form for the probability operator and shows that it is diagonal in the entropy representation. It also demonstrates the passage from the wave function and expectation value formulation of quantum mechanics to the statistical average formulation as a weighted sum over quantum states.

## 7.2 Uniform weight density of wave space

In chapter 5 equilibrium classical statistical mechanics was derived. It was shown in section 5.2 that the entropy of a point in the phase space of an isolated system was uniform. This result was first discussed in terms of the conventional ergodic hypothesis, section 5.2.1, then in terms of constant probability on a trajectory, section 5.2.2, and finally, and most satisfactorily, a proof was given based on uniformity in time and the density of energy hypersurfaces, section 5.2.3.

For the present quantum problem one can proceed analogously. For an isolated system, in section 7.2.1 it is simply asserted that quantum states are equally probable. In section 7.2.2 it is shown that the probability density is constant on a trajectory. And in section 7.2.3 it is proven that the probability density is uniform in wave space based on time uniformity and energy and norm hypersurface density.

### 7.2.1 Equal state probability hypothesis

The simplest presentation of the probability operator focuses on the quantum states, and, as in the classical case, it begins with the microstates of an isolated system. From their weight the required quantum state probabilities follow, first for an isolated system, and thence for a sub-system interacting with a reservoir. By analogy with the ergodic hypothesis of classical statistical mechanics, one can take it as an axiom that

$$\text{All quantum states of an isolated system have equal weight.} \tag{7.15}$$

As will be seen, the quantum states are the eigenstates of any complete operator. These are the diagonal elements in any representation.

In the two following subsections, this hypothesis is actually derived from more formal considerations; it appears as equation (7.40) below. Instead of quantum states, the following analysis in sections 7.2.2 and 7.2.3 is based on the Hilbert space of wave functions. The aim is to show that the wave space probability density for an isolated system is uniform.

### 7.2.2 Trajectory uniformity

Consider an isolated system with wave function $\psi$. Schrödinger's equation for the time rate of change of the latter is

$$i\hbar\psi^0 = \hat{\mathcal{H}}\psi. \tag{7.16}$$

Here $\hat{\mathcal{H}}$ is the Hamiltonian or energy operator. This is here taken to be independent of time. As in the earlier classical analysis, the superscript 0 is used for adiabatic quantities, which are those of the isolated system. In this section only the total isolated system appears.

One can construct a trajectory by integrating this

$$\psi(t) \equiv \psi^0(t|\psi_0, t_0) = \hat{U}^0(t - t_0)\psi_0. \tag{7.17}$$

Here the adiabatic time propagator is $\hat{U}^0(t) = e^{t\hat{\mathcal{H}}/i\hbar}$. For a time-dependent Hamiltonian, this is $\hat{U}^0(t - t_0) = e^{\int_{t_0}^{t} \mathrm{d}t' \, \hat{\mathcal{H}}(t')/i\hbar}$, with time-ordering assumed.

Each point $\psi$ in wave space has a real, non-negative weight density $\omega(\psi, t)$. As usual, the normalized form of this is the probability density $\wp(\psi, t)$. In section 7.3 these will be related to the probability operator and to the statistical average.

As for any density, $\mathrm{d}\psi \; \wp(\psi, t)$ is the probability of the system being within $|\mathrm{d}\psi|$ of $\psi$. This is a real non-negative number: $\mathrm{d}\psi \; \wp(\psi, t) = |\mathrm{d}\psi \; \wp(\psi, t)| = |\mathrm{d}\psi||\wp(\psi, t)|$. Since it is always the product that occurs, without loss of generality one may take each to be individually real.

The probability density is normalized

$$\int \mathrm{d}\psi \; \wp(\psi, t) = \int \mathrm{d}\underline{\psi} \; \wp(\underline{\psi}, t) = 1. \tag{7.18}$$

Here has been invoked an arbitrary representation, $\underline{\psi} = \{\psi_n\}$, $n = 1, 2, \ldots$. The coefficients are complex, $\psi_n = \psi_n^{\mathrm{r}} + i\psi_n^{\mathrm{i}}$, and the infinitesimal volume element can be written as

$$\mathrm{d}\underline{\psi} = \mathrm{d}\underline{\psi}^{\mathrm{r}} \, \mathrm{d}\underline{\psi}^{\mathrm{i}} \equiv \mathrm{d}\psi_1^{\mathrm{r}} \, \mathrm{d}\psi_1^{\mathrm{i}} \, \mathrm{d}\psi_2^{\mathrm{r}} \, \mathrm{d}\psi_2^{\mathrm{i}} \ldots \tag{7.19}$$

With this all the integrations are over the real line, $\psi_n^{\mathrm{r}} \in [-\infty, \infty]$, and $\psi_n^{\mathrm{i}} \in [-\infty, \infty]$, $n = 1, 2, \ldots$.

As is the case for Hamilton's classical equations of motion, Schrödinger's equation gives an incompressible trajectory,

$$\begin{aligned} \frac{\mathrm{d}\dot{\psi}^0}{\mathrm{d}\psi} &= \underline{\partial}_{\psi^{\mathrm{r}}} \cdot \underline{\dot{\psi}}^{0,\mathrm{r}} + \underline{\partial}_{\psi^{\mathrm{i}}} \cdot \underline{\dot{\psi}}^{0,\mathrm{i}} \\ &= \underline{\partial}_{\psi} \cdot \underline{\dot{\psi}}^{0} + \underline{\partial}_{\psi^*} \cdot \underline{\dot{\psi}}^{0*} \\ &= \frac{1}{i\hbar} \, \mathrm{TR} \, \underline{\underline{\mathcal{H}}} - \frac{1}{i\hbar} \, \mathrm{TR} \, \underline{\underline{\mathcal{H}}} \\ &= 0. \end{aligned} \tag{7.20}$$

The first two equalities are in fact the general expression for the trajectory compressibility. They hold as well for the total time derivative when the sub-system is open to a reservoir, heat bath, or environment. The final two equalities only hold when the evolution is given by Schrödinger's equation.

Again, as in classical phase space, the compressibility gives the logarithmic rate of change of the volume element. Since the compressibility vanishes, the volume element is a constant of the motion of the isolated system,

$$\mathrm{d}\psi^0(t) = \mathrm{d}\psi_0. \tag{7.21}$$

The total time derivative of the probability density on the adiabatic trajectory is

$$\begin{aligned}\frac{\mathrm{d}^0\wp(\psi,t)}{\mathrm{d}t} &= \frac{\partial\wp(\psi,t)}{\partial t} + \underline{\dot{\psi}}^{0,\mathrm{r}}\cdot\underline{\partial}_{\psi^{\mathrm{r}}}\wp(\psi,t) + \underline{\dot{\psi}}^{0,\mathrm{i}}\cdot\underline{\partial}_{\psi^{\mathrm{i}}}\wp(\psi,t)\\ &= \frac{\partial\wp(\psi,t)}{\partial t} + \underline{\dot{\psi}}^{0}\cdot\underline{\partial}_{\psi}\wp(\psi,t) + \underline{\dot{\psi}}^{0*}\cdot\underline{\partial}_{\psi^*}\wp(\psi,t)\\ &= \frac{\partial\wp(\psi,t)}{\partial t} + \underline{\partial}_{\psi}\cdot\left[\underline{\dot{\psi}}^{0}\wp(\psi,t)\right] + \underline{\partial}_{\psi^*}\cdot\left[\underline{\dot{\psi}}^{0*}\wp(\psi,t)\right].\end{aligned} \tag{7.22}$$

The first and second equalities are just the general definition of the total derivative. The final equality is valid on an adiabatic trajectory, upon which the compressibility vanishes. The probability flux can be seen to be $J_{\wp}(\psi,t)\equiv\dot{\psi}\wp(\psi,t)$.

For the isolated system, the probability density evolves adiabatically,

$$\wp(\psi_1,t_1) = \int\mathrm{d}\psi_0\,\wp(\psi_0,t_0)\,\delta(\psi_1-\psi^0(t_1|\psi_0,t_0)). \tag{7.23}$$

This invokes Bayes' theorem for the transition probability, essentially equation (3.14), $\wp(\psi_1,t_1;\psi_0,t_0)=\wp(\psi_1,t_1|\psi_0,t_0)\,\wp(\psi_0,t_0)$, together with the type 1 reduction condition, equation (3.17). The conditional transition probability in the present adiabatic case is $\wp(\psi_1,t_1|\psi_0,t_0)=\delta(\psi_1-\psi^0(t_1|\psi_0,t_0))$, which reflects the deterministic nature of Schrödinger's equation.

From the incompressibility of wave space under Schrödinger's equation, $\mathrm{d}\psi_0=\mathrm{d}\psi_1$. With $t_1=t_0+\Delta_t$, expanding to linear order in the time step yields

$$\begin{aligned}\wp(\psi_1,t_1) &= \int\mathrm{d}\psi_0\,\wp(\psi_0,t_0)\,\delta(\psi_1-\psi^0(t_1|\psi_0,t_0))\\ &= \wp(\psi_1-\Delta_t\dot{\psi}^0,t_0)\\ &= \wp(\psi_1,t_0) - \Delta_t\underline{\dot{\psi}}^{0}\cdot\underline{\partial}_{\psi}\wp(\psi,t_0) - \Delta_t\underline{\dot{\psi}}^{0*}\cdot\underline{\partial}_{\psi^*}\wp(\psi,t_0).\end{aligned} \tag{7.24}$$

Rearranging, this gives the partial time derivative as

$$\frac{\partial\wp(\psi,t)}{\partial t} = -\underline{\dot{\psi}}^{0}\cdot\underline{\partial}_{\psi}\wp(\psi,t) - \underline{\dot{\psi}}^{0*}\cdot\underline{\partial}_{\psi^*}\wp(\psi,t). \tag{7.25}$$

This holds for an isolated system evolving under Schrödinger's equation.

Combining this with the second equality of equation (7.22), one sees that the total adiabatic time derivative of the probability density vanishes, $\mathrm{d}^0\wp(\psi,t)/\mathrm{d}t=0$. Hence for an isolated system, the probability density is a constant of the motion,

$$\wp(\psi(t),t) = \wp(\psi_0,t_0), \tag{7.26}$$

where $\psi(t)\equiv\psi^0(t|\psi_0,t_0)$.

This is the quantum analogue of equation (5.20), in which the total time derivative of the probability density in the phase space of an isolated system vanishes under Hamilton's equations. This is not surprising as the physical picture underlying the original classical result is the same as that which holds for the present quantum result. Like Hamilton's equations, Schrödinger's equation is deterministic, which means that trajectories do not cross, and they are neither created nor destroyed. In both cases trajectories within a given volume remain inside throughout its evolution. In so far as the probability of the volume is proportional to the number of trajectories it contains, this is conserved by both Hamilton's and Schrödinger's equation. What the two equations additionally have in common is that they are incompressible, which means that the volume is also a constant of the motion. These two facts mean that the probability density must be conserved moving along the trajectory.

For the present equilibrium system, the Hamiltonian operator does not depend upon time. In this case the probability density must independent of time, $\wp(\psi)$. One has the stronger result

$$\wp(\psi^0(t|\psi_0, t_0)) = \wp(\psi_0). \tag{7.27}$$

This says that the probability density is the same everywhere on a trajectory.

Finally, by analogy with the classical ergodic hypothesis, one may suppose that a single trajectory passes sufficiently close to all relevant points of the state space. The hypothesis says that one can take any two points in state space, $\psi_1$ and $\psi_2$, with the same norm and energy expectation to lie on a single trajectory. (It will be shown explicitly shortly that Schrödinger's equation conserves the norm and energy of the wave function.) By the above, they therefore have the same probability density

$$\wp(\psi_2) = \wp(\psi_1), \text{ if } E(\psi_2) = E(\psi_1) \quad \text{and} \quad N(\psi_2) = N(\psi_1). \tag{7.28}$$

Hence the wave space probability density must be of the form

$$\wp(\psi|E, N) = \frac{\delta(E(\psi) - E)\delta(N(\psi) - N)}{W(E, N)}, \tag{7.29}$$

with the normalizing factor being

$$W(E, N) = \int \mathrm{d}\psi \; \delta(E(\psi) - E)\delta(N(\psi) - N). \tag{7.30}$$

These are the quantum analogues of equations (5.22) and (5.23). And so one is faced with an identical problem: in order to transform from the conditional wave space probability, $\wp(\psi|E, N)$, to the unconditional wave space probability, $\wp(\psi)$, one has to make some hypothesis about the dependence of the wave space weight density on the hypersurface, $w(E, N)$. If one assumes that this is constant or negligibly varying with energy and norm, then one can conclude that an isolated system has uniform weight in wave space,

$$\wp(\psi) = \frac{1}{W}, \quad W = \int \mathrm{d}\psi. \tag{7.31}$$

Obviously it is a little unsatisfactory to simply assume this. It is also unsatisfactory to assume that a single trajectory completely fills each hypersurface. Now an alternative derivation of the result is given, which avoids these assumptions. The approach is the direct analogue of the classical derivation of section 5.2.3.

### 7.2.3 Time average and hypersurface density

Schrödinger's equations conserve energy and norm. These are well-known results which are straightforward to prove directly.

The norm squared of the wave function is $N(\psi) \equiv \langle\psi|\psi\rangle$. Its adiabatic rate of change is

$$\begin{aligned} \dot{N}^0(\psi) &= \langle\dot{\psi}^0|\psi\rangle + \langle\psi|\dot{\psi}^0\rangle \\ &= \left\langle\frac{1}{i\hbar}\hat{\mathcal{H}}\psi|\psi\right\rangle + \left\langle\psi|\frac{1}{i\hbar}\hat{\mathcal{H}}\psi\right\rangle \\ &= \frac{-1}{i\hbar}\langle\psi|\hat{\mathcal{H}}^\dagger|\psi\rangle + \frac{1}{i\hbar}\langle\psi|\hat{\mathcal{H}}|\psi\rangle \\ &= 0. \end{aligned} \tag{7.32}$$

The final equality follows because the Hamiltonian operator, like all physical operators, is Hermitian conjugate, $\hat{\mathcal{H}}^\dagger = \hat{\mathcal{H}}$. In fact the Hamiltonian operator obeys a stronger symmetry: it is both real, $\hat{\mathcal{H}}^* = \hat{\mathcal{H}}$, and symmetric, $\hat{\mathcal{H}}^{\mathrm{T}} = \hat{\mathcal{H}}$.

The energy of the wave state $\psi$ is $E(\psi) \equiv \langle\psi|\hat{\mathcal{H}}|\psi\rangle/N(\psi)$. Hence by a similar argument it is also a constant of the adiabatic motion,

$$\begin{aligned} \dot{E}^0(\psi) &= \frac{1}{N(\psi)}\langle\dot{\psi}^0|\hat{\mathcal{H}}|\psi\rangle + \frac{1}{N(\psi)}\langle\psi|\hat{\mathcal{H}}|\dot{\psi}^0\rangle \\ &= \frac{1}{N(\psi)}\left\langle\frac{1}{i\hbar}\hat{\mathcal{H}}\psi|\hat{\mathcal{H}}|\psi\right\rangle + \frac{1}{N(\psi)}\left\langle\psi|\hat{\mathcal{H}}|\frac{1}{i\hbar}\hat{\mathcal{H}}\psi\right\rangle \\ &= \frac{-1}{i\hbar N(\psi)}\langle\psi|\hat{\mathcal{H}}^\dagger\hat{\mathcal{H}}|\psi\rangle + \frac{1}{i\hbar N(\psi)}\langle\psi|\hat{\mathcal{H}}\hat{\mathcal{H}}|\psi\rangle \\ &= 0. \end{aligned} \tag{7.33}$$

These results show that the trajectory of an isolated system, $\psi(t)$, is confined to a hypersurface of constant norm and energy. In view of this let $\tilde{\psi}$ denote a normalized wave function. Let $\chi$ denote a wave function on a hypersurface of constant energy $E$ and norm squared $N$. This hypersurface can be labeled $\{N, E\}$, and it is obviously a sub-space of the Hilbert space of all wave functions. In this notation the normalized wave function is $\tilde{\psi} = \tilde{\psi}(\chi, E)$, and the full wave function is $\psi = \psi(\tilde{\psi}, N) = \psi(\chi, E, N)$.

Ultimately the weight density of the full wave space of the isolated system, $\omega(\psi)$, is sought. Initially the weight density on the hypersurface, $\omega(\chi|N, E)$, is derived. This is then combined with the density of the $\{N, E\}$-hypersurface in the wave space to obtain the full weight density.

As in section 5.2.3, it is taken as axiomatic that time is homogeneous, which means that a statistical average is a simple time average over a trajectory. This implies that the weight density on the $\{N, E\}$-hypersurface must be inversely proportional to the speed of the trajectory,

$$\begin{aligned} \omega(\chi|N, E) &\propto |\dot{\psi}^0|^{-1} \\ &= \langle\psi|\hat{\mathcal{H}}\hat{\mathcal{H}}|\psi\rangle^{-1/2}, \end{aligned} \tag{7.34}$$

since the Hamiltonian operator is Hermitian. Notice that this depends upon $\psi$. This is just the time that the system spends in a volume element $|d\chi|$, which is the same as the time spent in the volume element of the full wave space $|d\psi|$. The proportionality factor is an immaterial constant on the hypersurface. As argued in section 5.2.3, this constant cannot vary with energy $E$ or with norm squared $N$ because this would violate the axiom that a statistical average is a simple time average over the trajectory irrespective of energy or norm squared.

Now we transform this hypersurface weight density to the full wave space. The Jacobean for the transformation $\chi \Rightarrow \tilde{\psi}$ is

$$|\tilde{\nabla}E| = \left[\frac{\partial E(\psi)}{\partial|\tilde{\psi}\rangle}\frac{\partial E(\psi)}{\partial\langle\tilde{\psi}|}\right]^{1/2} = \langle\tilde{\psi}|\hat{\mathcal{H}}\hat{\mathcal{H}}|\tilde{\psi}\rangle^{1/2}, \tag{7.35}$$

and that for the transformation $\tilde{\psi} \Rightarrow \psi$,

$$|\nabla N| = \left[\frac{\partial N(\psi)}{\partial|\psi\rangle}\frac{\partial N(\psi)}{\partial\langle\psi|}\right]^{1/2} = \langle\psi|\psi\rangle^{1/2}. \tag{7.36}$$

With these the full weight density is

$$\begin{aligned}\omega(\psi|N, E) &= \omega(\chi|N, E)\frac{|\tilde{\nabla}E|}{\Delta_E}\frac{|\nabla N|}{\Delta_N}, \quad |N(\psi) - N| < \Delta_N, \quad |E(\psi) - E| < \Delta_E \\ &\propto \frac{\langle\tilde{\psi}|\hat{\mathcal{H}}\hat{\mathcal{H}}|\tilde{\psi}\rangle^{1/2}\langle\psi|\psi\rangle^{1/2}}{\langle\psi|\hat{\mathcal{H}}\hat{\mathcal{H}}|\psi\rangle^{1/2}}\delta(N(\psi) - N)\,\delta(E(\psi) - E) \\ &= \delta(N(\psi) - N)\,\delta(E(\psi) - E).\end{aligned} \tag{7.37}$$

Integrating this over all energies and norm squareds yields the unconditional weight density for the wave space of the isolated system,

$$\omega(\psi) = 1. \tag{7.38}$$

The interpretation of the origin of this result is essentially the same as that for the result for classical phase space in section 5.2.3. The weight density is inversely proportional to the speed of the trajectory (i.e. low speed equates to more time in a volume element), and proportional to the hypersurface density, (i.e. for fixed spacing between discrete hypersurfaces, $\Delta_E$ and $\Delta_N$, large gradients correspond to more hypersurfaces per unit wave space volume). These ideas are the same as those depicted in the classical case in figure 5.3.

Just as in the classical case for Hamilton's equations of motion, it is a remarkable consequence of Schrödinger's equation that the speed is identical to the magnitude of the gradient of the hypersurface, so that these two cancel to give a weight density that is uniform in wave space. This result is the same as that given above as equation (7.31), which was obtained only by assuming that the hypersurface weight $w(E, N)$ was constant or negligibly varying.

This weight density is a physical observable, and as such there must be a corresponding weight operator whose expectation value it is. Hence one must have

$$\omega(\psi) \equiv \frac{\langle\psi|\hat{\omega}|\psi\rangle}{\langle\psi|\psi\rangle} = 1. \tag{7.39}$$

Since this must hold for all wave states, the weight operator for the isolated system must be the identity operator, and one must have

$$\hat{\omega} = \hat{\mathrm{I}}, \text{ or } \omega_{mn} = \delta_{m,n}. \tag{7.40}$$

The result holds in any representation.

One conclusion that follows from this is that the quantum microstates of the system are not unique. Rather any complete operator gives a collective of microstates. Also a particular set of microstates is the set of diagonal elements of a particular representation. Equivalently, they are the eigenstates of any complete operator. This result implies that these microstates all have equal weight, which is the axiom given as equation (7.15) at the start of this section.

A second conclusion is that the off-diagonal elements in any representation have zero weight. It should be clear from this that the weight operator is of no use for the expectation value, since the off-diagonal elements of the single wave function density matrix, the dyadic $\psi_m^*\psi_n$, $m \neq n$, certainly contribute to the expectation value. Nevertheless, it will be shown that for two interacting systems, the expectation value of an operator on one system has to be weighted by the states of the other system, and for this a weight operator is required (albeit one derived from the weight operator of the present isolated system).

## 7.3 Canonical equilibrium system

Now the canonical equilibrium system is analyzed in detail. As usual, the total isolated system consists of a sub-system and a thermal reservoir that can exchange energy with each other. In the literature the reservoir is often called the bath, or, particularly in the quantum field, the environment. Also the phrase ‘the system’ is often used to mean what is here called ‘the sub-system’. Here and throughout this book the terms used are total system, sub-system, and reservoir. These are denoted by superscripts or subscripts tot, s, and r where any ambiguity exists.

### 7.3.1 Entropy of energy states

As outlined in section 1.1.2, the classical theory of probability is formulated in terms of microstates and macrostates, and these now need to be defined for a quantum system. A microstate, which is labeled by a single lower case Roman letter, usually corresponds to an eigenstate of a complete set of commuting operators. In this case the eigenvalue is unique. However, it can also correspond to an eigenstate of an incomplete operator. In such a case the microstates are degenerate and microstates with different labels can share the same eigenvalue of the incomplete operator. A macrostate, which is labeled by a lower case Greek letter, corresponds to the principal quantum number of an operator. Each macrostate has a unique label and corresponds to a unique eigenvalue of the operator. A Greek and Roman letter paired together is used to signify a microstate, with the Greek letter labeling the principal quantum number, and the Roman letter labeling the degenerate quantum states.

For example, writing the eigenfunctions with a single microstate label, one has $\hat{O}|\zeta_n^{\rm O}\rangle = O_n|\zeta_n^{\rm O}\rangle$. In this case different values of $n$ may yield the same eigenvalue (assuming that $\hat{O}$ is an incomplete operator). Alternatively, the eigenfunctions may be written as the combination of principal and degenerate state labels, $\hat{O}|\zeta_{\alpha g}^{\rm O}\rangle = O_\alpha|\zeta_{\alpha g}^{\rm O}\rangle$. In this case different values of $\alpha$ necessarily yield different eigenvalues.

For the present canonical equilibrium system, it is energy that is exchanged between the sub-system and the reservoir. Hence the focus is here on energy states. Since the microstates of an isolated system must have equal weight, equation (7.40) the weight of the energy microstates is $w_{\alpha g}^{\rm E} = 1$, since without loss of generality this can be set to unity. This means that the energy microstates of the total system have no internal entropy, $S_{\alpha g}^{\rm E} = k_{\rm B} \ln w_{\alpha g}^{\rm E} = 0$.

In the general formulation of classical probability, section 1.1.2, the weight of a macrostate is the sum of the weights of the microstates contained within it. In the present case the energy macrostate weight is

$$w_\alpha^{\rm E} = \sum_g^{({\rm E},\alpha)} w_{\alpha g}^{\rm E} \equiv N_\alpha^{\rm E}. \tag{7.41}$$

This is just the number of degenerate states with energy $E_\alpha$. Hence the entropy of an energy macrostate is the logarithm of the degeneracy,

$$S_\alpha^{\rm E} = k_{\rm B} \ln N_\alpha^{\rm E}. \tag{7.42}$$

Analogous to the analysis for classical thermodynamics, section 2.2.1, the entropy of the reservoir when the sub-system is in a particular energy microstate will be required. Since energy is conserved one must have

$$E^{\rm tot} = E^{\rm s} + E^{\rm r}. \tag{7.43}$$

As mentioned, tot means the total system, s means the sub-system, and r means the reservoir. The interaction energy between the sub-system and the reservoir is included in the reservoir energy.

Now the reservoir entropy can be written in two different ways, which will be equated to each other below. The first expression directly invokes the energy degeneracy discussed above,

$$S^{\rm r}(E_\alpha^{\rm r}) = k_{\rm B} \ln N_\alpha^{\rm Er}, \quad \text{with } N_\alpha^{\rm Er} \equiv \sum_{g\in\alpha}^{({\rm Er})}. \tag{7.44}$$

This is the entropy of the isolated reservoir in this energy macrostate.

The second expression for the reservoir entropy invokes the definition of the thermodynamic temperature, equation (2.10), $T^{-1} \equiv \partial S(E)/\partial E$. By definition the reservoir is much larger than the sub-system, and so the sub-system energy may be used as an expansion variable. The Taylor expansion yields

$$S^{\rm r}(E^{\rm r}) = S^{\rm r}(E^{\rm tot} - E^{\rm s}) = \text{const.} - \frac{E^{\rm s}}{T^{\rm r}}. \tag{7.45}$$

The higher order terms are negligible in the thermodynamic limit. This gives the reservoir entropy as a function of the sub-system energy. Henceforth the superscript on the reservoir temperature will be dropped as it is the only temperature that will appear.

### 7.3.2 Wave function entanglement

For the total system, the wave function lies in the Hilbert space that comprises that of the sub-system and the reservoir. It is assumed that the sub-system and the reservoir interact so weakly that they may be treated as quasi-independent. This means that for any state of one all permitted states of the other are available. The total number of states is essentially the product of the two individual totals, subject to the conservation laws. Also the basis functions for the total system are the product of the basis functions of the sub-system and of the reservoir, each considered as isolated.

With $\{|\zeta_n^{\mathrm{s}}\rangle\}$ an orthonormal basis for the sub-system and $\{|\zeta_m^{\mathrm{r}}\rangle\}$ an orthonormal basis for the reservoir, the most general wave state of the total system can be expanded in terms of these basis functions,

$$|\psi_{\mathrm{tot}}\rangle = \sum_{n,m} c_{nm}|\zeta_n^{\mathrm{s}}, \zeta_m^{\mathrm{r}}\rangle. \tag{7.46}$$

If the coefficient matrix is dyadic, $c_{nm} = c_n^{\mathrm{s}} c_m^{\mathrm{r}}$, then the wave state is separable, $|\psi_{\mathrm{tot}}\rangle = |\psi_{\mathrm{s}}, \psi_{\mathrm{r}}\rangle$ with $|\psi_{\mathrm{s}}\rangle = \sum_n^{(\mathrm{s})} c_n^{\mathrm{s}}|\zeta_n^{\mathrm{s}}\rangle$, and $|\psi_{\mathrm{r}}\rangle = \sum_m^{(\mathrm{r})} c_m^{\mathrm{r}}|\zeta_m^{\mathrm{r}}\rangle$. Alternatively, if the coefficients are not dyadic, then the wave state is inseparable, which is called an entangled state (Messiah 1961, Merzbacher 1970).

In general terms entangled states arise from the conservation laws. This can be seen explicitly for the present case of energy exchange, in which the total energy is fixed, $E^{\mathrm{tot}} = E^{\mathrm{s}} + E^{\mathrm{r}}$. (Recall that the interaction energy is included in the reservoir energy.) Using the respective energy eigenfunctions as a basis for the sub-system and the reservoir, the most general expansion of the total wave function is

$$|\psi_{\mathrm{tot}}\rangle = \sum_{\alpha g, \beta h} c_{\alpha g, \beta h}|\zeta_{\alpha g}^{\mathrm{Es}}, \zeta_{\beta h}^{\mathrm{Er}}\rangle. \tag{7.47}$$

From energy conservation one must have that

$$c_{\alpha g, \beta h} = 0 \quad \text{if } E_\alpha^{\mathrm{s}} + E_\beta^{\mathrm{r}} \neq E_{\mathrm{tot}}. \tag{7.48}$$

This condition cannot be satisfied for a system able to exchange energy if the coefficient is in dyadic form. To see this, suppose the contrary, that $c_{\alpha g, \beta h} = c_{\alpha g}^{\mathrm{s}} c_{\beta h}^{\mathrm{r}}$. Denote one of the occupied sub-system energy macrostates by $\alpha_1$, so that $c_{\alpha_1 g}^{\mathrm{s}} \neq 0$. Because the macrostate labels are distinct, there can be only one non-zero reservoir macrostate, say $\beta_1$, that conserves energy, $E_{\alpha_1}^{\mathrm{s}} + E_{\beta_1}^{\mathrm{r}} = E_{\mathrm{tot}}$. Hence one must have that $c_{\beta h}^{\mathrm{r}} = 0$, $\beta \neq \beta_1$. But a unique occupied reservoir energy macrostate implies that $\alpha_1$ must also be unique, which is to say that there can be no other occupied sub-system energy macrostate.

This shows that if the total wave function is in product form, then the sub-system and reservoir each can only have a single fixed energy. This in turn means that neither wave function, $\psi_s$ or $\psi_r$, can contain a superposition of principal energy states (because these are non-degenerate). It follows from this that the sub-system is unable to change energy by exchange with the reservoir, because during such an exchange it would be in a superposition of the initial and final energy states. Therefore a dyadic form for the expansion coefficient corresponds to the sub-system and the reservoir being isolated from each other. But this contradicts the basis of a canonical equilibrium system, namely that the energy of the sub-system fluctuates over time as it exchanges energy with the reservoir. One concludes from this contradiction that if the sub-system and the reservoir can exchange energy, then the total wave function must be entangled.

Accordingly, for the present canonical equilibrium system, the total wave function cannot be factorized. Instead it must have a representation of the form

$$|\psi_{\text{tot}}\rangle = \sum_{\alpha,g,h} c_{\alpha g,h}|\zeta_{\alpha g}^{\text{Es}}, \zeta_{\beta_\alpha h}^{\text{Er}}\rangle. \tag{7.49}$$

Here the reservoir principal energy label $\beta_\alpha$ depends upon the sub-system principal energy $\alpha$ and is defined implicitly by $E_\alpha^{\text{s}} + E_{\beta_\alpha}^{\text{r}} = E_{\text{tot}}$. For brevity, the coefficients are written, $c_{\alpha g,h} \equiv c_{\alpha g}^{\text{s}} c_{\beta_\alpha h}^{\text{r}}$. Because $\beta_\alpha$ depends upon $\alpha$, this is entangled (i.e. the sum of the products is not equal to the product of the sums).

The most well known case of entanglement is that of particles with spin that result from radioactive decay (figure 7.2). It is the conservation of spin that is responsible for the entanglement. Measurement of the spin of either daughter particle collapses the superposed wave function into one or other of the two wave functions that form it. If it were not for spin conservation, the daughter wave function would be the superposition of four possible wave functions, which are the independent (i.e. unentangled) product of two spinors for each particle (figure 7.3). It is spin conservation that entangles the two daughter particles and that reduces the superposition basis set from four to two possibilities.

It should be clear in this example that the entangled and unentangled wave functions are very different. For the entangled case, a measurement of the product of the spins of the two daughter particles yields $\langle\psi|\hat{\sigma}_1\hat{\sigma}_r|\psi\rangle = -1$. The same measurement on the unentangled wave function yields $\langle\psi|\hat{\sigma}_1\hat{\sigma}_r|\psi\rangle = 0$. This example is but one of many that could be offered, since quite generally entanglement has far-reaching consequences.

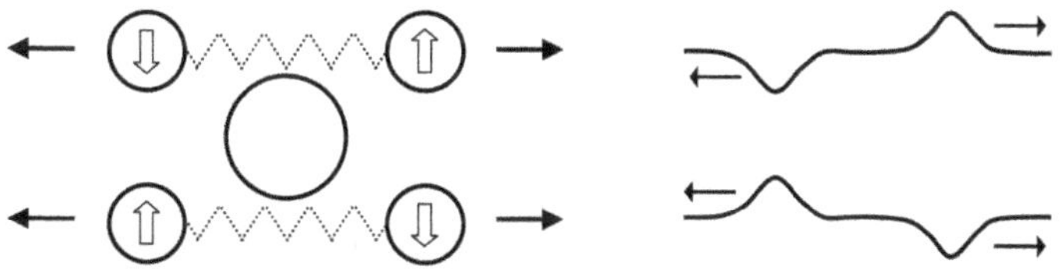

**Figure 7.2.** The two possible decays (upper and lower left) of a spin 0 particle (center left) into two spin half particles. On the right are the two wave functions that are superposed for the final state. Each represents the entanglement of two single particle wave functions.

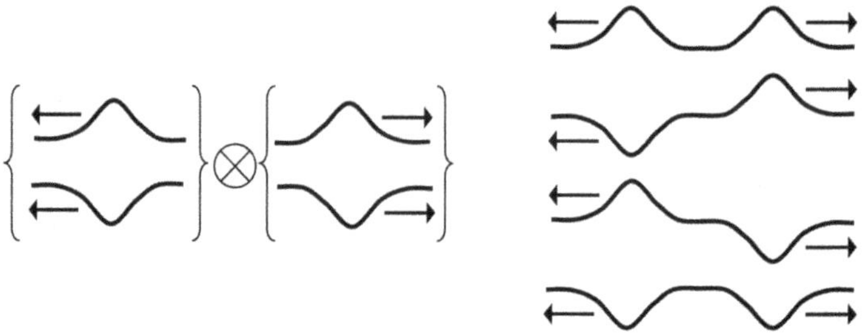

**Figure 7.3.** The dyadic product of two spin half particle wave functions (left), and the four wave functions this gives for superposition, each representing two unentangled single particle wave functions (right).

### 7.3.3 Expectation values and wave function collapse

For the present canonical equilibrium system, conservation of energy causes the entanglement of the sub-system and reservoir wave functions. It will now be shown that this causes the principal energy states of the sub-system to collapse, which is to say that a sub-system expectation value is diagonal in these. This is quite significant for quantum statistical mechanics in view of the discussion in section 7.1 that the single wave function density operator cannot be diagonalized. In contrast, if a probability operator existed in the usual sense of an observable operator, then it can be diagonalized. Therefore the collapse of the principal energy states due to entanglement that will now be established already goes a long way to proving the existence of the probability operator.

In view of the expansion for the wave function of the total system, equation (7.49), and assuming a normalized wave function, $\langle\psi_{\rm tot}|\psi_{\rm tot}\rangle = 1$, the expectation value of an operator on the sub-system is

$$\begin{aligned}\langle\psi_{\rm tot}|\hat{O}^{\rm s}|\psi_{\rm tot}\rangle &= \sum_{\alpha'g';h'}\sum_{\alpha g;h} c^{*}_{\alpha'g',h'}c_{\alpha g,h}\langle\zeta^{\rm Es}_{\alpha'g'},\,\zeta^{\rm Er}_{\beta_{\alpha'}h'}|\hat{O}^{\rm s}|\zeta^{\rm Es}_{\alpha g},\,\zeta^{\rm Er}_{\beta_\alpha h}\rangle\\ &= \sum_{\alpha'g';h'}\sum_{\alpha g;h} c^{*}_{\alpha'g',h'}c_{\alpha g,h}\langle\zeta^{\rm Es}_{\alpha'g'}|\hat{O}^{\rm s}|\zeta^{\rm Es}_{\alpha g}\rangle\langle\zeta^{\rm Er}_{\beta_{\alpha'}h'}|\zeta^{\rm Er}_{\beta_\alpha h}\rangle\\ &= \sum_{\alpha,g,g'}\sum_{h\in\beta_\alpha}^{\rm (Er)} c^{*}_{\alpha g',h}c_{\alpha g,h}\langle\zeta^{\rm Es}_{\alpha g'}|\hat{O}^{\rm s}|\zeta^{\rm Es}_{\alpha g}\rangle\\ &= \sum_{\alpha,g,g'}\sum_{h\in\beta_\alpha}^{\rm (Er)} c^{*}_{\alpha g',h}c_{\alpha g,h}O^{\rm s,E}_{\alpha g',\alpha g}.\end{aligned} \tag{7.50}$$

The third equality follows because the relationship between $\alpha$ and $\beta_\alpha$ is bijective (the macrostates are non-degenerate), $\langle\zeta^{\rm Er}_{\beta_{\alpha'}h'}|\zeta^{\rm Er}_{\beta_\alpha h}\rangle = \delta_{\alpha',\alpha}\delta_{h',h}$. Only the principal diagonal elements of the sub-system matrix operator in the energy representation contribute to this. This can be equivalently cast in terms of the principal diagonal elements of the density matrix. The projection onto the sub-system in this case has the effect of converting a superposition of the sub-system basis states into a mixture of pure sub-system basis states, as arises from a collapsed wave function.

In this result energy conservation and entanglement have reduced and collapsed the original four sums over independent principal energy states into a single sum over principal energy states, from $\alpha'\alpha$, $\beta'\beta$ then to two sums, $\alpha'\alpha$, and then to a single sum, $\alpha$. The degenerate energy states of the reservoir have also collapsed, since the orthogonality of the reservoir basis functions have reduced the sums over $h$ and $h'$ to a single sum over $h$. Specifically, whereas for an isolated system the non-diagonal entries of the operator matrix in the energy representation would contribute to the expectation value, when energy exchange with a reservoir occurs, orthogonality of the reservoir energy basis functions, together with energy conservation, eliminate the off-diagonal sub-system energy terms. The principal energy states have at this stage collapsed, while the degenerate energy states of the sub-system remain in superposition form.

Here collapse refers to the collapse of the wave function and to the collapse into principal energy states. Collapse means that the number of possible states has been reduced, and also that the interference between superposition states has been reduced by the elimination of off-diagonal energy terms.

*Random phase approach*

The microstates of the total isolated system have equal weight. This can be taken as a hypothesis, equation (7.15), or as a theorem that follows from the uniform density of wave space, equation (7.40). In view of this, the coefficients in the representation of the total wave function all have the same magnitude, which can be set to unity, $|c_{\alpha g,h}| = 1$. (With this choice, the expectation value, which is actually the statistical average, has to be explicitly normalized.) The coefficients therefore are of the form

$$c_{\alpha g,h} = e^{i\theta_{\alpha g,h}}, \tag{7.51}$$

with the phase $\theta$ being real and $i \equiv \sqrt{-1}$.

This form is reminiscent of the so-called EPR (Einstein–Podolsky–Rosen) state in which entangled qubits are often described in similar terms (see figure 7.2).

The given total wave functions correspond to specific sets of phases, $\{\theta_{\alpha g,h}\}$. Therefore, averaging over the phases makes the product of coefficients that appears in equation (7.50), zero unless $g' = g$,

$$\langle c^*_{\alpha g',h} c_{\alpha g,h} \rangle_{\text{stat}} = \left\langle e^{i\left[\theta_{\alpha g,h} - \theta_{\alpha g',h}\right]} \right\rangle_{\text{stat}} = \delta_{g,g'}. \tag{7.52}$$

This will be explicitly shown in the next section by averaging over the total wave space. With this, the reservoir sum in the expectation value becomes

$$\begin{aligned} \left\langle \sum_{h \in \beta_\alpha}^{(\text{Er})} c^*_{\alpha g',h} c_{\alpha g,h} \right\rangle_{\text{stat}} &= \delta_{g,g'} \sum_{h \in \beta_\alpha}^{(\text{Er})} \\ &= \delta_{g,g'}\, N^{\text{Er}}_{\beta_\alpha} \\ &= \delta_{g,g'}\, e^{S^{\text{r}}_{\beta_\alpha}/k_{\text{B}}} \\ &= \delta_{g,g'}\, e^{-E^{\text{s}}_\alpha/k_{\text{B}}T}. \end{aligned} \tag{7.53}$$

The statistical average on the left-hand side means the average over total wave functions, which means the average over randomly and uniformly distributed phases. The penultimate equality gives the entropy of the reservoir macrostate $\beta_\alpha$ as the logarithm of the number of microstates it contains, equation (7.44), and the final equality uses the thermodynamic relationship equation (7.45), with the constant part effectively incorporated into the normalization factor.

The effect of averaging the expectation value over the wave function is to collapse the degenerate sub-system energy states, $g = g'$. This both reduces the number of elements contributing to the expectation value and removes the interference between the degenerate energy states (i.e. they are no longer in superposition). Hence both the principal and degenerate states in the energy representation of an operator on the sub-system are in diagonal form. It will be recalled that this result had to be established in order to derive the mixture form for the statistical average, equation (7.9). Equivalently, the probability operator is fully diagonal in the energy representation. Since in the canonical equilibrium system, the entropy operator is proportional to the energy operator, $\hat{S} = -\hat{\mathcal{H}}/T$, this means that the probability operator is fully diagonal in the entropy representation. (In general, the entropy operator is derived from the variables that are exchangeable between the sub-system and the reservoir, and so in all equilibrium and non-equilibrium systems the probability operator will be diagonal in the entropy representation.)

The diagonal nature of the probability operator signifies the elimination of the interference between energy states, both principal and degenerate. This transforms the sub-system from a superposition of states to a mixture of states.

The statistical average also counts the degenerate reservoir energy states. The sum of these gives a weight to each sub-system energy state that is just the number of degenerate reservoir microstates. But by equation (7.44), this is just the exponential of the corresponding reservoir entropy. It is this second effect, combined with equation (7.45), that gives the Maxwell–Boltzmann form for the probability operator.

The average over the phases of the expectation value, equation (7.50), is the statistical average of the operator. With the above result it is

$$\langle \hat{O}^{\mathrm{s}} \rangle_{\mathrm{stat}} = \frac{1}{Z} \sum_{\alpha, g} e^{-E_\alpha / k_{\mathrm{B}} T} O^{\mathrm{s,E}}_{\alpha g, \alpha g}. \tag{7.54}$$

Here $Z$ is just the normalization constant (actually the partition function), and the sum is over the primary and degenerate energy states of the sub-system. This has the same functional form as the final equality in equation (7.9), which is the von Neumann trace form for the statistical average. This gives explicitly the probability matrix as the Maxwell–Boltzmann form, $\wp^{\mathrm{S}}_{\alpha g, \alpha g} = Z^{-1} e^{-E_\alpha / k_{\mathrm{B}} T}$.

*Wave space approach*

The statistical average of an operator on the sub-system can also be expressed as an integral over the total wave space of the expectation value with respect to the entangled wave function. The weight density of the total wave space is uniform, section 7.2. The integral over wave space is carried out by integrating the coefficients in the energy basis state expansion. The allowed coefficients respect energy conservation, equation (7.49), which means that $d\psi_{\text{tot}} \equiv d\underline{\underline{c}} \equiv \prod_{\alpha g,h} dc^{\text{r}}_{\alpha g,h}\, dc^{\text{i}}_{\alpha g,h}$. The real and imaginary parts of the coefficient each belong to the real line, $\in (-\infty, \infty)$. Entanglement has reduced the number of indices from four to three. With the expectation value given by equation (7.50), $O^{\text{s}}(\psi_{\text{tot}}) \equiv \langle \psi_{\text{tot}} | \hat{O}^{\text{s}} | \psi_{\text{tot}} \rangle / \langle \psi_{\text{tot}} | \psi_{\text{tot}} \rangle$, the statistical average is

$$
\begin{aligned}
\langle \hat{O}^{\text{s}} \rangle_{\text{stat}} &= \frac{1}{Z'} \int d\psi_{\text{tot}}\; O^{\text{s}}(\psi_{\text{tot}}) \\
&= \frac{1}{Z'} \sum_{\alpha,g,g'} \sum_{h \in \beta_\alpha} {}^{(\text{Er})}O^{\text{s,E}}_{\alpha g',\alpha g} \int d\underline{\underline{c}} \frac{c^*_{\alpha g',h} c_{\alpha g,h}}{N(\psi_{\text{tot}})} \\
&= \frac{1}{Z'} \sum_{\alpha,g} \sum_{h \in \beta_\alpha} {}^{(\text{Er})}O^{\text{s,E}}_{\alpha g,\alpha g} \int d\underline{\underline{c}} \frac{|c_{\alpha g,h}|^2}{N(\psi_{\text{tot}})} \\
&= \frac{1}{Z'} \sum_{\alpha,g} e^{S^{\text{r}}_{\beta_\alpha}/k_{\text{B}}} O^{\text{s,E}}_{\alpha g,\alpha g} \times \text{const.} \\
&= \frac{1}{Z} \sum_{\alpha,g} e^{-E^{\text{s}}_\alpha/k_{\text{B}}T} O^{\text{s,E}}_{\alpha g,\alpha g}.
\end{aligned}
\tag{7.55}
$$

Here the norm squared is $N(\psi_{\text{tot}}) = \sum_{\alpha g,h} |c_{\alpha g,h}|^2$. The third equality follows since the terms in which the integrand is odd with respect to any coefficient vanish, and so the only non-vanishing terms have $g = g'$. The fourth equality follows because all the integrations are identical and give a constant that is independent of $\alpha$, $g$, and $h$, which can be incorporated into the normalization factor. The fourth and fifth equalities use the two forms for the reservoir entropy, equations (7.44) and (7.45). The partition function ensures that $\langle \hat{\text{I}}^{\text{s}} \rangle_{\text{stat}} = 1$.

This agrees with the random phase approach, equation (7.54), which should not be surprising since the uniformity of wave space of an isolated system implies the equal weight of isolated system microstates. Both methods show that the reduction and collapse of the energy macrostates is due to energy conservation and the consequent entanglement of the total wave function. They both also show that the collapse of the sub-system degenerate energy states is due to the fact that the off-diagonal contributions average to zero.

This equivalence demonstrates that quantum statistical mechanics can be as well formulated in terms of the wave function as in terms of quantum states.

### 7.3.4 Maxwell–Boltzmann probability operator

The statistical average given by both approaches agree with the von Neumann trace form, specifically the final equality in equation (7.9). This is in the form of a classical average: it is the sum over states of the probability of a state times the value of the observable in the state. It is essential to the result that the off-diagonal elements in the entropy representation of the probability (density) matrix vanish, which is to say that the wave function in this basis is decoherent.

One can easily convert the present result for the average to the generic von Neumann trace form, the first equality in equation (7.9). Of course, the probability operator replaces the conventional many wave function density operator. In the energy basis, the representation of any function of the Hamiltonian operator is diagonal, and the diagonal entries are just the function of the eigenvalues,

$$e^{-E_\alpha^{\rm s}/k_{\rm B}T}\delta_{\alpha',\alpha}\delta_{g',g} = \{e^{-\hat{\mathcal{H}}/k_{\rm B}T}\}^{\rm E}_{\alpha'g',\alpha g}, \tag{7.56}$$

with $\hat{\mathcal{H}}$ being the sub-system Hamiltonian operator. For the canonical equilibrium system, the reservoir entropy operator is $\hat{S} = -\hat{\mathcal{H}}/T$, a consequence of equation (7.45). This means that in the present canonical equilibrium system, entropy eigenfunctions are identical to the energy eigenfunctions. More generally, the entropy representation is always the representation in which an open quantum system is diagonal. With this, the quantum statistical average of a sub-system operator can be written

$$\begin{aligned}\langle\hat{O}\rangle_{\rm stat} &= \frac{1}{Z}\sum_{\alpha,g} e^{-E_\alpha/k_{\rm B}T}O^{\rm s,E}_{\alpha g,\alpha g}\\ &= \frac{1}{Z}\sum_{\alpha g,\alpha' g'}\{e^{-\hat{\mathcal{H}}/k_{\rm B}T}\}^{\rm E}_{\alpha'g',\alpha g}\,O^{\rm s,E}_{\alpha g,\alpha' g'}\\ &= {\rm TR}\,\hat{\wp}\,\hat{O}.\end{aligned} \tag{7.57}$$

The final form is independent of any specific representation or basis. One can conclude that the probability operator for the canonical equilibrium system is just the Maxwell–Boltzmann operator,

$$\hat{\wp} \equiv \frac{1}{Z}e^{\hat{S}/k_{\rm B}} \equiv \frac{1}{Z}e^{-\hat{\mathcal{H}}/k_{\rm B}T}. \tag{7.58}$$

The normalizing partition function is $Z = {\rm TR}e^{-\hat{\mathcal{H}}/k_{\rm B}T}$. The first equality holds in general, whereas the second equality is specific for the canonical equilibrium system. This result for the canonical equilibrium probability operator (or many-wave function density operator) is well known, for example Feynman (1998).

The Maxwell–Boltzmann probability operator is the direct quantitative consequence of the sum over the degenerate energy microstates of the reservoir. This gives the exponential of the reservoir entropy for each particular sub-system energy macrostate. This is just the matrix representation of the Maxwell–Boltzmann probability operator in the energy basis. Although the energy representation was used to derive this result, the final expression as a trace of the product of the two operators is invariant with respect to the representation.

One sees in this derivation that it is the statistical average that causes the superposition of the degenerate energy microstates of the sub-system to collapse into a mixture. The reduction and collapse of the energy macrostates occurred at the level of the expectation value due to the energy conservation law and the entanglement of the reservoir and the sub-system. As has been mentioned above, both collapses are necessary for the mixture form for the statistical average, equation (7.9). The superposition entropy states, both principle and degenerate, had to collapse so that the probability operator was diagonal in the entropy representation. Although we do not work with a density operator, if we did, the result (i.e. equating it to the probability operator) would imply that the density operator is an average over multiple wave functions.

The Maxwell–Boltzmann probability operator can be written as a Feynman path integral for the time propagator (Feynman and Hibbs 1965). In this approach the temperature is interpreted as an imaginary time. Although this technique has been successfully exploited in many applications (Schulman 1981, Kleinert 2009) it is not analyzed or invoked in what follows.

For the present formulation, the statistical average can be written as an integral of the expectation value over the sub-system wave space

$$\langle \hat{O} \rangle_{\text{stat}} = \int \mathrm{d}\psi \; \frac{\langle \psi | \hat{\wp} \, \hat{O} | \psi \rangle}{\langle \psi | \psi \rangle}. \tag{7.59}$$

It is no longer necessary to show that these belong to the sub-system, the superscript s, because the reservoir has been integrated out. This is a continuum version of the trace. The equivalence of this with the above discrete expression for the trace can be shown by expanding the sub-system wave function in any basis, $|\psi\rangle = \sum_n \psi_n |\zeta_n\rangle$, and integrating,

$$\begin{aligned} \int \mathrm{d}\psi \; \frac{\langle \psi | \hat{\wp} \, \hat{O} | \psi \rangle}{\langle \psi | \psi \rangle} &= \int \mathrm{d}\underline{\psi} \frac{\sum_{m,n,l} \psi_m^* \psi_l \wp_{mn} O_{nl}}{\langle \psi | \psi \rangle} \\ &= \frac{1}{Z''} \sum_{m,n} \{ e^{\hat{S}/k_{\mathrm{B}}} \}_{mn} O_{nm} \int \mathrm{d}\underline{\psi} \; \frac{\psi_m^* \psi_m}{\langle \psi | \psi \rangle} \\ &= \frac{\text{const.}}{Z''} \sum_{m,n} \{ e^{\hat{S}/k_{\mathrm{B}}} \}_{mn} O_{nm} \\ &= \mathrm{TR} \, \hat{\wp} \, \hat{O}. \end{aligned} \tag{7.60}$$

In the second equality $l = m$ because odd powers of $\psi_l$ vanish. Since the integration is the same for all $m$, it is a constant that can be taken outside of the sums and incorporated into the partition function, as in the third equality.

In section 7.2, the probability density in wave space for an isolated total system was shown to be uniform in wave space. In contrast, the probability density in the wave space of a sub-system of a total system is non-uniform. For the present canonical equilibrium sub-system that can exchange energy with a reservoir, it is given by the expectation value of the Maxwell–Boltzmann probability operator. The probability density is the expectation value of the probability operator,

$$\wp(\psi) = \frac{\langle \psi | \hat{\wp} | \psi \rangle}{\langle \psi | \psi \rangle}, \tag{7.61}$$

where $\psi$ is the wave function of the sub-system. It should be clear that the statistical average is *not* the integral over the sub-system wave space of the probability density and the expectation value of the operator,

$$\langle \hat{O} \rangle_{\text{stat}} \neq \int \mathrm{d}\psi \ \wp(\psi) O(\psi). \tag{7.62}$$

The physical reason that this does not hold is that in general statistical averages are the value of a variable in a state times the probability of the state, summed over the states. The problem with this expression, is that both $\wp(\psi)$ and $O(\psi)$ are expectation values, which include the superposition of states.

*Environmental selection*

In quantum literature, the present sub-system that can exchange energy with a thermal reservoir is generally called an open quantum system. These have been treated in detail from a number of different perspectives (Davies 1976, Breuer and Petruccione 2002, Weiss 2008). One approach that is related to that developed here is environmental selection, or einselection (Zeh 2001, Zurek 2003, Schlosshauer 2005). Einselection includes the influence of the environment on the system of interest, and this is one obvious point in common with the above analysis of the sub-system and reservoir. The specific goal of einselection is to explain the apparent collapse and decoherence of the wave function upon measurement. Wave function collapse occurred several times in the derivation of equilibrium quantum statistical mechanics just given, and this is another point of similarity. Although the quantitative details differ, the basic conclusion of einselection—that the reservoir or environment suppresses the superposition states of the open sub-system rendering it diagonal—is essentially the same as that drawn here. A more detailed discussion of einselction in the context of the present approach is given in section 1.4 of Attard (2015).

## 7.4 Expansion for quantum statistical mechanics

### 7.4.1 Partition function

The probability operator derived above for the canonical equilibrium system is

$$\hat{\wp} = \frac{e^{-\beta \hat{\mathcal{H}}}}{Z(T)}, \tag{7.63}$$

where the canonical partition function that normalizes this is

$$Z(T) = \mathrm{TR}' e^{-\beta \hat{\mathcal{H}}} = \sum_n{}' e^{-\beta \mathcal{H}_n}. \tag{7.64}$$

In general, the partition function is the total number of states of the total system, and its logarithm gives the total entropy. The prime that has been added here signifies the restriction on the counting of states, so that forbidden states are not included, and so that the same state is not counted more than once. This restriction is essential, but it is usually not explicitly shown or averted to in the conventional expression for the partition function, equation (7.1).

The symmetrization of the wave function plays a role in the correct counting, as states that differ only by the permutation of the particle labels are not distinct. Further, states with repeated label values may be allowed for bosons but forbidden for fermions. The way to properly formulate the partition function to account for these effects is to define a characteristic function for the microstates.

*Characteristic function*

A basis wave function formed from the orthonormal set of unsymmetrized wave functions $\{\phi_{\mathbf{n}}\}$ has symmetrized form

$$\phi_{\mathbf{n}}^{\pm} = \frac{1}{\sqrt{N!\chi_{\mathbf{n}}^{\pm}}} \sum_{\hat{\mathrm{P}}} (\pm 1)^p \phi_{\hat{\mathrm{P}}\mathbf{n}}. \tag{7.65}$$

Here $\hat{\mathrm{P}}$ is the permutation operator, $p$ is its parity, the upper sign is for bosons, and the lower sign is for fermions. The characteristic function of the state, $\chi_{\mathbf{n}}^{\pm}$, is inversely proportional to the number of non-zero distinct permutations of the wave function. Specifically, normalization, $\langle \phi_{\mathbf{n}}^{\pm} | \phi_{\mathbf{n}}^{\pm} \rangle = 1$, gives

$$\chi_{\mathbf{n}}^{\pm} = \sum_{\hat{\mathrm{P}}} (\pm 1)^p \langle \phi_{\mathbf{n}} | \phi_{\hat{\mathrm{P}}\mathbf{n}} \rangle. \tag{7.66}$$

With the characteristic function, the sum over distinct, allowed states of some symmetric function $f_{\hat{\mathrm{P}}\mathbf{n}} = f_{\mathbf{n}}$ can be written as a sum over all states,

$$\sum_{\mathbf{n}}{}' f_{\mathbf{n}} = \frac{1}{N!} \sum_{\mathbf{n}} \chi_{\mathbf{n}}^{\pm} f_{\mathbf{n}}. \tag{7.67}$$

*Example: two particles in two states*

As an example, consider a two particle system, with characteristic function $\chi_{k_1 k_2}^{\pm}$, where $k_1$ is the state of the first particle and $k_2$ is that of the second. Suppose that there are just two one-particle states, $k_i \in \{1, 2\}$, $i = 1, 2$. One has

$$\begin{aligned} \chi_{11}^{\pm} &= \langle \phi_{11} | \phi_{11} \rangle \pm \langle \phi_{11} | \phi_{11} \rangle \\ &= \begin{cases} 2 \\ 0 \end{cases} \\ &= \chi_{22}^{\pm}. \end{aligned} \tag{7.68}$$

Since two fermions cannot be in the same state, the 11 and 22 states are forbidden for a fermionic system. Also,

$$\begin{aligned} \chi_{12}^{\pm} &= \langle \phi_{12} | \phi_{12} \rangle \pm \langle \phi_{12} | \phi_{21} \rangle \\ &= 1 \\ &= \chi_{21}^{\pm}. \end{aligned} \tag{7.69}$$

The state 12 is the same as the state 21.

Hence the sum over distinct, allowed states of some symmetric function $f_{\hat{\mathrm{P}}\mathbf{n}} = f_{\mathbf{n}}$ can be written as the sum over all states by using the characteristic function,

$$\begin{aligned} \sum_{\mathbf{n}}{}' f_{\mathbf{n}} &= \begin{cases} f_{11} + f_{12} + f_{22} \\ f_{12} \end{cases} \\ &= \begin{cases} f_{11} + \frac{1}{2}[f_{12} + f_{21}] + f_{22} \\ \frac{1}{2}[f_{12} + f_{21}] \end{cases} \\ &= \frac{1}{2}\{\chi_{11}^{\pm} f_{11} + \chi_{12}^{\pm} f_{12} + \chi_{21}^{\pm} f_{21} + \chi_{22}^{\pm} f_{22}\} \\ &= \frac{1}{N!}\sum_{\mathbf{n}} \chi_{\mathbf{n}}^{\pm} f_{\mathbf{n}}. \end{aligned} \tag{7.70}$$

For fermions, the 11 and 22 states are excluded. For bosons the 11 and 22 states are each counted with weight 1. The 12 = 21 state is counted once for both bosons and fermions.

*Grand partition function*

Let $\phi_{\mathbf{n}}$ be an entropy eigenfunction. In the canonical equilibrium case, entropy eigenfunctions are the same as energy eigenfunctions. With $\hat{\mathcal{H}}$ the Hamiltonian or energy operator, in the canonical equilibrium case the eigenvalue equation is $\hat{\mathcal{H}}|\phi_{\mathbf{n}}\rangle = \mathcal{H}_{\mathbf{n}}|\phi_{\mathbf{n}}\rangle$. Here the $\{\phi_{\mathbf{n}}\}$ form a complete orthonormal unsymmetrized set, and $\mathbf{n}$ labels entropy microstates.

The partition function is the total number of allowed, distinct states of the total system, which is the total number of reservoir-weighted allowed, distinct states of the sub-system. The emphasis is on the words allowed and distinct, because it would be wrong to count forbidden states, or to count the same state more than once. It is most convenient to work in the grand canonical system (constant chemical potential, variable particle number). With correct counting, the grand partition function is (Attard 2017)

$$\begin{aligned} \Xi^{\pm} &= \sum_{N=0}^{\infty} z^N \sum_{\mathbf{n}}{}' e^{-\beta\mathcal{H}_{\mathbf{n}}} \\ &= \sum_{N=0}^{\infty} \frac{z^N}{N!} \sum_{\mathbf{n}} \chi_{\mathbf{n}}^{\pm} e^{-\beta\mathcal{H}_{\mathbf{n}}} \\ &= \sum_{N=0}^{\infty} \frac{z^N}{N!} \sum_{\mathbf{n}} \sum_{\hat{\mathrm{P}}} (\pm 1)^p \langle \phi_{\hat{\mathrm{P}}\mathbf{n}} | \phi_{\mathbf{n}} \rangle e^{-\beta\mathcal{H}_{\mathbf{n}}} \\ &= \sum_{N=0}^{\infty} \frac{z^N}{N!} \sum_{\mathbf{n}} \sum_{\hat{\mathrm{P}}} (\pm 1)^p \langle \phi_{\hat{\mathrm{P}}\mathbf{n}} | e^{-\beta\hat{\mathcal{H}}} | \phi_{\mathbf{n}} \rangle. \end{aligned} \tag{7.71}$$

Here $N$ is the number of particles, the fugacity is $z \equiv e^{\beta\mu}$, where $\mu$ is the chemical potential, and $\beta = 1/k_B T$ is sometimes called the inverse temperature, with $k_B$ being Boltzmann's constant and $T$ the temperature.

The partition function (and statistical average) is usually written as the simple trace of the Maxwell–Boltzmann operator (Neumann 1932, Messiah 1961, Merzbacher 1970). Usually this does not explicitly exclude forbidden states or count duplicate states with correctly reduced weight. The present expression corrects this deficiency. It is dependent upon using entropy microstates and their characteristic function $\chi^{\pm}(\phi_{\mathbf{n}})$. If the partition function (and statistical average) is to be written as the trace of an operator, then the definition of the trace has to be altered to explicitly count only distinct, allowed states.

The partition function can be usefully cast in an arbitrary basis. Consider another orthonormal, complete, unsymmetrized but otherwise at this stage arbitrary basis $\{\zeta_{\mathbf{p}}\}$. (Here the subscript is used as both an index and a label.) Since this is complete, $\sum_{\mathbf{p}} |\zeta_{\mathbf{p}}\rangle\langle\zeta_{\mathbf{p}}| = \hat{\mathrm{I}}$, the identity operator in this form can be inserted to yield

$$\begin{aligned}
\Xi^{\pm} &= \sum_{N=0}^{\infty} \frac{z^N}{N!} \sum_{\mathbf{n}} \sum_{\mathbf{p}} \sum_{\hat{\mathrm{P}}} (\pm 1)^p \langle \phi_{\hat{\mathrm{P}}\mathbf{n}} | e^{-\beta\hat{\mathcal{H}}} | \zeta_{\mathbf{p}} \rangle \langle \zeta_{\mathbf{p}} | \phi_{\mathbf{n}} \rangle \\
&= \sum_{N=0}^{\infty} \frac{z^N}{N!} \sum_{\mathbf{n}} \sum_{\mathbf{p}} \sum_{\hat{\mathrm{P}}} (\pm 1)^p \langle \phi_{\mathbf{n}} | e^{-\beta\hat{\mathcal{H}}} | \zeta_{\hat{\mathrm{P}}\mathbf{p}} \rangle \langle \zeta_{\mathbf{p}} | \phi_{\mathbf{n}} \rangle \\
&= \sum_{N=0}^{\infty} \frac{z^N}{N!} \sum_{\mathbf{p}} \sum_{\hat{\mathrm{P}}} (\pm 1)^p \langle \zeta_{\mathbf{p}} | e^{-\beta\hat{\mathcal{H}}} | \zeta_{\hat{\mathrm{P}}\mathbf{p}} \rangle .
\end{aligned} \tag{7.72}$$

The second equality follows from the replacements $\mathbf{n} \Rightarrow \hat{\mathrm{P}}\mathbf{n}$ and $\mathbf{p} \Rightarrow \hat{\mathrm{P}}\mathbf{p}$, and the fact that $\langle \zeta_{\hat{\mathrm{P}}\mathbf{p}} | \phi_{\hat{\mathrm{P}}\mathbf{n}} \rangle = \langle \zeta_{\mathbf{p}} | \phi_{\mathbf{n}} \rangle$. The final equality has the same form as the final form of the sum over entropy states.

Finally, it will prove useful to express the partition function as a sum over a dual basis set. Consider another orthonormal, complete, unsymmetrized, but otherwise at this stage arbitrary basis $\{\zeta_{\mathbf{q}}\}$, with $\sum_{\mathbf{q}} |\zeta_{\mathbf{q}}\rangle\langle\zeta_{\mathbf{q}}| = \hat{\mathrm{I}}$. One can write

$$\begin{aligned}
\Xi^{\pm} &= \sum_{N=0}^{\infty} \frac{z^N}{N!} \sum_{\mathbf{q},\mathbf{p}} \sum_{\hat{\mathrm{P}}} (\pm 1)^p \langle \zeta_{\mathbf{p}} | \zeta_{\mathbf{q}} \rangle \, \langle \zeta_{\mathbf{q}} | e^{-\beta\hat{\mathcal{H}}} | \zeta_{\hat{\mathrm{P}}\mathbf{p}} \rangle \\
&= \sum_{N=0}^{\infty} \frac{z^N}{N!} \sum_{\mathbf{q},\mathbf{p}} \sum_{\hat{\mathrm{P}}} (\pm 1)^p \langle \zeta_{\hat{\mathrm{P}}\mathbf{p}} | \zeta_{\mathbf{q}} \rangle \, \langle \zeta_{\mathbf{q}} | e^{-\beta\hat{\mathcal{H}}} | \zeta_{\mathbf{p}} \rangle .
\end{aligned} \tag{7.73}$$

(One could replace here $\mathbf{q} \Rightarrow \mathbf{p}'$, in which case the final form is just the non-diagonal sum over a single basis set, analogous to the diagonal sum over the entropy basis set.)

### 7.4.2 Momentum and position states

*Basis functions*

Now the analysis is restricted to momentum and position basis functions. The eigenfunctions of the momentum operator in the position representation $\mathbf{r}$ are plane waves,

$$\zeta_{\mathbf{p}}(\mathbf{r}) = \frac{1}{V^{N/2}} e^{-\mathbf{p}\cdot\mathbf{r}/i\hbar} = \prod_{j=1}^{N} \frac{e^{-\mathbf{p}_j\cdot\mathbf{r}_j/i\hbar}}{V^{1/2}}. \tag{7.74}$$

Here $\mathbf{p}$ is the configuration momentum, and $V$ is the sub-system volume. These form a complete orthonormal set. Periodic boundary conditions give the width of the momentum state per particle per dimension as $\Delta_p = 2\pi\hbar/V^{1/3}$, assuming a cubic volume.

The position basis functions are Gaussians,

$$\zeta_{\mathbf{q}}(\mathbf{r}) = \frac{e^{-(\mathbf{r}-\mathbf{q})^2/4\xi^2}}{(2\pi\xi^2)^{3N/4}} = \prod_{j=1}^{N} \frac{e^{-(\mathbf{r}_j-\mathbf{q}_j)^2/4\xi^2}}{(2\pi\xi^2)^{3/4}}. \tag{7.75}$$

Here $\mathbf{q}$ is the configuration position. Normalization fixes the spacing of the configuration position states as $\Delta_q = \sqrt{8\pi\xi^2}$, which gives the completeness expression, $\sum_{\mathbf{q}} |\zeta_{\mathbf{q}}(\mathbf{r}')\rangle \langle\zeta_{\mathbf{q}}(\mathbf{r})| = e^{-(\mathbf{r}'-\mathbf{r})^2/8\xi^2}/(8\pi\xi^2)^{3N/2} \equiv \delta_\xi(\mathbf{r}' - \mathbf{r})$. In the limit $\xi \to 0$ the position basis functions form a complete orthonormal set. This limit will be taken in the final results below.

Note the distinction between the representation position $\mathbf{r}$ and the configuration position $\mathbf{q}$.

The transformation coefficient for the two basis sets is readily shown to be

$$\langle\zeta_{\mathbf{p}}|\zeta_{\mathbf{q}}\rangle \equiv \frac{(8\pi\xi^2)^{3N/4}}{V^{N/2}} e^{-\xi^2 p^2/\hbar^2} e^{\mathbf{q}\cdot\mathbf{p}/i\hbar} = \prod_{j=1}^{N} \frac{(8\pi\xi^2)^{3/4}}{V^{/2}} e^{-\xi^2 p_j^2/\hbar^2} e^{\mathbf{q}_j\cdot\mathbf{p}_j/i\hbar}. \tag{7.76}$$

Again, $N$ is the number of particles in the sub-system. It is important for the treatment of permutation loops below that this is the product of individual particle factors. The product of this and its complex conjugate is obviously

$$\langle\zeta_{\mathbf{p}}|\zeta_{\mathbf{q}}\rangle\langle\zeta_{\mathbf{q}}|\zeta_{\mathbf{p}}\rangle \equiv \frac{(8\pi\xi^2)^{3N/2}}{V^N} e^{-2\xi^2 p^2/\hbar^2}, \tag{7.77}$$

which product will shortly appear in the summand of the partition function.

*Non-commutativity function*

The position and momentum operators do not commute, which fundamental property can be accounted for by recasting the partition function to include what we shall call the 'non-commutativity' function.

Modifying slightly an argument due to Wigner (1932), one can formally commute the Maxwell–Boltzmann operator with the momentum basis function by writing

$$\begin{aligned} e^{-\beta\hat{\mathcal{H}}}\zeta_{\mathbf{p}}(\mathbf{r}) &= \zeta_{\mathbf{p}}(\mathbf{r})e^{-\beta\hat{\hat{\mathcal{H}}}}1 \\ &\equiv \zeta_{\mathbf{p}}(\mathbf{r})e^{-\beta\mathcal{H}(\mathbf{r},\mathbf{p})}e^{w(\mathbf{r},\mathbf{p})}. \end{aligned} \tag{7.78}$$

This defines the non-commutativity function $w(\mathbf{r}, \mathbf{p})$. The modified energy operator induced here is

$$\begin{aligned} \hat{\hat{\mathcal{H}}} &\equiv e^{\mathbf{p}\cdot\mathbf{r}/i\hbar}\hat{\mathcal{H}}e^{-\mathbf{p}\cdot\mathbf{r}/i\hbar} \\ &= e^{\mathbf{p}\cdot\mathbf{r}/i\hbar}\left[\frac{-\hbar^2}{2m}\nabla^2 + U(\mathbf{r})\right]e^{-\mathbf{p}\cdot\mathbf{r}/i\hbar} \\ &= \frac{p^2}{2m} + U(\mathbf{r}) - \frac{i\hbar}{m}\mathbf{p}\cdot\nabla - \frac{\hbar^2}{2m}\nabla^2. \end{aligned} \tag{7.79}$$

The non-commutativity function $w(\mathbf{r}, \mathbf{p})$ is extensive. This is one reason why the present formulation is preferable to those given by Wigner (1932) and by Kirkwood (1933). A recursion relation leading to an expansion for $w(\mathbf{r}, \mathbf{p})$ will be given in section 7.4.4 below. To leading order the non-commutativity function vanishes, $w(\hbar = 0) = w(\beta = 0) = 0$.

With $w(\mathbf{r}, \mathbf{p})$, the expectation value of the Maxwell–Boltzmann operator in the position-momentum space formulation of the partition function becomes

$$\begin{aligned} \langle\zeta_{\mathbf{q}}|e^{-\beta\hat{\mathcal{H}}}|\zeta_{\mathbf{p}}\rangle &= \langle\zeta_{\mathbf{q}}|e^{-\beta\hat{\mathcal{H}}}\zeta_{\mathbf{p}}\rangle \\ &= \langle\zeta_{\mathbf{q}}(\mathbf{r})|\zeta_{\mathbf{p}}(\mathbf{r})e^{-\beta\mathcal{H}(\mathbf{r},\mathbf{p})}e^{w(\mathbf{r},\mathbf{p})}\rangle \\ &= \langle\zeta_{\mathbf{q}}|\zeta_{\mathbf{p}}\rangle e^{-\beta\mathcal{H}(\mathbf{q},\mathbf{p})}e^{w(\mathbf{q},\mathbf{p})}. \end{aligned} \tag{7.80}$$

The final equality follows because $\lim_{\xi\to 0}\zeta_{\mathbf{q}}(\mathbf{r}) = \delta(\mathbf{q} - \mathbf{r})$.

*Symmetrization factor*

The obvious merit of formulating the grand partition function in terms of the asymmetric expectation value of the Maxwell–Boltzmann operator is that the sum over entropy states has become a sum over points in classical phase space. The continuum limit of this is

$$\sum_{\mathbf{q},\mathbf{p}} \Rightarrow \frac{1}{\left(\Delta_p\Delta_q\right)^{3N}}\int d\mathbf{\Gamma}. \tag{7.81}$$

The volume elements are $\Delta_p = 2\pi\hbar/V^{1/3}$ and $\Delta_q = \sqrt{8\pi\xi^2}$.

The present way of representing quantum mechanics in classical phase space is distinctly different to the way advocated by Wigner (1932) and by Kirkwood (1933). The Wigner–Kirkwood method has been followed and modified by many workers (for example, Groenewold 1946, Moyal 1949, Praxmeyer and Wódkiewicz 2002, Barnett and Radmore 2003, Gerry and Knight 2005, Zachos *et al* 2005, Dishlieva 2008, and Barker 2010). In particular, previous authors focus upon a transform of the wave function rather than the states themselves, they do not distinguish the position representation $\mathbf{r}$ from the position configuration $\mathbf{q}$ as here, and they do not account for wave function symmetrization. Finally, the Wigner–Kirkwood phase space representation has principally been applied or used in quantum optics, whereas the present focus is on the generic formulation of quantum statistical mechanics.

In view of the above the partition function becomes

$$\begin{aligned}
\Xi^{\pm} &= \sum_{N=0}^{\infty} \frac{z^N}{N!} \sum_{\mathbf{q},\mathbf{p}} \sum_{\hat{\mathrm{P}}} (\pm 1)^p \langle \zeta_{\hat{\mathrm{P}}\mathbf{p}} | \phi_{\mathbf{q}} \rangle \, \langle \zeta_{\mathbf{q}} | e^{-\beta \hat{\mathcal{H}}} | \zeta_{\mathbf{p}} \rangle \\
&= \sum_{N=0}^{\infty} \frac{z^N}{N! \left( \Delta_p \Delta_q \right)^{3N}} \int \mathrm{d}\mathbf{\Gamma} \sum_{\hat{\mathrm{P}}} (\pm 1)^p \langle \zeta_{\hat{\mathrm{P}}\mathbf{p}} | \zeta_{\mathbf{q}} \rangle \langle \zeta_{\mathbf{q}} | \zeta_{\mathbf{p}} \rangle e^{-\beta \mathcal{H}(\mathbf{q},\mathbf{p})} e^{w(\mathbf{q},\mathbf{p})} \\
&= \sum_{N=0}^{\infty} \frac{z^N}{N! \left( \Delta_p \Delta_q \right)^{3N}} \int \mathrm{d}\mathbf{\Gamma} \, \eta^{\pm}(\mathbf{q}, \mathbf{p}) \langle \zeta_{\mathbf{p}} | \zeta_{\mathbf{q}} \rangle \langle \zeta_{\mathbf{q}} | \zeta_{\mathbf{p}} \rangle e^{-\beta \mathcal{H}(\mathbf{q},\mathbf{p})} e^{w(\mathbf{q},\mathbf{p})} \\
&= \sum_{N=0}^{\infty} \frac{z^N}{N! h^{3N}} \int \mathrm{d}\mathbf{\Gamma} e^{-\beta \mathcal{H}(\mathbf{\Gamma})} e^{w(\mathbf{\Gamma})} \eta^{\pm}(\mathbf{\Gamma}).
\end{aligned} \tag{7.82}$$

Notice that the factors of $\xi$ cancel, which allows the limit $\xi \to 0$ to be taken. Here

$$\eta^{\pm}(\mathbf{\Gamma}) \equiv \frac{1}{\langle \zeta_{\mathbf{p}} | \zeta_{\mathbf{q}} \rangle} \sum_{\hat{\mathrm{P}}} (\pm 1)^p \langle \zeta_{\hat{\mathrm{P}}\mathbf{p}} | \zeta_{\mathbf{q}} \rangle \tag{7.83}$$

might be called the symmetrization factor.

*Statistical average*

The statistical average of an operator is

$$\begin{aligned}
\langle \hat{O} \rangle_{\mathrm{stat}} &= \frac{1}{\Xi^{\pm}} \mathrm{TR}' \, z^N \hat{O} e^{-\beta \hat{\mathcal{H}}} \\
&= \frac{1}{\Xi^{\pm}} \sum_{N=0}^{\infty} z^N \sum_{\mathbf{n}}{}' \, O_{\mathbf{n}} e^{-\beta \mathcal{H}_{\mathbf{n}}} \\
&= \frac{1}{\Xi^{\pm}} \sum_{N=0}^{\infty} \frac{z^N}{N!} \sum_{\mathbf{q},\mathbf{p}} \sum_{\hat{\mathrm{P}}} (\pm 1)^p \langle \zeta_{\hat{\mathrm{P}}\mathbf{p}} | \phi_{\mathbf{q}} \rangle \, \langle \zeta_{\mathbf{q}} | \hat{O} e^{-\beta \hat{\mathcal{H}}} | \zeta_{\mathbf{p}} \rangle.
\end{aligned} \tag{7.84}$$

One sees that compared to the grand partition function, the Maxwell–Boltzmann operator has simply been replaced by the product of it and the observable operator, $e^{-\beta \hat{\mathcal{H}}} \Rightarrow \hat{O} e^{-\beta \hat{\mathcal{H}}}$.

As for the non-commutativity factor, equation (7.78), and recalling that $\zeta_{\mathbf{p}}(\mathbf{r}) = V^{-N/2} e^{-\mathbf{p}\cdot\mathbf{r}/i\hbar}$, one can make the transformation

$$\begin{aligned}
\hat{O} e^{-\beta \hat{\mathcal{H}}} \zeta_{\mathbf{p}}(\mathbf{r}) &= \zeta_{\mathbf{p}}(\mathbf{r}) e^{\mathbf{p}\cdot\mathbf{r}/i\hbar} \hat{O} e^{-\mathbf{p}\cdot\mathbf{r}/i\hbar} e^{\mathbf{p}\cdot\mathbf{r}/i\hbar} e^{-\beta \hat{\mathcal{H}}} e^{-\mathbf{p}\cdot\mathbf{r}/i\hbar} 1 \\
&\equiv \zeta_{\mathbf{p}}(\mathbf{r}) \hat{\hat{O}} e^{-\beta \hat{\hat{\mathcal{H}}}} 1 \\
&= \zeta_{\mathbf{p}}(\mathbf{r}) \hat{\hat{O}} \{ e^{-\beta \mathcal{H}(\mathbf{r},\mathbf{p})} e^{w(\mathbf{r},\mathbf{p})} \} \\
&= \zeta_{\mathbf{p}}(\mathbf{r}) O(\mathbf{r}, \mathbf{p}) \{ e^{-\beta \mathcal{H}(\mathbf{r},\mathbf{p})} e^{w_O(\mathbf{r},\mathbf{p})} \}.
\end{aligned} \tag{7.85}$$

The final equality defines the function $w_O(\mathbf{r}, \mathbf{p})$.

The momentum operator is $\hat{\mathbf{p}} \equiv -i\hbar\nabla_r$, and its transform is $\hat{\tilde{\mathbf{p}}} = \mathbf{p} + \hat{\mathbf{p}}$. If the observable operator is a function of the position and momentum operators, $\hat{O} = O(\mathbf{r}, \hat{\mathbf{p}})$, then its transform is

$$\hat{\tilde{O}} = O(\tilde{\mathbf{r}}, \hat{\tilde{\mathbf{p}}}) = O(\mathbf{r}, \mathbf{p} + \hat{\mathbf{p}}). \tag{7.86}$$

This may be confirmed for the transform of the Hamiltonian operator given above, equation (7.79).

With this transformation, one sees that the expression for the statistical average of an operator observable is essentially the same as that for the grand partition function with $w(\mathbf{\Gamma}) \Rightarrow w_O(\mathbf{\Gamma})$. This is

$$\langle \hat{O} \rangle_{\text{stat}} = \frac{1}{\Xi^\pm} \sum_{N=0}^{\infty} \frac{z^N}{N! h^{3N}} \int \mathrm{d}\mathbf{\Gamma} e^{-\beta \mathcal{H}(\mathbf{\Gamma})} e^{w_O(\mathbf{\Gamma})} \eta^\pm(\mathbf{\Gamma}) O(\mathbf{\Gamma}). \tag{7.87}$$

In the event that the operator observable does not depend upon the momentum operator, $\hat{O} = O(\mathbf{r})$, then the modified operator is just the observable operator itself, and the non-commutability function is unchanged,

$$\hat{\tilde{O}} = \hat{O}(\mathbf{r}), \quad \text{and } w_O(\mathbf{\Gamma}) = w(\mathbf{\Gamma}). \tag{7.88}$$

In this case the statistical average reduces to

$$\langle \hat{O}(\mathbf{r}) \rangle_{\text{stat}} = \frac{1}{\Xi^\pm} \sum_{N=0}^{\infty} \frac{z^N}{N! h^{3N}} \int \mathrm{d}\mathbf{\Gamma} e^{-\beta \mathcal{H}(\mathbf{\Gamma})} e^{w(\mathbf{\Gamma})} \eta^\pm(\mathbf{\Gamma})\, O(\mathbf{q}). \tag{7.89}$$

### 7.4.3 Expansions for non-commutativity and symmetrization

*Monomer and classical grand potential*

Shortly it will be shown that the symmetrization factor, $\eta^\pm$, breaks up into loops of particles, $l = 1, 2, \ldots$, which may be called monomers, dimers, etc. The monomer term is unity, $\eta^{(1)} = 1$, and the monomer grand partition function is

$$\Xi_1 = \sum_{N=0}^{\infty} \frac{z^N}{N! h^{3N}} \int \mathrm{d}\mathbf{\Gamma} e^{-\beta \mathcal{H}(\mathbf{\Gamma})} e^{w(\mathbf{\Gamma})}. \tag{7.90}$$

The ratio of the full partition function to the monomer partition function is just the monomer average of the symmetrization factor,

$$\begin{aligned} \frac{\Xi^\pm}{\Xi_1} &= \frac{1}{\Xi_1} \sum_{N=0}^{\infty} \frac{z^N}{N! h^{3N}} \int \mathrm{d}\mathbf{\Gamma} e^{-\beta \mathcal{H}(\mathbf{\Gamma})} e^{w(\mathbf{\Gamma})} \eta^\pm(\mathbf{\Gamma}) \\ &= \langle \eta^\pm \rangle_1. \end{aligned} \tag{7.91}$$

Since the leading order of the non-commutativity factor vanishes, $w(\hbar = 0) = w(\beta = 0) = 0$ (see section 7.4.4 below), the leading order of the monomer grand potential is just the classical grand potential,

$$\Xi_{1,0} \equiv \Xi_1(w = 0) = \sum_{N=0}^{\infty} \frac{z^N}{N! h^{3N}} \int \mathrm{d}\boldsymbol{\Gamma} e^{-\beta \mathcal{H}(\boldsymbol{\Gamma})}. \tag{7.92}$$

The logarithm of this gives the classical equilibrium grand potential,

$$\Omega_{\mathrm{cl}}(\mu, V, T) \equiv \Omega_{1,0} = -k_{\mathrm{B}} T \ln \Xi_{1,0}. \tag{7.93}$$

With this result, classical statistical mechanics has been derived from quantum statistical mechanics.

The quantum correction from monomers is the classical average

$$e^{-\beta[\Omega_1 - \Omega_{1,0}]} = \frac{\Xi_1}{\Xi_{1,0}} = \langle e^w \rangle_{1,0}, \tag{7.94}$$

where $\Omega_1$ is the full monomer grand potential. The subscript 1,0 is synonymous with the classical equilibrium average or thermodynamic potential.

*Permutation loop expansion of the grand potential*

In the treatment of wave function symmetrization, the basis functions can be composed from single particle functions,

$$\zeta_{\mathbf{n}}(\mathbf{r}) = \zeta_{\mathbf{n}_1}(\mathbf{r}_1)\zeta_{\mathbf{n}_2}(\mathbf{r}_2)\ldots\zeta_{\mathbf{n}_N}(\mathbf{r}_N). \tag{7.95}$$

This factorized form holds for the position and momentum basis functions, equations (7.74) and (7.75). The permutation operator can be applied to either the arguments, $\zeta_{\mathbf{n}}(\hat{\mathrm{P}}\mathbf{r})$ or else the indices $\zeta_{\hat{\mathrm{P}}\mathbf{n}}(\mathbf{r})$.

In general, any permutation of objects can be factored as the product of permutation loops. In such a loop, each successive element is the one permuted into the position originally occupied by the preceding element (see figure 7.4).

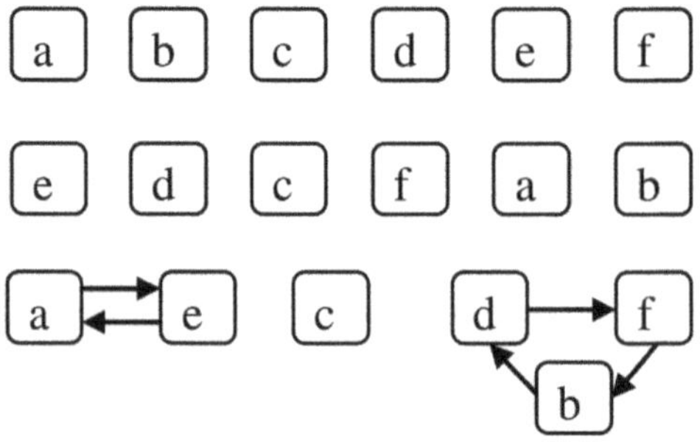


**Figure 7.4.** Six objects (top row) and a permutation of their order (second row). The bottom row shows the permutation factored into the product of the 2-loop a→e→a, the 1-loop c→c, and the 3-loop d→f→b→d.

Accordingly, any particular particle permutation operator can be factored into loop permutation operators. Hence the sum over all permutation operators can be written as the sum over all possible factors of loop permutations,

$$\sum_{\hat{P}}(\pm 1)^p\hat{P} = \hat{I} \pm \sum_{i,j}{}' \hat{P}_{ij} + \sum_{i,j,k}{}' \hat{P}_{ij}\hat{P}_{jk} + \sum_{i,j,k,l}{}' \hat{P}_{ij}\hat{P}_{kl} \pm \cdots \tag{7.96}$$

Here $\hat{P}_{jk}$ is the transpose of particles $j$ and $k$. The prime on the sums restrict them to unique loops, with each index being different. The first term is just the identity. The second term is a dimer loop, the third term is a trimer loop, and the fourth term shown is the product of two different dimers.

The symmetrization factor, $\eta^{\pm}(\mathbf{\Gamma}) = \sum_{\hat{P}}(\pm 1)^p \langle \zeta_{\hat{P}\mathbf{p}}|\zeta_{\mathbf{q}}\rangle / \langle \zeta_{\mathbf{p}}|\zeta_{\mathbf{q}}\rangle$, is the sum of the expectation values of these loops.

The monomer symmetrization factor comes from the unpermuted expectation value,

$$\eta^{(1)}(\mathbf{\Gamma}) \equiv \frac{\langle \zeta_{\mathbf{p}}|\zeta_{\mathbf{q}}\rangle}{\langle \zeta_{\mathbf{p}}|\zeta_{\mathbf{q}}\rangle} = 1. \tag{7.97}$$

The dimer overlap factor in the microstate $\mathbf{\Gamma}$ for particles $j$ and $k$ is

$$\begin{aligned}\eta_{jk}^{\pm(2)}(\mathbf{\Gamma}) &= \frac{\pm\langle \zeta_{\hat{P}_{jk}\mathbf{p}}|\zeta_{\mathbf{q}}\rangle}{\langle \zeta_{\mathbf{p}}|\zeta_{\mathbf{q}}\rangle} \\ &= \frac{\pm\langle \zeta_{\mathbf{p}_k}|\zeta_{\mathbf{q}_j}\rangle\langle \zeta_{\mathbf{p}_j}|\zeta_{\mathbf{q}_k}\rangle}{\langle \zeta_{\mathbf{p}_j}|\zeta_{\mathbf{q}_j}\rangle\langle \zeta_{\mathbf{p}_k}|\zeta_{\mathbf{q}_k}\rangle} \\ &= \pm\, e^{(\mathbf{q}_k-\mathbf{q}_j)\cdot\mathbf{p}_j/i\hbar} e^{(\mathbf{q}_j-\mathbf{q}_k)\cdot\mathbf{p}_k/i\hbar}.\end{aligned} \tag{7.98}$$

Note that since the basis functions are the product of single particle functions, equations (7.74) and (7.75), the expectation value factorizes leaving only the permuted particles to contribute.

The symmetrization factors are localized in the sense that they are only non-zero when the separations between consecutive neighbors around the loop are all small. This will be shown explicitly below (see equation (7.128)), but here it can be noted that the exponents give highly oscillatory and therefore canceling behavior unless the differences in configuration positions are all close to zero.

Similarly, the trimer symmetrization factor for particles $j$, $k$, and $l$ is

$$\begin{aligned}\eta_{jkl}^{\pm(3)}(\mathbf{\Gamma}) &= \frac{\langle \zeta_{\hat{P}_{jk}\hat{P}_{kl}\mathbf{p}}|\zeta_{\mathbf{q}}\rangle}{\langle \zeta_{\mathbf{p}}|\zeta_{\mathbf{q}}\rangle} \\ &= \frac{\langle \zeta_{\mathbf{p}_k}|\zeta_{\mathbf{q}_j}\rangle\langle \zeta_{\mathbf{p}_j}|\zeta_{\mathbf{q}_l}\rangle\langle \zeta_{\mathbf{p}_l}|\zeta_{\mathbf{q}_k}\rangle}{\langle \zeta_{\mathbf{p}_j}|\zeta_{\mathbf{q}_j}\rangle\langle \zeta_{\mathbf{p}_k}|\zeta_{\mathbf{q}_k}\rangle\langle \zeta_{\mathbf{p}_l}|\zeta_{\mathbf{q}_l}\rangle} \\ &= e^{(\mathbf{q}_j-\mathbf{q}_k)\cdot\mathbf{p}_k/i\hbar} e^{(\mathbf{q}_k-\mathbf{q}_l)\cdot\mathbf{p}_l/i\hbar} e^{(\mathbf{q}_l-\mathbf{q}_j)\cdot\mathbf{p}_j/i\hbar}.\end{aligned} \tag{7.99}$$

Continuing in this fashion, the symmetrization factor can be written as a series of loop products,

$$\eta^{\pm}(\boldsymbol{\Gamma}) = 1 + \sum_{ij}{}' \eta_{ij}^{\pm(2)}(\boldsymbol{\Gamma}) + \sum_{ijk}{}' \eta_{ijk}^{\pm(3)}(\boldsymbol{\Gamma}) + \sum_{ijkl}{}' \eta_{ij}^{\pm(2)}(\boldsymbol{\Gamma})\eta_{kl}^{\pm(2)}(\boldsymbol{\Gamma}) + \cdots \tag{7.100}$$

Here the superscript is the order of the loop, and the subscripts are the atoms involved in the loop.

This gives the ratio of the full to the monomer partition function as

$$\begin{aligned}
\frac{\Xi^{\pm}}{\Xi_1} &= \langle \eta^{\pm} \rangle_1 \\
&= 1 + \left\langle \sum_{ij}{}' \eta_{ij}^{\pm(2)} \right\rangle_1 + \left\langle \sum_{ijk}{}' \eta_{ijk}^{\pm(3)} \right\rangle_1 + \left\langle \sum_{ijkl}{}' \eta_{ij}^{\pm(2)} \eta_{kl}^{\pm(2)} \right\rangle_1 + \cdots \\
&= 1 + \left\langle \frac{N}{(N-2)!2} \eta^{\pm(2)} \right\rangle_1 + \left\langle \frac{N!}{(N-3)!3} \eta^{\pm(3)} \right\rangle_1 \\
&\quad + \frac{1}{2} \left\langle \frac{N!}{(N-2)!2} \eta^{\pm(2)} \right\rangle_1^2 + \cdots \\
&= \sum_{\{m_l\}} \frac{1}{m_l!} \prod_{l=2}^{\infty} \left\langle \frac{N!}{(N-l)!l} \eta^{\pm(l)} \right\rangle_1^{m_l} \\
&= \prod_{l=2}^{\infty} \sum_{m_l=0}^{\infty} \frac{1}{m_l!} \left\langle \frac{N!}{(N-l)!l} \eta^{\pm(l)} \right\rangle_1^{m_l} \\
&= \prod_{l=2}^{\infty} \exp \left\langle \frac{N!}{(N-l)!l} \eta^{\pm(l)} \right\rangle_1 .
\end{aligned} \tag{7.101}$$

The third and following equalities write the average of the product as the product of the averages. This is valid in the thermodynamic limit, since the product of the average of two loops scales as $V^2$, whereas the correlated interaction of two loops scales as $V$. The combinatorial factor accounts for the number of unique loops in each term; $\eta^{\pm(l)}$ without subscripts refers to any one set of $l$ particles, since all sets give the same average. Explicitly, the $l$-loop overlap factor is

$$\begin{aligned}
\tilde{\eta}^{\pm(l)}(\boldsymbol{\Gamma}^l) &= (\pm 1)^{l-1} \prod_{j=1}^{l} e^{(\mathbf{q}_j - \mathbf{q}_{j+1})\cdot \mathbf{p}_j / i\hbar} \\
&\equiv (\pm 1)^{l-1} e^{(\mathbf{q}-\mathbf{q}')\cdot \mathbf{p}/i\hbar}.
\end{aligned} \tag{7.102}$$

The subscripts for particles in a loop are to be understood mod $l$; in the first equality, $\mathbf{q}_{l+1} \equiv \mathbf{q}_1$. In the second equality, $\mathbf{q} = \{\mathbf{q}_1, \mathbf{q}_2, \ldots, \mathbf{q}_l\}$ and $\mathbf{q}' = \{\mathbf{q}_l, \mathbf{q}_1, \mathbf{q}_2, \ldots, \mathbf{q}_{l-1}\}$.

The grand potential is essentially the logarithm of the grand partition function, $\Omega \equiv -k_{\mathrm{B}}T \ln \Xi$. Hence the difference between the full grand potential and the monomer grand potential is just the series of loop potentials,

$$\begin{aligned} -\beta[\Omega^{\pm} - \Omega_1] &= \ln \frac{\Xi^{\pm}}{\Xi_1} \\ &= \sum_{l=2}^{\infty} \left\langle \frac{N!}{(N-l)!l} \eta^{\pm(l)} \right\rangle_1 \\ &\equiv -\beta \sum_{l=2}^{\infty} \Omega_l^{\pm}. \end{aligned} \tag{7.103}$$

The monomer grand potential is of course $\Omega_1 \equiv -k_{\mathrm{B}}T \ln \Xi_1$, with the monomer grand partition function being given by equation (7.90).

### 7.4.4 Expansion of the non-commutativity function

Now several expansions for the non-commutativity function $w(\mathbf{\Gamma})$ are given. The approach is similar to that of Kirkwood (1933). Whereas in essence he expanded the function that is here denoted $e^w$, here the function $w$ itself is expanded. This is an extensive function, which is an advantage.

The defining equation (7.78),

$$e^{-\beta\hat{\mathcal{H}}}1 = e^{-\beta\mathcal{H}(\mathbf{\Gamma})}e^{w(\mathbf{\Gamma})}, \tag{7.104}$$

has temperature derivative

$$\begin{aligned} \left[\frac{\partial w}{\partial \beta} - \mathcal{H}\right] e^{-\beta\mathcal{H}}e^{w} &= \hat{\mathcal{H}}e^{-\beta\hat{\mathcal{H}}}1 \\ &= \hat{\mathcal{H}}e^{-\beta\mathcal{H}}e^{w}. \end{aligned} \tag{7.105}$$

With the modified Hamiltonian operator, equation (7.79),

$$\hat{\mathcal{H}} \equiv \frac{p^2}{2m} + U(\mathbf{r}) - \frac{i\hbar}{m}\mathbf{p}\cdot\nabla - \frac{\hbar^2}{2m}\nabla^2, \tag{7.106}$$

this may be rearranged as

$$\begin{aligned} \frac{\partial w}{\partial \beta} &= \frac{i\hbar}{m} e^{\beta U - w}\mathbf{p}\cdot\nabla\{e^{w-\beta U}\} + \frac{\hbar^2}{2m} e^{\beta U - w}\nabla^2\{e^{w-\beta U}\} \\ &= \frac{i\hbar}{m}\mathbf{p}\cdot\nabla(w-\beta U) + \frac{\hbar^2}{2m}\{\nabla(w-\beta U)\cdot\nabla(w-\beta U) + \nabla^2(w-\beta U)\}. \end{aligned} \tag{7.107}$$

The non-commutativity function expanded in powers of Planck's constant is

$$w \equiv \sum_{n=1}^{\infty} w_n \hbar^n. \tag{7.108}$$

(Actually an expansion in powers of inverse temperature is slightly simpler.) This begins at $n = 1$ because it must reduce to the classical Maxwell–Boltzmann factor when Planck's constant is zero, $w(\hbar = 0) = 0$. This expansion leads to the recursion relation for $n > 2$,

$$\begin{aligned} \frac{\partial w_n}{\partial \beta} = & \frac{i}{m}\mathbf{p} \cdot \nabla w_{n-1} + \frac{1}{2m}\sum_{j=0}^{n-2} \nabla w_{n-2-j} \cdot \nabla w_j \\ & - \frac{\beta}{m}\nabla w_{n-2} \cdot \nabla U + \frac{1}{2m}\nabla^2 w_{n-2}. \end{aligned} \tag{7.109}$$

It is straightforward to derive the first several coefficient functions explicitly. One has for $n = 1$,

$$w_1 = \frac{-i\beta^2}{2m}\mathbf{p} \cdot \nabla U, \tag{7.110}$$

for $n = 2$,

$$w_2 = \frac{\beta^3}{6m^2}\mathbf{pp} : \nabla\nabla U + \frac{1}{2m}\left\{\frac{\beta^3}{3}\nabla U \cdot \nabla U - \frac{\beta^2}{2}\nabla^2 U\right\}, \tag{7.111}$$

for $n = 3$,

$$w_3 = \frac{i\beta^4}{24m^3}\mathbf{ppp} \vdots \nabla\nabla\nabla U + \frac{5i\beta^4}{24m^2}\mathbf{p}(\nabla U) \colon \nabla\nabla U - \frac{i\beta^3}{6m^2}\mathbf{p} \cdot \nabla\nabla^2 U, \tag{7.112}$$

and for $n = 4$,

$$\begin{aligned} w_4 = & \frac{-i^4\beta^5}{5!m^4}(\mathbf{p} \cdot \nabla)^4 U - \frac{\beta^5}{30m^3}(\nabla U)\mathbf{pp} \vdots \nabla\nabla\nabla U \\ & - \frac{\beta^5}{15m^2}(\nabla U)(\nabla U) : \nabla\nabla U + \frac{\beta^4}{16m^2}\nabla U \cdot \nabla\nabla^2 U \\ & + \frac{\beta^4}{48m^3}\mathbf{pp} : \nabla\nabla\nabla^2 U + \frac{\beta^4}{48m^2}\nabla^2(\nabla U \cdot \nabla U) \\ & - \frac{\beta^3}{24m^2}\nabla^2\nabla^2 U - \frac{\beta^5}{40m^3}(\mathbf{p} \cdot \nabla\nabla U) \cdot (\mathbf{p} \cdot \nabla\nabla U). \end{aligned} \tag{7.113}$$

Notice that the odd coefficient functions are pure imaginary and odd in momentum. Because $\mathcal{H}(\mathbf{\Gamma})$ is an even function of momentum, these terms in the quantum weight $e^w$ average out to real oscillatory (cosine) contributions. (The $\eta^{\pm(l)}$ contain terms that are either real and even, or else imaginary and odd in momentum.)

*Monomer expansion A*

The quantum correction to the classical grand potential due to the monomers is just a classical average of the quantum weight due to non-commutativity, equation (7.94). To fourth order in $\hbar$ this is

$$
\begin{aligned}
& -\beta[\Omega_1 - \Omega_{1,0}] \\
& = \ln \langle e^{w} \rangle_{1,0} \\
& = \ln \left\langle 1 + w + \frac{1}{2!} w^2 + \frac{1}{3!} w^3 + \frac{1}{4!} w^4 \right\rangle_{1,0} \\
& = \left\langle w + \frac{1}{2} w^2 + \frac{1}{3!} w^3 + \frac{1}{4!} w^4 \right\rangle_{1,0} - \frac{1}{2} \left\langle w + \frac{1}{2} w^2 \right\rangle_{1,0}^2 \\
& = \hbar^2 \langle w_2 \rangle_{1,0} + \frac{\hbar^2}{2} \langle w_1^2 \rangle_{1,0} \\
& \quad + \frac{\hbar^4}{2} \left\langle w_2^2 - \langle w_2 \rangle_{1,0}^2 \right\rangle_{1,0} + \frac{\hbar^4}{4!} \langle w_1^4 \rangle_{1,0} - \frac{\hbar^4}{8} \langle w_1^2 \rangle_{1,0}^2 \\
& \quad + \hbar^4 \langle w_4 \rangle_{1,0} + \hbar^4 \langle w_1 w_3 \rangle_{1,0} + \frac{\hbar^4}{2} \langle w_1^2 w_2 \rangle_{1,0} - \frac{\hbar^4}{2} \langle w_1^2 \rangle_{1,0} \langle w_2 \rangle_{1,0}.
\end{aligned}
\tag{7.114}
$$

All higher order contributions have been set to zero, and the odd imaginary terms have explicitly canceled here. It is $w_j(\mathbf{p}, \mathbf{q})$ whose average is taken. Recall that the subscript 1,0 denotes the classical average over position and momentum configurations. Note that the right-hand side must be extensive, which requires the cancelation of products of extensive terms This expansion terminated at $\mathcal{O}(\hbar^2)$ may be called A2, and at $\mathcal{O}(\hbar^4)$ it may be called A4.

*Monomer expansion B*

One can define the cumulative weight as

$$
w^{(n)} = \hbar w_1 + \hbar^2 w_2 + \cdots + \hbar^n w_n, \tag{7.115}
$$

and the $n$th approximation to the quantum correction to the monomer grand potential as

$$
\Delta\Omega_1^{(n)} = -k_B T \ln \langle e^{w^{(n)}} \rangle_{1,0}. \tag{7.116}
$$

One has $\lim_{n\to\infty} \Delta\Omega_1^{(n)} = \Omega_1 - \Omega_{1,0}$.

The classical average $\langle \cdots \rangle_{1,0}$ includes an average over the momenta $\langle \cdots \rangle_{1,0,p}$ as well as one over the position configurations. The classical monomer probability distribution for momentum is a Gaussian,

$$
\wp(\mathbf{p}) = \frac{e^{-\beta p^2/2m}}{[2\pi m k_B T]^{3N/2}}. \tag{7.117}
$$

Hence $\langle \mathbf{p}\mathbf{p} \rangle_{1,0,p} = m k_B T \, \underline{\underline{\mathrm{I}}}$.

Because the probability distribution over momenta is a Gaussian, it is straightforward, if perhaps a little tedious, to perform the momentum integrals analytically. To a particular order one has

$$
\langle e^{w^{(n)}(\mathbf{q},\mathbf{p})} \rangle_{1,0,p} \equiv e^{w^{(n)}(\mathbf{q})}. \tag{7.118}
$$

One can show (details are given in Attard (2017); the same method is used in section 7.4.6 below) that to fourth order the result is

$$\begin{aligned} w^{(4)}(\mathbf{q}) =& \frac{\hbar^2\beta^3}{24m}\nabla U \cdot \nabla U - \frac{\hbar^2\beta^2}{12m}\nabla^2 U \\ &+ \frac{2\hbar^4\beta^4}{45m^2}(\nabla\nabla U) : (\nabla\nabla U) - \frac{\hbar^4\beta^5}{240m^2}\nabla U \nabla U : \nabla\nabla U \\ &+ \frac{\hbar^4\beta^4}{20m^2}(\nabla U) \cdot \nabla\nabla^2 U - \frac{11\hbar^4\beta^3}{240m^2}\nabla^2\nabla^2 U + \mathcal{O}(\hbar^6) \\ \equiv& \hbar^2\tilde{w}_2(\mathbf{q}) + \hbar^4\tilde{w}_4(\mathbf{q}). \end{aligned} \tag{7.119}$$

This is the exponent that one has to use to weight the position configurations (in addition to the Boltzmann factor of the potential energy). This expansion terminated at $\mathcal{O}(\hbar^2)$ may be called B2, and at $\mathcal{O}(\hbar^4)$ it may be called B4.

*Monomer expansion C*

Expansion A has the merit of requiring less analysis, and of requiring the classical average of extensive terms and their products. It has the additional advantage of being explicitly dependent upon the momenta, which means it can be used for higher order loops (dimer, trimer, etc). It has the disadvantage of requiring the average over the momentum configurations to be taken numerically. Also, as a fluctuation expression, it requires relatively high accuracy in the individual averages of products to get the necessary cancelation between these super-extensive terms to end up with the final extensive result for $\Omega_1 - \Omega_{1,0}$. The larger the system size, the greater the number of configurations that need to be generated to get acceptable statistical accuracy.

Expansion B does not require numerical momentum averaging, which leads to higher accuracy. In a computer simulation, this can make it 1–2 orders of magnitude more efficient than expansion A. Also, exponentiating the expansion, as in B, should give faster convergence than expanding the final expression, as in A. A disadvantage of expansion B is that it requires more explicit algebra. Also, more problematic, since $w$ is an extensive variable, taking the average of $e^{w^{(n)}}$ can lead to computational overflow problems for large systems.

One way to avoid numerical overflow in evaluating the exponent in expansion B, but to preserve the advantage of analytical momentum average, is to use the momentum averaged $w^{(4)}$ in the expansion A. That is, writing equation (7.119) as explicit powers of Planck's constant, $w^{(4)} \equiv \hbar^2\tilde{w}_2 + \hbar^4\tilde{w}_4$, this can be inserted into expansion A, equation (7.114), with $w_1 = w_3 = 0$,

$$\begin{aligned} -\beta[\Omega_1 - \Omega_{1,0}] &= \ln\left\langle e^{w^{(4)}}\right\rangle_{1,0} \\ &= \hbar^2\langle\tilde{w}_2\rangle_{1,0} + \frac{\hbar^4}{2}\left\langle\tilde{w}_2^2 - \langle\tilde{w}_2\rangle_{1,0}^2\right\rangle_{1,0} + \hbar^4\langle\tilde{w}_4\rangle_{1,0}. \end{aligned} \tag{7.120}$$

This may be called expansion C4. The averages here are classical averages over configuration positions. The neglected terms are $\mathcal{O}(\hbar^6)$. The difference from expansion A4 is that the momentum averages have been performed analytically to leading order before expanding the exponential.

*Results for Lennard-Jones Ar, Ne, and He*

Classical Monte Carlo simulations for Lennard-Jones models of argon, neon, and helium have been carried out to test the three expansions given above for the change in the monomer grand potential due to non-commutativity, $\Omega_1 - \Omega_{1,0}$ (see Attard (2017) for full details). The Lennard-Jones pair potential is $u(r) = 4\varepsilon[(\sigma/r)^{12} - (\sigma/r)^6]$, and the thermal wave length is $\Lambda = [2\pi\hbar^2/mk_BT]^{1/2}$. Comparing this difference with the classical virial pressure, $p_{cl} = -\Omega_{1,0}/V$, measures the quantum correction due to non-commutativity, since symmetrization does not affect the monomer term. Basically, the Metropolis algorithm was used to generate the position configurations for the averages, and, for expansion A, momentum configurations (typically 32 for each position configuration used for the average) were drawn directly from the Gaussian distribution. There was generally quite good agreement between the three expansions, although A had the worst statistical error. There was good agreement between A2 and the Wigner–Kirkwood second order expansion (Wigner 1932, Kirkwood 1933).

Results on a typical sub-critical isotherm are shown in figure 7.5. Regions where the pressure has negative slope are unstable in the thermodynamic limit. From the classical virial pressure, there is a gas phase for $\rho\sigma^3 \lesssim 0.1$, a liquid phase $0.8 \lesssim \rho\sigma^3 \lesssim 0.95$, and a solid phase $\rho\sigma^3 \gtrsim 1$. Inclusion of the quantum correction may shift these phase boundaries. The C4 quantum correction is qualitatively similar for all three noble elements, with it being positive and mainly increasing with increasing density. Compared to the classical virial pressure, the quantum correction is substantially less for argon, somewhat larger for neon, and very much larger for helium. Judging by the change between the second and the fourth order expansions, the fourth order quantum correction calculated here as an estimate of the total quantum correction appears reliable for argon, marginal for neon, and unreliable for helium.

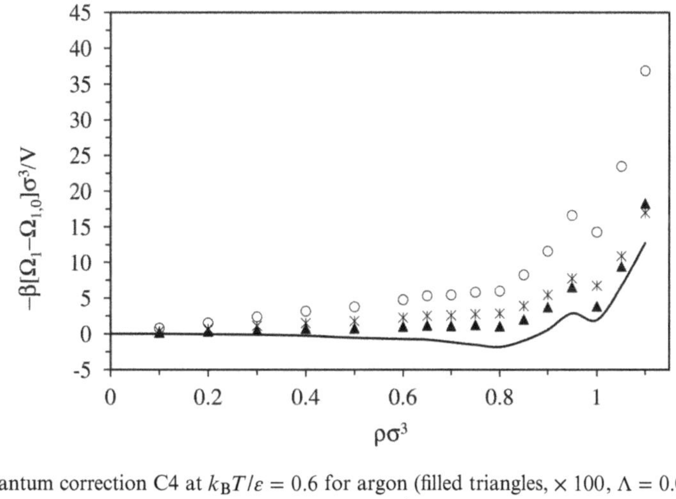


**Figure 7.5.** Quantum correction C4 at $k_BT/\varepsilon = 0.6$ for argon (filled triangles, × 100, $\Lambda = 0.0937\sigma$), for neon (open circles, $\Lambda = 0.3011\sigma$), and for helium (asterisks, × $10^{-3}$, $\Lambda = 1.3787\sigma$). The solid curve is the classical virial pressure. The standard deviation is less than 1%. Data from Attard (2017).

### 7.4.5 Expansion of the symmetrization function

The difference between the full grand potential and the monomer grand potential may be expanded in terms of loop potentials

$$\Omega^{\pm} - \Omega_1 = \sum_{l=2}^{\infty} \Omega_l^{\pm}. \tag{7.121}$$

The loop potentials arise from factoring the permutations involved in symmetrizing the wave function, with the $l$-loop potential being given by equation (7.103),

$$\begin{aligned} -\beta\Omega_l^{\pm} &\equiv \left\langle \frac{N!}{(N-l)!l}\eta^{\pm(l)} \right\rangle_1 \\ &= \frac{1}{\Xi_1}\sum_{N=0}^{\infty} \frac{z^N}{N!h^{3N}} \int \mathrm{d}\mathbf{\Gamma} e^{-\beta\mathcal{H}(\mathbf{\Gamma})} e^{w(\mathbf{\Gamma})} \frac{N!}{(N-l)!l}\eta^{\pm(l)}(\mathbf{\Gamma}). \end{aligned} \tag{7.122}$$

The monomer grand partition function is given by equation (7.90),

$$\Xi_1 = \sum_{N=0}^{\infty} \frac{z^N}{N!h^{3N}} \int \mathrm{d}\mathbf{\Gamma} e^{-\beta\mathcal{H}(\mathbf{\Gamma})} e^{w(\mathbf{\Gamma})}. \tag{7.123}$$

The monomer grand potential is of course $\Omega_1 \equiv -k_{\mathrm{B}}T \ln \Xi_1$.

*Commuting part of the expansion*

In the preceding sub-section, three expansions for the weight $w(\mathbf{\Gamma})$ were discussed. Of particular interest is what might be called the commuting part of the expansion, in which case one sets $w(\mathbf{\Gamma}) = 0$. This can be denoted by an additional subscript 0, so that the commuting part of the loop potential is

$$\begin{aligned} -\beta\Omega_{l,0}^{\pm} &\equiv \left\langle \frac{N!}{(N-l)!l}\eta^{\pm(l)} \right\rangle_{1,0} \\ &= \frac{1}{\Xi_{1,0}}\sum_{N=0}^{\infty} \frac{z^N}{N!h^{3N}} \int \mathrm{d}\mathbf{\Gamma} e^{-\beta\mathcal{H}(\mathbf{\Gamma})} \frac{N!}{(N-l)!l}\eta^{\pm(l)}(\mathbf{\Gamma}). \end{aligned} \tag{7.124}$$

Since this $l$-loop overlap factor depends only on the first $l$ particles, its classical equilibrium average can be written as an integral over their configuration momenta and positions. The standard definition in classical equilibrium statistical mechanics of the $l$-particle density is (Attard 2002)

$$\begin{aligned} &\rho^{(l)}(\mathbf{q}^l) \\ &= \frac{1}{\Xi_{1,0}}\sum_{N=l}^{\infty} \frac{\Lambda^{-3N}z^N}{(N-l)!} \int \mathrm{d}\mathbf{q}_{l+1}\cdots \mathrm{d}\mathbf{q}_N\, e^{-\beta U(\mathbf{q}^N)} \\ &= \frac{1}{\Xi_{1,0}}\sum_{N=l}^{\infty} \frac{z^N\Lambda^{-3l}e^{\beta\mathcal{K}(\mathbf{p}^l)}}{(N-l)!h^{3(N-l)}} \int \mathrm{d}\mathbf{\Gamma}_{l+1}\cdots \mathrm{d}\mathbf{\Gamma}_N\, e^{-\beta\mathcal{H}(\mathbf{\Gamma}^N)}. \end{aligned} \tag{7.125}$$

Recall that the thermal wave length is $\Lambda = [2\pi\hbar^2/mk_BT]^{1/2}$. With this the commuting part of the $l$-loop grand potential can be written

$$
\begin{aligned}
-\beta\Omega^{\pm}_{l,0} &= \frac{1}{\Xi_{1,0}} \sum_{N=l}^{\infty} \frac{z^N}{l(N-l)!h^{3N}} \int \mathrm{d}\boldsymbol{\Gamma} e^{-\beta\mathcal{H}(\boldsymbol{\Gamma})} \eta^{\pm(l)}(\boldsymbol{\Gamma}) \\
&= \frac{\Lambda^{3l}}{lh^{3l}} \int \mathrm{d}\boldsymbol{\Gamma}^l e^{-\beta\mathcal{K}(\mathbf{p}^l)} \rho^{(l)}(\mathbf{q}^l) \eta^{\pm(l)}(\boldsymbol{\Gamma}) \\
&= \frac{(\pm 1)^{l-1}\Lambda^{3l}}{lh^{3l}} \int \mathrm{d}\boldsymbol{\Gamma}^l e^{-\beta\mathcal{K}(\mathbf{p}^l)} \rho^{(l)}(\mathbf{q}^l) e^{\mathbf{p}\cdot(\mathbf{q}-\mathbf{q}')/i\hbar}.
\end{aligned}
\tag{7.126}
$$

The final equality uses the $l$-loop overlap factor, equation (7.102). In this $\mathbf{q} = \{\mathbf{q}_1, \mathbf{q}_2, \ldots, \mathbf{q}_l\}$ and $\mathbf{q}' = \{\mathbf{q}_l, \mathbf{q}_1, \mathbf{q}_2, \ldots, \mathbf{q}_{l-1}\}$.

The part of the exponent that depends upon the momenta is

$$
\frac{-\Lambda^2 p^2}{4\pi\hbar^2} + \frac{\mathbf{p}\cdot(\mathbf{q}-\mathbf{q}')}{i\hbar} = \frac{-\Lambda^2}{4\pi\hbar^2}\left[\mathbf{p} - \frac{2\pi\hbar^2}{\Lambda^2 i\hbar}(\mathbf{q}-\mathbf{q}')\right]^2 - \frac{\pi}{\Lambda^2}(\mathbf{q}-\mathbf{q}')^2. \tag{7.127}
$$

Hence the momentum contributes just a Gaussian integral, which is readily evaluated, reducing the loop potential to

$$
\begin{aligned}
-\beta\Omega^{\pm}_{l,0} &= \frac{(\pm 1)^{l-1}\Lambda^{3l}}{lh^{3l}} (4\pi^2\hbar^2/\Lambda^2)^{3l/2} \int \mathrm{d}\mathbf{q}^l e^{-\pi(\mathbf{q}-\mathbf{q}')^2/\Lambda^2} \rho^{(l)}(\mathbf{q}) \\
&= \frac{(\pm 1)^{l-1}}{l} \int \mathrm{d}\mathbf{q}^l e^{-\pi(\mathbf{q}-\mathbf{q}')^2/\Lambda^2} \rho^{(l)}(\mathbf{q}) \\
&= \frac{(\pm 1)^{l-1} V}{l} \int \mathrm{d}\mathbf{q}^{l-1} e^{-\pi(\mathbf{q}-\mathbf{q}')^2/\Lambda^2} \rho^{(l)}(\mathbf{q}), \quad \mathbf{q}_l = \mathbf{0}.
\end{aligned}
\tag{7.128}
$$

Since this is homogeneous in space, in the final equality particle $l$ has been fixed at the origin, and a factor of $V$ has replaced the integration over this coordinate.

The thermal wavelength $\Lambda = [2\pi\hbar^2/mk_BT]^{1/2}$ provides the length scale for the symmetrization Gaussian. If any nearest neighbors around the loop are separated by much more than this, then the Gaussian is zero and the configuration does not contribute to the loop potential. Therefore one needs $q_{j,j+1} \lesssim \Lambda$. Conversely, it is generally the case that particles have finite size, say $\sigma$, and at low and moderate densities and the $l$-particle density is zero if any two particles are separated by less than this. Therefore, at low and moderate densities one needs

$$
\Lambda \gtrsim q_{j,j+1} \gtrsim \sigma \tag{7.129}
$$

for symmetrization effects to be measurable. For the noble elements on the isotherm $k_BT/\varepsilon = 0.6$, in figure 7.5, this suggests that symmetrization is negligible for argon and neon.

In fact of course there is never such a thing as a perfectly impenetrable and rigid hard core; $\sigma$ is really a measure of the onset of a steep inter-particle repulsion. Hence if the density is high, then the separation between nearest neighbors is approximately the cube root of the particle volume, $\rho^{-1/3}$. In this case, for symmetrization effects to be measurable, one requires

$$\rho\Lambda^3 \gtrsim 1. \tag{7.130}$$

By this criterion, only helium is affected by wave function symmetrization over the density range of figure 7.5. But even in this case one suspects that the effects of non-commutativity, which are $\mathcal{O}(10^4)$ times the classical pressure, dominate those of particle symmetrization, since $\rho\Lambda_{\mathrm{He}}^3 \lesssim 3$. (Explicit computer simulation results confirm this suspicion.)

It is worth pointing out that the Gaussian exponent can be written as a quadratic form. With $\mathbf{q}_l = \mathbf{0}$, and $\alpha = x, y, z$, one simply has

$$\begin{aligned}(\mathbf{q} - \mathbf{q}')_\alpha^2 &= q_{1;\alpha}^2 + q_{12;\alpha}^2 + q_{23;\alpha}^2 + \cdots + q_{l-2,l-1;\alpha}^2 + q_{l-1;\alpha}^2 \\ &= \underline{\underline{A}}^{(l-1)} : \underline{q}_\alpha^{l-1} \underline{q}_\alpha^{l-1},\end{aligned} \tag{7.131}$$

where $q_{jk,\alpha}^2 = (q_{j,\alpha} - q_{k,\alpha})^2$. Here $\underline{\underline{A}}^{(l-1)}$ is an $(l-1) \times (l-1)$ tridiagonal matrix with 2 on the main diagonal and $-1$ immediately above and below the main diagonal, and all other entries 0. It is readily shown that this has determinant

$$\begin{aligned}|\underline{\underline{A}}^{(l-1)}| &= 2|\underline{\underline{A}}^{(l-2)}| - |\underline{\underline{A}}^{(l-3)}| \\ &= l.\end{aligned} \tag{7.132}$$

This result will be used in section 7.5 below.

### 7.4.6 Second order non-commuting dimer contribution

For the case of a dimer loop, $l = 2$, the overlap factor, equation (7.98), is

$$\begin{aligned}\frac{N!}{(N-2)!2}\eta^{\pm(2)}(\mathbf{\Gamma}^N) &= \pm \sum_{j<k}^{N} e^{-\mathbf{q}_{jk}\cdot\mathbf{p}_{jk}/i\hbar} \\ &= \frac{\pm N(N-1)}{2} e^{-\mathbf{q}_{12}\cdot\mathbf{p}_{12}/i\hbar},\end{aligned} \tag{7.133}$$

since the averaging makes all particle pairs equivalent. From the general expression, equation (7.103), the dimer loop potential is

$$\begin{aligned}-\beta\Omega_2^\pm &\equiv \left\langle \frac{N!}{(N-2)!2}\eta^{\pm(2)} \right\rangle_1 \\ &= \frac{1}{\Xi_1^\pm} \sum_{N=0}^{\infty} \frac{\pm z^N}{2h^{3N}(N-2)!} \int \mathrm{d}\mathbf{\Gamma}^N e^{-\beta\mathcal{H}(\mathbf{\Gamma})} e^{w(\mathbf{\Gamma})} e^{-\mathbf{q}_{12}\cdot\mathbf{p}_{12}/i\hbar}.\end{aligned} \tag{7.134}$$

Note that the integrand is highly oscillatory unless $\mathbf{q}_1 \approx \mathbf{q}_2$. Hence the integral vanishes unless the two particles are close together, which means that the integral has effectively lost a factor of volume compared to $\Xi_1^\pm$. Conversely, the denominator of $(N-2)!$ is a factor of $N^2$ smaller than the denominator $N!$ of $\Xi_1^{\pm}$. Hence in total $\Omega_2^{\pm} \sim \mathcal{O}(V^{-1}N^2)$, which is extensive, as it must be.

The second order approximation for non-commutativity, using equations (7.110) and (7.111), is

$$\begin{aligned} w^{(2)} &= \hbar w_1 + \hbar^2 w_2 \\ &= \frac{-i\hbar\beta^2}{2m}\mathbf{p}\cdot\nabla U + \frac{\hbar^2\beta^3}{6m^2}\mathbf{pp}:\nabla\nabla U + \frac{\hbar^2}{2m}\left\{\frac{\beta^3}{3}\nabla U\cdot\nabla U - \frac{\beta^2}{2}\nabla^2 U\right\} \\ &\equiv w_p^{(2)} + w_q^{(2)}. \end{aligned} \tag{7.135}$$

The two momenta terms have here been separated out.

In the integrand of the loop potential, it is simplest to deal with the Gaussian exponent for the momenta directly by completing the squares. Write $\mathbf{U}' \equiv \nabla U$, $\mathbf{U}'' \equiv \nabla\nabla U$, and

$$\{\tilde{\mathbf{U}}'\}_{j\alpha} \equiv \nabla_{j\alpha}U - \frac{2m}{\hbar^2\beta^2}q_{12;\alpha}\{\delta_{j,1} - \delta_{j,2}\}. \tag{7.136}$$

The final term here comes from the exponent of the dimer loop overlap factor. With these, that part of the exponent in the integrand that depends upon the momenta is

$$\begin{aligned} e_p &\equiv \frac{-\beta}{2m}\mathbf{p}\cdot\mathbf{p} - \frac{i\hbar\beta^2}{2m}\mathbf{p}\cdot\nabla U + \frac{\hbar^2\beta^3}{6m^2}\mathbf{pp}:\nabla\nabla U - \frac{\mathbf{q}_{12}\cdot\mathbf{p}_{12}}{i\hbar} \\ &= \frac{-\beta}{2m}\mathbf{p}\cdot\mathbf{p} - \frac{i\hbar\beta^2}{2m}\mathbf{p}\cdot\tilde{\mathbf{U}}' + \frac{\hbar^2\beta^3}{6m^2}\mathbf{pp}:\mathbf{U}'' \\ &= \frac{-\beta}{2m}\mathbf{A}:\left[\mathbf{p} + \frac{2m}{2\beta}\frac{i\hbar\beta^2}{2m}\mathbf{A}^{-1}\tilde{\mathbf{U}}'\right]^2 - \frac{\hbar^2\beta^3}{8m}\mathbf{A}^{-1}:\tilde{\mathbf{U}}'\tilde{\mathbf{U}}', \end{aligned} \tag{7.137}$$

where $\mathbf{A} \equiv \mathrm{I} - \hbar^2\beta^2\mathbf{U}''/3m$. The final equality is correct to $\mathcal{O}(\hbar^2)$. Clearly,

$$\mathbf{A}^{-1} = \mathrm{I} + \frac{\hbar^2\beta^2}{3m}\mathbf{U}'' + \mathcal{O}(\hbar^4). \tag{7.138}$$

Since the sum of the squares of the eigenvalues is the trace of the square of the matrix, one has

$$\begin{aligned} \ln|\mathbf{A}|^{1/2} &= \frac{1}{2}\sum_j \ln[1+\lambda_j] \\ &= \frac{1}{2}\sum_j\left[\lambda_j - \frac{1}{2}\lambda_j^2 + \cdots\right] \\ &= \frac{-\hbar^2\beta^2}{6m}\nabla^2 U - \frac{1}{4}\frac{\hbar^4\beta^4}{9m^2}(\nabla\nabla U):(\nabla\nabla U). \end{aligned} \tag{7.139}$$

With these the integral over the momenta for the dimer loop grand potential with second order expansion for the non-commutativity factor yields

$$\begin{aligned}
-\beta\Omega_{2,2}^{\pm} &= \frac{1}{\Xi_{1,2}^{\pm}} \sum_{N=0}^{\infty} \frac{\pm z^N/2h^{3N}}{(N-2)!} \int \mathrm{d}\boldsymbol{\Gamma}^N e^{-\beta\mathcal{H}(\boldsymbol{\Gamma})} e^{w^{(2)}(\boldsymbol{\Gamma})} e^{-\mathbf{q}_{12}\cdot\mathbf{p}_{12}/i\hbar} \\
&= \frac{1}{\Xi_{1,2}^{\pm}} \sum_{N=0}^{\infty} \frac{\pm z^N/2h^{3N}}{(N-2)!} \int \mathrm{d}\mathbf{q}^N e^{-\beta U(\mathbf{q})} e^{w_q^{(2)}(\mathbf{q})} \\
&\quad \times |2\pi m k_{\mathrm{B}} T \mathbf{A}(\mathbf{q})^{-1}|^{1/2} e^{(-\hbar^2\beta^3/8m)\mathbf{A}(\mathbf{q})^{-1}:\tilde{\mathbf{U}}'(\mathbf{q})\tilde{\mathbf{U}}'(\mathbf{q})} \\
&= \frac{1}{\Xi_{1,2}^{\pm}} \sum_{N=0}^{\infty} \frac{\pm z^N \Lambda^{-3N}}{2(N-2)!} \int \mathrm{d}\mathbf{q}^N e^{-\beta U(\mathbf{q})} e^{w_q^{(2)}(\mathbf{q})} \\
&\quad \times e^{(\hbar^2\beta^2/6m)\nabla^2 U(\mathbf{q}) + (\hbar^4\beta^4/36m^2)(\nabla\nabla U(\mathbf{q})):(\nabla\nabla U(\mathbf{q}))} \\
&\quad \times e^{(-\hbar^2\beta^3/8m)\mathbf{A}(\mathbf{q})^{-1}:\tilde{\mathbf{U}}'(\mathbf{q})\tilde{\mathbf{U}}'(\mathbf{q})}.
\end{aligned} \tag{7.140}$$

Arguably, only terms to $\mathcal{O}(\hbar^2)$ should be retained in the exponent. The second order approximation to the monomer grand partition function, the denominator $\Xi_{1,2}^{\pm}$, is identical to this with the replacements $\tilde{\mathbf{U}}'(\mathbf{q}) \Rightarrow \mathbf{U}'(\mathbf{q})$ and $2(N-2)! \Rightarrow N!$. The full result may be called 'the' $\Omega_{2,2}^{\pm}$ term, although of course there is more than one way of expanding the non-commuting weight. As mentioned above, $\Omega_{2,2}^{\pm}$ is extensive.

## 7.5 Quantum ideal gas

For an ideal gas the potential energy is zero, $U(\mathbf{r}) = 0$. In this case the modified Hamiltonian operator, equation (7.79), is

$$\hat{\tilde{\mathcal{H}}}^{\mathrm{id}} = \mathcal{K}(\mathbf{p}) - \frac{i\hbar}{m}\mathbf{p}\cdot\nabla - \frac{\hbar^2}{2m}\nabla^2, \tag{7.141}$$

where the kinetic energy for a momentum configuration is $\mathcal{K}(\mathbf{p}) = p^2/2m$. With this the defining equation for the non-commutativity function, equation (7.78), becomes

$$\begin{aligned}
\zeta_{\mathbf{p}}(\mathbf{r}) e^{-\beta\mathcal{H}^{\mathrm{id}}(\mathbf{r},\mathbf{p})} e^{w^{\mathrm{id}}(\mathbf{r},\mathbf{p})} &= \zeta_{\mathbf{p}}(\mathbf{r}) e^{-\beta\hat{\tilde{\mathcal{H}}}^{\mathrm{id}}} 1 \\
&= \zeta_{\mathbf{p}}(\mathbf{r}) e^{-\beta\mathcal{K}(\mathbf{p})}.
\end{aligned} \tag{7.142}$$

One can see that this is satisfied by

$$w^{\mathrm{id}}(\mathbf{r}, \mathbf{p}) = 0. \tag{7.143}$$

This is to be expected since the $\zeta_{\mathbf{p}}(\mathbf{r})$ are momentum eigenfunctions. Therefore, one need only retain the commuting part of the expansion, section 7.4.5, $\Omega_l^{\pm,\mathrm{id}} \equiv \Omega_{l,0}^{\pm,\mathrm{id}}$.

The commuting part of the $l$-loop grand potential, equation (7.128), requires the classical $l$-particle density. For the case of the ideal gas this is (Attard 2002)

$$\rho^{(l),\mathrm{id}}(\mathbf{q}) = \Lambda^{-3l} z^l. \tag{7.144}$$

This assumes a homogeneous system.

With this and using equation (7.132), the $l$-loop grand potential is

$$\begin{aligned} -\beta\Omega_l^{\pm,\mathrm{id}} &= \frac{(\pm 1)^{l-1} V \Lambda^{-3l} z^l}{l} \int \mathrm{d}\mathbf{q}^{l-1} e^{-\pi(\mathbf{q}-\mathbf{q}')^2/\Lambda^2} \\ &= \frac{(\pm 1)^{l-1} V \Lambda^{-3l} z^l}{l} \left( \frac{2\pi\Lambda^2}{2\pi} \right)^{3(l-1)/2} |\underline{\underline{A}}^{(l-1)}|^{-3/2} \\ &= \frac{(\pm 1)^{l-1} \Lambda^{-3} V z^l}{l^{5/2}}. \end{aligned} \tag{7.145}$$

The upper sign is for bosons, and the lower for fermions. This holds for $l \geqslant 2$. The monomer case, $l = 1$, is the classical case, and direct calculation shows that $-\beta\Omega_1^{\mathrm{id}} = zV/\Lambda^3$. This is just this expression for $-\beta\Omega_l^{\pm,\mathrm{id}}$ with $l = 1$.

The thermodynamic relation between the pressure and the grand potential is given by equation (2.65), $p = -\Omega/V$. The classical ideal gas pressure is $p^{\mathrm{cl,id}} = zk_{\mathrm{B}}T/\Lambda^3 = -\Omega_1^{\mathrm{id}}/V$. With these and the above result, the pressure of the quantum ideal gas is given by

$$\begin{aligned} \beta p^{\pm,\mathrm{id}}(z, T)\Lambda^3 &= \frac{-\beta\Lambda^3}{V} \sum_{l=1}^{\infty} \Omega_l^{\pm,\mathrm{id}} \\ &= \sum_{l=1}^{\infty} (\pm 1)^{l-1} z^l l^{-5/2}. \end{aligned} \tag{7.146}$$

This is the known result (Pathria 1972).

The relationship between density and fugacity for the quantum ideal gas is readily obtained from the thermodynamic result, equation (2.60),

$$\overline{N}(z, V, T) = \frac{-\partial\Omega(z, V, T)}{\partial\mu}. \tag{7.147}$$

Applying this to the loop expansion gives

$$\begin{aligned} \rho^{\pm,\mathrm{id}}(z, T) &= \frac{1}{\beta} \frac{\partial\beta p^{\pm,\mathrm{id}}(z, T)}{\partial z} \frac{\partial z}{\partial\mu} \\ &= \Lambda^{-3} \sum_{l=1}^{\infty} (\pm 1)^{l-1} z^l l^{-3/2}. \end{aligned} \tag{7.148}$$

## 7.6 The classical world

In view of the above formulation of quantum statistical mechanics as an expansion with classical statistical mechanics as the leading term, it is of interest to discuss the origin of the classical world around us. The puzzle is that the underlying laws of the Universe must be quantum mechanical, but the behavior that we actually observe, and the equations of motion that have historically been developed to describe the world quantitatively, are classical in nature. How in actual fact does quantum mechanics give rise to classical behavior in general, and to classical mechanics in particular, on the terrestrial sphere?

The usual answer to this question, with which teachers have fobbed off students for generations, is that quantum mechanics applies to the very small, and that classical mechanics applies to macroscopic objects. This answer is as unsatisfactory as it is uninformative. It is nothing more than a restatement of the original question and it provides no mechanistic explanation as to why the everyday world is classical.

It seems to me that a genuine attempt to address this issue should proceed in two stages. First qualitatively, namely to explain why quantum phenomena are absent from everyday experience. And second quantitatively, namely to demonstrate how Newton's (equivalently, Hamilton's) equations of motion account for the movement that we observe and measure.

The following discussion invokes the quantum statistical mechanical results of sections 7.3 and 7.4 for a sub-system interacting with a reservoir. We shall apply those results to part of a macroscopic object, taking it as axiomatic that if any part behaves classically, then the object as a whole must behave classically. The results also imply the classical behavior of individual microscopic particles and molecules that interact with each other and with an environment.

The non-classical world is defined by three quantum phenomena: lack of simultaneity for certain variables (non-commutativity), the superposition of states (coherence, or interference), and the complete symmetrization or anti-symmetrization of the wave function with respect to particle interchange. These are absent in the classical universe, where in the first place there is no impediment in principle to measuring any two properties at the same time, which is to say that the system can be simultaneously in macrostates of different collectives. Also, a classical measurement of a property yields only one value at a time, which is to say that it is in one and only one state of a collective at a time, and that these states cannot interfere with each other. Finally, classically identical particles can be interchanged without measurable consequences.

The results in section 7.3 explain the absence of the superposition of states in the classical world around us. There it was shown that the conservation law for the exchange of energy and material between a sub-system and a reservoir (equivalently, an open system and its environment) caused the total wave function to become entangled, which caused the suppression of the superposition of the principle quantum states, and the cancelation of the superposition of the degenerate quantum states. The result of this collapse of the total wave function is that the sub-system is manifest as a mixture of pure quantum states, which do not interfere with each other and which defines a classical statistical system.

This result explains the corresponding absence of superposition states in our classical world. Any macroscopic particle, can be considered as the sum total of sub-systems that can exchange energy and material with each other. Hence each such sub-system acts as a mixture of pure states, and the total wave function of such an object has collapsed. Measurements of the mass, energy, position, or momentum of the object yield results compatible with classical experience.

The formulation of quantum statistical mechanics in section 7.4 showed that the effects of non-commutativity of the position and momentum operators could be dealt with by a particular expansion, the leading order term of which corresponded to the classical case in which both could be measured simultaneously. Higher order terms, which are weighted by powers of Planck's constant and of inverse temperature, are negligible at terrestrial temperatures and pressures. As above, any macroscopic object must therefore behave classically in so far as it simultaneously possesses a well-defined position and momentum.

Also in section 7.4 wave function symmetrization was expressed as an expansion, with the higher order term depending on the spacing between particles relative to their thermal wave length. Again for terrestrial temperatures and pressures, these are practically negligible. Hence any macroscopic object must behave classically in so far as insensitivity to identical particle interchange is concerned.

The analysis of sections 7.3 and 7.4 explains qualitatively the origins of the classical nature of the world around us given the underlying quantum properties of the atoms and molecules. The three unique features of a quantum system—the superposition of states, the lack of simultaneity, and particle interchange total symmetry or asymmetry—are absent in the observed universe because of the collapse of the total wave function due to entanglement, and because the effects of non-commutativity and of interchange symmetry are quantitatively negligible at terrestrial temperatures and pressures. This argument, which was made explicitly for macroscopic particles, also holds for microscopic particles and molecules that interact with each other and their environment.

It remains to explain how the quantitative characterization of classical motion, namely Hamilton's equations of motion, arise from Schrödinger's equation. An elementary result in quantum mechanics is Ehrenfest's theorem (Messiah 1961, Merzbacher 1970). For a system in the wave state $\psi$, the particle positions and momenta are given by the expectation values

$$\mathbf{q}(\psi) = \langle \psi | \hat{\mathbf{q}} | \psi \rangle, \quad \text{and} \quad \mathbf{p}(\psi) = \langle \psi | \hat{\mathbf{p}} | \psi \rangle. \tag{7.149}$$

Ehrenfest's theorem says that the (adiabatic) rate of change of these is given by the expected derivatives of the Hamiltonian operator,

$$\dot{\mathbf{q}}^0(\psi) = \left\langle \psi \left| \frac{\partial \hat{\mathcal{H}}}{\partial \hat{\mathbf{p}}} \right| \psi \right\rangle, \quad \text{and} \quad \dot{\mathbf{p}}^0(\psi) = \left\langle \psi \left| \frac{-\partial \hat{\mathcal{H}}}{\partial \hat{\mathbf{q}}} \right| \psi \right\rangle. \tag{7.150}$$

The Hamiltonian operator is just the classical Hamiltonian function of the position and momentum operators, $\hat{\mathcal{H}} \equiv \mathcal{H}_{\rm cl}(\hat{\mathbf{p}}, \hat{\mathbf{q}})$. Ehrenfest's theorem derives directly from Schrödinger's equation (Messiah 1961, Merzbacher 1970).

The derivation of the first of these from Schrödinger's equation is

$$\begin{aligned}\dot{\mathbf{q}}^0(\psi) &= \langle\psi|\hat{\mathbf{q}}|\dot{\psi}^0\rangle + \langle\dot{\psi}^0|\hat{\mathbf{q}}|\psi\rangle \\ &= \frac{1}{i\hbar}\langle\psi|\{\hat{\mathbf{q}}\hat{\mathcal{H}} - \hat{\mathcal{H}}\hat{\mathbf{q}}\}|\psi\rangle \\ &= \frac{-\hbar^2/2m}{i\hbar}\int \mathrm{d}\mathbf{r}\ \psi(\mathbf{r})^*\{\mathbf{r}\nabla^2 - \nabla^2\mathbf{r}\}\psi(\mathbf{r}) \\ &= \frac{-i\hbar}{m}\int \mathrm{d}\mathbf{r}\ \psi(\mathbf{r})^*\nabla\psi(\mathbf{r}) \\ &= \frac{1}{m}\langle\psi|\hat{\mathbf{p}}|\psi\rangle \\ &= \left\langle\psi\left|\frac{\partial\hat{\mathcal{H}}}{\partial\hat{\mathbf{p}}}\right|\psi\right\rangle.\end{aligned} \tag{7.151}$$

The Hermitian nature of the Hamiltonian operator gives the second equality, and integration by parts gives the fourth equality after some cancelation. The derivation of the second of these from Schrödinger's equation is

$$\begin{aligned}\dot{\mathbf{p}}^0(\psi) &= \langle\psi|\hat{\mathbf{p}}|\dot{\psi}^0\rangle + \langle\dot{\psi}^0|\hat{\mathbf{p}}|\psi\rangle \\ &= \frac{1}{i\hbar}\langle\psi|\{\hat{\mathbf{p}}\hat{\mathcal{H}} - \hat{\mathcal{H}}\hat{\mathbf{p}}\}|\psi\rangle \\ &= -\int \mathrm{d}\mathbf{r}\ \psi(\mathbf{r})^*\{\nabla U(\mathbf{r}) - U(\mathbf{r})\nabla\}\psi(\mathbf{r}) \\ &= -\int \mathrm{d}\mathbf{r}\ \psi(\mathbf{r})^*\psi(\mathbf{r})\nabla U(\mathbf{r}) \\ &= \left\langle\psi\left|\frac{-\partial\hat{\mathcal{H}}}{\partial\hat{\mathbf{q}}}\right|\psi\right\rangle.\end{aligned} \tag{7.152}$$

The fourth equality follows from the product rule for differentiation, since the gradient operator acts on everything to its right.

For the case of momentum eigenfunctions, $\hat{\mathbf{p}}|\zeta_\mathbf{p}\rangle = \mathbf{p}|\zeta_\mathbf{p}\rangle$, the first of these is just

$$\dot{\mathbf{q}}^0 = \left\langle\zeta_\mathbf{p}\left|\frac{\partial\hat{\mathcal{H}}}{\partial\hat{\mathbf{p}}}\right|\zeta_\mathbf{p}\right\rangle = \frac{\partial\mathcal{H}_{\mathrm{cl}}(\mathbf{p},\mathbf{q})}{\partial\mathbf{p}}, \tag{7.153}$$

since in an eigenstate, the expectation of a function of an operator is just that function of the eigenvalue. For the case of position eigenfunctions, $\hat{\mathbf{q}}|\zeta_\mathbf{q}\rangle = \mathbf{q}|\zeta_\mathbf{q}\rangle$, the second of these is just

$$\dot{\mathbf{p}}^0 = \left\langle\zeta_\mathbf{q}\left|\frac{-\partial\hat{\mathcal{H}}}{\partial\hat{\mathbf{q}}}\right|\zeta_\mathbf{q}\right\rangle = \frac{-\partial\mathcal{H}_{\mathrm{cl}}(\mathbf{p},\mathbf{q})}{\partial\mathbf{q}}. \tag{7.154}$$

These are just Hamilton's equations of motion. Since we have just seen that any macroscopic object or microscopic particle that interacts with its environment is simultaneously in a well-defined position and momentum state, this shows how in the terrestrial sphere Hamilton's (equivalently, Newton's) classical equations of motion apply.

## Summary

- The wave space of an isolated quantum system has uniform weight. This implies that the microstates of an isolated system have equal weight.
- In an open quantum system, the conservation laws entangle the wave functions of the sub-system and reservoir, which causes them to collapse into principle entropy states. Statistical averaging over the reservoir causes the degenerate entropy states to also collapse. These mean that an open sub-system is a mixture of pure entropy microstates.
- The probability operator is the exponential of the entropy operator, which, for the canonical equilibrium system, is the negative of the sub-system Hamiltonian operator divided by the reservoir temperature.
- The partition function and statistical average are the weighted sum over distinct, allowed entropy microstates. Accounting for symmetrization, transform this into a sum over all entropy microstates, and thence into a double sum over simultaneous position and momentum states.
- Written as an integral over classical phase space, the partition function and the statistical average include the classical Maxwell–Boltzmann weight, a non-commutativity factor for the position and momentum states, and a permutation loop factor from wave function symmetrization.
- Expansion of these factors in powers of Planck's constant shows that classical statistical mechanics is the leading order contribution to quantum statistical mechanics. Classical mechanics is just quantum statistical mechanics for an open system.

## References

Attard P 2002 *Thermodynamics and Statistical Mechanics: Equilibrium by Entropy Maximisation* (London: Academic)

Attard P 2015 *Quantum Statistical Mechanics: Equilibrium and Non-Equilibrium Theory from First Principles* (Bristol: IOP Publishing)

Attard P 2017 Quantum statistical mechanics results for argon, neon, and helium using classical Monte Carlo arXiv:1702.00096v1

Barker J R 2010 Quantum phase-space distributions with compact support *Physica* E **42** 491

Barnett S M and Radmore P M 2003 *Methods in Theoretical Quantum Optics* (Oxford: Oxford University Press)

Breuer H-P and Petruccione F 2002 *The Theory of Open Quantum Systems* (Oxford: Oxford University Press)

Davies E B 1976 *Quantum Theory of Open Systems* (London: Academic)

Dishlieva K G 2008 Kirkwood and Wigner distribution functions: graphical imaging *Int. J. Pure Appl. Math.* **42** 583

Dubos R L 1950 *Louis Pasteur Freelance of Science* (Boston, MA: Little Brown & Co.)

Feynman R P and Hibbs A R 1965 *Quantum Mechanics and Path Integrals* (New York: McGraw-Hill)

Feynman R P 1998 *Statistical Mechanics* 2nd edn (Perseus: Westview Press)

Gerry C and Knight P 2005 *Introductory Quantum Optics* (Cambridge: Cambridge University Press)

Groenewold H J 1946 On the principles of elementary quantum mechanics *Physica* **12** 405

Kirkwood J 1933 Quantum statistics of almost classical particles *Phys. Rev.* **44** 31

Kleinert H 2009 *Path Integrals in Quantum Mechanics, Statistics, Polymer Physics, and Financial Markets* (Singapore: World Scientific)

Messiah A 1961 *Quantum Mechanics* vols I and II (Amsterdam: North-Holland)

Merzbacher E 1970 *Quantum Mechanics* 2nd edn (New York: Wiley)

Moyal J E 1949 Quantum mechanics as a statistical theory *Proc. Camb. Phil. Soc.* **45** 99

Pathria R K 1972 *Statistical Mechanics* (Oxford: Pergamon Press)

Praxmeyer L and Wódkiewicz K 2002 Quantum interference in the Kirkwood–Rihaczek representation arXiv:quant-ph/0207127v1

Schlosshauer M 2005 Decoherence, the measurement problem, and interpretations of quantum mechanics arXiv:quant-ph/0312059v4

Schulman L S 1981 *Techniques and Applications of Path Integration* (New York: Wiley)

von Neumann J 1932 *Mathematische Grundlagen der Quantenmechanik* (Berlin: Springer)

Weinberg S 2015 *To Explain the World. The Discovery of Modern Science* (New York: Harper) p xiii

Weiss U 2008 *Quantum Dissipative Systems* 3rd edn (Singapore: World Scientific)

Wigner E 1932 On the quantum correction for thermodynamic equilibrium *Phys. Rev.* **40** 749

Zeh H D 2001 *The Physical Basis of the Direction of Time* 4th edn (Berlin: Springer)

Zachos C, Fairlie D and Curtright T 2005 *Quantum Mechanics in Phase Space* (Singapore: World Scientific)

Zurek W H 2003 Decoherence, einselection and the quantum origins of the classical arXiv:quant-ph/0105127v3

Lightning Source UK Ltd.
Milton Keynes UK
UKHW051856270122
397741UK00002B/115

9 780750 315883